Advanced Nanotechnology in Plants

Nanotechnology uses nanomaterials/nanoparticles that can penetrate plant cells and interact with intracellular organelles and metabolites, impacting plant growth, development, physiology, and biochemistry. *Advanced Nanotechnology in Plants: Methods and Applications* explores emerging plant nanotechnology, covering advanced methods and applications with an emphasis on the mitigation of plant diseases and environmental stressors. This technology can lead to the improvement of crop quality and yield to face the challenge of global climate change with an expanding global population.

Features:

- Summarizes advanced methods and current applications of nanotechnology to mitigate plant stress
- Supports the Paris Agreement, which tackles three main objectives for sustainably increasing agricultural productivity and incomes, adapting and building resilience to climate change, and reducing and/or removing greenhouse gas emissions
- Discusses potential uses and future directions in green nanotechnology for smart and sustainable agriculture
- The content fits the goals of the UN Sustainable Development Goals (SDGs), contributing to goals 12 and 15 for responsible consumption and production and sustainable use of terrestrial ecosystems
- Provides current research findings on engineered nanoparticles for phytoremediation
- This book is a reference for students, researchers, and scientists in the field of plant sciences and nanotechnology. It is also useful for those in green chemistry and environmental sciences, and will be a practical handbook for academics, including teachers, students, and agricultural experts.

Jen-Tsung Chen is currently a professor at the National University of Kaohsiung in Taiwan. He teaches cell biology, genomics, proteomics, medicinal plant biotechnology, and plant tissue culture. Dr. Chen's research interests include bioactive compounds, chromatography techniques, plant tissue culture, phytochemicals, and plant biotechnology. He serves as an editorial board member for *Plant Methods*, *GM Crops & Food*, *Plant Nano Biology*, *Biomolecules*, and *International Journal of Molecular Sciences*, and a guest editor for *Frontiers in Plant Science*, *Frontiers in Pharmacology*, and *Journal of Fungi*.

Advanced Nanotechnology in Plants

Methods and Applications

Edited by
Jen-Tsung Chen

CRC Press
Taylor & Francis Group
Boca Raton London New York

CRC Press is an imprint of the
Taylor & Francis Group, an **informa** business

First edition published 2024
by CRC Press
2385 NW Executive Center Drive, Suite 320, Boca Raton FL 33431

and by CRC Press
4 Park Square, Milton Park, Abingdon, Oxon, OX14 4RN

CRC Press is an imprint of Taylor & Francis Group, LLC

Library of Congress Cataloging-in-Publication Data
Names: Chen, Jen-Tsung, editor.
Title: Advanced nanotechnology in plants : methods and applications / edited by Jen-Tsung Chen
Description: First edition | Boca Raton, FL : CRC Press, 2024 | Includes bibliographical references and index.
Identifiers: LCCN 2023033296 (print) | LCCN 2023033297 (ebook) | ISBN 9781032429458 (hardback) | ISBN 9781032437194 (paperback) | ISBN 9781003368571 (ebook)
Subjects: LCSH: Plant biotechnology. | Nanobiotechnology. | Agricultural innovations.
Classification: LCC TP248.27.P55 A35 2024 (print) | LCC TP248.27.P55 (ebook) | DDC 630--dc23/eng/20230729
LC record available at https://lccn.loc.gov/2023033296
LC ebook record available at https://lccn.loc.gov/2023033297

ISBN: 9781032429458 (hbk)
ISBN: 9781032437194 (pbk)
ISBN: 9781003368571 (ebk)

DOI: 10.1201/b23308

Typeset in Times
by Deanta Global Publishing Services, Chennai, India

Contents

Contributors

Hajra Abid
Department of Botany
University of Poonch Rawalakot
Azad Jammu and Kashmir
Rawalakot, Pakistan

Muhammad Adil
Pharmacology & Toxicology Section
University of Veterinary & Animal Sciences
Jhang campus
Lahore, Pakistan

Faheem Ahmad
Department of Botany
Aligarh Muslim University
Aligarh, India

Khawaja Shafique Ahmad
Department of Botany
University of Poonch Rawalakot
Azad Jammu and Kashmir
Rawalakot, Pakistan

Mukhtiar Ahmed
Institute of Process Engineering
Chinese Academy of Sciences
Beijing, PR China

Fatima El Amerany
Laboratory of Sustainable Development and Health Research
Faculty of Science and Technology of Marrakech
Department of Chemistry
Cadi Ayyad University
Marrakech, Morocco

Arshi Anees
Centre for Agricultural Education
Faculty of Agricultural Sciences
Aligarh Muslim University
Aligarh, India

Sundas Asghar
Department of Zoology
The University of Lahore
Sargodha campus
Sargodha, Pakistan

Rabiya Basri
School of Agriculture
ITM University
Gwalior, India

Tesleem Taye Bello
Department of Agricultural Science Education
Federal College of Education
Abeokuta, Nigeria

Renu Bhardwaj
Plant Stress Physiology Laboratory
Department of Botanical and Environmental Sciences
Guru Nanak Dev University
Amritsar, India

Tamanna Bhardwaj
Plant Stress Physiology Laboratory
Department of Botanical and Environmental Sciences
Guru Nanak Dev University
Amritsar, India

Subir Kumar Bose
Department of Biotechnology
Meerut Institute of Engineering and Technology
Meerut, India

José Juvencio Castañeda-Nava
Department of Plant Biotechnology
Center for Research and Assistance in Technology and Design of the State of Jalisco (CIATEJ)
Zapopan, Mexico

Kishore Chand
College of Material Science and Chemical Engineering
Harbin Engineering University
Harbin, PR China

Sunbal Khalil Chaudhari
Department of Botany
The University of Lahore
Sargodha campus
Sargodha, Pakistan

Celia Chávez-Mendoza
Research Center in Food and Development (CIAD)
Delicias, Mexico

Jen-Tsung Chen
Department of Life Sciences
National University of Kaohsiung
Kaohsiung, Taiwan

Anam Choudhary
School of Agricultural Sciences
Sri Venkateshwara University
Amroha, India

Soleyman Dayani
Department of Biotechnology
Faculty of Agriculture and Natural Resources
Imam Khomeini International University
Qazvin, Iran

Anuvab Dey
Department of Biological Sciences and Bioengineering
Indian Institute of Technology Guwahati
North Guwahati, India

Laurent Dufosse
University of Reunion
CHEMBIOPRO Laboratory (Chemistry and Biotechnology of Natural Products)
Agri-food Department
University of Reunion Island's Engineering School
Saint-Denis, France

Oluwatoyin Adenike Fabiyi
Department of Crop Protection
Faculty of Agriculture
University of Ilorin
Ilorin, Nigeria

Soledad García-Morales
Department of Plant Biotechnology
CONAHCYT-Center for Research and Assistance in Technology and Design of the State of Jalisco (CIATEJ)
Zapopan, Mexico

Subhrojyoti Ghosh
Department of Biotechnology
Indian Institute of Technology Madras
Chennai, India

Rishil Gupta
Department of Botany
Aligarh Muslim University
Aligarh, India

Abdul Hanan
College of Material Science and Chemical Engineering
Harbin Engineering University
Harbin, PR China

Ramin Hosseini
Department of Biotechnology
Faculty of Agriculture and Natural Resources
Imam Khomeini International University
Qazvin, Iran

Altaf Hussain
State Key Laboratory of Electroanalytical Chemistry
Changchun Institute of Applied Chemistry
Chinese Academy of Sciences
Changchun, PR China

Sajjad Hussain
Department of Botany
University of Poonch Rawalakot
Azad Jammu and Kashmir
Sargodha, Pakistan

Arshi Jamil
Centre of Agricultural Education
Faculty of Agricultural Sciences
Aligarh Muslim University
Aligarh, India

Srinivasan Kameswaran
Department of Botany
Vikrama Simhapuri University College
SPSR Nellore district, India

Masudulla Khan
Botany Section
Women's College
Aligarh Muslim University
Aligarh, India

Kanika Khanna
Plant Stress Physiology Laboratory
Department of Botanical and Environmental Sciences
Guru Nanak Dev University
Amritsar, India

Jaspreet Kour
Plant Stress Physiology Laboratory
Department of Botanical and Environmental Sciences
Guru Nanak Dev University
Amritsar, India

Sandeep Kour
Department of Zoology
Guru Nanak Dev University
Amritsar, India

Muhammad Nazim Lakhan
College of Material Science and Chemical Engineering
Harbin Engineering University
Harbin, PR China

Janet María León-Morales
Academic Coordination West Altiplano Campus
Autonomous University of San Luis Potosi
Salinas de Hidalgo, Mexico

Sourav Majhi
Department of Biological Sciences and Bioengineering
Indian Institute of Technology Guwahati
North Guwahati, India

Madiha Manzoor
Department of Botany
The University of Lahore
Sargodha campus
Sargodha, Pakistan

Maryam Mazaheri-Tirani
Department of Biology
Faculty of Science
University of Jiroft
Jiroft, Iran

Ansar Mehmood
Department of Botany
University of Poonch Rawalakot
Azad Jammu and Kashmir
Rawalakot, Pakistan

Anamika Mishra
Department of Biotechnology
School of Biosciences & Technology
Vellore Institute of Technology
Vellore, India

Oishi Mitra
Department of Biotechnology
School of Biosciences & Technology
Vellore Institute of Technology
Vellore, India

Abhijit Mondal
University Institute of Biotechnology
Chandigarh University
Sahibzada Ajit Singh Nagar, India

Iqra Munir
Department of Botany
The University of Lahore
Sargodha campus
Sargodha, Pakistan

Nasreen Musheer
School of Agricultural Science
Glocal University
Saharanpur, India

Amna Mustafa
Department of Botany
University of Poonch Rawalakot
Azad Jammu and Kashmir
Rawalakot, Pakistan

Ghazala Mustafa
Department of Botany
Quaid-i-Azam University
Islamabad, Pakistan

Sagnik Nag
Department of Biotechnology
School of Biosciences & Technology
Vellore Institute of Technology
Vellore, India

Fahim Nawaz
Research School of Biology
The Australian National University
Australian Capital Territory
Canberra, Australia

Saiqa Nazir
Department of Botany
University of Poonch Rawalakot
Azad Jammu and Kashmir
Rawalakot, Pakistan

Puja Ohri
Department of Zoology
Guru Nanak Dev University
Amritsar, India

Alejandro Palacio-Márquez
Research Center in Food and Development (CIAD)
Delicias, Mexico

Mandala Ramakrishna
Department of Botany
Vikrama Simhapuri University College
Nellore, India

Carlos A. Ramírez-Estrada
Research Center in Food and Development (CIAD)
Delicias, Mexico

Anisa Ratnasari
Department of Environmental Engineering
Sepuluh Nopember Institute of Technology
Surabaya, Indonesia

Mohammed Rhazi
Interdisciplinary Laboratory in Bio-Resources, Environment and Materials
Higher Normal School
Department of Biology
Cadi Ayyad University
Marrakech, Morocco

Sabiha Saeed
Department of Plant Protection
Faculty of Agricultural Sciences
Aligarh Muslim University
Aligarh, India

Esteban Sánchez
Research Center in Food and Development (CIAD)
Delicias, Mexico

Adina-Elena Segneanu
Institute for Advanced Environmental Research-West University of Timisoara
Timisoara, Romania

Mehreen Shah
Department of Environmental Science and Engineering
Jamia Millia Islamia
New Delhi, India

Nandni Sharma
Department of Zoology
Guru Nanak Dev University
Amritsar, India

Neerja Sharma
Plant Stress Physiology Laboratory
Department of Botanical and Environmental Sciences
Guru Nanak Dev University
Amritsar, India

Junaid Shehzad
Department of Plant Sciences
Quaid-i-Azam University
Islamabad, Pakistan

Muhammad Asif Shehzad
Institute of Plant Breeding and Biotechnology
MNS University of Agriculture
Multan, Pakistan

Arun Dev Singh
Plant Stress Physiology Laboratory
Department of Botanical and Environmental Sciences
Guru Nanak Dev University
Amritsar, India

Irfan Ali Soomro
State Key Laboratory of Chemical Resource Engineering
Institute of Computational Chemistry
College of Chemistry
Beijing University of Chemical Technology
Beijing, PR China

Battana Swapna
Department of Botany
Vikrama Simhapuri University College
Nellore, India

Uday Kumar Thera
Department of Plant Sciences and Landscape Architecture
University of Maryland, College Park
College Park, USA

Pragya Tiwari
Department of Biotechnology
Yeungnam University
Gyeongsan, Republic of Korea

Phanindra P V
Department of Mycology and Plant Pathology
Institute of Agricultural Sciences
Banaras Hindu University
Varanasi, India

Gilberto Velázquez-Juárez
Department of Chemistry
Center for Exact Sciences and Engineering (CUCEI)
University of Guadalajara
Jalisco, Mexico

Saira Zaheer
Department of Plant Sciences
Quaid-i-Azam University
Islamabad, Pakistan

Preface

Nanotechnology has been intensively explored in biomedical science, and a paradigm shift was going well in the past decade in the fields of plant science and agriculture, which provides a promising tool for the development of a range of engineered nanoparticles and might greatly benefit climate-smart and sustainable agriculture in the future. This book comprehensively summarizes recent advances in plant nanotechnology and systematically refines methods and applications with an emphasis on the mitigation of plant diseases and environmental stressors and the improvement of crop quality and yield to face the challenges in the scenario of global climate change and the expanding global human population. It supports the Paris Agreement, which tackles three main objectives: sustainably increasing agricultural productivity and incomes, adapting and building resilience to climate change, and reducing and/or removing greenhouse gas emissions, where possible. The content of this book fits the goals of the SDGs (Sustainable Development Goals) and particularly contributes to goals 12 and 15 for responsible consumption and production and for sustainable use of terrestrial ecosystems.

This book covers emerging nano-biotechnological methods for enhancing crop production when facing biotic/pathogenic and abiotic/environmental stresses, which may be worse and more complicated in the future due to climate change, particularly with rising temperatures. The use of nano-biofertilizers and nanopesticides is a powerful aid to achieving smart agricultural practices for easy and efficient application and reducing waste of resources while making crops more resilient to multiple stressors. These eco-friendly technologies not only increase crop yield but also can improve nutritional quality and are indispensable for supporting global food security, sustainable agriculture, and the protection of the environment. With the application of green chemistry, plant nanotechnology has entered a new era of smart manipulation of green-synthesized nanomaterials/nanoparticles, using natural resources of plants and microorganisms such as endophytes, to save energy and reduce waste and therefore minimize damage to the environment. They are promising agents and/or carriers for a certain range of applications, including abiotic stress mitigation, growth promotion, nutritional quality enhancement, and plant protection, and additionally, are able to alleviate climate change effects in crops and might play a critical role in the phytoremediation of heavy metal–contaminated soils and water pollution.

This book offers a brand new perspective on plant nanotechnology by covering updated theory, recent advances in method development, and practical approaches. It explores and discusses emerging areas of green chemistry for nanotechnology in plant research, crop improvement, agriculture, and environmental protection. This book is a suitable reference for scholars in multiple research fields, including plant physiology, plant pathology, plant biotechnology, agronomy, agriculture, green chemistry, environmental science, and material science, and will be a practical handbook for academics, including teachers, students, and agricultural experts.

The editor is very grateful to the staff of the publisher for their guidance and assistance, and to all the chapter authors for their intelligence and efforts. This book would not have been finalized without your valuable contributions.

1 Nanotechnology in Plant Sciences and Agriculture
An Overview

Masudulla Khan

1.1 INTRODUCTION

In 2020, around 720 million people faced chronic hunger, and globally, 149.2 million children under five years of age suffer from stunting. To address this situation, the 2030 Agenda for Sustainable Development aims to achieve food security. The new strategy to achieve this vision is through sustainability for a better environment, better production, and a better life.[1]

Conventionally, farmers mainly use toxic chemicals to control plant diseases. The increasing use of herbicides leads to risks to the environment. Around two million tonnes of pesticides are used worldwide per year. The high use of these chemicals gives rise to environmental deterioration, and the residues of pesticides in food cause harmful diseases like cancer and asthma in humans. For instance, DDT (dichlorodiphenyltrichloroethane) pesticide causes genotoxicity in humans.[2] Also, 3,6-dichloro-2-methoxy benzoic acid and 2,4-dichlorophenoxyacetic acid (2,4-D) are the herbicides that belong to the Auxinic group, and it has been reported that they can cause DNA damage in humans. Nowadays, glyphosate is the most popular herbicide, and it contributes to 60% of herbicide sales.[3] In 1999, it was reported that around one million deaths annually occur in the world due to pesticides. According to the World Health Organization (WHO), three million poisoning cases per year are caused by pesticides.[4] Herbicides can cause acute and chronic toxicity, and many of them have chronic effects on human health. In addition, pesticides cause prostate cancer in farmers, decreased sperm count, endocrine disruption, or increased abnormalities in sperm.[5,6]

There is a need to cut down on the utilization of toxic pesticides in agriculture. Moreover, in the future, we need to reduce the toxicity of chemical pesticides by using alternative and advanced ways to control plant pathogens. Among these alternatives, some of the most effective are nanomaterials. It is well known that using nanomaterials to manage plant disease can be eco-friendly as well as targeted to the host. Nanotechnology has high potential to bring a revolution in agriculture, food systems, and crop yield, and has given birth to a new era of "Agronanotechnology". The word "nano" means one-billionth part. Different types of nanomaterials have been obtained, such as zinc, copper, magnesium, titanium, gold, alginate, and silver. In 2010, the global production of nanoparticles reached around 260,000–309,000 metric tons. Although nanotechnology bears the potential to revolutionize agriculture, it is currently in the budding stage and still demands intensive investigation into its methods and applications. In conclusion, at present, managing diseases and improving yields are the key applications in the field of agriculture.[7,8]

1.2 NANOMATERIALS IN MANAGEMENT OF PHYTOPARASITIC NEMATODES

Nematodes are omnipresent tiny roundworms, and some are parasitic to critical food crops. Phytoparasitic nematodes (PNs) are soil-borne, obligate biotrophs. To date, around 4,100 nematode species have been reported to infect crop plants, around 15% of the total known nematode species.

DOI: 10.1201/b23308-1

The yield loss caused by nematodes was approximately US$70 billion in 1987;[9] in 2003, the loss was around US$125 billion,[10] and it had risen to US$173 billion in 2013.[11] PNs may be ectoparasitic or endoparasitic. Ectoparasitic nematodes use their stylets to feed on the plant roots from the outer surface. These nematodes have long stylets, which help them in feeding. Endoparasitic nematodes remain inside the host plant and feed there. Endoparasitic nematodes may be sedentary or migratory. The most destructive nematodes belong to the sedentary endoparasitic lifestyle category. Two main nematodes fall into this category:root-knot nematodes and cyst nematodes. In these nematodes, the J2 larval stage invades the root. Among all phytoparasitic nematodes, root-knot nematodes and cyst nematodes cause the highest economic losses to crops.[12] Root-knot nematodes (*Meloidogynne* sp.) may cause the loss of upto 80% of vegetable crops in subtropical and tropical countries.[13] In addition to this, *Pratylenchus* spp., *Rotylenchulus reniformis*, and *Aphelenchoides besseyi* drastically harm crop plants.[14,15] Furthermore, phytoparasitic nematodes enhance the severity of fungal and bacterial infections; for instance, the presence of *Meloidogyne incognita* increases the severity of bacterial wilting and fungal Phomopsis blight diseases in eggplant. PNs puncture the roots and wound the surface with the stylet, and this wound provides away for other bacterial and fungal pathogens to enter host plants.[13–15]

Many previous studies have reported the nematicidal activity of nanoparticles. Khan et al. (2022a) reported that silicon dioxide and titanium dioxide (SiO_2 and TiO_2) nanoparticles have an inhibitory effect on egg hatching of the plant-parasitic nematode *M. incognita*. Both nanoparticles were found to be effective against *M. incognita*, but SiO_2nanoparticles showed higher mortality than TiO_2 nanoparticles. Among all three concentrations (0.05, 0.10, and 0.20 mg/mL), the most effective concentration was 0.20 mg/mL. The effect of both nanoparticles on *M. incognita* was concentrationbased.[16] Khan et al. (2022b) reported that plant medium–synthesized silver nanoparticles (AgNPs) have an inhibitory effect on the plant-parasitic nematode *M. incognita*. The effect of AgNPs on *M. incognita* was concentrationbased.[17] Ahamad et al. (2021) screened three different nanoparticles (SiO_2, ZnO, and TiO_2). All three nanoparticles showed an inhibitory effect against the plant-parasitic nematode *M. incognita*. Among all the screened nanoparticles, SiO_2nanoparticles were found to be most effective against *M. incognita*.[18] Khan and Siddiqui (2018) observed the impact of zinc oxide nanoparticles (ZnONPs) on *M. incognita*. ZnONPs at 200 ppm concentration showed the highest mortality of *M. incognita*.[19] Nour El-Deen and El-Deeb (2018 reported a reduction in the number of *M. incognita* root-knot nematodes after the application of silver nanoparticles in a tomato field.[20] Kalaiselvi et al. (2017) screened two types of silver nanoparticles (AgNPs) synthesized from *Azadirachta indica* and *Curcuma longa* plants. They studied the effect on J2 of *M. incognita*. The results showed that both AgNPs had nematicidal activity, but the ones produced using *C. longa* showed greater nematicidal activity.[21] Thakur and Shirkot (2017) reported the efficacy of gold nanoparticles against phytoparasitic nematodes.[22] Abdellatif et al. (2016) found that silver nanoparticles at 17 mg/ml reduced the population of J2 of *M. javanica* in the soil by 69.44%.[23] Cromwell et al. (2014) reported inactivity in 99% of nematodes after exposing the J2 of *M. incognita* to AgNPs in water at a concentration of 30–150 mg/ml for six hours.[24] Ardakani (2013) reported toxicity of silver, siliconoxide, and titanium oxide against *M. incognita*. He found that treatments of 200, 400, and 800 mg/ml silver nanoparticles caused 100% J2 mortality (Figures 1.1 and 1.2).[25]

1.3 IMPACT OF DIFFERENT NANOPARTICLES ON PLANT GROWTH

Nanotechnology can play a key role in improving crop yield.[26] Khan et al. (2022a) reported that titanium dioxide and silicon dioxide nanoparticles could improve the growth of eggplant, but silicon dioxide nanoparticles were found better for plant growth in comparison to titanium dioxide.[16]It has been reported that zinc oxide nanoparticles were able to enhance plant growth.[19]Additionally, the application of titanium dioxide (TiO_2) could promote plant growth parameters;[27] silicon has a beneficial effect on plant growth and physiological activities;[28] SiO_2 nanoparticles increased the growth

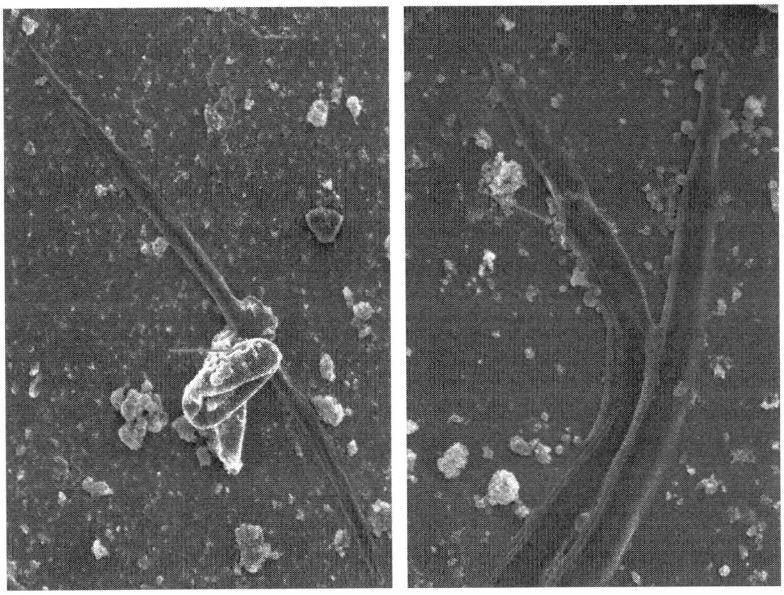

FIGURE 1.1 Silicon dioxide and titanium dioxide nanoparticles affect *M. incognita*. (From Khan et al. *Nanotechnological Review*, 2022, *11*(1), 1606–1619.)

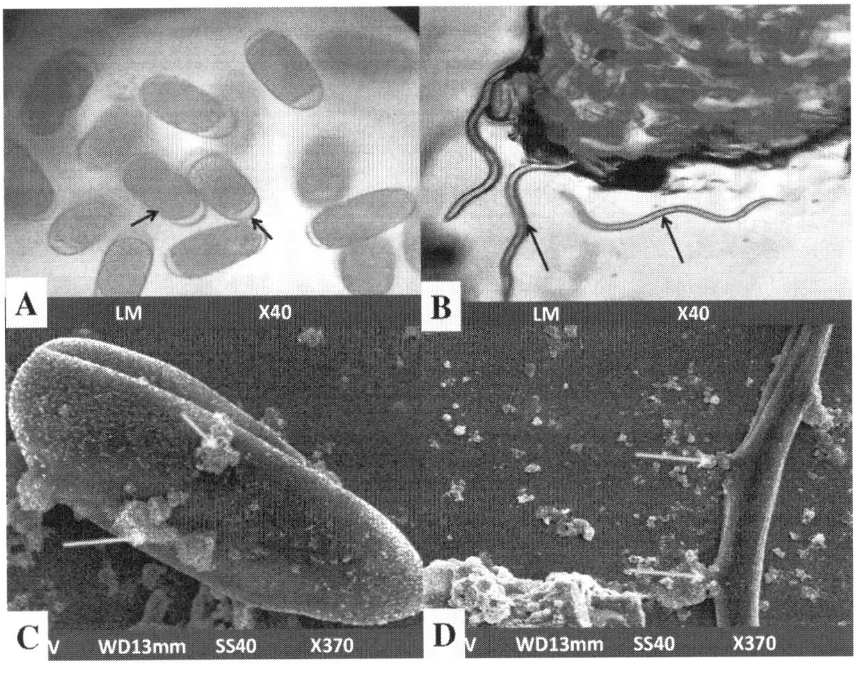

FIGURE 1.2 Light microscopy images: (A) eggs of *Meloidogyne incognita* (untreated); (B) second-stage (J2) juveniles of *M. incognita* (untreated). Scanning electron microscopy images of AgNPs-treated *M. incognita*: (C) eggs; (D) J2 of *M. incognita*. (From Khan et al. *Biomolecules*, 2022, 12, 74)

of *Agropyron elongatum* seedlings;[29] nano-SiO_2 has the potential to act as a fertilizer and could reduce fertilizer loss through release in a controlled and slow manner;[30] silica nanoparticles improve the growth of soybean (*Glycine max*) and mitigate the toxicity of mercury (Table 1.1).[31]

1.4 NANOPARTICLES IN REDUCING PLANT DISEASES

Due to the toxic effect of pesticides, the search for an alternative approach to disease management has increased. It is reported that these chemicals can harm environmental factors plus human health.[49] Nanomaterials may be applied to plants via two different mechanisms: (i) applied directly on crops and (ii) as carriers of pesticides. The controlled release of agrochemicals can provide them in the necessary quantities for increased efficiency and therefore, reduces harmful effects.[50]

Silicon NPs provide plants with the capability to reduce infections.[51,52] Silicon is effective in inducing host resistance.[53] Silicon dioxide nanoparticles (SiO_2NPs) were found effective against the phytopathogenic fungus *Phomopsis vexans* and the bacterium *Ralstonia solanacearum*.[16] SiO_2NPs at 0.2 mg/L reduced the growth of *Pseudomonas synringae*, *Xanthomonas campestris*, *R. solanacearum*, *Pectobacterium carotovorum*, *Alternaria solani*, and *Fusarium oxysporum*.[54]

Zinc oxide (ZnO) NPs recorded a global annual production of 33,400 tons and become the third most often utilized NPs with potential for plant disease management.[55,56] George et al. (2014) screened the antifungal action of ZnO and TiO_2NPs on the fungal pathogens *A. flavus*, *Aspergillus niger*, *Rhizoctonia oryzae*, *Trichophyton*, *Fusarium*, *Fonsecaea*, *Ramichloridium schulzeri*, and *Cladosporium*. Both nanoparticles were highly effective against test pathogens.[57] The application of ZnO and TiO_2 NPs could potentially reduce bacterial diseases in plants.[58] Silver NPs exhibit antimicrobial action and could be applied for the management of plant diseases.[59,60]Namburi et al. (2021) observed the antimicrobial activity of biogenic AgNPs against the pathogenic bacterium *Xanthomonas oryzae* pv. *oryzae*, which causes leaf blight inrice.[60] AgNPs also showed antifungal action against *Fusarium oxysporum*, which causes wilting in tomato.[61] Upto 79–98% inhibitions in the growth of *F. oxysporum* was recorded *in vitro* in comparison to control. The growth of molds, rot, fungi, and various other plant pathogens could be prevented through the application of silver NPs.[62] Antifungal actions of AgNPs were reported against *Fusarium oxysporum*, *Macrophomina phaseolina*, and *A. alternata*.[63] Zirconium dioxide nanoparticles (ZrONPs) could serve as an antibacterial, antifungal, and antimicrobial agent.[64,65] ZrONPs inhibited the growth of the root rot pathogen and enhanced resistance against disease, possibly due to their nano size.[26]ZrONPs were found to be inhibitory against the growth and pathogenicity of the fungal pathogen (Table 1.2, Figures 1.3–1.5).[66]

1.5 NANOPESTICIDES

Around 90% of pesticide is lost during application, and less than 0.1% of pesticide applied acts against the target organisms.[76,77] As an effective alternative, nanopesticides are emerging products, developed with novel nanomaterials, used in agriculture to protect crops. Controlled release of pesticides could reduce the loss. Nanoparticles may serve as more effective vehicles for pesticide delivery due to their easy attachment, larger surface area, and rapid transfer. Nanoparticles could deliver insecticides and increase their solubility. Some common target pests are *Tetranychus urticae*, *Spodoptera litura*, and *Helicoverpa armigera*. Insecticides with low solubility in water have been effectively loaded into chitosan and porous silica.[78–80] Silica (SiO_2) nanoparticles (NPs) have pesticidal activity. SiO_2 NPs showed pesticidal effects on the insect *Spodoptera littoralis*,[81] the aphid *Aphis craccivora*,[82] and the leaf-mining fly *Litura trifolii*.[82] The WHO approved the use of amorphous silica as ananobiopesticide.[81–83] Nano-insecticides improve the stability of active compounds. Yang et al. (2009) observed 80% mortality of *Tribolium castaneum* when treated with garlic essential oil encapsulated in polyethylene glycol (PEG).[84] Yang et al. (2018) used chitosan as a carrier of nicotine insecticide against house flies. The delivery of herbicides by nano-based carriers mainly focuses on diminishing the toxicity of herbicides on the environment.[85] Oliveira et al.

TABLE 1.1
Impact of Different Nanoparticles on the Growth of Plants

Nanoparticles	Concentrations	Plants	Effects	References
GA-Ag NPs (silver NPs coated in Gum Arabic)	5, 10, 20, and60 mg/L	*Phaseolusvulgaris*	Increase in leaf area and number of leaves	[32]
Zn and FeO NPs	100 mg/L	Wheat	Increased wheat growth	[34]
Ag	0 and 44 mg/L	*Solanumlycopersicum*	Increase in plant biomass and shoot length	[33]
ZnO NPs	600 mg/L	*Vigna mungo*	Improved germination and seedling vigor	[35]
ZnO NPs	300 ppm	*Arachis hypogaea* L.	Chlorophyll, reducing sugar, and total soluble sugar levels were increased	[36]
ZnO NPs	8 mg/L	Tomato	Increased growth and photosynthetic efficiency	[37]
ZnO NPs	<400 mg/kg	*Coriandrum sativum*	Improved photosynthesis pigments	[38]
ZnO NPs	50 mg/L	*Triticum aestivum*	Positive effect on seed germination	[39]
ZnO NPs	20 ppm	*Vigna radiata*	A 6.47% increase in plant height	[40]
Nano-Si	1 mM	Tomato	Si NPs increased photosynthesis rate and transpiration under salinity stress	[42]
SiO_2NPs	300 mg/L	*Indocalamus barbatus*	Increased net photosynthetic rate	[43]
Si NPs	5 and 10 mM	*Avena sativa*	Increased growth in plants	[41]
Nano-Si	1 mM	Tomato	Enhancement in dry weight, germination rate, and root length	[44]
SiO_2NPs	500 μL/L	*Larix olgensis*	Augmented chlorophyll content and plant height	[28]
TiO_2 NPs	0.05%	*Glycine max* L.	The yield of oil and seed was positively affected	[45]
TiO_2 NPs	10 mg/L	*Vigna radiata* L.	The length of the shoot and root was increased significantly	[46]
TiO_2 NPs	100 mg/L	*Mentha piperita*	Significant increase in length of roots	[47]
TiO_2 NPs	5 ppm	*Agropyron desertorum*	Improved seed germination	[48]

(2015) formed poly(epsilon-caprolactone) polymeric nanoparticles loaded with atrazine and found that nano-encapsulated atrazine was a more efficient herbicide at a lower concentration.[86]

The carbon-based nanomaterials also showed antifungal activity. Six types of carbon nano materials (reduced graphene oxide [rGO], graphene oxide [GO], activated carbon [AC], fullerenes, single-walled carbon nanotubes [SWCNTs], and multi-walled carbon nanotubes [MWCNTs]) were screened for antifungal action against *Fusarium poae* and *F. graminearum*. The results showed that MWCNTs, SWCNTs, rGO, and GO have antifungalbehavior.[87]

TABLE 1.2
Impact of Nanoparticles on Plant Pathogens

Nanoparticles	Concentrations	Tested Pathogen	Effects	References
SiO_2 and TiO_2 NPs	0.2 mg/ml	*Meloidogyne incognita*, *Ralstonia solanacearum*, and *Phomopsis vexans*	Inhibitory	[16]
AgNPs	100 ppm	*M. incognita* and *Aspergillus flavus*	Inhibitory	[17]
AgNPs	100 ppm	*M. incognita*, *R. solanacearum*, and *F. oxysporum*	Inhibitory	[67]
SiO_2NPs	0.20 g/L	*Pseudomonas synringae*, *Xanthomonas campestris*, *Pactobacterium carotovorum*, *R. solanacearum*, *Fusarium oxysporum*, *and Alternaria solani*	Inhibitory	[54]
TiO_2 NPs	40 mg/L	*Bipolaris sorokiniana*	Inhibitory	[62]
ZnO NPs	0.10 g/L,0.20 g/L	*Pseudomonas syringae*, *X. campestris*, *P. carotovorum*, *A. solani*, and *F. oxysporum f. sp. lycopersici*	Inhibitory	[68]
AgNPs	100 ppm	*Ralstonia* and *Phomopsis*	Inhibitory	[69]
TiO_2 NPs	250 ppm	*Podosphaera xanthii*	Inhibitory	[70]
Se NPs	100 ppm	*P. grisea*, *C. capsica*, and *A. solani*	Inhibitory	[71]
Si	2 mmol/L	*Colletotrichum sublineolum*	Inhibitory	[72]
Nano-Si-Ag	100 ppm	*P. ultimum*, *M. grisea*, *C. gloeosporioides*, and*X. compestris* pv. vesicatoria	Inhibitory	[73]

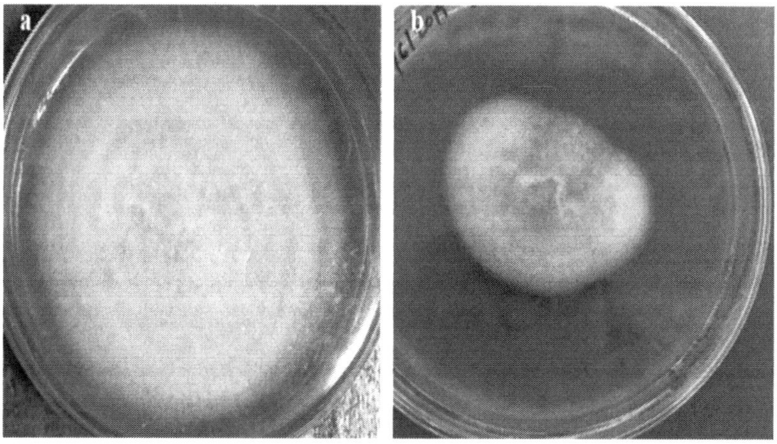

FIGURE 1.3 Antifungal activity of *F. oxysporum* on potato dextrose agar (PDA) medium (a) without AgNPs; (b) treated with AgNPs showed a decrease in mycelium. (From Khan, M. et al., *Molecules,* 26, 2462, 2021.)

1.6 CONCLUSION

Synthetic pesticides have toxic effects on human health and the environment. Due to this, the search for new eco-friendly products to reduce plant diseases has increased. Currently, nanotechnology is a promising alternative, which can enhance crop yield and productivity by reducing plant diseases. These biocontrol agents are termed biopesticides, and the demand for nano-based pesticides will increase in the future due to awareness of the need to use less ecologically toxic pesticides. Additionally, it has been wellestablished that a range of nanoparticles can promote plant growth

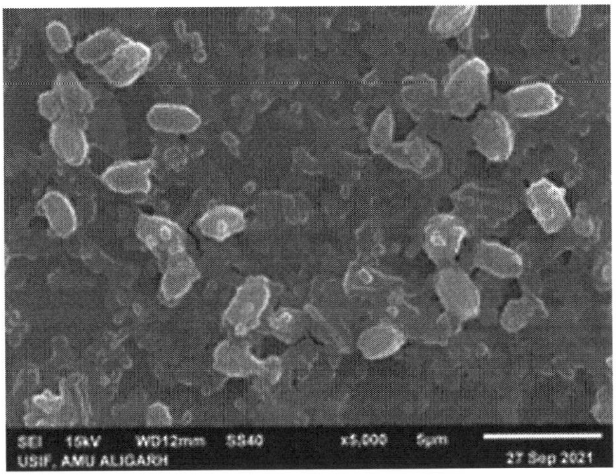

FIGURE 1.4 SEM micrograph of bacteria (*Ralstonia solanacearum*) treated with MgO-NPs. (From Khan, A. U. et al., *Bioinorganic Chemistry and Applications*, 321, 2022.)

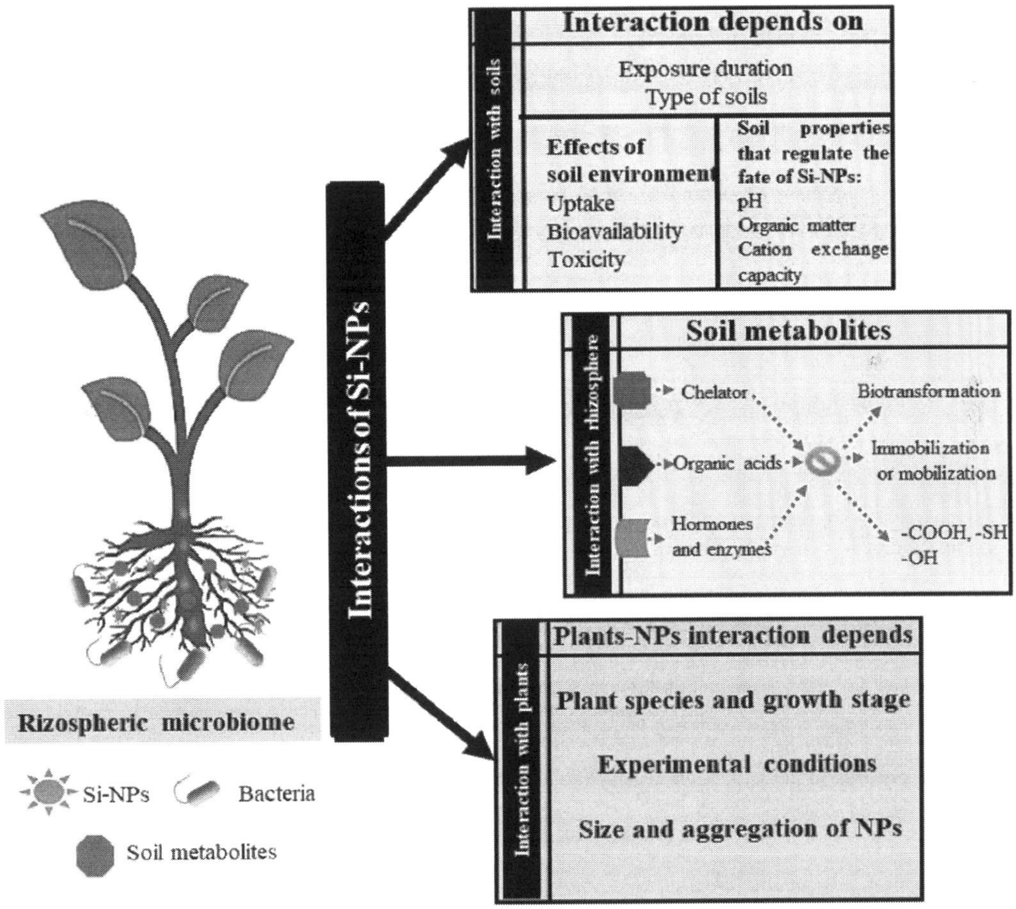

FIGURE 1.5 Schematic representations of interactions of Si-NPs with the rhizospheric microbiome. (From Rajput, V. D. et al., *Biology*, 2021, 10, 791.)

and lead to better crop yields. Altogether, this chapter provides knowledge for designing research programs that aim to develop nano-based products as potential biopesticides for managing a range of crop diseases and as growth-enhancing nanoparticles for better crop yields to achieve sustainable agriculture and global food security in the future.

REFERENCES

1. World Health Organization (WHO). The State of Food Security and Nutrition in the World Transforming food systems for food security, improved nutrition and affordable healthy diets for all. *Food and Agriculture Organization*, 2021July 12.
2. Cohn, B.; Wolff, M.; Cirillo, P.; Sholtz, R.DDT and breast cancer in young women: New data on the significance of age at exposure. *Environmental Health Perspectives*, 2007, *115*(10), 1406–1412.
3. Atwood, D.; Paisley-Jones, C.2008-2012 Market estimates. *Pesticides Industry Sales and Usage*, 2017.
4. World Health Organization (WHO). *Public Health Impact of Pesticides Used in Agriculture*, 2016.
5. Weisenburger, D. D.Human health effects of agrichemical use. *Human Pathology*, 1993, *24*(6), 571–576.
6. Kortekamp, A.*Herbicides and Environment*. BoD–Books on Demand Jan 8, 2011.
7. Khan, A. U.; Khan, M.; Malik, N.; Cho, M. H.; Khan, M. M.Recent progress of algae and blue–green algae-assisted synthesis of gold nanoparticles for various applications. *Bioprocess and Biosystems Engineering*, 2019, *42*(1), 1–15.
8. Khan, M.; Khan, A. U.; Hasan, M. A.; Yadav, K. K.; Pinto, M. M. C.; Malik, N.; Yadav, V. K.; Khan, A. H.; Islam, S.; Sharma, G. K. et al. Agro-nanotechnology an emerging field: A novel sustainable approach for improving the plant growth by reducing biotic stress. *Applied Sciences*, 2021, *11*(5), 2282.
9. Sasser, J. N.; Freckman, D. W.A world perspective on nematology: The role of the society. In: *Vistas on Nematology* (J.A.Veech and D.W.Dickerson, eds), pp. 7–14. Hyatsville: Society of Nematologists, 1987.
10. Chitwood, D. J.Research on plant-parasitic nematode biology conducted by the United States Department of Agriculture–Agricultural Research Service. *Pest Management Science: Formerly Pesticide Science*, 2003, *59*(6–7), 748–753.
11. Elling, A. A.Major emerging problems with minor meloidogyne species. *Phytopathology*, 2013, *103*(11), 1092–1102.
12. Tamilarasan, S.; Rajam, M.Engineering crop plants for nematode resistance through host-derived RNA interference. *Cell and Developmental Biology*, 2013, *2*, 2–5.
13. Khan, M.; Siddiqui, Z. A.Interaction of *Meloidogyne incognita*, *Ralstonia solanacearum* and *Phomopsis vexans* on eggplant in the sand mix and fly ash mix soils. *Scientia Horticulturae*, 2017, *225*, 177–184.
14. Khan, M.; Khan, A. U.Plant parasitic nematodes effectors and their crosstalk with defense response of host plants: A battle underground. *Rhizosphere*, 2021, *17*, 100288.
15. Jones, J. T.; Haegeman, A.; Danchin, E. G.; Gaur, H. S.; Helder, J.; Jones, M. G.; Perry, R. N.Top 10 plant-parasitic nematodes in molecular plant pathology. *Molecular Plant Pathology*, 2013, *14*(9), 946–961.
16. Khan, M.; Siddiqui, Z. A.; Parveen, A.; Khan, A. A.; Moon, I. S.; Alam, M.Elucidating the role of silicon dioxide and titanium dioxide nanoparticles in mitigating the disease of the eggplant caused by Phomopsis vexans, Ralstonia solanacearum, and root-knot nematode *Meloidogyne incognita*. *Nanotechnology Reviews*, 2022a, *11*(1), 1606–1619.
17. Khan, M.; Khan, A. U.; Rafatullah, M.; Alam, M.; Bogdanchikova, N.; Garibo, D.Search for effective approaches to fight microorganisms causing high losses in agriculture: Application of P. lilacinum metabolites and Mycosynthesised silver nanoparticles. *Biomolecules*, 2022b, *12*(2), 74.
18. Ahamad, L.; Siddiqui, Z. A.Effects of silicon dioxide, zinc oxide and titanium dioxide nanoparticles on *Meloidogyne incognita*, *Alternaria dauci and Rhizoctonia solani* disease complex of carrot. *Experimental Parasitology*, 2021, *230*, 108176.
19. Khan, M.; Siddiqui, Z. A.Zinc oxide nanoparticles for the management of *Ralstonia solanacearum*, *Phomopsis vexans* and *Meloidogyne incognita* incited disease complex of eggplant. *Indian Phytopathology*, 2018, *71*(3), 355–364.
20. Nour El-Deen, A. H.; El-Deeb, B. A.Effectiveness of silver nanoparticles against root-knot nematode, *Meloidogyne incognita* infecting tomato under greenhouse conditions. *Journal of Agricultural Science*, 2018, *10*(2), 148–156.
21. Kalaiselvi, D.; Sundararaj, P.; Premasudha, P.; Hafez, S. L.Nematicidal activity of green synthesized silver nanoparticles using plant extracts against root-knot nematode meloidogyne incognita. *International Journal of Nematology*, 2017, *22*(1 and 2), 81–94.

22. Thakur, R. K.; Shirkot, P.Potential of biogold nanoparticles to control plant pathogenic nematodes. *Journal of Bioanalysis and Biomedicine*, 2017, *9*, 220–222.
23. Abdellatif, K. F.; Hamouda, R. A.; El-Ansary, M. S. M.Green nanoparticles engineering on root-knot nematode infecting eggplants and their effect on plant DNA modification. *Iranian Journal of Biotechnology*, 2016, *14*(4), 250–259.
24. Cromwell, W. A.; Yang, J.; Starr, J. L.; Jo, Y. K.Nematicidal effects of silver nanoparticles on root-knot nematode in Bermudagrass. *Journal of Nematology*, 2014, *46*(3), 261–266.
25. Ardakani, A. S.Toxicity of silver, titanium and silicon nanoparticles on the root-knot nematode, *Meloidogyne incognita* and growth parameters of tomato. *Nematology*, 2013, *15*(6), 671–677.
26. Derbalah, A.; Elsharkawy, M. M.; Hamza, A.; El-Shaer, A.Resistance induction in cucumber and direct antifungal activity of zirconium oxide nanoparticles *againstRhizoctonia solani*. *Pesticide Biochemistry and Physiology*, 2019, *157*, 230–236.
27. Chao, S. H. L.; Choi, H. S.Method for providing enhanced photosynthesis. Korea Research Institute of Chemical Technology: Jeonju, South Korea. *Bulletin*, 2005, 10.
28. Bao-shan, L.; Chun-hui, L.; Li-jun, F.; Shu-chun, Q.; Min, Y.Effect of TMS (nanostructured silicon dioxide) on growth of Changbai larch seedlings. *Journal of Forestry Research*, 2004, *15*(2), 138–140.
29. Azimi, R.; Borzelabad, M. J.; Feizi, H.; Azimi, A.Interaction of SiO_2 nanoparticles with seed prechilling on germination and early seedling growth of tall wheatgrass (*Agropyron elongatum* L.). *Polish Journal of Chemical Technology*, 2014, *16*(3), 25–29.
30. Nair, R.; Varghese, S. H.; Nair, B. G.; Maekawa, T.; Yoshida, Y.; Kumar, D. S.Nanoparticulate material delivery to plants. *Plant Science*, 2010, *179*(3), 154–163.
31. Li, Y.; Zhu, N.; Liang, X.; Bai, X.; Zheng, L.; Zhao, J.; Li, Y.-F.; Zhang, Z.; Gao, Y.Silica nanoparticles alleviate mercury toxicity via immobilization and inactivation of Hg(ii) in soybean (Glycine max). *Environmental Science: Nano*, 2020, *7*(6), 1807–1817.
32. El-Batal, A. I.; Gharib, F. A. E.-L.; Ghazi, S. M.; Hegazi, A. Z.; Hafz, A. G. M. A. E.Physiological responses of two varieties of common bean (Phaseolus vulgaris L.) to foliar application of silver nanoparticles. *Nanomaterials and Nanotechnology*, 2016, *6*, 13.
33. Noshad, A.; Hetherington, C.; Iqbal, M.Impact of AgNPs on seed germination and seedling growth: A focus study on its antibacterial potential against Clavibacter michiganensis subsp. michiganensis infection in *Solanum lycopersicum*. *Journal of Nanomaterials*, 2019, 6316094. https://doi.org/10.1155/2019/6316094.
34. Rizwan, M.; Ali, S.; Ali, B.; Adrees, M.; Arshad, M.; Hussain, A.; Zia Ur Rehman, M.; Waris, A. A.Zinc and iron oxide nanoparticles improved the plant growth and reduced the oxidative stress and cadmium concentration in wheat. *Chemosphere*, 2019a, *214*, 269–277. https://doi.org/10.1016/J.CHEMOSPHERE.2018.09.120.
35. Raja, K.; Sowmya, R.; Sudhagar, R.; Moorthy, P. S.; Govindaraju, K.; Subramanian, K. S.Biogenic ZnO and Cu nanoparticles to improve seed germination quality in black gram (*Vigna mungo*). *Materials Letters*, 2019, *235*, 164–167.
36. Rajiv, P.; Vanathi, P.Effect of Parthenium based vermicompost and zinc oxide nanoparticles on growth and yield of *Arachis hypogaea* L. in zinc deficient soil. *Biocatalysis and Agricultural Biotechnology*, 2018, *13*, 251–257.
37. Faizan, M.; Faraz, A.; Yusuf, M.; Khan, S. T.; Hayat, S.Zinc oxide nanoparticle-mediated changes in photosynthetic efficiency and antioxidant system of tomato plants. *Photosynthetica*, 2018, *56*(2), 678–686.
38. Pullagurala, R. V. L.; Adisa, I. O.; Rawat, S.; Kalagara, S.; Hernandez-Viezcas, J. A.; Peralta-Videa, J. R.; Gardea-Torresdey, J. L.ZnO nanoparticles increase photosynthetic pigments and decrease lipid peroxidation in soil grown cilantro (*Coriandrum sativum*). *Plant Physiology and Biochemistry: PPB*, 2018, *132*, 120–127.
39. Awasthi, A.; Bansal, S.; Jangir, L. K.; Awasthi, G.; Awasthi, K. K.; Awasthi, K.Effect of ZnO nanoparticles on germination of *Triticum aestivum* Seeds. *Macromolecular Symposium*, 2017, *376*(1), 1700043. https://doi.org/10.1002/masy.201700043.
40. Dhoke, S. K.; Mahajan, P.; Kamble, R.; Khanna, A.Effect of nanoparticles suspension on the growth of mung (*Vigna radiata*) seedlings by foliar spray method. *Nanotechnology Development*, 2013, *3*(1), 1–5.
41. Asgari, F.; Majd, A.; Jonoubi, P.; Najafi, F.Effects of silicon nanoparticles on molecular, chemical, structural and ultrastructural characteristics of oat (*Avena sativa* L.). Plant Physiology and Biochemistry: PPB, *127*, 152–160.
42. Haghighi, M.; Pessarakli, M.Influence of silicon and nanosilicon on salinity tolerance of cherry tomatoes (*Solanum lycopersicum* L.) at early growth stage. *Scientific Horticulture*, 2013, *161*, 111–117.

43. Yinfeng, X.; Bo, L.; Gongsheng, T. A. O.; Qianqian, Z.; Chunxia, Z.Effects of nano-silicon dioxide on photosynthetic fluorescence characteristics of *Indocalamus barbatus* McClure. Journal of Nanjing Forestry University (Natural Sciences Edition), 2012, 36, 59–63.
44. Haghighi, M.; Afifipour, Z.; Mozafarian, M.The effect of N-Si on tomato seed germination under salinity levels. *Journal of Biological & Environmental Sciences*, 2012, *6*, 87–90.
45. Rezaei, F.; Moaveni, P.; Mozafari, H.Effect of Different Concentrations and Time of Nano TiO2 Spraying on Quantitative and Qualitative Yield of Soybean (*Glycine max L.*) at Shahr-e-Qods. *Biological Forum*, 2015, *7*, 957–964.
46. Raliya, R.; Biswas, P.; Tarafdar, J.TiO_2 nanoparticle biosynthesis and its physiological effect on mung bean (*Vigna radiata* L.). *Biotechnology Reports*, 2015, *5*, 22–26.
47. Samadi, N.; Yahyaabadi, S.; Rezayatmand, Z.Effect of TiO2 and TiO2 nanoparticle on germination, root and shoot length and photosynthetic pigments of *Mentha piperita*. *International Journal of Plant and Soil Science*, 2014, *3*(4), 408–418.
48. Reyhaneh, A.; Hassan, F.; Hosseini, M. K.Can bulk and nanosized titanium dioxide particles improve seed germination features of wheatgrass (*Agropyron desertorum*). *Notulae Scientia Biologicae*, 2013, *5*(3), 325–331.
49. Mall, D.; Larsen, A.; Martin, E.Investigating the (mis) match between natural pest control knowledge and the intensity of pesticide use. *Insects*, 2018, *9*(1), 2.
50. Tsuji, K.Microencapsulation of pesticides and their improved handling safety. *Journal of Microencapsulation*, 2001, *18*(2), 137–147.
51. Strout, G.; Russell, S. D.; Pulsifer, D. P.; Erten, S.; Lakhtakia, A.; Lee, D. W.Silica nanoparticles aid in structural leaf coloration in the Malaysian tropical rainforest understorey herb *Mapania caudata*. *Annals of Botany*, 2013, *112*(6), 1141–1148.
52. Rastogi, A.; Tripathi, D. K.; Yadav, S.; Chauhan, D. K.; Zivcak, M.; Ghorbanpour, M.; El-Sheery, N. I.; Brestic, M.Application of silicon nanoparticles in agriculture. *3 Biotech*, 2019, *9*(3), 90.
53. Ye, M.; Song, Y. Y.; Long, J.; Wang, R. L.; Baerson, S. R.; Pan, Z.; Zhu-Salzman, K., Xie, J., Cai, K., Luo, S.; Zeng, R. Priming of jasmonate-mediated antiherbivore defense responses in rice by silicon. *Proceedings of the National Academy of Sciences of the United States of America*, 2013, *110*(38), E3631–E3639.
54. Parveen, A.; Siddiqui, Z. A.Impact of silicon dioxide nanoparticles on growth, photosynthetic pigments, proline, activities of defense enzymes and some bacterial and fungal pathogens of tomato. *Vegetos*, 2022, *35*, 83–93.
55. Sabir, S.; Arshad, M.; Chaudhari, S. K.Zinc oxide nanoparticles for revolutionizing agriculture: Synthesis and applications. *The Scientific World Journal*, 2014,*925494*, 1–8. https://doi.org/10.1155/2014/925494
56. Bondarenko, O.; Juganson, K.; Ivask, A.; Kasemets, K.; Mortimer, M.; Kahru, A.Toxicity of Ag, CuO and ZnO nanoparticles to selected environmentally relevant test organisms and mammalian cells in vitro: A critical review. *Archives of Toxicology*, 2013, *87*(7), 1181–1200.
57. George, S. A.; Raj, M. S.; Solomon, D.; Roselin, P.Comparative study of the anti-fungal activity of zinc oxide and titanium dioxide Nano and bulk particles with anti-fungals against fungi isolated from infected skin and dandruff flakes. *ResearchReview and Journal of Microbiology and Biotechnology*, 2014, *3*(3), 23–30.
58. Siddiqui, Z. A.; Khan, M. R.; Abd Allah, E. F.; Parveen, A.Titanium dioxide and zinc oxide nanoparticles affect some bacterial diseases, and growth and physiological changes of beetroot. *International Journal of Vegetable Science*, 2018, *25*(5), 409–430.
59. Kaur, J.; Singh, J.; Rawat, M.An efficient and blistering reduction of 4-nitrophenol by green synthesized silver nanoparticles. *SN Applied Sciences*, 2019, *1*(9), 1–6.
60. Namburi, K. R.; Kora, A. J.; Chetukuri, A.; Kota, V. S. M. K.Biogenic silver nanoparticles as an antibacterial agent against bacterial leaf blight causing rice phytopathogen *Xanthomonas oryzae* pv. *oryzae*. *Bioprocess and Biosystems Engineering*, 2021, 1–14.
61. Khan, M.; Khan, A. U.; Bogdanchikova, N.; Garibo, D.Antibacterial and antifungal studies of biosynthesized silver nanoparticles against plant parasitic nematode meloidogyne incognita, Plant pathogens Ralstonia solanacearum and Fusarium oxysporum. *Molecules*, 2021, *26*(9), 2462.
62. Satti, S. H.; Raja, N. I.; Javed, B.; Akram, A.; Mashwani, Z. U. R.; Ahmad, M. S.; Ikram, M.Titanium dioxide nanoparticles elicited agro-morphological and physicochemical modifications in wheat plants to control *Bipolaris sorokiniana*. *PLOS ONE*, 2021, *16*(2), e0246880.

63. Parveen, A.; Siddiqui, Z. A. Zinc oxide nanoparticles affect growth, photosynthetic pigments, proline content and bacterial and fungal diseases of tomato. *Archives of Phytopathology and Plant Protection*, 2021, *54*(17–18), 1519–1538.
64. Renuka, L.; Anantharaju, K. S.; Sharma, S. C.; Nagaswarupa, H. P.; Prashantha, S. C.; Nagabhushana, H.; Vidya, Y. S.Hollow microspheres Mg-doped ZrO2 nanoparticles: Green assisted synthesis and applications in photocatalysis and photoluminescence. *Journal of Alloys and Compounds*, 2016, *672*, 609–622.
65. Gowri, S.; Gandhi, R. R.; Sundrarajan, M.Structural, optical, antibacterial and antifungal properties of zirconia nanoparticles by biobased protocol. *Journal of Materials Science and Technology*, 2014, *30*(8), 782–790.
66. Jalill, R. D. A.; Numan, R. S.Silver nitrate and zirconium oxide nanoparticles as management of wheat damping-off caused by *Fusarium graminearum*. *Journal of Genetic and Environmental Resources Conservation*, 2016, *4*(2), 85–93.
67. Khan, M.; Khan, A. U.; Bogdanchikova, N.; Garibo, D.Antibacterial and antifungal studies of biosynthesized silver nanoparticles against plant parasitic nematode meloidogyne incognita, Plant pathogens Ralstonia solanacearum and Fusarium oxysporum. *Molecules*, 2021, *26*(9), 2462.
68. Parveen, A.; Siddiqui, Z. A.Zinc oxide nanoparticles affect growth, photosynthetic pigments, proline content and bacterial and fungal diseases of tomato. *Archives of Phytopathology and Plant Protection*, 2021, 1–20.
69. Khan, M.; Khan, A. U.; Park, J.; Alam, M.; Alam, M.Biosynthesis of silver nanoparticles and its application against phytopathogenic bacterium and fungus. *International Journal of Environmental Analytica Chemistry*, 2020, *100*(12), 1390–1401.
70. Hafez, Y. M.; Attia, K. A.; Kamel, S.; Alamery, S. F.; El-Gendy, S.; Al-Dosse, A., Mehiar, F., Ghazy, A.; Abdelaal, K. A.Bacillus subtilis as a bio-agent combined with Nano molecules can control powdery mildew disease through histochemical and physiobiochemical changes in cucumber plants. *Physiological and Molecular Plant Pathology*, 2020, *111*, 101489.
71. Joshi, S. M.; De Britto, S.; Jogaiah, S.; Ito, S. I.Mycogenic selenium nanoparticles as potential new generation broad spectrum antifungal molecules. *Biomolecules*, 2019, *9*(9), 419.
72. Resende, R. S.; Rodrigues, F. A.; Gomes, R. J.; Nascimento, K. J. T.Microscopic and biochemical aspects of sorghum resistance to anthracnose mediated by silicon. *Annals of Applied Biology*, 2013, *163*(1), 114–123.
73. Park, H. J.; Kim, S. H.; Kim, H. J.; Choi, S. H.A new composition of nano sized silica- silver for control of various plant diseases. *Plant Pathology Journal*, 2006, *22*(3), 295–302.
74. Khan, A. U.; Khan, M.; Parveen, A.; Ansari, S.; Alam, M.; Alam, M.Effect of phyto-assisted synthesis of magnesium oxide nanoparticles (MgO-NPs) on bacteria and the root-knot nematode. *Bioinorganic Chemistry and Applications*, 2022, 321.
75. Rajput, V. D.; Minkina, T.; Feizi, M.; Kumari, A.; Khan, M.; Mandzhieva, S., et al Effects of silicon and silicon-based nanoparticles on rhizosphere microbiome, plant stress and growth. *Biology*, 2021, *10*(8), 791.
76. Liang, Y.; Guo, M.; Fan, C.; Dong, H.; Ding, G.; Zhang, W.Development of novel urease responsive pendimethalin microcapsules using silica- IPTS-PEI as controlled release carrier materials. *ACS Sustainable Chemistry and Engineering*, 2017, *5*(6), 4802–4810.
77. Nguyen, H.; Hwang, I.; Park, J.; Park, H.Enhanced payload and photo-protection for pesticides using nanostructured lipid carriers with corn oil as liquid lipid. *Journal of Microencapsulation*, 2012, *29*(6), 596–604.
78. Wang, Y.; Cui, H.; Sun, C.; Zhao, X.; Cui, B.Construction and evaluation of controlled-release delivery system of abamectin using porous silica nanoparticles as carriers. *Nanoscale Research Letters*, 2014, *9*(1), 2490.
79. Campos, E. V. R.; Proença, P. L. F.; Oliveira, J. L.; Melville, C. C.; Della Vechia, J. F.; De Andrade, D. J.; Fraceto, L. F.Chitosan nanoparticles functionalized with β-cyclodextrin: A promising carrier for botanical pesticides. *Scientific Reports*, 2018, *8*(1), 2067.
80. Zhang, J.; Li, M.; Fan, T.; Xu, Q.; Wu, Y.; Chen, C.; Huang, Q.Construction of novel amphiphilic chitosan copolymer nanoparticles for chlorpyrifos delivery. *Journal of Polymer Research*, 2013, *20*(3), 107.
81. El-Samahy, M.; Galal, O. A.Evaluation of silica nanoparticles as a new approach to control faba bean (Vicia faba L.) insects and its genotoxic effect on M2 plants. *Egyptian Journal of Agricultural Research*, 2012, *90*(2), 869–888.

82. El-Samahy, M.; Khafagy, I.; El-Ghobary, A.Efficiency of silica nanoparticles, two bioinsecticides, peppermint extract and insecticide in controlling cotton leafworm, Spodoptera littoralis Boisd. and their effects on some associated natural enemies in sugar beet fields. *Journal of Plant Protection and Pathology*, 2015, *6*(9), 1221–1230.
83. Deka, B.; Babu, A.; Baruah, C.; Barthakur, M.Nanopesticides: A systematic review of their prospects with special reference to tea pest management. *Frontiers in Nutrition*, 2021, *8*, 686131.
84. Yang, F. L.; Li, X. G.; Zhu, F.; Lei, C. L.Structural characterization of nanoparticles loaded with garlic essential oil and their insecticidal activity against Tribolium castaneum (Herbst) (Coleoptera: Tenebrionidae). *Journal of Agriculture and Food Chemistry*, 2009, *57*(21), 10156–10162.
85. Yang, Y.; Cheng, J.; Garamus, V. M.; Li, N.; Zou, A.Preparation of an environmentally friendly formulation of the insecticide nicotine hydrochloride through encapsulation in chitosan/tripolyphosphate nanoparticles. *Journal of Agriculture and Food Chemistry*, 2018, *66*(5), 1067–1074.
86. Oliveira, H. C.; Stolf-Moreira, R.; Martinez, C. B. R.; Grillo, R.; de Jesus, M. B.; Fraceto, L. F.Nanoencapsulation enhances the post-emergence herbicidal activity of atrazine against mustard plants. *PLOS ONE*, 2015, *10*(7), e0132971.
87. Wang, X.; Liu, X.; Chen, J.; Han, H.; Yuan, Z.Evaluation and mechanism of antifungal effects of carbon nanomaterials in controlling plant fungal pathogen. *Carbon*, 2014, *68*, 798–806.

2 The Application of Nanotechnology to Accelerate Plant Adaptation to Abiotic Stresses

Saira Zaheer, Junaid Shehzad, Sunbal Khalil Chaudhari, and Ghazala Mustafa

2.1 INTRODUCTION

Changing climate and the resulting stresses on plants are major restraints that negatively affect plant growth and cause a decline in plant yield (Rasheed *et al.*, 2020). There will be different types of stresses, like drought, salt, changing patterns of rainfall, floods, temperature, UV, and heavy metals (Shinozaki and Yamaguchi-Shinozaki, 2007). As a result of the changing climate, global hunger and food insecurity have been constantly increasing in the last few decades (Usman *et al.*, 2020). Moreover, about 90% of cultivated land areas are exposed to many types of abiotic pressures (Dos Reis *et al.*, 2021), which may ultimately reduce the harvest of chief food crops by 70% (Mantri *et al.*, 2012). Therefore, the development of new and improved technologies in the agricultural division is compulsory to deal with such challenges. Although earlier skills, like the use of synthetic chemicals, genetic engineering, and plant breeding, enhanced agricultural efficiency, these techniques have some limitations (Fincheira *et al.*, 2020). Therefore, further inventions are highly encouraged to meet the increasing need for food and provide economic and environmental health benefits.

Nanotechnology and the nano-sciences are the most auspicious and innovative tools in the current situation and have emerged broadly into the "agri-food sector" (Khan *et al.*, 2019). Improvement in plant growth under different diseases and abiotic stresses is one of the most important objectives of biotechnological research. Nanotechnology also includes the transfer of targeted essential components, tracking devices, nanosensors, product developments, and accurate processing (Tariq *et al.*, 2021). Respective findings have been documented to describe the whole mechanism by which NPs improve plant safety against several abiotic stresses. NPs improve the morpho-physiological parameters of plants by moderating their molecular and biochemical mechanisms. Among the biochemical attributes, NPs enhanced antioxidant enzyme activities and the buildup of osmolytes (Abdoli *et al.*, 2020). Moreover, NPs also enhance nitrogen metabolism, chlorophyll content, and photosynthetic actions (Alsaeedi *et al.*, 2019). Additionally, NPs help to upregulate the expression of several genes under stress conditions. All these features affect the secondary metabolism and plant growth regulators that produce the bioactive compounds that protect plants against abiotic stresses.

NPs can be applied for refining agricultural productivity and for abiotic stress management. This chapter summarizes and explains the molecular, physiological, and biochemical mechanisms by which plants cope with abiotic stress using nanotechnology. This frame of knowledge identifies the key gaps to assess and minimize the properties of different stresses in plants through nanotechnology.

DOI: 10.1201/b23308-2

2.2 NANOPARTICLE SYNTHESIS, MECHANISM OF INCORPORATION, AND TRANSLOCATION IN PLANTS

Recently, the synthesis of biocompatible NPs by eco-friendly technologies has become significant for its applications in various fields. For NPs synthesis, several biological, physical, and chemical methods have been proposed that significantly contribute to distinctive nanoparticle properties (Mustafa *et al.*, 2022). Commonly, the chemical and biological approaches are the best to obtain stable and consistently sized NPs (Tan *et al.*, 2019). Furthermore, the physiological parameters of NPs (e.g., shape and size) can also be modulated by changing the reaction conditions and chemical concentrations (Joris *et al.*, 2013). This goal is achieved by using natural resources and ideal solvent systems. Microorganisms, fungi, and plants can be used for biological synthesis. Among the different biological synthesis methods, the utilization of plant extracts is a reasonably easy and simple procedure to produce metal-based NPs (Yashni *et al.*, 2020). The occurrence of macromolecules in medicinal plants also improves the functional properties of the green synthesized NPs (Murthy *et al.*, 2020). These types of nanomaterials have applications in biomedical diagnostics, catalysis, labeling of biological systems, optical imaging, antimicrobials, and molecular sensing.

The uptake and transport of NPs are dependent on their size and the presence of the plant cell wall, which constitutes an extra layer that they must cross to enter the cell. The uptake of NPs by plants depends primarily on the means of exposure. This is why the selection of efficient methods for NP incorporation is the first and foremost step toward achieving the desired results (De Almeida *et al.*, 2021). NPs can be incorporated through the incubation of seeds, pollen, roots, and protoplast or by foliar spray on plants. Hydroponic treatment, biolistic delivery of NPs, and irrigation with NPs are some other accessible methods of application (Noori and Kareem, 2019). Several researchers have proposed that NPs, due to their lower toxicity, might help plants to overcome abiotic stresses when compared with their bulk and ionic counterparts (Khan *et al.*, 2022). When NPs are applied to the soil, they gradually move through the soil pores through physical pressure and water movement. After that, they adhere to the plant surfaces via different forces such as van der Waals, hydrophobic, and electrostatic forces (Zhang, 2015). When NPs are applied to the aerial plant parts, they directly insert themselves through wounds, stigmata, stomata, lenticels, hydathodes, trichomes, and cuticles (Gelaw and Sanan-Mishra, 2022). After the uptake of NPs by the roots of plants, they are translocated to the aerial parts, including newly developed seeds (Ali *et al.*, 2021). The accumulation and translocation of NPs in plants mostly depend on the structure and physiology of the plant cells. NPs take both symplastic and apoplastic routes for cell-to-cell transportation (Zahedi *et al.*, 2020). These NPs also travel via some exclusive pathways (for instance, by generating pores or via the endocytosis pathway). Passive mechanisms of NPs transport include translocation through the endodermis by the apoplast, while transport through the plasma membrane is an active mechanism (Usman *et al.*, 2020).

Recently, several studies were performed to describe the mechanism of NP translocation in different plants. CuO NPs are translocated from the origin to sprouts via the xylem and back to the root via the phloem in *Z. mays* (Wang *et al.*, 2012). The absorbance and transfer of conjugated TiO_2 NPs exhibited tissue- and cell-specific distribution in *Arabidopsis thaliana* (Kurepa *et al.*, 2020). Furthermore, Lee *et al.* (2018) detected the probable uptake and translocation of CuNPs in *Phaseolus radiatus* and *Triticum aestivum* by using an agar growth medium. The transformation ability of NPs in the plants was observed using synchrotron X-ray absorption spectroscopy. Despite all these studies, still, a well-defined study related to the uptake and transport of NPs is essential to better understand the mechanisms by which they are translocated.

2.3 INTERACTION MECHANISMS OF PLANTS AND NANOPARTICLES

2.3.1 Nanoparticles' Impact on Seed Germination and Growth

Numerous studies have verified that the application of non-toxic NPs treatments can increase yield by promoting seed germination and plant growth. Furthermore, NPs can also be used to defend plants from environmental stresses such as salt, heat, cold, or drought and lessen the accretion and toxicity of heavy metals (Landa, 2021). In *Brassica napus* and *Camelina* seedlings, the growth percentage of each treatment (0.1 to 1000 ppm) was measured for 6 days. After the growth test, these plantlets were implanted in the Hoagland solution and again treated with different doses of ZnO NPs for a week. The results exhibited a 41% decrease in plantlets of *B. napus* and *Camelina* when treated with ZnO NPs (50 ppm) as compared with the control. Only a 15% increase in root length of *Camelina* seedlings was observed when treated with ZnO NPs (1 ppm) as compared with control plants (Sarkhosh *et al.*, 2022). The germination rate was significantly improved in zucchini and watermelon in response to Ag NPs as compared with nontreated seeds (Almutairi and Alharbi, 2015). Nair *et al.* (2012) exposed *Oryza sativa* seedlings to graphene and found better growth and yield relative to untreated seedlings. Likewise, in tomato, the germination capability was improved by graphene; this may be due to graphene's capability to progress water uptake through the seed coat (Zhang, 2015). The germination rate of lettuce and cucumber seeds was encouraged by treatment with a 62 μg mL^{-1} solution of Au NPs for a week (Barrena *et al.*, 2009). In pearl millet (*Pennisetum glaucum*), parallel results were also observed when seeds were soaked for 2 hours in 20 and 50 μg mL^{-1} Au NPs solution (Parveen *et al.*, 2016). Lee *et al.* (2008) verified significant effects of Cu NPs on mung bean and wheat, and at comparatively high doses (200–1000 mg L^{-1}), a significant reduction is observed in both root and seedling growth in a concentration-dependent manner. Markedly significant effects of Cu NPs (0.2–1-mg L^{-1} doses) were observed in the germination assessment of wheat (Pakistani wheat variety Millat-2011) (Hafeez *et al.*, 2015). In plants, seed priming with ZnO NPs increases the grain vintage as compared with non-primed seeds in response to salinity stress. Nutrient uptake, especially K, N, and P, in both grains and straw were significantly increased at high salinity levels in the nano-primed seeds. While in field training, we may enhance the effectiveness of NPs in seed priming for salinity tolerance (Al-Salama, 2022).

2.3.2 Nanoparticles' Effect on Photosynthesis

The interaction of NPs with plants results in several physiological and genotoxic changes that alter plant growth. NPs promote both negative and positive effects on plant growth, depending upon the properties of the nanomaterials as well as the plant species. The interaction involves oxidative stress responses, mitochondrial respiration, cell wall metabolism, and photosynthetic electron transport (Da Costa and Sharma, 2016). It has been reported that a low concentration (50 ppm) of CuO NPs significantly increased the growth characteristics, photosynthetic pigments, and antioxidant enzymes of wheat plants, while no significant changes occurred in carbohydrate or total protein content (Badawy *et al.*, 2021). Silver NPs (Ag NPs) can disrupt the photosynthetic system by reducing the synthesis of chlorophyll in leaves (Tripathi *et al.*, 2017). In *Arabidopsis* leaves, Ag NPs inhibit growth and decrease chlorophyll content by disrupting the structure of the thylakoid membrane (Qian *et al.*, 2013). Similarly, Nair and Chung (2014) also reported that 1 week of exposure to Ag NPs decreased the carotenoids and total chlorophyll content in *Oryza sativa* L. After accumulation in *Brassica* species, Ag NPs caused severe inhibition of photosynthesis (Vishwakarma *et al.*, 2017). Exposure of *Lupinus termis* L. seedlings to Ag NPs reduced the root and shoot elongation, protein, and total chlorophyll content (Al-Huqail *et al.*, 2018). Liang *et al.* (2018) stated that exposure to Ag NPs changed the structure of thylakoids in *Physcomitrella patens* and reduced the chlorophyll b content. The rate of transpiration was significantly reduced in *Cucurbita pepo* after exposure to Ag NPs (Hawthorne *et al.*, 2012). Another important nutrient, Fe_3O_4 NPs, significantly increased plant growth (9%–32%) and chlorophyll a content (1.94–2.8-fold) (Kokina *et al.*, 2020). The findings of Tirani *et al.* (2019) highlight that tobacco responded positively to bulk ZnO and all nano ZnO levels by adjustment at the morphological, physiological, and anatomical levels. Van Nguyen *et al.* (2022) showed that application of Cu NPs significantly amplified the water content and photosynthesis in *Zea mays* L. under drought stress.

2.4 NANOPARTICLES AND PLANTS' STRESS TOLERANCE

2.4.1 Extreme Temperature

Plant growth, productivity, and development are compromised and suffer due to extreme heat and cold temperatures. These extreme temperatures cause denaturation of biomolecules, low germination rate, the disintegration of membrane lipids in crops, decreased photosynthetic rate, and slow growth rate (Ding *et al.*, 2020). NPs can lessen the effects of these extreme temperatures primarily by increased production of reactive oxygen species (ROS) and reducing oxidative stress (Aref *et al.*, 2013). Se NPs at a size of 10–40 nm increased the hydration of plants, chlorophyll content, and growth of tomatoes (*Solanum lycopersicum* L.) (Djanaguiraman *et al.*, 2018). In tomato seedlings, TiO_2 NPs stimulate the photosynthetic system by decreasing the energy dissolution of nonregulated Photosystem II (PS II) and increasing the energy degeneracy of regulated PS II in response to high-temperature stress (Qi *et al.*, 2013). In wheat, Ag NPs (10–20 nm) improved plant growth significantly under high heat stress (Iqbal *et al.*, 2019). In chickpeas (*Cicer arietinum* L.), the application of TiO_2 NPs not only prevented membrane damage but also improved oxidative damage and maintained the stability of chlorophyll and carotenoid accretions under cold stress (Mohammadi *et al.*, 2014; Mohammadi *et al.*, 2013) enhancing the chlorophyll content and expression of the gene of Rubisco-binding proteins (Hasanpour *et al.*, 2015). Khodakovkaya *et al.* (2012) showed that multi-walled carbon nanotubes (MWCNTs) prompted the stress response in plants by upregulating the different stress-linked gene expressions comprising *HSP90*. Similarly, CeO_2 NPs reduce the H_2O_2 content and upregulate the gene expression of *HSP70*. Song *et al.* (2021) reported that ZnO NPs mitigate cold stress by moderating the chilling response transcription factors in plant leaves.

2.4.2 Salinity

Advances in nanotechnology and engineering indicate that nano fertilizers can expand crop yield in the current atmosphere. Salt stress strongly restricts about 23% of crop production in the world's agricultural areas (Onaga and Wydra, 2016). In crops, NPs can improve the damage caused due to salinity stress in many ways, such as altering metal accumulation in plants and restoring the damaged photosynthetic system. In tomatoes and squash, treatment with SiO_2 NPs increased proline accumulation, chlorophyll content, seed germination, and dry weight under NaCl stress (Siddiqui and Al-Whaibi, 2014). In a sunflower cultivar, $FeSO_4$ NPs (foliar application) enhanced chlorophyll content and maximized the photochemical efficacy of PS II (Fv/Fm), shoot dry weight, net CO_2 assimilation rate, and leaf area, but reduced the sodium content in sunflower leaves under salinity stress (Torabian *et al.*, 2018). CeO_2 NPs improve tolerance in cotton by enabling the plant to maintain the cytosolic ratio of K^+/Na^+ (Liu *et al.*, 2021). In tomatoes (*S. lycopersicum* L.), 20 nm SiO_2 NPs reduces the effects of salinity on photosynthetic rate, chlorophyll, and fresh weight (Haghighi and Pourkhaloee, 2013), while in squash (*Cucurbita pepo* L.), 12 nm SiO_2 NPs reduced H_2O_2, electrolyte leakage, and levels of malondialdehyde (Siddiqui *et al.*, 2014). In rapeseed, seed priming improves salinity tolerance by moderating homeostasis of ROS and α-amylase actions (Khan *et al.*, 2021).

2.4.3 Drought

Phyto-nanotechnology has enormous potential to improve the effects of drought on different crop seedlings, restoring the damage to plant growth triggered by severe drought, encouraging seed water absorption, and improving water accumulation in crops. In barley, the application of MWCNTs and sodium nitroprusside (SN) NPs boosted seed water absorption under drought stress (Karami and Sepehri, 2017). Other findings stated that the application of TiO_2 NPs (0.02%) improved plant agronomic features, i.e., ear number, plant height, number of seeds, ear weight, gluten-included harvest index, biomass, starch content, and final yield under drought stress (Khan and Bano, 2018). Potentially, silicon can be used to reduce drought stress in plants. Previous findings demonstrated that in wheat, a 1.0 mM concentration of sodium silicate showed moderately antagonistic effects on drought stress. Though the detailed mechanism is still mysterious, silicon moderately increases leaf chlorophyll content, maintains the water capacity of the leaf, promotes shoot growth and decreases lipid peroxidation of the membrane in stressed plants (Pei *et al.*, 2010). Cumulative shortage of agronomic water badly affects agricultural production and destroys plants in the arid and semiarid ranges of the world. In soybean, the use of micronutrient NPs and CeO_2 NPs improved crop growth and development under drought stress (Dimkpa *et al.*, 2017).

2.4.4 Waterlogging and Flooding

Exposure of soybeans under flooding stress to Al_2O_3 NPs and Ag NPs reduces the adverse effects of stress. Interestingly, Poly(A)-Binding Protein 2 (PABP2) is the most interactive protein in soybean root under flooding stress when treated with NPs. This PABP of mRNA plays a significant role in translation initiation (Mustafa *et al.*, 2015a). Another study by Mustafa *et al.* (2016) analyzes the proteome of soybean in response to different sizes of Ag NPs under flooding stress. Morphological analysis revealed the modifying effect of 15 nm Ag NPs to mitigate flooding stress. All the reports together suggest that flooding stress mitigation by NPs involves limiting cell death in roots and reducing O_2 deprivation stress as well as the detoxification of toxic products and regulation of protein synthesis. Hasanuzzaman *et al.* (2017) showed that the introduction of Ag NPs to flooded plants reflects the inhibition of *Glyoxalase II 3*, thereby decreasing the cytotoxicity of plants. (Table 2.1).

2.5 MECHANISMS OF STRESS MITIGATION IN PLANTS BY NANOPARTICLE APPLICATION

2.5.1 Antioxidant Enzymes

Ag NPs significantly amplified the antioxidant activities in pearl millet leaves in response to salinity stress, and this increased with higher NPs doses. Salinity stress reduced the antioxidant enzymes' activities, but a 20 mM dose of NPs enhanced the activity of superoxide dismutase (SOD) by 25%, catalase (CAT) by 24%, and peroxidase (POD) by 49%. A similar trend at a similar dose was detected in the case of glutathione peroxidase (GPX) and glutathione reductase (GR) activities (Khan *et al.*, 2020). The most significant enhancement in antioxidant enzymes over control plants when treated with 100 ppm silver NPs in response to *Meloidogyne incognita* attack on plant roots (Danish *et al.*, 2021). However, the supreme improvement in the action of SOD (44%), CAT (57%), and POD (59%) was noted in plants subjected to ZnO NPs along with NaCl (Faizan *et al.*, 2021). In pomegranate trees under intense drought, POD, ascorbate peroxidase (APX), CAT, and SOD activities increased after foliar spraying with 10 nm Se NPs as compared with other treatments (Zahedi *et al.*, 2021). APX and POD were decreased, while SOD and CAT increased, when treated with lead and SiO_2 NPs (Ullah *et al.*, 2022). The highest activity of all antioxidant enzymes was documented at 5 mg L^{-1} CS-Se NPs in plants grown in a normal environment and under Cd stress (Azimi *et al.*, 2021). A rise in POD, CAT, and SOD activities of about 55.81%, 52.72%, and 39.02%, respectively, over the control was observed when plant leaves were sprayed with ZnO NPs (Singh *et al.*, 2021). Seed priming with 150 mM NaCl and Zn NPs alone mostly regulated all tested antioxidant activities, and this regulation was more apparent in plants treated with Zn NPs than others (Abdel Latef *et al.*, 2017). In addition, SOD displayed treatment-dependent increases, and POD and CAT activity were significantly induced, though antioxidant enzymes were increased at lower TiO_2 NPs (10–30 μg mL^{-1}) and reduced at the higher concentrations (Laware and Rasker, 2014).

TABLE 2.1
Plants' Responses to Nanomaterials Under Different Abiotic Stresses

Plant	Abiotic stress	Nanomaterial	Findings	References
Tomato	Salinity	Si NPs	Regulated genes *CRK*; *TAS1*; *AREB*; *NCED3*	Almutairi (2016a)
Soybean	Flooding	Al_2O_3 NPs	Protein synthesis/degradation, lipid metabolism. *FQR1*; *PABP2*; *NmrA-Like*	Mustafa *et al.* (2015a)
Tomato	Heat	TiO_2 NPs	Enhanced photosynthesis and stomatal opening, regulated energy dissipation	Qi *et al.* (2013)
Soybean	Salinity	SiO_2 NPs	Reduced damage because of expression of antioxidative enzymes	Farhangi-Abriz and Torabian (2018)
Maize	Heat	CeO_2 NPs	Decreased production of hydrogen peroxide and upregulation of *HSP70*	Zhao *et al.* (2012)
Cabbage	Salinity	MWCNTs	Reduced injurious effects of salinity stress	Martinez-Ballesta *et al.* (2016)
Sunflower	Drought	Fe_2O_3 NPs	Countered drought stress has less effect on amino acids and proline	Martinez-Fernandez *et al.* (2015)
Wheat	Drought	TiO_2 NPs	Increased yield, development, and starch content	Jaberzadeh *et al.* (2013)
Strawberry	Drought	Fe NPs	Boosted resistance and acclimation against stress	Havas and Ghaderi (2018)
Rapeseed	Salinity	CeO_2 NPs	Empowered capability to uphold cytosolic ratio of K^+/Na^+	Liu *et al.* (2021)
Tomato	Salinity	SiO_2 NPs	Suppressed the effect of salinity on the germination rate	Almutairi (2016)
Cucumber	Salinity	SiO_2 NPs	Increased plant growth and characteristics like plant height, chlorophyll, and fruit yield.	Alsaeedi *et al.* (2019)
Tomato	Heat	TiO_2 NPs	Regulated energy dissipation, enhanced photosynthetic rate, and induced stomatal opening	Qi *et al.* (2013)
Soybean	Flooding	Ag NPs	Root proteins for stress signaling and cell metabolism	Mustafa *et al.* (2015b)
Barley	Drought	MWCNTs	Increased seed water content and boosted seed absorption capacity	Karami and Sepehri (2017)

2.5.2 Interaction of NPs with Molecular Mechanisms

Molecular mechanisms such as suitable alterations of gene expression have been adopted by plants to cope with different stresses. The impact induced by defined doses of NPs stimulates the plant's defensive system, which activates the molecular mechanisms in response to particular abiotic stress situations. MicroRNAs (miRNAs) act as regulators of animal and plant biological processes, and play a significant role in plants under different biotic and abiotic stresses. Treatment with Al_2O_3 NPs upregulated the miRNA genes. With an increasing level of Al_2O_3 NPs, the expression of nine miRNAs (miR167, miR159, miR395, miR162, miR398, miR396, miR399, miR397, and miR169) was significantly upregulated (Singh *et al.*, 2017). *O. sativa* roots treated with Ag NPs revealed the NPs' response in transcription, responsive protein regulation, cell division, protein degradation, etc. Whole-genome cDNA microarray analysis revealed that 286 genes are upregulated, mainly related to oxidative and high metal stress (e.g., cytochrome P450-dependent oxidase [vacuolar cation/proton exchanger], POD, and SOD), while 81 genes are downregulated, including hormone stimulus and plant defense genes (Myint *et al.*, 2021). Calcium ions act as a second messenger involved in signal transduction under several stress situations (Ali *et al.*, 2020). Ag NPs influence cell metabolism and bind to Ca_I/Na_I ion pumps or calcium ion channels through calcium ion receptors.

Furthermore, the combination of nanocrystals and C60 promoted the regulation of Ca_I/calmodulin-dependent protein kinase II (Maghsoodi *et al.*, 2020). *Solanum lycopersicum* L. seedlings exhibited results that reduce the effect of salt stress under Ag NP treatment. Four genes (*AREB* [ABA Response Element-Binding Protein], *MAPK2*, *P5CS1* [Delta-1-Pyrroline-5-Carboxylate Synthetase 1], and *CRK1* [Cysteine-Rich Receptor Like Protein Kinase 42-Like]) are upregulated, and three genes (*TAS14* [Environmental Stress-Inducible Protein and ABA], *ZFHD1* [Zinc Finger Homeodomain Transcription Factor], and *DDF2* [Dwarf and Delayed Flowering 2]) are downregulated by NPs under salt stress (Almutairi., 2016b). In *Arabidopsis thaliana*, cadmium sulfide promoted overexpression of calcium-binding proteins (such as calcium-dependent protein kinase 23 and CML45) to control stress responses and increased plants' acceptance of different abiotic stresses (Boudsocq and Sheen, 2013). Transcriptional analysis for Al_2O_3 NPs–responsive genes under flooding stress revealed that the *NmrALike* (negative transcriptional regulator) gene is strongly activated, and the *FQR1* (Flavodoxin-Like Quinone Reductase) gene is downregulated (Mustafa *et al.*, 2015a). The activation of genes like *CRK1* and *P5CS1* by nanoparticles involved in ROS sensing, reduction, and abscisic acid (ABA) signaling is described by Almutairi (2016a, b).

2.5.3 Genetic Engineering and Crop Breeding

Nanotechnology is a novel, significant approach to increasing the accuracy and efficiency of genetic engineering and crop breeding. Nanobiotechnology enhances the efficiency of crop breeding by transporting exogenous nucleotides and biochemicals into plant cells (Joga *et al.*, 2016). The grouping of nanotechnology and biotechnology suggests more opportunities as a new method to transport biomolecules into the cells via the cell wall, because passage through the plant cell wall and size of introduced hereditary material are major barriers to external molecules moving into cells of plants (Zhao *et al.*, 2020). With no negative side effects, SiO_2 NPs have been created to transfer DNA into crops like maize and tobacco (Galbraith *et al.*, 2007). A technique for delivering genes through nanomaterials has recently been invented. Without the use of external physical or chemical means, high transformation efficiency is acquired in plant cells, with nanomaterials demonstrating outstanding submissions of genetic material into the plant (Yan *et al.*, 2022). It has been found that silicon carbide–assisted plant transformation is an efficient way to assign DNA fragments and sequences in a variety of *calli*, including cotton (*Gossypium hirsutum* L.), soybean, tobacco, and rice (Lau *et al.*, 2017). Non-toxic and biodegradable clay nanosheets carrying double-stranded RNAs (dsRNAs) protect against the cauliflower mosaic virus in tobacco leaves (Mitter *et al.*, 2017). Now, NPs are also being used to transport RNA into the plant cell in addition to the DNA delivery already mentioned. Implanting a composite of chitosan NPs and small interfering RNA (siRNA) into the plant cell has provided a new approach to improve crop growth and development, because chitosan can bind to RNA and penetrate cell membranes (Zhang *et al.*, 2010).

Traditional transformation and regeneration processes have made genome editing by CRISPR/Cas beneficial in plants. Recent advances in nanomaterial-mediated precise transport of single guide RNA (sgRNA) through CRISPR/Cas9 have paved the way for a new era of genetic engineering techniques. Furthermore, mesoporous silica nanoparticles (MSNs) have been used as carriers to deliver Cre-recombinase into the adolescent embryo of maize, loading the recombined loxP sites into the chromosomal DNA. The loxP fragment was successfully integrated to deliver improved MSNs into the plant cell (Martin-Ortigosa *et al.*, 2014). With their small size, high throughput, high tensile strength, and difference in charge, nanomaterials can improve efficiency and specificity by reducing the possibility of off-target effects of the CRISPR/Cas9 (Demirer *et al.*, 2021). Cas9En (RNP) (E-tag)-ribonucleoproteins transported the sgRNA in cultivated cell lines by cationic arginine gold. This provided highly efficient (around 30%) active cytoplasmic and nuclear gene editing, which can markedly advance crop research (Mout *et al.*, 2017). Although current phyto-nanotechnology

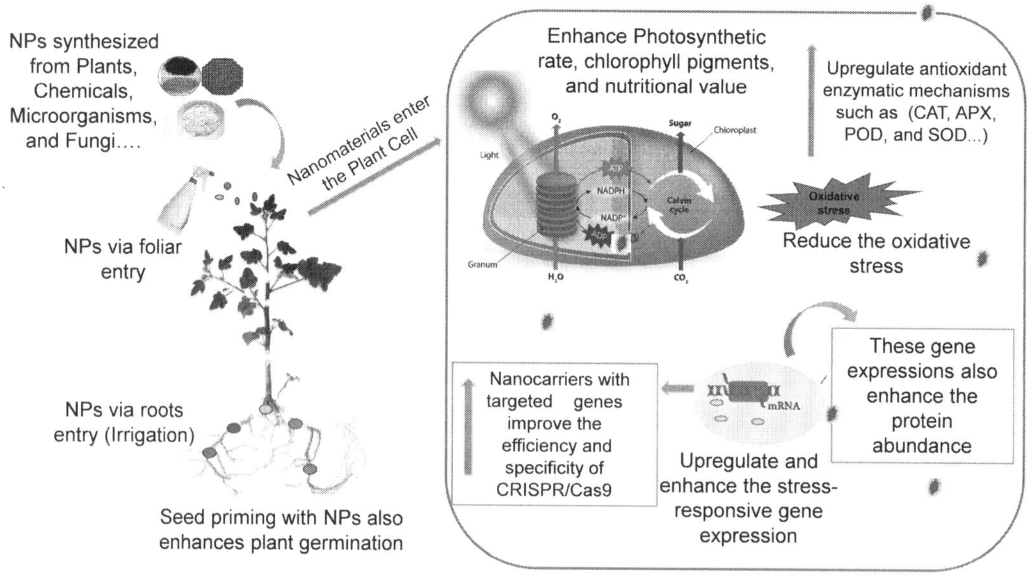

FIGURE 2.1 Mechanisms and pathways that illustrate the positive effects of NPs on plant development and growth.

approaches have promoted the CRISPR/Cas9 method in many plants, to overcome the challenges in plant genome editing, there are still many challenges to overcome, such as low transport efficiency, the number of plant species that can be improved by genetic engineering, and the types of CRISPR/Cas that can be effectively used in several plants (Figure 2.1).

2.6 CONCLUSION AND FUTURE PERSPECTIVE

Phyto-nanotechnology (the progress of NPs) has been recommended as an influential technique to reduce complications that are caused by traditional systems and conventional fertilizers. Among all the syntheses, green synthesis is eco-friendly, cost-effective, safe, and time-saving, which eradicates all the detrimental effects of the NPs in numerous applications. The unpredictable results of NPs biosynthesis mechanisms, their uptake and utilization in plant systems, and their assembly with biotic systems have enhanced the need for complete understanding to design effective and safe NPs. Despite their relative insufficiency, the findings in this chapter describe how NPs trigger molecular resistance to overcome stress. A limited number of studies exist that examine the molecular mechanisms activated by NPs against biotic and abiotic stress. The presented reports also validate that NPs can increase plant tolerance in response to environmental stresses by increasing the capability of defense mechanisms and reducing the severity of the damage. The basic dosage, translocation, exposure duration, accumulation, and action mechanism of NPs on plants are significant approaches to be established.

REFERENCES

Abdel Latef, A. A. H., Abu Alhmad, M. F., & Abdelfattah, K. E. (2017). The possible roles of priming with ZnO nanoparticles in mitigation of salinity stress in lupine (*Lupinus terms*) plants. *Journal of Plant Growth Regulation*, *36*(1), 60–70.

Abdoli, S., Ghassemi-Golezani, K., & Alizadeh-Salteh, S. (2020). Responses of ajowan (*Trachyspermum ammi* L.) to exogenous salicylic acid and iron oxide nanoparticles under salt stress. *Environmental Science and Pollution Research International*, *27*(29), 36939–36953.

Al-Huqail, A. A., Hatata, M. M., Al-Huqail, A. A., & Ibrahim, M. M. (2018). Preparation, characterization of silver phyto nanoparticles and their impact on growth potential of Lupinus termis L. seedlings. *Saudi Journal of Biological Sciences*, *25*(2), 313–319.

Ali, S., Hayat, K., Iqbal, A., & Xie, L. (2020). Implications of abscisic acid in the drought stress tolerance of plants. *Agronomy*, *10*(9), 1323.

Ali, S., Mehmood, A., & Khan, N. (2021). Uptake, translocation, and consequences of nanomaterials on plant growth and stress adaptation. *Journal of Nanomaterials*, *2021*.

Almutairi, Z. M. (2016). Effect of nano-silicon application on the expression of salt tolerance genes in germinating tomato (*Solanum lycopersicum* L.) seedlings under salt stress. *Plant Omics*, *9*(1), 106–114.

Almutairi, Z. M. (2016). Influence of silver nanoparticles on the salt resistance of tomato (*Solanum lycopersicum*) during germination. *International Journal of Agriculture and Biology*, *18*(2), 449–457.

Almutairi, Z. M., & Alharbi, A. (2015). Effect of silver nanoparticles on seed germination of crop plants. *International Journal of Nuclear and Quantum Engineering*, *9*(6), 689–693.

Alsaeedi, A., El-Ramady, H., Alshaal, T., El-Garawany, M., Elhawat, N., & Al-Otaibi, A. (2019). Silica nanoparticles boost growth and productivity of cucumber under water deficit and salinity stresses by balancing nutrients uptake. *Plant Physiology and Biochemistry*, *139*, 1–10.

Al-Salama, Y. (2022). Effect of seed priming with ZnO nanoparticles and saline irrigation water in yield and nutrients uptake by wheat plants. *Environmental Sciences Proceedings*, *16*(1), 37.

Aref, I., El Atta, H., El Obeid, M., Ahmed, A., Khan, P., & Iqbal, M. (2013). Effect of water stress on relative water and chlorophyll contents of *Juniperus procera* Hochst. ex Endlicher in Saudi Arabia. *Life Science Journal*, *10*(4), 681–685.

Azimi, F., Oraei, M., Gohari, G., Panahirad, S., & Farmarzi, A. (2021). Chitosan-selenium nanoparticles (Cs–Se NPs) modulate the photosynthesis parameters, antioxidant enzymes activities and essential oils in *Dracocephalum moldavica* L. under cadmium toxicity stress. *Plant Physiology and Biochemistry*, *167*, 257–268.

Badawy, A. A., Abdelfattah, N. A., Salem, S. S., Awad, M. F., & Fouda, A. (2021). Efficacy assessment of biosynthesized copper oxide nanoparticles (CuO-NPs) on stored grain insects and their impacts on morphological and physiological traits of wheat (Triticum aestivum L.) plant. *Biology*, *10*(3), 233.

Barrena, R., Casals, E., Colón, J., Font, X., Sánchez, A., & Puntes, V. (2009). Evaluation of the ecotoxicity of model nanoparticles. *Chemosphere*, *75*(7), 850–857.

Boudsocq, M., & Sheen, J. (2013). CDPKs in immune and stress signaling. *Trends in Plant Science*, *18*(1), 30–40.

Da Costa, M. V. J., & Sharma, P. K. (2016). Effect of copper oxide nanoparticles on growth, morphology, photosynthesis, and antioxidant response in Oryza sativa. *Photosynthetica*, *54*(1), 110–119.

Danish, M., Altaf, M., Robab, M. I., Shahid, M., Manoharadas, S., Hussain, S. A., & Shaikh, H. (2021). Green synthesized silver nanoparticles mitigate biotic stress induced by meloidogyne incognita in Trachyspermum ammi (L.) by improving growth, biochemical, and antioxidant enzyme activities. *ACS Omega*, *6*(17), 11389–11403.

De Almeida, M. S., Susnik, E., Drasler, B., Taladriz-Blanco, P., Petri-Fink, A., & Rothen-Rutishauser, B. (2021). Understanding nanoparticle endocytosis to improve targeting strategies in nanomedicine. *Chemical Society Reviews*, *50*(9), 5397–5434.

Demirer, G. S., Silva, T. N., Jackson, C. T., Thomas, J. B., W. Ehrhardt, Rhee, D. S. Y., & Landry, M. P. (2021). Nanotechnology to advance CRISPR–Cas genetic engineering of plants. *Nature Nanotechnology*, *16*(3), 243–250.

Dimkpa, C. O., Bindraban, P. S., Fugice, J., Agyin-Birikorang, S., Singh, U., & Hellums, D. (2017). Composite micronutrient nanoparticles and salts decrease drought stress in soybean. *Agronomy for Sustainable Development*, *37*(1), 1–13.

Ding, Y., Shi, Y., & Yang, S. (2020). Molecular regulation of plant responses to environmental temperatures. *Molecular Plant*, *13*(4), 544–564.

Djanaguiraman, M., Belliraj, N., Bossmann, S. H., & Prasad, P. V. (2018). High-temperature stress alleviation by selenium nanoparticle treatment in grain sorghum. *ACS Omega*, *3*(3), 2479–2491.

Dos Reis, M., de Alencastro Graça, P. M. L., Yanai, A. M., Ramos, C. J. P., & Fearnside, P. M. (2021). Forest fires and deforestation in the central Amazon: Effects of landscape and climate on spatial and temporal dynamics. *Journal of Environmental Management*, *288*, 112310.

Faizan, M., Bhat, J. A., Chen, C., Alyemeni, M. N., Wijaya, L., Ahmad, P., & Yu, F. (2021). Zinc oxide nanoparticles (ZnO-NPs) induce salt tolerance by improving the antioxidant system and photosynthetic machinery in tomato. *Plant Physiology and Biochemistry*, *161*, 122–130.

Farhangi-Abriz, S., & Torabian, S. (2018). Nano-silicon alters antioxidant activities of soybean seedlings under salt toxicity. *Protoplasma*, *255*(3), 953–962.

Fincheira, P., Tortella, G., Duran, N., Seabra, A. B., & Rubilar, O. (2020). Current applications of nanotechnology to develop plant growth inducer agents as an innovation strategy. *Critical Reviews in Biotechnology, 40*(1), 15–30.

Galbraith, C. G., Yamada, K. M., & Galbraith, J. A. (2007). Polymerizing actin fibers position integrins primed to probe for adhesion sites. *Science, 315*(5814), 992–995.

Gelaw, T. A., & Sanan-Mishra, N. (2022). Nanomaterials coupled with microRNAs for alleviating plant stress: A new opening towards sustainable agriculture. *Physiology and Molecular Biology of Plants, 28*(4), 791–818.

Hafeez, A., Razzaq, A., Mahmood, T., & Jhanzab, H. M. (2015). Potential of copper nanoparticles to increase growth and yield of wheat. *Journal of Nanoscience with Advanced Technology, 1*(1), 6–11.

Haghighi, M., & Pourkhaloee, A. (2013). Nanoparticles in agricultural soils: Their risks and benefits for seed germination. *Minerva Biotecnologica, 25*(2), 123–132.

Hasanpour, H., Maali-Amir, R., & Zeinali, H. (2015). Effect of TiO_2 nanoparticles on metabolic limitations to photosynthesis under cold in chickpea. *Russian Journal of Plant Physiology, 62*(6), 779–787.

Hasanuzzaman, M., Nahar, K., Hossain, M. S., Mahmud, J. A., Rahman, A., Inafuku, M., Oku, H., & Fujita, M. (2017). Coordinated actions of glyoxalase and antioxidant defense systems in conferring abiotic stress tolerance in plants. *International Journal of Molecular Sciences, 18*(1), 200.

Havas, F., & Ghaderi, N. (2018). Application of iron nanoparticles and salicylic acid in in vitro culture of strawberries (Fragaria× ananassa Duch.) to cope with drought stress. *Plant Cell, Tissue and Organ Culture (PCTOC), 132*(3), 511–523.

Hawthorne, J., Musante, C., Sinha, S. K., & White, J. C. (2012). Accumulation and phytotoxicity of engineered nanoparticles to Cucurbita pepo. *International Journal of Phytoremediation, 14*(4), 429–442.

Iqbal, M., Raja, N. I., Mashwani, Z. U. R., Wattoo, F. H., Hussain, M., Ejaz, M., & Saira, H. (2019). Assessment of AgNPs exposure on physiological and biochemical changes and antioxidative defence system in wheat (*Triticum aestivum* L) under heat stress. *IET nanobiotechnology, 13*(2), 230–236.

Jaberzadeh, A., Moaveni, P., Moghadam, H. R. T., & Zahedi, H. (2013). Influence of bulk and nanoparticles titanium foliar application on some agronomic traits, seed gluten and starch contents of wheat subjected to water deficit stress. *Notulae Botanicae Horti Agrobotanici Cluj-Napoca, 41*(1), 201–207.

Joga, M. R., Zotti, M. J., Smagghe, G., & Christiaens, O. (2016). RNAi efficiency, systemic properties, and novel delivery methods for pest insect control: What we know so far. *Frontiers in Physiology, 7*, 553.

Joris, F., Manshian, B. B., Peynshaert, K., De Smedt, S. C., Braeckmans, K., & Soenen, S. J. (2013). Assessing nanoparticle toxicity in cell-based assays: Influence of cell culture parameters and optimized models for bridging the in vitro–in vivo gap. *Chemical Society Reviews, 42*(21), 8339–8359.

Karami, A., & Sepehri, A. (2017). Multiwalled carbon nanotubes and nitric oxide modulate the germination and early seedling growth of barley under drought and salinity. *Agriculturae Conspectus Scientificus, 82*(4), 331–339.

Khan, I., Raza, M. A., Awan, S. A., Shah, G. A., Rizwan, M., Ali, B., Tariq, R., Hassan, M. J., Alyemeni, M. N., Brestic, M. Zhang, X., Ali, S., & Huang, L. (2020). Amelioration of salt induced toxicity in pearl millet by seed priming with silver nanoparticles (AgNPs): The oxidative damage, antioxidant enzymes, and ions uptake are major determinants of salt tolerant capacity. *Plant Physiology and Biochemistry, 156*, 221–232.

Khan, M. N., Li, Y., Khan, Z., Chen, L., Liu, J., Hu, J., Wu, H., & Li, Z. (2021). Nanoceria seed priming enhanced salt tolerance in rapeseed by modulating ROS homeostasis and α-amylase activities. *Journal of Nanobiotechnology, 19*(1), 1–19.

Khan, N. A., Khan, S. U., Ahmed, S., Farooqi, I. H., Dhingra, A., Hussain, A., & Changani, F. (2019). Applications of nanotechnology in water and wastewater treatment: A review. *Asian Journal of Water, Environment and Pollution, 16*(4), 81–86.

Khan, N., & Bano, A. (2018). Effects of exogenously applied salicylic acid and putrescine alone and in combination with rhizobacteria on the phytoremediation of heavy metals and chickpea growth in sandy soil. *International Journal of Phytoremediation, 20*(5), 405–414.

Khan, I., Hasan, M., Kausar, R., Shehzad, J., & Mustafa, G. (2022). Plant molecular responses to nanoparticle stress. In *Plant and Nanoparticles* (pp. 239–264). Springer, Singapore.

Khodakovskaya, M. V., De Silva, K., Biris, A. S., Dervishi, E., & Villagarcia, H. (2012). Carbon nanotubes induce growth enhancement of tobacco cells. *ACS Nano, 6*(3), 2128–2135.

Kokina, I., Plaksenkova, I., Jermaļonoka, M., & Petrova, A. (2020). Impact of iron oxide nanoparticles on yellow medick (Medicago falcata L.) plants. *Journal of Plant Interactions, 15*(1), 1–7.

Kurepa, J., Shull, T. E., & Smalle, J. A. (2020). Metabolomic analyses of the bio-corona formed on TiO2 nanoparticles incubated with plant leaf tissues. *Journal of Nanobiotechnology, 18*(1), 1–10.

Landa, P. (2021). Positive effects of metallic nanoparticles on plants: Overview of involved mechanisms. *Plant Physiology and Biochemistry, 161*, 12–24.

Lau, H. Y., Wu, H., Wee, E. J., Trau, M., Wang, Y., & Botella, J. R. (2017). Specific and sensitive isothermal electrochemical biosensor for plant pathogen DNA detection with colloidal gold nanoparticles as probes. *Scientific Reports, 7*(1), 1–7.

Laware, S. L., & Raskar, S. (2014). Effect of titanium dioxide nanoparticles on hydrolytic and antioxidant enzymes during seed germination in onion. *International Journal of Current Microbiology and Applied Sciences, 3*(7), 749–760.

Lee, S. M., Raja, P. M., Esquenazi, G. L., & Barron, A. R. (2018). Effect of raw and purified carbon nanotubes and iron oxide nanoparticles on the growth of wheatgrass prepared from the cotyledons of common wheat (Triticum aestivum). *Environmental Science: Nano, 5*(1), 103–114.

Lee, W. M., An, Y. J., Yoon, H., & Kweon, H. S. (2008). Toxicity and bioavailability of copper nanoparticles to the terrestrial plant's mung bean (Phaseolus radiates) and wheat (Triticum aestivum): Plant agar test for water-insoluble nanoparticles. *Environmental Toxicology and Chemistry: An International Journal, 27*(9), 1915–1921.

Liang, L., Tang, H., Deng, Z., Liu, Y., Chen, X., & Wang, H. (2018). Ag nanoparticles inhibit the growth of the bryophyte, Physcomitrella patens. *Ecotoxicology and Environmental Safety, 164*, 739–748.

Liu, J., Li, G., Chen, L., Gu, J., Wu, H., & Li, Z. (2021). Cerium oxide nanoparticles improve cotton salt tolerance by enabling a better ability to maintain cytosolic K+/Na+ ratio. *Journal of Nanobiotechnology, 19*(1), 1–16.

Maghsoodi, M. R., Ghodszad, L., & Lajayer, B. A. (2020). The dilemma of hydroxyapatite nanoparticles as phosphorus fertilizer: Potentials, challenges, and effects on plants. *Environmental Technology and Innovation, 19*, 100869.

Mantri, N., Patade, V., Penna, S., Ford, R., & Pang, E. (2012). Abiotic stress responses in plants: Present and future. In *Abiotic Stress Responses in Plants* (pp. 1–19). Springer, New York.

Martin-Ortigosa, S., Peterson, D. J., Valenstein, J. S., Lin, V. S. Y., Trewyn, B. G., Lyznik, L. A., & Wang, K. (2014). Mesoporous silica nanoparticle mediated intracellular Cre protein delivery for maize genome editing via *loxP* site excision. *Plant Physiology, 164*(2), 537–547.

Martínez-Ballesta, M., Zapata, L., Chalbi, N., & Carvajal, M. (2016). Multiwalled carbon nanotubes enter broccoli cells enhancing growth and water uptake of plants exposed to salinity. *Journal of Nanobiotechnology, 14*(1), 1–14.

Martínez-Fernández, D., Vítková, M., Bernal, M. P., & Komárek, M. (2015). Effects of Nano-maghemite on trace element accumulation and drought response of Helianthus annuus L. in a contaminated mine soil. *Water, Air, and Soil Pollution, 226*(4), 1–9.

Mitter, N., Worrall, E. A., Robinson, K. E., Li, P., Jain, R. G., Taochy, C., Fletcher, S. J., Carroll, B. J., Lu, G. Q., & Xu, Z. P. (2017). Clay nanosheets for topical delivery of RNAi for sustained protection against plant viruses. *Nature Plants, 3*(2), 1–10.

Mohammadi, H., Esmailpour, M., Gheranpaye, A., & Al Sahli, A. A. (2014). Effects of TiO_2 nanoparticles and water-deficit stress on morpho-physiological characteristics of dragonhead (*Dracocephalum moldavica L.*) plants. *Environmental Toxicology and Chemistry, 33*(11), 2429–2437.

Mohammadi, R., Amiri, N. M., & Mantri, L. (2013). Effect of TiO_2 nanoparticles on oxidative damage and antioxidant defense systems in chickpea seedlings during cold stress. *Russian Journal of Plant Physiology, 61*, 768–775.

Mout, R., Ray, M., Yesilbag Tonga, G., Lee, Y. W., Tay, T., Sasaki, K., & Rotello, V. M. (2017). Direct cytosolic delivery of CRISPR/Cas9-ribonucleoprotein for efficient gene editing. *ACS Nano, 11*(3), 2452–2458.

Murthy, H. C., Desalegn, T., Kassa, M., Abebe, B., & Assefa, T. (2020). Synthesis of green copper nanoparticles using medicinal plant *Hagenia abyssinica* (Brace) JF. Gmel. leaf extract: Antimicrobial properties. *Journal of Nanomaterials, 2020*.

Mushtaq, M., Hassan, S. M., & Mughal, S. S. (2022). Synthesis, characterization and biological approach of Nano oxides of calcium by Piper nigrum. *American Journal of Chemical Engineering, 10*(4), 79–88.

Mustafa, G., Sakata, K., & Komatsu, S. (2015). Proteomic analysis of flooded soybean root exposed to aluminum oxide nanoparticles. *Journal of Proteomics, 128*, 280–297.

Mustafa, G., Sakata, K., Hossain, Z., & Komatsu, S. (2015). Proteomic study on the effects of silver nanoparticles on soybean under flooding stress. *Journal of Proteomics, 122*, 100–118.

Mustafa, G., Sakata, K., & Komatsu, S. (2016). Proteomic analysis of soybean root exposed to varying sizes of silver nanoparticles under flooding stress. *Journal of Proteomics, 148*, 113–125.

Mustafa, G., Farooq, A., Riaz, Z., & Hasan, M. (2022). Nano-proteomics of stress tolerance in crop plants. *Sustainable Agriculture Reviews, 53*, 373–397.

Myint, K. Z., Yu, Q., Xia, Y., Qing, J., Zhu, S., Fang, Y., & Shen, J. (2021). Bioavailability and antioxidant activity of nanotechnology-based botanic antioxidants. *Journal of Food Science, 86*(2), 284–292.

Nair, P. M. G., & Chung, I. M. (2014). Physiological and molecular level effects of silver nanoparticles exposure in rice (*Oryza sativa* L.) seedlings. *Chemosphere, 112*, 105–113.

Nair, R., Mohamed, M. S., Gao, W., Maekawa, T., Yoshida, Y., Ajayan, P. M., & Kumar, D. S. (2012). Effect of carbon nanomaterials on the germination and growth of rice plants. *Journal of Nanoscience and Nanotechnology, 12*(3), 2212–2220.

Noori, A. J., & Kareem, F. A. (2019). The effect of magnesium oxide nanoparticles on the antibacterial and antibiofilm properties of glass-ionomer cement. *Heliyon, 5*(10), e02568.

Onaga, G., & Wydra, K. (2016). Advances in plant tolerance to abiotic stresses. *Plant Genomics, 10*, 229–272.

Parveen, A., Mazhari, B. B. Z., & Rao, S. (2016). Impact of bio-nanogold on seed germination and seedling growth in Pennisetum glaucum. *Enzyme and Microbial Technology, 95*, 107–111.

Pei, Z. F., Ming, D. F., Liu, D., Wan, G. L., Geng, X. X., Gong, H. J., & Zhou, W. J. (2010). Silicon improves the tolerance to water-deficit stress induced by polyethylene glycol in wheat (*Triticum aestivum* L.) seedlings. *Journal of Plant Growth Regulation, 29*(1), 106–115.

Qi, M., Liu, Y., & Li, T. (2013). Nano-TiO_2 improves the photosynthesis of tomato leaves under mild heat stress. *Biological Trace Element Research, 156*(1), 323–328.

Qian, H., Peng, X., Han, X., Ren, J., Sun, L., & Fu, Z. (2013). Comparison of the toxicity of silver nanoparticles and silver ions on the growth of terrestrial plant model Arabidopsis thaliana. *Journal of Environmental Sciences, 25*(9), 1947–1956.

Rasheed, R., Ashraf, M. A., Iqbal, M., Hussain, I., Akbar, A., Farooq, U., & Shad, M. I. (2020). Major constraints for global rice production: Changing climate, abiotic and biotic stresses. In *Rice Research for Quality Improvement: Genomics and Genetic Engineering* (pp. 15–45). Springer, Singapore.

Sarkhosh, S., Kahrizi, D., Darvishi, E., Tourang, M., Haghighi-Mood, S., Vahedi, P., & Ercisli, S. (2022). Effect of zinc oxide nanoparticles (ZnO-NPs) on seed germination characteristics in two Brassicaceae Family species: Camelina sativa and Brassica napus L. *Journal of Nanomaterials, 2022*, 1–15.

Shinozaki, K., & Yamaguchi-Shinozaki, K. (2007). Gene networks involved in drought stress response and tolerance. *Journal of Experimental Botany, 58*(2), 221–227.

Siddiqui, M. H., & Al-Whaibi, M. H. (2014). Role of nano-SiO2 in the germination of tomato (*Lycopersicum esculentum* seeds Mill.). *Saudi Journal of Biological Sciences, 21*(1), 13–17.

Singh, P., Arif, Y., Siddiqui, H., Sami, F., Zaidi, R., Azam, A., Alam, P., & Hayat, S. (2021). Nanoparticles enhance the salinity toxicity tolerance in *Linum usitatissimum* L. by modulating the antioxidative enzymes, photosynthetic efficiency, redox status, and cellular damage. *Ecotoxicology and Environmental Safety, 213*, 112020.

Singh, S., Vishwakarma, K., Singh, S., Sharma, S., Dubey, N. K., Singh, V. K., Liu, S., Tripathi, D.K., & Chauhan, D. K. (2017). Understanding the plant and nanoparticle interface at the transcriptomic and proteomic level: A concentric overview. *Plant Gene, 11*, 265–272.

Song, Y., Jiang, M., Zhang, H., & Li, R. (2021). Zinc oxide nanoparticles alleviate chilling stress in rice (Oryza sativa L.) by regulating the antioxidative system and chilling response transcription factors. *Molecules, 26*(8), 2196.

Tan, M., Wang, Y., Taguchi, A., Abe, T., Yang, G., Wu, M., & Tsubaki, N. (2019). Highly dispersed Ru nanoparticles sputtered on graphene for hydrogen production. *International Journal of Hydrogen Energy, 44*(14), 7320–7325.

Tariq, M., Choudhary, S., Singh, H., Siddiqui, M. A., Kumar, H., Amir, A., & Kapoor, N. (2021). Role of nanoparticles in abiotic stress. *Technology in Agriculture, 323*.

Tirani, M. M., Haghjou, M. M., & Ismaili, A. (2019). Hydroponically grown tobacco plants respond to zinc oxide nanoparticles and bulk exposures by morphological, physiological, and anatomical adjustments. *Functional Plant Biology, 46*(4), 360–375.

Torabian, S., Farhangi-Abriz, S., & Zahedi, M. (2018). Efficacy of $FeSO_4$ Nano formulations on osmolytes and antioxidative enzymes of sunflower under salt stress. *Indian Journal of Plant Physiology, 23*(2), 305–315.

Tripathi, D. K., Singh, S., Singh, S., Srivastava, P. K., Singh, V. P., Singh, S., Prasad, S. M., Singh, P. K., Dubey, N. K., Pandey, A. C., & Chauhan, D. K. (2017). Nitric oxide alleviates silver nanoparticles (AgNps)-induced phytotoxicity in Pisum sativum seedlings. *Plant Physiology and Biochemistry, 110*, 167–177.

Ullah, H., Khan, I., Mustafa, G., Shehzad, J., Hasan, M., Saeed, U., & Chaudhari, S. K. (2022). Molecular characterization and mitigative role of silicon dioxide nanoparticles in *Ocimum basilicum* under lead (Pb) stress. *Silicon*, 1–15.

Usman, M., Farooq, M., Wakeel, A., Nawaz, A., Cheema, S. A., ur Rehman, H., Ashraf, I., & Sanaullah, M. (2020). Nanotechnology in agriculture: Status, challenges, and future opportunities. *Science of the Total Environment*, *721*, 137778.

Usman, M., Qamar, M. U., Becker, R., Zaman, M., Conrad, C., & Salim, S. (2020). Numerical modelling and remote sensing-based approaches for investigating groundwater dynamics under changing land-use and climate in the agricultural region of Pakistan. *Journal of Hydrology*, *581*, 124408.

Van Nguyen, D., Nguyen, H. M., Le, N. T., Nguyen, K. H., Nguyen, H. T., Le, H. M., Nguyen, A. T., Dinh, N. T. T., Hoang, S. A., & Van Ha, C. (2022). Copper nanoparticle application enhances plant growth and grain yield in maize under drought stress conditions. *Journal of Plant Growth Regulation*, *41*(1), 364–375.

Vishwakarma, K., Upadhyay, N., Singh, J., Liu, S., Singh, V. P., Prasad, S. M., Chauhan, D. K., Tripathi, D. K., & Sharma, S. (2017). Differential phytotoxic impact of plant mediated silver nanoparticles (AgNPs) and silver nitrate (AgNO3) on Brassica sp. *Frontiers in Plant Science*, *8*, 1501.

Wang, Z., Xie, X., Zhao, J., Liu, X., Feng, W., White, J. C., & Xing, B. (2012). Xylem-and phloem-based transport of CuO nanoparticles in maize (Zea mays L.). *Environmental Science and Technology*, *46*(8), 4434–4441.

Yan, Y., Zhu, X., Yu, Y., Li, C., Zhang, Z., & Wang, F. (2022). Nanotechnology strategies for plant genetic engineering. *Advanced Materials*, *34*(7), 2106945.

Yashni, G., Willy, K. B., Al-Gheethi, A. A., Mohamed, R. M. S. R., Salleh, S. M., & Hashim, M. A. (2020). A review on green synthesis of ZnO nanoparticles using Coriandrum sativum leaf extract for degrading dyes in textile wastewater: A prospect towards green chemistry. In *IOP Conference Series. IOP Conference Series: Materials Science and Engineering* (Vol. 736, No. 4, p. 042003). IOP Publishing.

Zahedi, S. M., Hosseini, M. S., Daneshvar Hakimi Meybodi, N., & Peijnenburg, W. (2021). Mitigation of the effect of drought on growth and yield of pomegranates by foliar spraying of different sizes of selenium nanoparticles. *Journal of the Science of Food and Agriculture*, *101*(12), 5202–5213.

Zahedi, S. M., Karimi, M., & Teixeira da Silva, J. A. (2020). The use of nanotechnology to increase quality and yield of fruit crops. *Journal of the Science of Food and Agriculture*, *100*(1), 25–31.

Zhao, L., Peng, B., Hernandez-Viezcas, J. A., Rico, C., Sun, Y., Peralta-Videa, J. R., Tang, X., Niu, G., Jin, L., Varela-Ramirez, A., Zhang, J. Y., & Gardea-Torresdey, J. L. (2012). Stress response and tolerance of Zea mays to CeO_2 nanoparticles: Cross talk among H_2O_2, heat shock protein, and lipid peroxidation. *ACS Nano*, *6*(11), 9615–9622.

Zhang, B. (2015). *Journal of Experimental Botany*, *66*(7), 1749–1761.

Zhang, M., Gao, B., Chen, J., & Li, Y. (2015). Effects of graphene on seed germination and seedling growth. *Journal of Nanoparticle Research*, *17*(2), 1–8.

Zhang, X., Zhang, J., & Zhu, K. Y. (2010). Chitosan/double-stranded RNA nanoparticle-mediated RNA interference to silence chitin synthase genes through larval feeding in the African malaria mosquito (Anopheles gambiae). *Insect Molecular Biology*, *19*(5), 683–693.

Zhao, L., Lu, L., Wang, A., Zhang, H., Huang, M., Wu, H., Xing, B., Wang, Z., & Ji, R. (2020). Nano-biotechnology in agriculture: Use of nanomaterials to promote plant growth and stress tolerance. *Journal of Agricultural and Food Chemistry*, *68*(7), 1935–1947.

3 Methods for Nanotechnology in Plant Biotic Stress Management

Sagnik Nag, Subhrojyoti Ghosh, Anuvab Dey, Sourav Majhi, Abhijit Mondal, Anamika Mishra, Oishi Mitra, and Faheem Ahmad

3.1 INTRODUCTION

Major obstacles to crop productivity include biotic stresses caused by fungi, bacteria, viruses, viroids, phytoplasmas, nematodes, insects, and weeds. Agricultural yield losses resulting from biotic stressors can be as high as 50% in some circumstances (Martinez-Antonio et al., 2017), which accounts for a total loss of around \$2 trillion annually (Oluwaseun, 2017). Moreover, significant disparities exist between supply and demand for agricultural products, making it difficult to balance the growing demand for food due to an expanding global population and agricultural output (Savary et al., 2012). Rapid plant disease diagnosis techniques with high sensitivity and accuracy are crucial to stop the disease from spreading over a vast area with little loss to crop productivity and food quality assurance (Kashyap et al., 2016). The traditional methods for detecting plant pathogens take a long time, are less accurate, and demand complicated sample handling. Although molecular (nucleic acid–based) and immunological techniques have advanced, they still have certain problems with speed and signal strength (Kashyap et al., 2016).

Originally coined by Taniguchi in 1974, nanotechnology is the manipulation of matter at the nanoscale size (1–100 nm or 1.0×10^{-9} m) to produce new and distinctive nanomaterials (NMs) with the potential to drastically alter current processes (Chinnamuthu and Boopathi, 2009). Compared with their bulk forms and individual molecules, NMs exhibit very peculiar physical, chemical, and biological features (Li et al., 2001) due to their unique size, shape, surface-to-volume ratio, crystal phase, and chemical composition. High-throughput analysis, which can be utilized for high-quality monitoring and crop protection, is made possible by the potential for enhancing sensitivity, specificity, and speed of detection using nanoimaging, nanopore DNA sequencing tools, biosensors, quantum dots (QDs), and nanostructured platforms (Khiyami et al., 2014). Crop diseases can be identified and managed using nanotechnology based on microRNA (miRNA). According to Prasad, 2014 and Mishra et al., 2017, integrating nanotechnology-based technologies in agriculture has significant potential to address the world's concerns related to food production or security, sustainability, and climate change.

3.2 NANOTECHNOLOGY AND BIOTIC STRESS MANAGEMENT

Nanotechnology makes the development of quick, precise, time-efficient, simple-to-use detection methods for biotic stressors and effective disease management possible. Nano-phytopathology is a field of study that employs nanotechnology for the early detection, diagnosis, and management of plant diseases and associated pathogens (Hussain et al., 2019). Nanotechnological instruments

DOI: 10.1201/b23308-3

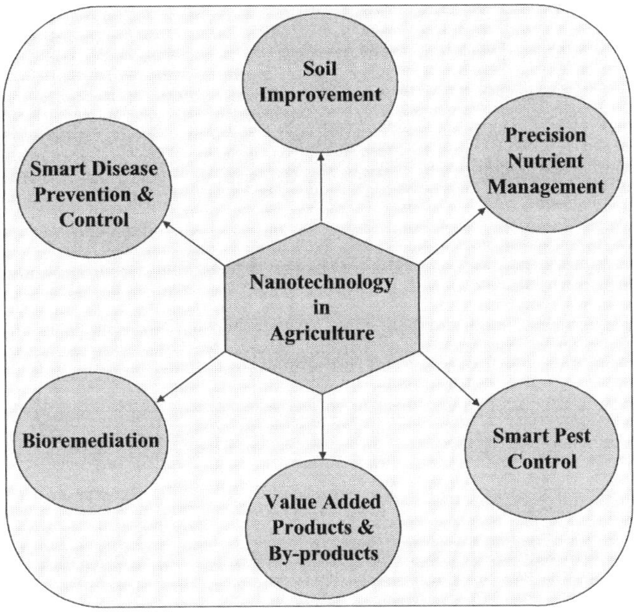

FIGURE 3.1 Factors responsible for plant biotic stress and their effect on plant growth, and different types of nanoparticles that can help manage those problems.

such as nano-biosensors, nanofabrication, nano-barcodes, and other portable diagnostic devices can identify several plant diseases. Moreover, nano-diagnostic kits can swiftly and easily identify possible plant infections, helping to stop the spread of epidemic illnesses. The genetics of the pathogen population and the transmission of genes between pathogens and hosts are studied and monitored using nanomolecular methods. Nanoparticles (NPs) and QDs are the most useful biological markers for rapidly detecting bacteria, fungi, and viruses. These tiny particles can be engineered to bind specifically to certain pathogens, making them an excellent tool for identifying the presence of disease-causing agents. They can also be used to track the movement of pathogens within plant tissues, helping researchers to understand how diseases spread and how they can be contained (Boonham et al., 2008; Chartuprayoon et al., 2010; Yao et al., 2009) (Figure 3.1).

3.3 DETECTION AND DIAGNOSIS OF BIOTIC STRESSES

3.3.1 Biosensors

Biosensors are fascinating analytical devices that detect chemical compounds in biological samples using specific biochemical reactions. They combine a biological component and a physicochemical detector, which work together to detect an analyte. This innovative technology has revolutionized how we detect and quantify biological processes and analytes. Researchers use various transduction mechanisms to create plant biosensors, including electrochemiluminescence, surface plasmon resonance, loop-mediated isothermal amplification, fluorescence, lateral flow, and electrochemical approaches. These biosensors detect, quantify, transmit, and record information on a biological analyte or process using a biological recognition element, transducer, and processor (Kumar and Arora, 2020). The biological recognition element interacts with the analyte under study, such as tissue, microorganisms, organelles, cell receptors, enzymes, antibodies, nucleic acids, or biomimetic components. The transducer then converts the molecular biological signal into an electric or digital signal proportional to the analyte's concentration or physical, chemical, and spatial characteristics. A processor can amplify, measure, visualize, and process this signal. Biosensors are crucial in various fields, including fundamental sciences (physics, chemistry, biology), engineering, biochemical, chemical, and material science, and medicine. These innovative devices come as either catalytic or affinity (non-catalytic) biosensors. The bioreceptor produces a new biochemical reaction product in catalytic biosensors upon contact with the analyte (Cao et al., 2008; Galandova et al., 2008). In contrast, the analyte is permanently bonded to the receptor in affinity biosensors, and no new biochemical reaction product is produced. Aside from enzyme-based and affinity-based biosensors, there are whole-cell, aptamer-based, and nanomaterial-based biosensors. The transduction mechanism determines the biosensor's analytical properties, including signal stability, reproducibility, detection limit, operational stability, and selectivity. Depending on the transduction mechanism, various biosensors are available, including optical, colorimetric, electric, and electrochemical biosensors (amperometric, impedimetric, potentiometric, and conductometric) and gravimetric biosensors (quartz crystal microbalance [QCM] and piezoelectric) (Tortella et al., 2023). Biosensors are groundbreaking devices that have revolutionized how we detect and quantify biological analytes and processes. Their applications are vast, ranging from medical diagnoses to environmental monitoring.

Nano-biosensors are revolutionizing the field of analytical chemistry thanks to their superior performance and remarkable capabilities. Nanostructured materials–based transducers can increase sensitivity by more than one order of magnitude compared with conventional materials. Nanotubes, nanorods, and nanofibers enable more significant loading of recognition elements and superior electrical communication ability, enhancing analytical performance. One of the defining aspects of nano-biosensors is their ability to increase both specificity and sensitivity while enabling miniaturization and increased mobility. Compared with conventional biosensors, nano-biosensors can estimate biological interactions to a much more significant and reliable degree due to the efficient use of surface plasmon resonance phenomena, which enables them to detect even the most minor interacting phenomena. Nanomaterials, including carbon-based materials, metals, organic compounds, and inorganic materials, can all be used to create nano-biosensors. Common structural nanomaterials include nanotubes, dendrimers, QDs, nanoparticles, nanowires, and fullerenes. Nano-biosensors offer many advantages, including stable support for immobilizing bio-molecular receptors, ease of immobilization, effective signal transduction, and the ability to deal with various signals depending on the desired phenomena being measured. Their uses have been extensively researched in life sciences applications such as human diagnostics, pathogen recognition, food quality control, and pollution monitoring. However, their application in plant sciences has been limited until recently. Recent studies by Elmer et al. have demonstrated the potential of various nanomaterials, including nano-copper, nano-zinc, metallic and non-metallic oxides, carbon nanomaterials, liposomes, dendrimers, and nano-silver, in plant pathology. These nanomaterials have been shown to effectively combat biotic stressors, including fungi, bacteria, and viruses, that negatively impact plant growth and yield (Elmer et al., 2018). One exciting application of nanomaterials in plant science is the development of plant biosensors. Plants are natural sensors that react to their surroundings, and with the help of nanomaterials, researchers can develop biosensors to monitor plant physiology and intracellular communication signals. For example, Cao et al. (2008) showed that carbon nanotubes could be used to detect particular DNA oligonucleotides through several optical nano-biosensors, while Galandova et al. (2008) developed an electrochemical biosensor that can recognize DNA damage using a carbon electrode screen printed with multi-walled carbon nanotubes and chitosan (Cao et al., 2008; Galandova et al., 2008). Mycotoxins, hazardous secondary metabolites generated by fungi such as *Aspergillus*, *Penicillium*, *Rhizopus*, *Alternaria*, and *Fusarium*, can also be detected using nanosensors, as per Otles and Yalcin (2010). Additionally, nanomaterials such as ZnO and Ag NPs generate reactive oxygen species (ROS), known to have antimicrobial activity. Silicon has been shown to increase plant disease and stress resistance. Chitosan nanoparticles have also been shown to modulate innate immune responses in plants (Chandra et al., 2015; Otles and Yalcin, 2010; Rémus-Borel et al., 2005).

3.3.2 Nano-barcodes

Barcodes are visible markers frequently employed to monitor and identify important data in the macrocosmic realm. This technique allows the detection of bacteria in compost and tracking numerous illnesses that may be detected by fluorescence using nano-barcodes. Nano-barcodes are being utilized successfully for bioimaging due to their nanoscale dimension(s), safety, and combinatorial bioassay use. Micron- and nanoscale nanostructures are the encoding carriers for diverse codes that may be prepared using different components to create micro- and nanoscale barcodes with various characteristics, such as variable sizes, shapes, and colors. Nano-barcodes' smaller size enables them to recognize tiny targets for utilization in imaging applications (Li et al., 2005). The encoding carriers can be magnetic, optical, or nanoparticles and can be buried inside a nanostructure or ornamented on the outside with self-assembly. An encoding approach that produces the requisite number of stable nano-barcodes is necessary for detecting nano-barcodes (optical encoding, phase change encoding, fluorescent spectroscopy). Several encoding methods are addressed, including optically encoded organic dye nano-barcodes, QD-encoded nano-barcodes, fluorescent carbon dot–encoded nano-barcodes, fluorescence kinetics–based nano-barcodes, and emission color–based fluorescent nano-barcodes. As well as optical and fluorescent encoding, there are also magnetic encoding, phase change–based encoding, morphological encoding, sequence, and graphical encoding; these are also encoding principles for the preparation of nano-barcodes, described by Shikha et al. (2017).

One such application, developed by Li et al. in 2005, uses nano-barcodes to recognize pathogen DNA and unknown individuals. With advancements in nanotechnology-based gene sequencing, nano-barcodes are projected to bring quick and affordable capability to identify and use plant gene trait resources effectively (Li et al., 2005). Husale et al. (2009) conducted a study using barcodes to identify target genomic ssDNA, mRNA, or cDNA in biological samples, particularly plant material. This method involves hybridizing these targets with known bait DNA molecules adsorbed on a nanomaterial, allowing digital and quantitative identification of even femtomolar concentrations without prior amplification. This approach can be a valuable tool in the early detection and diagnosis of plant diseases caused by biotic stressors (Husale et al., 2009). Overall, using nano-barcodes in plant biotic stress management can aid in the early identification and treatment of plant diseases, ultimately leading to healthier and more productive crops. By utilizing nanotechnology-based gene sequencing and nano-barcodes, researchers can identify and use plant gene trait resources efficiently and cost-effectively, leading to more effective management of biotic stressors in plants.

3.3.3 Nanofabrication

Nanofabrication refers to fabricating microscopic structures (1–100 nm). Two methods, known as top-down and bottom-up nanofabrication, may be used to shape, position, and arrange matter at the nanoscale. Self-assembling building blocks require proper control over their size, shape, and surface characteristics to form well-organized constructions. There are two conventional approaches to self-assembly: one is a top-down and another is a bottom-up approach. Other than that, we can categorize fabrication into three major divisions: physical, chemical, and biological methods. The standard physical methods are ball milling, lithography, laser ablation, spray pyrolysis, etc. Physical methods work under a top-down approach; in other cases, chemical methods like vapor deposition, vapor pyrolysis, sol-gel techniques, etc., follow the same approach. Green synthesis, where bacteria and fungi take part, predominantly follows the bottom-up approach, and accuracy increases (Shoala, 2023).

Lithography is well known in most industrial nanofabrication techniques. The semiconductor industry's typical fabrication method entails optical projection lithography followed by electron beam lithography for high-throughput replication of the specified patterns on a series of masks, along with other material pattern transfer processes. In traditional lithography, the necessary material is typically covered by a mask, and the exposed material is removed using an etching process.

The final product's feature resolutions are determined by chemical or mechanical etching utilizing acids or ultraviolet, X-ray, or electron beams, respectively. E-beam lithography, optical lithography, soft lithography, nanoimprint lithography, block co-polymer lithography, and scanning probe lithography are examples of lithography types that use the top-down approach; on the other hand, atomic layer deposition, sol-gel manufacturing, molecular self-assembly, vapor phase deposition, and DNA nanostructures are bottom-up techniques. The method used to create multifunctional nanostructures and gadgets from the bottom up requires the self-assembly of atoms or molecules. In bottom-up nanofabrication, the accuracy and precision are increased, and the amount of material waste is decreased. Alternatives to lithography include micromachining, wet and dry etching, optoelectronics, and microelectronics. Nanofabrication, a technique that involves the manipulation of matter at the nanoscale level, has been utilized in plant biotic stress management. Scientists have used nanofabrication to create synthetic versions of stomata and xylem vessels, as noted by Meng et al. (2005). This innovation has led to the development of highly sensitive and precise diagnostic tools that can detect plant diseases early, facilitating quick and effective management of biotic stressors (Meng et al., 2005). Furthermore, nanofabrication has been identified as a powerful nano-diagnostic imaging technique in plant pathology. In one study, researchers used nanofabrication to examine plant tissue and cells and identify plant ailments. The ability to accurately and rapidly diagnose plant diseases using nanofabrication can aid in the early detection and treatment of biotic stressors (El-Saadony et al., 2022). Nanofabrication presents promising solutions for managing plant biotic stressors by developing highly sensitive and precise diagnostic tools. By utilizing nanofabrication in plant pathology, researchers can detect and diagnose plant diseases early, leading to more effective management and ultimately, healthier and more productive crops.

3.4 DIFFERENT NANOPARTICLES IN BIOTIC STRESS MANAGEMENT

Biotic stress can significantly impact our daily lives in today's fast-paced world. Fortunately, advances in nanotechnology have paved the way for novel solutions to these problems. Nanoparticles, in particular, have shown great potential in detecting and treating biotic stress–related issues. To help readers navigate this exciting field, we present Table 3.1, a comprehensive guide to the most promising nanoparticles and their unique properties. With information on size, detection limit, and target moiety, this table is an essential resource for anyone interested in harnessing the power of nanotechnology to combat biotic stress.

3.4.1 Zinc Oxide NPs

ZnO NPs show promising results in alleviating biotic stress in plants. They increase plant growth, development, and resistance against pathogens. ZnO NPs enter pathogenic (bacterial) cells, release zinc ions, and generate ROS, which can cause membrane dysfunction and cell death when the anti-oxidative capacity of the cells is exceeded (Dizaj et al., 2014; Tortella et al., 2023). ZnO NPs have also shown agricultural and food safety applications because of their bactericidal and fungicidal activities. The antimicrobial activity of ZnO NPs is even better than that of large zinc particles, as the smaller size and higher surface-to-volume ratio of nanoparticles allow better interaction with the pathogenic cells (Xie et al., 2011). ZnO NPs on crops have enhanced production by reducing pest activity and stopping spoilage of harvested fruits during storage for up to 8 days (Luksiene et al., 2020). Inoculation of ZnO NPs into the soil reduces many pathogenic populations that cause bacterial wilting and other plant diseases. ZnO NPs supplemented with media like Murashige Skoog (MS) show increased activity of antioxidative enzymes like catalase, superoxide dismutase (SOD), and peroxidases, inducing proline synthesis and promoting somatic embryogenesis, shooting, and regeneration in plants, thus increasing their overall tolerance against biotic stress (Helaly et al., 2014).

TABLE 3.1
Various Biosensors Used for the Detection of Biotic Pathogens Infesting the Plant Body

Nanoparticle	Size	Principle used	Detection limit	Target moiety	References
Gold nanoparticles (Au NPs)	2–100 nm	Surface-enhanced Raman scattering (SERS)	1 ng/mL	Pathogen-derived DNA/RNA	Kulabhusan et al., 2022; Liu et al., 2017
Carbon dots (C-dots)	<10 nm	Fluorescence quenching	100 ng/mL	Reactive oxygen species	Anju Maholiya et al., 2023; Zulfajri et al., 2020
Quantum dots (QDs)	2–10 nm	Fluorescence resonance energy transfer (FRET)	0.01 ng/mL	Bacterial lipopolysaccharides	Zou et al., 2010
Magnetic nanoparticles (MNPs)	10–100 nm	Magnetoresistance	100 ng/mL	Bacterial cells	Singh et al., 2021
Silica nanoparticles (Si NPs)	10–100 nm	Colorimetry	1 μg/mL	Fungal mycotoxins	L. Wang et al., 2022
Titanium dioxide nanoparticles (TiO_2 NPs)	5–100 nm	Impedance spectroscopy	10 ng/mL	Phytopathogenic bacteria	Satti et al., 2022
Zinc oxide nanoparticles (ZnO NPs)	5–100 nm	Colorimetry	1 ng/mL	*Botrytis cinerea*	Kalia et al., 2020
Iron oxide nanoparticles (IO NPs)	5–100 nm	Magnetization	1 μg/mL	Fungal spores	Xue et al., 2014
Copper nanoparticles (Cu NPs)	10–100 nm	Fluorescence quenching	1 ng/mL	*Phytophthora infestans*	Rai et al., 2018
Graphene oxide nanoparticles (GO)	5–100 nm	Colorimetry	10 ng/mL	*Fusarium oxysporum*	Ocsoy et al., 2013
Silver nanoparticles (Ag NPs)	2–100 nm	Surface plasmon resonance (SPR)	0.1 ng/mL	Bacterial endotoxins	Gupta et al., 2018
Cerium oxide nanoparticles (CeO_2 NPs)	5–100 nm	Electrochemical impedance spectroscopy (EIS)	10 ng/mL	Phytopathogenic fungi	An et al., 2020
Palladium nanoparticles (Pd NPs)	2–100 nm	Fluorescence quenching	1 ng/mL	*Phytophthora capsici*	Maryška et al., 2023
Manganese oxide nanoparticles (MnOx NPs)	5–100 nm	Chemiluminescence (CL)	10 ng/mL	*Fusarium graminearum*	Kasote et al., 2021

3.4.2 Silver NPs

Silver nanoparticles are the most studied and widely used nanoparticles of all. The antimicrobial and plant disease management properties of Ag NPs are well established. Green synthesis of Ag NPs using plant derivatives produces less toxicity than physical and chemical techniques, while their antimicrobial properties increase (Danish et al., 2021). Ag NPs are more effective than chemical antimicrobial agents as they exhibit multiple killing mechanisms like DNA unwinding, changing membrane permeability, and membrane disruption, which makes it difficult for pathogens to develop resistance against them (Kumari et al., 2017). Ag NPs can efficiently penetrate microbial cells, such that lower concentrations are sufficient for their antimicrobial activity, thus reducing the danger of high-concentration toxicity (Samuel and Guggenbichler, 2004). Ag NPs are also known to increase seed germination and modify the biochemical profile of some plants (Khan et al., 2016). ROS are produced in plants under biotic stress condition, which enhances the production of antioxidative enzymes like SOD, POD (peroxidase), and CAT (catalase). The application of Ag NPs has been found to reduce stress significantly, and the production of these antioxidative enzymes means that the production of ROS and the collateral damage they caused is also minimized (Hussain et al., 2019; Kumari et al., 2020).

3.4.3 Copper NPs

Copper is a significant micronutrient, which plays a vital role in plant growth and development. It is also crucial in preventing and treating many plant diseases (Borkow and Gabbay, 2005). Thus, copper-based products (chemicals, nanoparticles, etc.) have found a massive role in the agricultural sector. Copper is well known to protect plants against diseases caused by fungi, bacteria, insects, and other pests; hence, Cu NPs can serve as fungicides, pesticides, insecticides, and micronutrients. The small size of the Cu NPs facilitates easy absorption by plants. Being cost-effective and eco-friendly, Cu NPs can also serve as a potent antimicrobial agent in food packaging (Bhagat et al., 2021). Under stress conditions, the production of ROS increases, which can cause damage to the DNA, proteins, and lipids of the cell. However, plants treated with Cu NPs show reduced levels of ROS due to the increased activity of SOD and ascorbate peroxidase (Apx) enzymes (Choudhury et al., 2017). Thus, the Cu NPs help the plants to develop tolerance against stress. Finally, some studies have shown that Cu NPs have adverse effects like phytotoxicity and bioaccumulation when used at higher concentrations. Hence, for safer use of these in agriculture, further studies must be conducted to understand the interaction mechanism between the plant and the Cu NPs (Rai et al., 2018).

3.4.4 Silica NPs

Although a non-essential element, Silicon has proven beneficial to most plant species (Dhakate et al., 2022; Sharma et al., 2022). Plant growth, development, and nutrient uptake are positively affected by silicon (Mitani and Jian, 2005). Disease and stress resistance in plants are also induced by silicon (Brecht et al., 2004). Silica nanoparticles display unique physiochemical characteristics compared with bulk silica material (Bhat et al., 2021). Due to their unique physical and chemical properties, SiO_2 NPs have found major effectiveness in modern agriculture (Seleiman et al., 2020). They can serve as a tool for the targeted delivery of nucleotides, proteins, and different chemicals into plant cells, thus providing a way to manipulate plant genes (Gad et al., 2020). The property of SiO_2 NPs in promoting tolerance against stress in plants is well documented. The uptake and translocation of silicon are increased by SiO_2 NPs, which in turn lowers the production and accumulation of ROS and lipid peroxidation, conferring greater tolerance to biotic stress (Mathur and Roy, 2020). SiO_2 NPs are also known to induce plant immune response (acquired resistance) by expressing defense genes, contributing to protection under biotic stress (Kandhol et al., 2022; Wang et al., 2022). SiO_2 NPs can also be combined with other nanoparticles to alleviate biotic stress in plants (Tortella et al., 2023).

3.4.5 Polymer-based NPs

Over the last decade, polymer-based nanoparticles have been studied extensively due to their potential use in medicine and biology (Tortella et al., 2023). Polymeric nanoparticles are colloidal systems composed of natural or synthetic polymers (Gagliardi et al., 2021). They have various applications due to their biodegradable, non-toxic, and antifungal properties. The potential of polymeric nanoparticles to protect active compounds from unfavorable environmental conditions and to release them in a specific zone of the target plant has led to their use in sustainable agriculture and crop protection against biotic stress (Mosaddik et al., 2018; Tortella et al., 2023). Chitosan is an example of one such polymeric nanomaterial that is very commonly used in agriculture due to its cost-effectiveness and biocompatible and biodegradable nature (Saberi Riseh et al., 2022). Chitosan also stimulates plant growth and induces tolerance to biotic stress (Malerba and Cerana, 2016). However, it has been observed that when chitosan is used in nanoform, it is more effective than bulk chitosan (Chun and Chandrasekaran, 2019). Chitosan nanoparticles can load active molecules, such as nitric oxide (NO) donors, to enhance plant growth. They can also be used as nanocarriers of traditional agrochemicals to increase their effectiveness (Seabra et al., 2022; Tortella et al., 2023). Moreover, chitosan nanoparticle treatment improves the innate immune response of plants by upregulating their defense genes, producing antioxidant enzymes, and increasing phenolic and NO levels. Thus, chitosan nanoparticles can also be used as a better disease control agent than bulk chitosan for sustainable agriculture.

3.5 ROLE OF EMERGING NANO-FORMULATIONS AGAINST BIOTIC STRESS

Emerging nano-formulations have shown great potential in mitigating the effects of biotic stress in agriculture caused by pests, diseases, and other living organisms that can harm crops. Nanoparticles can deliver active compounds like pesticides more effectively to the target site, thereby reducing the amount of pesticide needed, which is beneficial for both the environment and human health (Dhakate et al., 2022). Nanoparticles can also enhance plant defense mechanisms. For instance, chitosan nanoparticles can stimulate the plant's immune system and increase the production of plant hormones that can help combat stress (Chun and Chandrasekaran, 2019). Other nanoparticles have also been shown to improve plant resistance to fungal diseases by inducing the expression of defense-related genes (Pande and Arora, 2019). In addition, some nano-formulations can directly interact with pests themselves, disrupting their physiological processes and reducing their ability to cause damage. Silver nanoparticles, for example, have insecticidal properties against various problems, including aphids, spider mites, and whiteflies (Mishra et al., 2017). Overall, emerging nano-formulations offer promising results in combating biotic stress in agriculture. However, further research is necessary to fully understand their potential benefits and ensure environmental and human health safety (Table 3.2).

3.6 MECHANISM OF NP–PLANT INTERACTION AGAINST BIOTIC STRESS

Nanoparticles regulate bacteria, fungi, insects, pests, and nematodes. A complex process is involved in the interaction of nanoparticles with plant systems. The properties of nanoparticles are unique and facilitate the transportation of nanoparticles within the plants. Due to their high reactive potential and large surface area, they can efficiently interrelate with various plant parts and tissues. Moreover, the interaction between nanoparticles and plants is described as being based on the nanoparticle size, chemical structure, amount, and consistency (Awasthi et al., 2020). According to various research groups, the chemical movement and attachment of NPs to the plant parts are determined by different forces such as gravity, Brownian motion, double layer forces, and van der Waals forces. Some metal oxide NPs have been shown to improve plant nutrient and water uptake, increase the biomass of shoots and roots, and promote seed germination (Rajput et al., 2021).

TABLE 3.2
Tabulated Representation of Different Nanoformulations Active against Different Pathogens

Nanoformulation	Type	Target Species	Mode of Action	References
Chitosan nanoparticles	Biopolymer-based	Wheat, paddy, maize, soybean	Induce systemic acquired resistance (SAR)	Ma et al., 2020
Liposomes	Lipid-based	Tomato, wheat, crabgrass	Deliver bioactive molecules to the target site	Chaudhary et al., 2018
Silver nanoparticles	Metal-based	Tomato, rice, soybean	Inhibit microbial growth and suppress plant diseases	Shang et al., 2019a
Zinc oxide nanoparticles	Metal-based	Tomato plants	Enhance plant growth and inhibit pathogen infection	Faizan et al., 2021
Graphene oxide nanoparticles	Carbon-based	Rice plants	Induce pathogen resistance and activate defense mechanisms	Elmer et al., 2018
Carbon nanotubes	Carbon-based	Wheat, tomato, radish	Enhance plant growth and induce systemic resistance	Cao et al., 2008
Polymeric nanoparticles	Polymer-based	Tomato plants	Deliver bioactive molecules and induce plant growth	Kalia et al., 2020

Additionally, using nanoparticles may alter how plant genes are expressed and regulated, improving the plant's ability to withstand biotic stressors (Pande and Arora, 2019). Furthermore, nanoparticles can improve plant growth and development by improving plant productivity and nutrient uptake. However, there are also worries about environmental and health risks associated with using nanoparticles in agriculture. More research is required to determine the safe and efficient use of nanoparticles in agriculture. Rules and guidelines must also be developed to ensure their secure and accountable application in agriculture (Ali et al., 2021).

Physical adsorption, diffusion, and endocytosis are some of the processes involved in the mechanism of nanoparticle uptake by plants under abiotic stress. Plant cell membranes become more permeable under abiotic stress conditions, allowing nanoparticles to enter cells via diffusion or endocytosis. The uptake of the nanoparticles by plants is also influenced by their size and surface charge (Awasthi et al., 2020). Nanoparticles can interact with various cellular elements that have entered the cells, including organelles, proteins, and enzymes. These interactions can change these elements' structure and function, enhancing plant growth and abiotic stress tolerance. Additionally, nanoparticles can transport nutrients and bioactive substances like fertilizers and pesticides to the target cells, increasing their efficiency (Pande and Arora, 2019). When NPs are given to plants, their absorption and transportation depend on size. Yet, an additional barrier is created by the formation of the cell wall over the plant cell. This obstructs the passage of NPs into cells. Therefore, specific methods are required for the distribution of NPs in plants. NPs have been delivered via foliar spraying or culturing of various tissues such as seeds, roots, pollen, isolated cells, and protoplasts with NPs (Manzoor et al., 2022). The different methods of transferring the NPs into plants are spraying, injecting the NPs directly, the hydroponic method, and the gene gun method. The application of NPs has been attempted by pollen and root incubation. The barrier to NPs, that is, the cell wall, is removed by cell wall–degrading enzymes, and the protoplasts are also removed for the better delivery of the NPs. Further, the biolistic method transfers the NPs into the plant cell. The root system uses a complex pathway to take up the nanoparticles because of its interaction with the soil (Worrall et al., 2018). The NPs have the potential to overcome biotic stress, depending on the interaction that occurs due to their uptake and translocation by the plant system. Different species of plants have various hindrances that stop the transport of NPs in the plant. The uptake and transport of NPs in plants take place with the help of morphological changes and chemical composition of roots and leaf tissues. As the nanoparticles enter the plant system, they can cause changes in the plant's morphology, biochemistry, and physiology, and improve the biotic stress caused in the plants (Awasthi et al., 2020).

The other technique employed in the uptake of NPs by plants is hydroponics. Accurate parameters should be maintained in this process for efficient uptake of the NPs. In this process, Ag NPs, SiO_2 NPs, ZnO NPs, etc., are applied as multi-walled carbon nanotubes; the NPs are transported more readily in the plants. After the nanoparticles have infiltrated plants, their tracking is initiated, enabling the observation of their uptake, absorption, and subsequent transfer within the plants, thereby showcasing their impact. The apoplastic and symplastic paths are highly involved in the uptake of NPs by plants; the root involves the epidermis, and the aerial part involves stomata, rhizomes, stigma, hydathodes, cuticles, and lenticels for the transfer of NPs through the phloem (Awasthi et al., 2020). The NPs are transferred in the xylem and phloem through the tip of the root meristem, where the Casparian strips are not completely developed, and are transported in the shoot. Endocytosis transfers the NPs when the cell wall, cell membrane, and Casparian strips are present. Other methods, such as carrier proteins, aquaporins, etc., transport the NPs. NPs with a size of 4–100 nm can pass through the wax layer and cuticle barrier and accumulate in the substomatal cavity. The aquaporins can transfer the NPs through the intercellular parenchymatous space in the seed coat. The amounts of NPs taken up and accumulated within plant cells are determined by their charge and other physicochemical properties. The outermost layers of a plant cell, the cell wall and cell membrane, contain both hydrophobic and hydrophilic components and are negatively charged (Sood et al., 2021). As a result, a cell's negatively charged wall facilitates the uptake of positively charged NPs. Positively charged Au NPs and CeO NPs are absorbed in higher concentrations and exhibit charge-dependent transport efficiency (Ali et al., 2021). NPs enter cells via osmotic pressure and capillary action. The interaction of NPs with the outer surface of the cell modifies its receptor, transporter, and other membrane proteins. Plants adsorb NPs. NPs larger than the cellular pores are not absorbed; NPs smaller than the cellular pores are absorbed, transmitted, and consequently accumulate in various parts. The transport of NPs is influenced by the cell wall, cytoplasmic membrane, plasmodesmata, Casparian strips, vascular tissue, cuticle, stomata, and hydathodes based on size exclusion limits (SELs). Nanoparticles as small as 36–50 nm have been determined to gain entry and move within plants; however, this size excludes the SELs of barriers. According to one theory, the SELs of barriers are active and dynamic. They can alter in response to NP influence, changes in ionic concentrations, and various biotic and abiotic stresses (Shang et al., 2019b).

Conclusively, larger NPs are constrained by the semipermeable epidermal cell wall. In contrast, small NPs (3–5 nm in diameter) can enter plant roots by osmotic pressure, capillary forces, or passing through the root epidermal cells. Some NPs cause the cell wall to develop new pores to allow entry. Once inside, NPs are transported to the central vascular cylinder via extracellular spaces, passing through the Casparian strip barrier to enter the xylem. Through endocytosis, the formation of pores, and transportation, NPs may bind to carrier proteins. NPs that cannot internalize gather on the Casparian strip, while those that have entered the xylem are moved to the shoots and then returned to the roots via the phloem. NPs can be found in the nuclei, cytoplasm, and cell walls of cortical cells (Ali et al., 2021) (Figure 3.2).

3.7 LIMITATIONS AND CHALLENGES OF NANOTECHNOLOGICAL INTERVENTIONS

The application of nanotechnology has the potential to completely transform the agriculture sector, resulting in more productive and sustainable farming methods, higher crop yields, and better environmental health. However, it's crucial to ensure that these technologies are both environmentally and humanly safe and that their advantages outweigh any possible risks (Ali et al., 2021). The use of nanotechnology in agriculture has the potential to completely change how we produce food and take

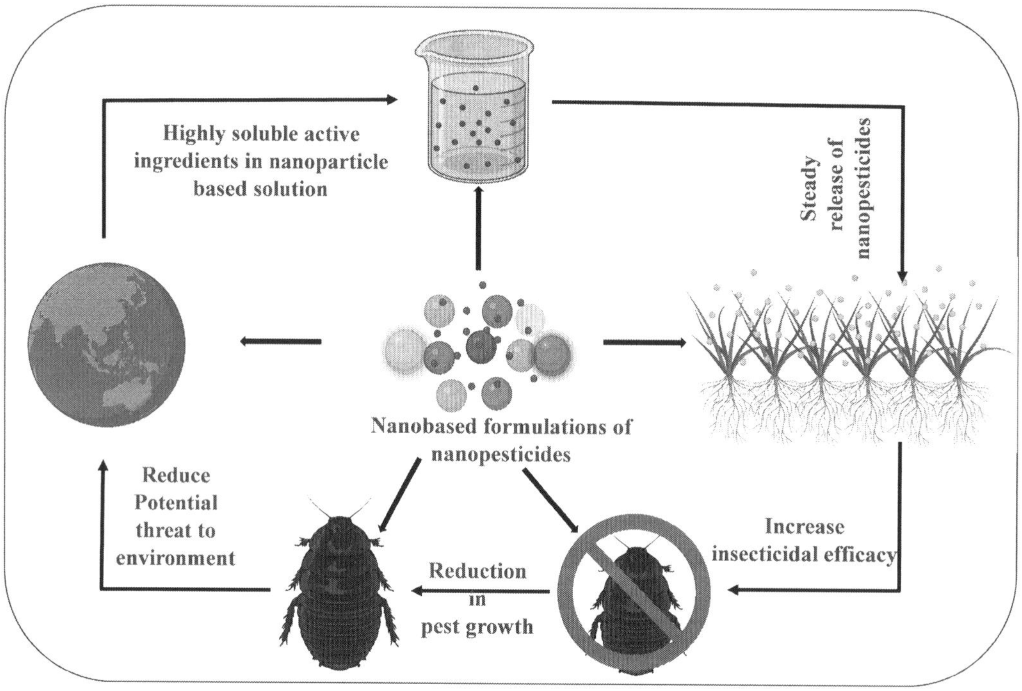

FIGURE 3.2 Mechanisms of nanoparticle transfer in the plant. NPs such as gold NPs, multi-walled carbon nanotubes, and carbon nanotubes are transferred by the roots into the plant tissues, such as cuticles, stomata, phloem, xylem, etc. Endocytosis, hydroponics, and biolistic methods are used.

care of our natural resources. To ensure the safe and responsible application of these technologies, there are a number of issues and restrictions that must be resolved. One of the biggest worries is the potential environmental effects of nanotechnological interventions in agriculture. In soil, water, and air, nanoparticles have the potential to build up, which could be harmful to ecosystems and human health (Awasthi et al., 2020). If nanoparticles are ingested, inhaled, or come into contact with the skin, they may harm human health. There is evidence that nanoparticles can cause cell damage and inflammation, among other problems, though the health effects of nanoparticles are still being researched (Usman et al., 2020).

Since agriculture is a relatively new field, there are few rules governing nanotechnological interventions. There is a chance that these technologies will be abused or lead to unintended consequences if there are no appropriate regulations in place (Mishra et al., 2017). If consumers are not fully informed about these technologies' possible risks and advantages, they may be wary of products treated with nanotechnology. This might make it challenging for companies to market their goods and prevent the agricultural industry from adopting these technologies quickly (Usman et al., 2020). Nanotechnological interventions in agriculture can be pricey, which may prevent them from being used outside of larger farms or more affluent nations. This might worsen significant inequalities in the world food system and restrict access to these potentially game-changing technologies (Mishra et al., 2017).

3.8 FUTURE PROSPECTS

Nanotechnology can potentially be a sustainable method of crop protection and stress management in plants (Farooq et al., 2022). However, the negative impact of using nanotechnology in the agricultural sector cannot be overlooked. As such, detailed study must be done on the phytotoxicity effects of nanoparticles on plants. Several features, such as the size of the nanoparticles, their chemical nature, dose, shape, concentration, duration of treatment, route of application, the nature of the target plant, and the environmental microbiota, determine the effects of the nanoparticles on plants (Y. Wang et al., 2022). The molecular mechanism of the plant–nanoparticle interaction also needs to be studied thoroughly. To apply accurate and non-toxic doses of nanoparticles, a study on their concentration and bioaccumulation must be performed. The use of synthetic nanoparticles should be minimized as much as possible. Green chemistry should be applied to synthesize nanoparticles using living cells and plant extracts (Dhakate et al., 2022). Another challenge would be the acceptance of new technology by people. Smart methods of delivering nutrients to plants via nanoparticles need to be developed. Furthermore, ecotoxicological experiments must be performed in natural settings for a better understanding of nanoparticles' environmental impacts (Farooq et al., 2022). There is also a need to perform futuristic research to fill the knowledge gaps and build new regulatory frameworks in agriculture to deal with the environmental risks of using these technologies (Grillo et al., 2021). Collaborative studies must be performed among researchers from different fields (agriculture, plant physiology, and nanotechnology) to improve future research.

3.9 CONCLUSION

Nanotechnology is emerging as a fast-growing field that has the potential to revolutionize agriculture along with many other fields. Although nanotechnology has already found its application in the medical field, its application in sustainable agriculture and crop improvement is still in its juvenile stage. Thus, to use this technology to its full potential, a thorough understanding of the interaction of nanoparticles with plants at the cellular and molecular levels is essential. However, scientists should also consider the risk involved in using this technology. Bioaccumulation of silver, copper, silica, zinc oxide nanoparticles, etc. in the food chain may have many detrimental and toxic effects. It can also cause phytotoxicity at higher concentrations and toxicity for non-targeted organisms. Therefore, nanotechnology-based products must be used with caution. More research is needed to screen and utilize nanomaterials to stimulate plant development in stress conditions effectively. The nanoparticle size and concentration must be optimized before application to reduce the potential adverse effects on the crops and the environment. Biosynthesized nanoparticles should be used instead of synthetic ones to reduce the generation of toxic materials. The development of environment-friendly methods should be the focus for the synthesis and delivery of nanoparticles in the future. Nano-phytopathology is an exciting and growing field that requires more research to make it a safer and environment-friendly tool for sustainable agriculture.

REFERENCES

Ali, S., Mehmood, A., & Khan, N. (2021). Uptake, translocation, and consequences of nanomaterials on plant growth and stress adaptation. *Journal of Nanomaterials*. https://doi.org/10.1155/2021/6677616

An, J., Hu, P., Li, F., Wu, H., Shen, Y., White, J. C., Tian, X., Li, Z., & Giraldo, J. P. (2020). Emerging investigator series: molecular mechanisms of plant salinity stress tolerance improvement by seed priming with cerium oxide nanoparticles. *Environmental Science: Nano*, *7*(8), 2214–2228. https://doi.org/10.1039/D0EN00387E

Maholiya, A., Ranjan, P., Khan, R., Murali, S., Nainwal, C. R., Chauhan, S. P., Sathish, N., Chaurasia, P. J., & Srivastava, K. A. (2023). An insight into the role of carbon dots in the agriculture system: A review. *Environmental Science: Nano*. https://doi.org/10.1039/D2EN00954D

Awasthi, G., Singh, T., Tiwari, Y., Awasthi, A., Tripathi, R. D., Shrivastava, S., Vajpayee, P., Kumar, A., & Awasthi, K. K. (2020). A review on nanotechnological interventions for plant growth and production. *Materials Today: Proceedings*, *31*, 685–693. https://doi.org/10.1016/J.MATPR.2020.07.255

Bhagat, M., Anand, R., Sharma, P., Rajput, P., Sharma, N., & Singh, K. (2021). Review—Multifunctional copper nanoparticles: Synthesis and applications. *ECS Journal of Solid State Science and Technology*, *10*(6), 063011. https://doi.org/10.1149/2162-8777/AC07F8

Bhat, J. A., Rajora, N., Raturi, G., Sharma, S., Dhiman, P., Sanand, S., Shivaraj, S. M., Sonah, H., & Deshmukh, R. (2021). Silicon nanoparticles (SiNPs) in sustainable agriculture: Major emphasis on the practicality, efficacy and concerns. *Nanoscale Advances*, *3*(14), 4019–4028. https://doi.org/10.1039/D1NA00233C

Boonham, N., Glover, R., Tomlinson, J., & Mumford, R. (2008). Exploiting generic platform technologies for the detection and identification of plant pathogens. https://doi.org/10.1007/s10658-008-9284-3

Borkow, G., & Gabbay, J. (2005). Copper as a biocidal tool. *Current Medicinal Chemistry*, *12*(18), 2163–2175. https://doi.org/10.2174/0929867054637617

Brecht, M. O., Datnoff, L. E., Kucharek, T. A., & Nagata, R. T. (2004). Influence of silicon and chlorothalonil on the suppression of gray leaf spot and increase plant growth in St. Augustinegrass. *Plant Disease*, *88*(4), 338–344. https://doi.org/10.1094/PDIS.2004.88.4.338

Cao, C., Kim, J. H., Yoon, D., Hwang, E. S., Kim, Y. J., & Baik, S. (2008). Optical detection of DNA hybridization using absorption spectra of single-walled carbon nanotubes. *Materials Chemistry and Physics*, *112*(3), 738–741. https://doi.org/10.1016/J.MATCHEMPHYS.2008.07.129

Chandra, S., Chakraborty, N., Dasgupta, A., Sarkar, J., Panda, K., & Acharya, K. (2015). Chitosan nanoparticles: A positive modulator of innate immune responses in plants. *Scientific Reports*, *5*(1), 1–14. https://doi.org/10.1038/srep15195

Chartuprayoon, N., Rheem, Y., Chen, W., & Myung, N. (2010). Detection of plant pathogen using LPNE grown single conducting polymer nanoribbon. *ECS Meeting Abstracts*, MA2010-02(40), 2278–2278. https://doi.org/10.1149/MA2010-02/40/2278

Chaudhary, V., Jangra, S., & Yadav, N. R. (2018). Nanotechnology based approaches for detection and delivery of microRNA in healthcare and crop protection. *Journal of Nanobiotechnology*, *16*(1), 1–18. https://doi.org/10.1186/S12951-018-0368-8

Chinnamuthu, C. R., & Boopathi, P. M. (2009). Nanotechnology and agroecosystem. *Madras Agricultural Journal*, *96*(1/6), 17–31.

Choudhury, F. K., Rivero, R. M., Blumwald, E., & Mittler, R. (2017). Reactive oxygen species, abiotic stress and stress combination. *The Plant Journal: For Cell and Molecular Biology*, *90*(5), 856–867. https://doi.org/10.1111/TPJ.13299

Chun, S. C., & Chandrasekaran, M. (2019). Chitosan and chitosan nanoparticles induced expression of pathogenesis-related proteins genes enhances biotic stress tolerance in tomato. *International Journal of Biological Macromolecules*, *125*, 948–954. https://doi.org/10.1016/J.IJBIOMAC.2018.12.167

Danish, M., Altaf, M., Robab, M. I., Shahid, M., Manoharadas, S., Hussain, S. A., & Shaikh, H. (2021). Green synthesized silver nanoparticles mitigate biotic stress induced by meloidogyne incognita in Trachyspermum ammi (L.) by improving growth, biochemical, and antioxidant enzyme activities. *ACS Omega*, *6*(17), 11389–11403. https://doi.org/10.1021/ACSOMEGA.1C00375/ASSET/IMAGES/MEDIUM/AO1C00375_M004.GIF

Dhakate, P., Kandhol, N., Raturi, G., Ray, P., Bhardwaj, A., Srivastava, A., Kaushal, L., Singh, A., Pandey, S., Chauhan, D. K., Dubey, N. K., Sharma, S., Singh, V. P., Sahi, S., Grillo, R., Peralta-Videa, J., Deshmukh, R., & Tripathi, D. K. (2022). Silicon nanoforms in crop improvement and stress management. *Chemosphere*, *305*, 135165. https://doi.org/10.1016/J.CHEMOSPHERE.2022.135165

Dizaj, S. M., Lotfipour, F., Barzegar-Jalali, M., Zarrintan, M. H., & Adibkia, K. (2014). Antimicrobial activity of the metals and metal oxide nanoparticles. *Materials Science and Engineering C: Biomimetic Materials Sensors and Systems*, *44*, 278–284. https://doi.org/10.1016/J.MSEC.2014.08.031

Elmer, W., Ma, C., & White, J. (2018). Nanoparticles for plant disease management. *Current Opinion in Environmental Science & Health*, *6*, 66–70. https://doi.org/10.1016/J.COESH.2018.08.002

El-Saadony, M. T., Saad, A. M., Soliman, S. M., Salem, H. M., Desoky, E. S. M., Babalghith, A. O., El-Tahan, A. M., Ibrahim, O. M., Ebrahim, A. A. M., Abd El-Mageed, T. A., Elrys, A. S., Elbadawi, A. A., El-Tarabily, K. A., & AbuQamar, S. F. (2022). Role of nanoparticles in enhancing crop tolerance to abiotic stress: A comprehensive review. *Frontiers in Plant Science*, *13*. https://doi.org/10.3389/FPLS.2022.946717

Faizan, M., Faraz, A., Hayat, S., Bhat, J. A., & Yu, F. (2021). Zinc oxide nanoparticles and epibrassinolide enhanced growth of tomato via modulating antioxidant activity and photosynthetic performance. *Biologie Cellulaire*, *45*(4), 1081–1093. https://doi.org/10.32604/BIOCELL.2021.015363

Farooq, M. A., Hannan, F., Islam, F., Ayyaz, A., Zhang, N., Chen, W., Zhang, K., Huang, Q., Xu, L., & Zhou, W. (2022). The potential of nanomaterials for sustainable modern agriculture: Present findings and future perspectives. *Environmental Science: Nano*, *9*(6), 1926–1951. https://doi.org/10.1039/D1EN01124C

Gad, M. A., & Li, M. J., Ahmed, F. K., & Almoammar, H. (2020). Nanomaterials for gene delivery and editing in plants: Challenges and future perspective. *Multifunctional Hybrid Nanomaterials for Sustainable Agri-Food and Ecosystems*, 135–153. https://doi.org/10.1016/B978-0-12-821354-4.00006-6

Gagliardi, A., Giuliano, E., Venkateswararao, E., Fresta, M., Bulotta, S., Awasthi, V., & Cosco, D. (2021). Biodegradable polymeric nanoparticles for drug delivery to solid tumors. *Frontiers in Pharmacology*, *12*, 17. https://doi.org/10.3389/FPHAR.2021.601626/BIBTEX

Galandova, J., Ziyatdinova, G., & Labuda, J. (2008). Disposable electrochemical biosensor with multiwalled carbon nanotubes—Chitosan composite layer for the detection of deep DNA damage. *Analytical Sciences: The International Journal of the Japan Society for Analytical Chemistry*, *24*(6), 711–716. https://doi.org/10.2116/ANALSCI.24.711

Grillo, R., Fraceto, L. F., Amorim, M. J. B., Scott-Fordsmand, J. J., Schoonjans, R., & Chaudhry, Q. (2021). Ecotoxicological and regulatory aspects of environmental sustainability of nanopesticides. *Journal of Hazardous Materials*, *404*(A), 124148. https://doi.org/10.1016/J.JHAZMAT.2020.124148

Gupta, N., Upadhyaya, C. P., Singh, A., Abd-Elsalam, K. A., & Prasad, R. (2018). Applications of silver nanoparticles in plant protection. *Nanotechnology in the Life Sciences*, 247–265. https://doi.org/10.1007/978-3-319-91161-8_9/COVER

Helaly, M. N., Ahmed El-Metwally, M., El-Hoseiny, H., Omar, S. A., & El-Sheery, I. (2014). Effect of nanoparticles on biological contamination of in vitro cultures and organogenic regeneration of banana. *AJCS*, *8*(4), 612–624.

Husale, S., Persson, H. H. J., & Sahin, O. (2009). DNA nanomechanics allows direct digital detection of complementary DNA and microRNA targets. *Nature 462*(7276), 1075–1078. https://doi.org/10.1038/nature08626

Hussain, M., Raja, N. I., Mashwani, Z. U. R., Iqbal, M., Ejaz, M., & Aslam, S. (2019). Green synthesis and evaluation of silver nanoparticles for antimicrobial and biochemical profiling in Kinnow (Citrus reticulata L.) to enhance fruit quality and productivity under biotic stress. *IET Nanobiotechnology*, *13*(3), 250–256. https://doi.org/10.1049/IET-NBT.2018.5049

Kalia, A., Abd-Elsalam, K. A., & Kuca, K. (2020). Zinc-based nanomaterials for diagnosis and management of plant diseases: Ecological safety and future prospects. *Journal of Fungi*, *6*(4), 1–29. https://doi.org/10.3390/JOF6040222

Kandhol, N., Singh, V. P., Peralta-Videa, J., Corpas, F. J., & Tripathi, D. K. (2022). Silica nanoparticles: The rising star in plant disease protection. *Trends in Plant Science*, *27*(1), 7–9. https://doi.org/10.1016/J.TPLANTS.2021.10.007

Kashyap, P. L., Rai, P., Sharma, S., Chakdar, H., Kumar, S., Pandiyan, K., & Srivastava, A. K. (2016). *Nanotechnology for the Detection and Diagnosis of Plant Pathogens*, 253–276. https://doi.org/10.1007/978-3-319-39306-3_8

Kasote, D. M., Lee, J. H. J., Jayaprakasha, G. K., & Patil, B. S. (2021). Manganese oxide nanoparticles as safer seed priming agent to improve chlorophyll and antioxidant profiles in watermelon seedlings. *Nanomaterials*, *11*(4), 1016. https://doi.org/10.3390/NANO11041016/S1

Khan, M. S., Zaka, M., Abbasi, B. H., Rahman, L., & Shah, A. (2016). Seed germination and biochemical profile of Silybum marianum exposed to monometallic and bimetallic alloy nanoparticles. *IET Nanobiotechnology*, *10*(6), 359–366. https://doi.org/10.1049/IET-NBT.2015.0050

Khiyami, M. A., Almoammar, H., Awad, Y. M., Alghuthaymi, M. A., & Abd-Elsalam, K. A. (2014). Plant pathogen nanodiagnostic techniques: forthcoming changes? *Biotechnology and, Biotechnological Equipment*, *28*(5), 775. https://doi.org/10.1080/13102818.2014.960739

Kulabhusan, P. K., Tripathi, A., Kant, K., Gold, N., Kulabhusan, P. K., Tripathi, A., & Kant, K. (2022). Gold nanoparticles and plant pathogens: An overview and prospective for biosensing in forestry. *Sensors*, *22*(3), 1259. https://doi.org/10.3390/S22031259

Kumar, V., & Arora, K. (2020). Trends in nano-inspired biosensors for plants. *Materials Science for Energy Technologies*, *3*, 255–273. https://doi.org/10.1016/j.mset.2019.10.004

Kumari, M., Pandey, S., Giri, V. P., Bhattacharya, A., Shukla, R., Mishra, A., & Nautiyal, C. S. (2017). Tailoring shape and size of biogenic silver nanoparticles to enhance antimicrobial efficacy against MDR bacteria. *Microbial Pathogenesis*, *105*, 346–355. https://doi.org/10.1016/J.MICPATH.2016.11.012

Kumari, M., Pandey, S., Mishra, S. K., Giri, V. P., Agarwal, L., Dwivedi, S., Pandey, A. K., Nautiyal, C. S., & Mishra, A. (2020). Omics-based mechanistic insight into the role of bioengineered nanoparticles for biotic stress amelioration by modulating plant metabolic pathways. *Frontiers in Bioengineering and Biotechnology*, *8*, 242. https://doi.org/10.3389/FBIOE.2020.00242/BIBTEX

Li, L. S., Hu, J., Yang, W., & Alivisatos, A. P. (2001). Band gap variation of size- and shape-controlled colloidal CdSe quantum rods. *Nano Letters*, *1*(7), 349–351. https://doi.org/10.1021/NL015559R/3DBANDGAP.WRL

Li, Y., Cu, Y. T. H., & Luo, D. (2005). Multiplexed detection of pathogen DNA with DNA-based fluorescence nanobarcodes. *Nature Biotechnology*, *23*(7), 885–889. https://doi.org/10.1038/nbt1106

Liu, Y., Zhou, H., Hu, Z., Yu, G., Yang, D., & Zhao, J. (2017). Label and label-free based surface-enhanced Raman scattering for pathogen bacteria detection: A review. *Biosensors and Bioelectronics*, *94*, 131–140. https://doi.org/10.1016/J.BIOS.2017.02.032

Luksiene, Z., Rasiukeviciute, N., Zudyte, B., & Uselis, N. (2020). Innovative approach to sunlight activated biofungicides for strawberry crop protection: ZnO nanoparticles. *Journal of Photochemistry and Photobiology, Part B: Biology*, *203*, 111656. https://doi.org/10.1016/J.JPHOTOBIOL.2019.111656

Ma, Y., Chen, S., Liao, W., Zhang, L., Liu, J., & Gao, Y. (2020). Formation, physicochemical stability, and redispersibility of curcumin-loaded rhamnolipid nanoparticles using the pH-driven method. *Journal of Agricultural and Food Chemistry*, *68*(27), 7103–7111. https://doi.org/10.1021/ACS.JAFC.0C01326/ASSET/IMAGES/LARGE/JF0C01326_0004.JPEG

Malerba, M., & Cerana, R. (2016). Chitosan effects on plant systems. *International Journal of Molecular Sciences*, *17*(7), 996. https://doi.org/10.3390/IJMS17070996

Manzoor, N., Ali, L., Ahmed, T., Noman, M., Adrees, M., Shahid, M. S., Ogunyemi, S. O., Radwan, K. S. A., Wang, G., & Zaki, H. E. M. (2022). Recent advancements and development in nano-enabled agriculture for improving abiotic stress tolerance in plants. *Frontiers in Plant Science*, *13*, 2401. https://doi.org/10.3389/FPLS.2022.951752/BIBTEX

Martinez-Antonio, A., Pestovsky, Y. S., & Martínez-Antonio, A. (2017). The use of nanoparticles and nanoformulations in agriculture biological compute in FPGAs view project bioeconomy based on waste and residual materials view project the use of nanoparticles and nanoformulations in agriculture. *Article in Journal of Nanoscience and Nanotechnology*, *17*, 8699–8730. https://doi.org/10.1166/jnn.2017.15041

Maryška, L., Jindřichová, B., Siegel, J., Záruba, K., & Burketová, L. (2023). Impact of palladium nanoparticles on plant and its fungal pathogen. A case study: Brassica napus–Plenodomus lingam. *AoB Plants*, *15*(2), 1–11. https://doi.org/10.1093/AOBPLA/PLAD004

Mathur, P., & Roy, S. (2020). Nanosilica facilitates silica uptake, growth and stress tolerance in plants. *Plant Physiology and Biochemistry*, *157*, 114–127. https://doi.org/10.1016/J.PLAPHY.2020.10.011

Meng, Y., Li, Y., Galvani, C. D., Hao, G., Turner, J. N., Burr, T. J., & Hoch, H. C. (2005). Upstream migration of Xylella fastidiosa via pilus-driven twitching motility. *Journal of Bacteriology*, *187*(16), 5560–5567. https://doi.org/10.1128/JB.187.16.5560-5567.2005/SUPPL_FILE/HOCHMOVIE3.ZIP

Mishra, S., Keswani, C., Abhilash, P. C., Fraceto, L. F., & Singh, H. B. (2017). Integrated approach of agri-nanotechnology: Challenges and future trends. *Frontiers in Plant Science*, *8*, 471. https://doi.org/10.3389/FPLS.2017.00471/BIBTEX

Mitani, N., & Jian, F. M. (2005). Uptake system of silicon in different plant species. *Journal of Experimental Botany*, *56*(414), 1255–1261. https://doi.org/10.1093/JXB/ERI121

Mosaddik, A., Ravinayagam, V., Elaanthikkal, S., Fessi, H., Badri, W., & Elaissari, A. (2018). Development and use of polymeric nanoparticles for the encapsulation and administration of plant extracts. *Natural Products as Source of Molecules with Therapeutic Potential: Research and Development, Challenges and Perspectives*, 391–463. https://doi.org/10.1007/978-3-030-00545-0_11/COVER

Ocsoy, I., Paret, M. L., Ocsoy, M. A., Kunwar, S., Chen, T., You, M., & Tan, W. (2013). Nanotechnology in plant disease management: DNA-directed silver nanoparticles on graphene oxide as an antibacterial against Xanthomonas perforans. *ACS Nano*, *7*(10), 8972–8980. https://doi.org/10.1021/NN4034794/SUPPL_FILE/NN4034794_SI_001.PDF

Oluwaseun, A. C. (2017). Impacts of biogenic nanoparticle on the biological control of plant pathogens. *Advances in Biotechnology & Microbiology*, *7*(3). https://doi.org/10.19080/AIBM.2017.07.555711

Otles, S., & Yalcin, B. (2010). Nano-biosensors as new tool for detection of food quality and safety. *LogForum*, *6*(4), 7.

Pande, A., & Arora, S. (2019). Nanotechnological interventions for improving plant health and productivity. *Plant Biotechnology: Progress in Genomic Era*, 375–395. https://doi.org/10.1007/978-981-13-8499-8_17/COVER

Prasad, R. (2014). Synthesis of silver nanoparticles in photosynthetic plants. *Journal of Nanoparticles*, *2014*, 1–8. https://doi.org/10.1155/2014/963961

Rai, M., Ingle, A. P., Pandit, R., Paralikar, P., Shende, S., Gupta, I., Biswas, J. K., & Da Silva, S. S. (2018). Copper and copper nanoparticles: Role in management of insect-pests and pathogenic microbes. *Nanotechnology Reviews*, *7*(4), 303–315. https://doi.org/10.1515/NTREV-2018-0031/ASSET/GRAPHIC/J_NTREV-2018-0031_CV_008.JPG

Rajput, V. D., Minkina, T., & Kumari, A., Harish, Singh, V. K., Verma, K. K., Mandzhieva, S., Sushkova, S., Srivastava, S., & Keswani, C. (2021). Coping with the challenges of abiotic stress in plants: New dimensions in the field application of nanoparticles. *Plants*, *10*(6), 1221. https://doi.org/10.3390/PLANTS10061221

Rémus-Borel, W., Menzies, J. G., & Bélanger, R. R. (2005). Silicon induces antifungal compounds in powdery mildew-infected wheat. *Physiological and Molecular Plant Pathology*, *66*(3), 108–115. https://doi.org/10.1016/J.PMPP.2005.05.006

Saberi Riseh, R., Hassanisaadi, M., Vatankhah, M., Soroush, F., & Varma, R. S. (2022). Nano/microencapsulation of plant biocontrol agents by chitosan, alginate, and other important biopolymers as a novel strategy for alleviating plant biotic stresses. *International Journal of Biological Macromolecules*, *222*, 1589–1604. https://doi.org/10.1016/J.IJBIOMAC.2022.09.278

Samuel, U., & Guggenbichler, J. P. (2004). Prevention of catheter-related infections: The potential of a new nano-silver impregnated catheter. *International Journal of Antimicrobial Agents*, *23*(Suppl. 1), 75–78. https://doi.org/10.1016/J.IJANTIMICAG.2003.12.004

Satti, S. H., Raja, N. I., Ikram, M., Oraby, H. F., Mashwani, Z. U. R., Mohamed, A. H., Singh, A., & Omar, A. A. (2022). Plant-based titanium dioxide nanoparticles trigger biochemical and proteome modifications in Triticum aestivum L. under biotic stress of Puccinia striiformis. *Molecules (Basel, Switzerland)*, *27*(13). https://doi.org/10.3390/MOLECULES27134274

Savary, S., Ficke, A., Aubertot, J. N., & Hollier, C. (2012). Crop losses due to diseases and their implications for global food production losses and food security. *Food Security*, *4*(4), 519–537. https://doi.org/10.1007/S12571-012-0200-5/TABLES/3

Seabra, A. B., Silveira, N. M., Ribeiro, R. V., Pieretti, J. C., Barroso, J. B., Corpas, F. J., Palma, J. M., Hancock, J. T., Petřivalský, M., Gupta, K. J., Wendehenne, D., Loake, G. J., Durner, J., Lindermayr, C., Molnár, Á., Kolbert, Z., & Oliveira, H. C. (2022). Nitric oxide-releasing nanomaterials: from basic research to potential biotechnological applications in agriculture. *New Phytologist*, *234*(4), 1119–1125. https://doi.org/10.1111/NPH.18073

Seleiman, M. F., Almutairi, K. F., Alotaibi, M., Shami, A., Alhammad, B. A., & Battaglia, M. L. (2020). Nano-fertilization as an emerging fertilization technique: Why can modern agriculture benefit from its use? *Plants*, *10*(1). https://doi.org/10.3390/PLANTS10010002

Shang, Y., Kamrul Hasan, M., Ahammed, G. J., Li, M., Yin, H., & Zhou, J. (2019a). Applications of nanotechnology in plant growth and crop protection: A review. *Molecules*, *24*(14). https://doi.org/10.3390/MOLECULES24142558

Shang, Y., Kamrul Hasan, M., Ahammed, G. J., Li, M., Yin, H., & Zhou, J. (2019b). Applications of nanotechnology in plant growth and crop protection: A review. *Molecules*, *24*(14), 2558. https://doi.org/10.3390/MOLECULES24142558

Sharma, A., Vishwakarma, K., Singh, N. K., Prakash, V., Ramawat, N., Prasad, R., Sahi, S., Singh, V. P., Tripathi, D. K., & Sharma, S. (2022). Synergistic action of silicon nanoparticles and indole acetic acid in alleviation of chromium (CrVI) toxicity in Oryza sativa seedlings. *Journal of Biotechnology*, *343*, 71–82. https://doi.org/10.1016/J.JBIOTEC.2021.09.005

Shikha, S., Salafi, T., Cheng, J., & Zhang, Y. (2017). Versatile design and synthesis of nano-barcodes. *Chemical Society Reviews*, *46*(22), 7054–7093. Royal Society of Chemistry. https://doi.org/10.1039/c7cs00271h

Shoala, T. (2023). The dual role of nanotechnology in the management of biotic and abiotic stresses in plants. In *Nanotechnology in Agriculture and Agroecosystems* (pp. 51–70). Elsevier. https://doi.org/10.1016/B978-0-323-99446-0.00011-8

Singh, J., Cobb-Smith, D., Higgins, E., & Khan, A. (2021). Comparative evaluation of lateral flow immunoassays, LAMP, and quantitative PCR for diagnosis of fire blight in apple orchards. *Journal of Plant Pathology*, *103*(Suppl. 1), 131. https://doi.org/10.1007/S42161-020-00644-W

Sood, M., Kapoor, D., Kumar, V., Kalia, N., Bhardwaj, R., Sidhu, G. P. S., & Sharma, A. (2021). Mechanisms of plant defense under pathogen stress: A review. *Current Protein and Peptide Science*, *22*(5), 376–395. https://doi.org/10.2174/1389203722666210125122827

Tortella, G., Rubilar, O., Pieretti, J. C., Fincheira, P., de Melo Santana, B., Fernández-Baldo, M. A., Benavides-Mendoza, A., & Seabra, A. B. (2023). Nanoparticles as a promising strategy to mitigate biotic stress in agriculture. *Antibiotics*, *12*(2), 338. https://doi.org/10.3390/ANTIBIOTICS12020338

Usman, M., Farooq, M., Wakeel, A., Nawaz, A., Cheema, S. A., & ur Rehman, H., Ashraf, I., & Sanaullah, M. (2020). Nanotechnology in agriculture: Current status, challenges and future opportunities. *Science of the Total Environment*, *721*, 137778. https://doi.org/10.1016/J.SCITOTENV.2020.137778

Wang, L., Pan, T., Gao, X., An, J., Ning, C., Li, S., & Cai, K. (2022). Silica nanoparticles activate defense responses by reducing reactive oxygen species under Ralstonia solanacearum infection in tomato plants. *NanoImpact*, *28*, 100418. https://doi.org/10.1016/J.IMPACT.2022.100418

Wang, Y., Xu, X., Fang, X., Yao, N., Lei, H., Yang, G., Wang, Z., Dong, Y., & Hua, Z. (2022). Rationally designing renewable plant oil-based polymers as efficient nanocarriers for sustained pesticide delivery. *Chemical Engineering Journal*, *450*, 138294. https://doi.org/10.1016/J.CEJ.2022.138294

Worrall, E. A., Hamid, A., Mody, K. T., Mitter, N., & Pappu, H. R. (2018). Nanotechnology for plant disease management. *Agronomy*, *8*(12), 285. https://doi.org/10.3390/AGRONOMY8120285

Xie, Y., He, Y., Irwin, P. L., Jin, T., & Shi, X. (2011). Antibacterial activity and mechanism of action of zinc oxide nanoparticles against Campylobacter jejuni. *Applied and Environmental Microbiology*, *77*(7), 2325–2331. https://doi.org/10.1128/AEM.02149-10/ASSET/26983837-6597-4BEB-96EB-4D13136F758A/ASSETS/GRAPHIC/ZAM9991019360004.JPEG

Xue, S., Wang, Y., Wang, M., Zhang, L., Du, X., Gu, H., & Zhang, C. (2014). Iodinated oil-loaded, fluorescent mesoporous silica-coated iron oxide nanoparticles for magnetic resonance imaging/computed tomography/fluorescence trimodal imaging. *International Journal of Nanomedicine*, *9*(1), 2527. https://doi.org/10.2147/IJN.S59754

Yao, K. S., Li, S. J., Tzeng, K. C., Cheng, T. C., Chang, C. Y., Chiu, C. Y., Liao, C. Y., Hsu, J. J., & Lin, Z. P. (2009). Fluorescence silica nanoprobe as a biomarker for rapid detection of plant pathogens. *Advanced in Materials Research*, *79–82*, 513–516. https://doi.org/10.4028/WWW.SCIENTIFIC.NET/AMR.79-82.513

Zou, Z., Du, D., Wang, J., Smith, J. N., Timchalk, C., Li, Y., & Lin, Y. (2010). Quantum dot-based immunochromatographic fluorescent biosensor for biomonitoring trichloropyridinol, a biomarker of exposure to chlorpyrifos. *Analytical Chemistry*, *82*(12), 5125–5133. https://doi.org/10.1021/AC100260M/ASSET/IMAGES/LARGE/AC-2010-00260M_0008.JPEG

Zulfajri, M., Abdelhamid, H. N., Sudewi, S., Dayalan, S., Rasool, A., Habib, A., & Huang, G. G. (2020). Plant part-derived carbon dots for biosensing. *Biosensors*, *10*(6), 68. https://doi.org/10.3390/BIOS10060068

4 The Use of Nanopesticides in Crop Protection

Anam Choudhary, Nasreen Musheer, Rabiya Basri, Arshi Jamil, and Sabiha Saeed

4.1 INTRODUCTION

The livelihood of 70% of the population of developed and under-developed countries depends on agriculture (Gupta and Kumar, 2018). Crops are infected by various pests, pathogens, nematodes, and weeds under field conditions, causing economic loss to the farmers (Kitherian, 2017). Pests and diseases causing crop losses are a key threat to food security and the farming community (Pandey et al. 2017). To protect the crops from biological disasters and enhance crop productivity, pesticides are used as a vital input (Zhao et al., 2017). The Food and Agriculture Organization of the United States (FAO) reported that about 30% of agricultural product output worldwide is protected by monitoring pathogens and pests (Lamberth et al., 2013). All around the world, 4.6 million tonnes of pesticides are used as a plant protection input. But during their application, only 1% of pesticide reaches the targeted site, and the residues run off into the environment and pass through the ecological cycle (Memarizadeh et al., 2014). The harmful effect of chemical pesticides has disrupted crop production and food security in sustainable agriculture (Joginder et al., 2017). Using synthetic specialized pesticides, genetically modified disease-resistant crops and botanical extracts have helped to combat such problems for five decades. Their usage has improved food production, but food quality and soil fertility have declined. With the decrease in soil biodiversity, pathogen and pest resistance start emerging (Ghormade et al., 2011). These chemical pesticides also have negative effects on the environment and non-target organisms. Therefore, a novel technology that is cost-effective, species-specific, with low toxicity to non-target species, and producing fewer pesticide residues is required for sustainable agriculture.

Nanotechnology has emerged as the most promising tool to reduce the hazardous effects of pesticides on the environment and increase the quality and productivity of crops. It is used to develop antimicrobial agents for crop protection (Banik and Sharma, 2011; Ocsoy et al., 2013). Nanoparticles (NPs) comprise a size range of 1–100 nm with a larger surface area (Rajan, 2004). Nanoparticles have enhanced activity due to their greater chemical and physical properties as compared with their precursors (Shobha et al., 2014). These nanoparticles can penetrate biological barriers and infuse into the plant tissue by foliar and root application due to their small size (Poddar et al., 2018). Nanopesticides and nanoherbicides are used to effectively manage pests, herbs, and weeds, whereas nanofertilizers enhance soil fertility and supply nutrients (Chippa, 2017). Nanosensors are used for disease detection and to monitor nutrient concentration and soil moisture (Omanovic-Miklicanina and Maksimovic, 2016). This chapter deals with the current method used in nanotechnology for the management of pests and diseases.

4.2 ROLE OF NANOTECHNOLOGY IN CROP PROTECTION

Excessive use of chemical pesticides can be avoided by using nanotechnology that is safe and eco-friendly for the environment. It has emerged as an effective and potential approach to managing

DOI: 10.1201/b23308-4

pests and diseases in farm applications (Abd-Elsalam and Prasad, 2018). Nanoparticles are advantageous to the plant as well as farmers and have no significant adverse effects on the soil microflora, germination of seeds, and root-to-shoot ratio. Numerous nanoparticles are synthesized and used as pesticides, such as gold, silver, iron oxide, and polymeric nanoparticles. Nanoparticles can be applied directly via seed, soil, and foliage to defend the plant from foreign pests and pathogens. These NPs directly affect the pathogens and are also used in the formulation of pesticides (Khan and Rizvi, 2014). Nanopesticides formed using nanomaterials have provided properties such as solubility, biodegradability, thermal stability, and permeability (Bouwmeester et al., 2009). The active ingredients of targeted nanoparticles have shown broad-spectrum activity against pests and diseases. Broad-spectrum characteristics are exhibited by metal and plant oils. Various types of inorganic nanomaterials such as Ag (silver), Al (aluminum), Au (gold), Cu (copper), Fe (iron), Mg (magnesium), Ti (titanium), Zn (zinc), and metal oxide polymers such as zinc oxide (ZnO) and titanium dioxide (TiO_2) are used for pest and disease management. Significant research and application have been done mainly on silver and gold against pests and pathogens. Still, silver nanoparticles have been shown as the most effective and potential antimicrobial agent against pests, fungi, bacteria, nematodes, and viruses. Nanoparticles control targeted nematodes by inducing oxidative stress in their cells (Lim et al., 2012). The active ingredients of nanomaterials affect the functioning of membrane-bound enzymes in the respiratory chain and inhibit oxygen metabolism, causing suffocation and cell death of fungi and bacteria (Puebla et al., 2004). They control viruses by preventing the replication of nucleic acid (Jain and Kothari 2014). Metal nanoparticles are synthesized using physical, chemical, and biological methods. The biosynthesis of nanoparticles using microorganisms and plants has become prevalent recently, and these are also used as stabilizing and reducing agents (Abdellatif et al., 2016; Prasad et al., 2016). Active ingredients present in nanopesticides show target delivery, slow degradation, and controlled release of active ingredients for longer periods, which are less toxic and more environmentally safe. Nanopesticides are also used for the early detection of diseases. Various types of nanopesticides have been developed, such as nanoemulsions, nanoencapsulated formulations, nanospheres, nanogels, and metal and metal oxide nanoparticles. Improved properties such as solubility, permeability, thermal properties, crystallinity, and photostability have been developed in the majority of nanopesticides (Chippa et al., 2017; Nuruzzaman et al., 2016). Microbes and plant metabolites have been used in the development of nanobiopesticides by conjugation or capping of nanoparticles. The nanopesticides are developed with precise release of pesticides and deliberate field application to minimize the impact on the environment.

Among microbial pathogens, fungi cause about 70–80% of loss (Moore et al., 2011). They affect the crop at the growth, flowering, and post-harvest stages, causing enormous damage (Pusztahelyi et al., 2015). Among plant pathogens, 200 species of bacteria infect plants (Schaad et al., 2001). An effective control strategy to control plant diseases caused by bacteria urgently needs to be developed.

About 47% of plant diseases are caused by viruses, as reported by Anderson et al. (2004). Diseases caused by viruses are difficult to control and sometimes are incurable. Viral diseases cause harmful consequences for agriculture, and metal nanoparticles are used as potential antiviral agents by intervening in certain steps of viral replication (Vargas-Hernandez et al., 2020).

Plant-pathogenic nematodes affect major crops such as grain, vegetable, and fruit crops (Khan et al., 2021). Nematode infection causes a significant risk to crop production, leading to crop loss. Environment-friendly alternatives are necessary to control plant-parasitic nematodes (Fabiyi et al., 2020).

Crops are also attacked by numerous insects and pests, affecting the quality, quantity, and commercial value of the products (Taghizadeh and Mohammadkhani, 2017). Chemicals are used to kill and inhibit the reproduction and feeding habits of insects (Mogul et al., 1996). Pesticides harm the environment; therefore, innovative pest-resistive and advanced technology has been developed. Pest management with nanoparticles includes insecticides and insect repellents.

Largely because of the negative impact of pesticides, much effort has been devoted to finding new pest control strategies through the introduction of innovative pest-resistive concepts and advanced technologies for pest management. NPs help to produce new pesticides, insecticides, and insect repellents.

Traditionally, chemical pesticides, genetically modified crops, water management, crop rotation, and planting clean seeds and plant parts are used commonly to control diseases (Sundin et al., 2016). Antimicrobial compounds used for bacterial diseases have been banned due to antibiotic resistance (Almoneafy et al., 2014). Therefore, an environment-friendly and effective management strategy such as using nanoparticles in laboratory, greenhouse, and field application for controlling bacterial diseases plays a significant role (Figure 4.1; Table 4.1).

4.3 ANTIMICROBIAL MECHANISM AGAINST FUNGI/BACTERIA

The mechanisms involved in the biocidal activity of nanoparticles are (i) translocation and internalization of metal ions, resulting in intracellular ATP degradation and disruption of DNA duplication (Lok et al., 2007); (ii) generation of reactive oxygen species (ROS) damaging the cellular structure (Kim et al., 2007); and (iii) the permeability of the cell membrane is altered, and proton motive force is depleted (Mc Quillan, 2010).

In fungi, the target site for the nanoparticles is the cell membrane, causing an inhibitory effect, as reported by Mallman et al. (2015). Nanoparticles form pores in the cell membrane of the fungus, releasing biomolecules and leading to fungal cell death (Kim et al., 2009). In the study, membrane depolarization and formation of pits and pores in the cell wall of *Candida albicans* were observed on treatment with silver nanoparticles, causing liberation of intracellular glucose and trehalose into the medium. Growth of the pathogen was prevented by cell cycle arrest at the G2/M phase of the pathogen (Kim et al., 2009).

In bacteria, two different antimicrobial methods are observed: bacterial plasmolysis and bacteriolysis. In bacterial plasmolysis, there is a contraction of the plasma membrane and degradation, alteration, and reduction of cytoplasm, causing loss of intracellular components, whereas in bacteriolysis, the rigid cell wall is substantially degraded due to the stimulated autolytic wall enzymes, particularly after the death of the cell (Ginsburg, 1989).

Nanoparticles affect the survival of cells, causing cell death and affecting cell signaling, producing ROS (Mueller et al., 2005) comprising free radicals: singlet oxygen ($^1O_2$), superoxide (O_2^-), carbonate (CO_3^-), hydroxyl ($HO^\cdot$), hydroperoxyl (HO_2), peroxyl ($RO_2^\cdot$), carbonate (CO_3^-), carbon dioxide radical ($CO_2^{\cdot\,-}$), alkoxyl ($RO^\cdot$), and non-radicals: ozone (O_3), nitric oxide (NO), hydrogen peroxide (H_2O_2), hypochlorite (OCl^-), hypochlorous acid (HOCl), hypobromous acid (HOBr),

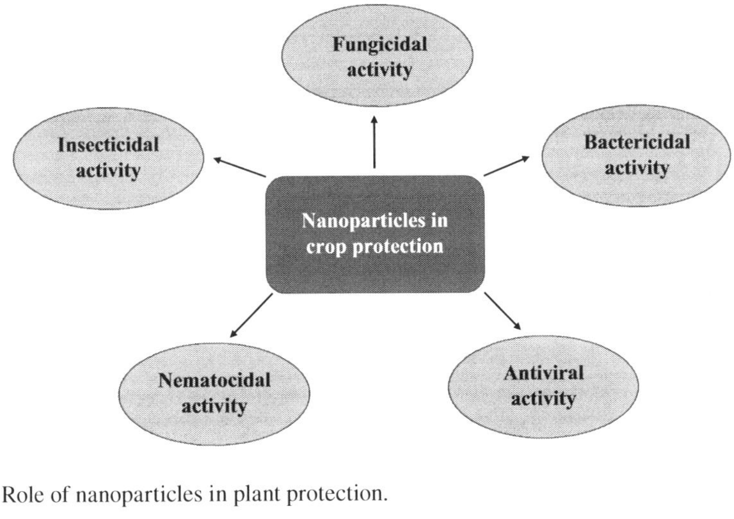

FIGURE 4.1 Role of nanoparticles in plant protection.

TABLE 4.1
Nanopesticides Used against Plant Pathogenic Fungi, Bacteria, Viruses, and Nematodes

Nanoparticles	Pathogen	References
	Fungi	
AgNPs	*Alternaria solani*, *A. alternata*	Narware et al., 2019
AgNPs	*Aspergillus* spp. and *Penicillium digitatum*	Al-Zubaidi et al., 2019
AgNPs	*A. niger*, *Fusarium udum*, and *Rhizoctonia solani*	Akhter and Hemlatha, 2019
AgNPs	*Fusarium oxysporum*, *Rhizoctonia solani*, *Sclerotium rolfsii*, and *Sclerotinia sclerotiorum*	Kaman and Dutta, 2019
AgNPs	*Bipolaris sorghicola*	Almaary et al., 2020
AgNPs	*Cladosporium fulvum*	Elgorban et al., 2017
AgNPs	*Fusarium fujikuroi*, *Macrophomina phaseolina*	Zaki et al., 2022
AgNPs	*Fusarium graminearum*	Ibrahim et al., 2020
AgNPs	*Fusarium verticilloides*	Marathe et al., 2020
AgNPs	*Magnaporthe oryzae*	Ibrahim et al., 2020
AgNPs	*M. oryzae triticum*	Islam et al., 2018
AgNPs	*S. sclerotiorum*	Tomah et al., 2020
AgNPs	*Ustilaginoidea virens*	Roseline et al., 2019
Chitosan NPs	*F. oxysporum*, *R. solani*, *S. rolfsii*	Boruah and Dutta, 2021
CuONPs	*A. alternata*, *A. niger*, *Fusarium oxysporum f.sp. chrysanthemi*, *Pythium ultimum*	Hassan et al., 2019
CuONPs and ZnONPs	*Alternaria citri*	Lasso-Robledo et al., 2022
CuONPs	*F. oxysporum f.sp. chrysanthemi*	Elmer et al., 2021
CuONPs	*Phytophthora nicotianae*	Juan-ni et al., 2022
MgONPs	*P. nicotianae*, *Thielaviopsis basicola*	Chen et al., 2020
MgONPs	*Phytophthora infestans*	Wang et al., 2022
CuONPs	*F. solani*, *R. solani*	Bukhari et al., 2021
SiNPs	*Botrytis fabae*	Hasan et al., 2020
ZnONPs	*F. solani*	Malandrakis et al., 2021
	Bacteria	
AgNPs	*Acidovorax oryzae*	Masum et al., 2019
AgNPs	*Agrobacterium tumefacians*, *Erwinia amylovora*, *Pectobacterium carotovorum subsp. carotovorum*, *Pseudomonas lachrymans*, *Ralstonia solanacearum*	Mohammed et al., 2020
AgNPs, ZnONPs, TiO_2NPs	*Dickeya dedantii*	Hossain et al., 2019
AgNPs	*Erwinia carotovora subsp. atroseptica*	Abbas et al., 2019
AgNPs and AuNPs	Gram-negative bacteria	Costa et al., 2020
AgNPs	*Pseudomonas aeruginosa*	Bhuyar et al., 2020
AgNPs	*Xanthomonas oryzae pv. oryzae*	Ahmed et al., 2020
CuNPs, Cu-Cs nanocomposite	*Clavibacter michiganensis subsp. Michiganensis*, *E. amylovora*, *P. syringae pv. syringae*, and *X. perforans*	Shahryari et al., 2020
Caprylate-capped cobalt ferrite NPs	*E. carotovora*	Johnson et al., 2020
Chromium NPs	*Erwinia amylovora*	Koth et al., 2020
CuONPs	Gram-positive and gram-negative bacteria	Mohamed et al., 2020
CuO	*R. solanacearum*	Chen et al., 2019
CuNPs, CuONPs	*Xanthomonas axonopodis pv. punicae*	Chikte et al., 2019
CuNPs and FeNPs	*Xanthomonas campestris*	Moradian and Biparva, 2018
CuNPs	*Xanthomonas euvesucatoria*	Fan et al., 2021
FeONPs, ZnONPs	*R. solanacearum*	Jiang et al., 2021

(Continued)

TABLE 4.1 (CONTINUED)
Nanopesticides Used against Plant Pathogenic Fungi, Bacteria, Viruses, and Nematodes

Nanoparticles	Pathogen	References
Fe_2O_3	*R. solanacearum*	Alam et al., 2019
MgO, MnO_2	*Xanthomonas oryzae* pv. *oryzae*	Ogunyemi et al., 2020
SiO_2	*Pseudomonas syringae* pv. *tomato*, *Pectobacterium carotovorum* subsp. *carotovorum*, *R. solanacearum*, *Xanthomonas campestris* pv. *vesicatoria*	Parveen and Siddiqui, 2021
ZnONPs	*Bacillus cereus*, *Pseudomonas aeruginosa*, and *Staphylococcus aureus*	Sumanth et al., 2020
ZnONPs	*Escherichia coli*	El-Belely et al., 2021
ZnONPs	Gram-positive and Gram-negative bacteria	Taghizadeh et al., 2020
ZnONPs	*Xanthomonas oryzae* pv. *oryzae*	Abdallah et al., 2020
	Virus	
AgNPs	Potato virus Y	Hao et al., 2017
AgNPs	Tobacco mosaic virus	Wang et al., 2016
AgNPs	Tomato mosaic virus	El-Dougdoug et al., 2018
AgNPs	Tomato spotted wilt virus	Shafie et al., 2018
Au	Barley yellow dwarf virus	Alkubaisi and Aref, 2016
Carbon nanotubes	Tobacco mosaic virus	Adeel et al., 2021
CeO_2NPs	Tobacco mosaic virus	Eugene and Zholobak, 2016
Fe_2O_3NPs	Tobacco mosaic virus	Cai et al., 2019
GO-AgNPs	Tomato bushy stunt virus	Elazzazy et al., 2017
NiONPs	Cucumber mosaic virus	Derbalah and Elsharkawy, 2019
SiO_2NPs	Papaya ringspot virus	Tan et al., 2018
SiO_2NPs	Tobacco mosaic virus	Cai et al., 2019
SiO_2NPs	Tomato yellow leaf curl virus	El-Shazly et al., 2017
TiO_2NPs	Broad bean stain virus	Elsharkawy and Derbalah, 2018
TiO_2NPs	Turnip mosaic virus	Hao et al., 2018
ZnONPs	Cucumber mosaic virus	Hao et al., 2017
ZnONPs	Tobacco mosaic virus	Cai et al., 2019
	Nematode	
AgNPs	*Meloidogyne javanica*	Ghareeb et al., 2020
AgNPs	*Meloidogyne incognita*	Khan et al., 2021
AgNPs	Cyst nematodes	Fabiyi et al., 2020
CuONPs	*M. incognita*	Tauseef et al., 2021
FeONPs	*M. incognita*, *M. javanica*	Gkanatsiou et al., 2019
ZnONPs	*M. incognita*	Elansary et al., 2022 ~~2021~~
	Insects	
AgNPs	*Holotrichia serrata*	Shukla et al., 2020
AgNPs	*Coptotermes* spp.	Iqtedar et al., 2020
AgNPs	*Helicoverpa armigera* and *Spodoptera litura*	Manimegalai et al., 2022
CuNPs	*Rhyzopertha dominica*, *Sitophilus granarius*	Badawy et al., 2021
CuNPs	*Tribolium castenium*	El-Saadony et al., 2020
SeNPs	*S. litura*	Arunthirumeni et al., 2022 ~~2021~~
TiNPs	*Galleria mellonella*	Yosri et al., 2018
TiO_2NPs, ZnONPs	*Bactericera cockerelli*	Gutierrez-Ramirez et al., 2021
TiO_2NPs	*H. armigera*	Chinnaperumal et al., 2018
TiO_2NPs, ZnONPs	*Spodoptera frugiperda*	Kumaravel et al., 2021

peroxomonocarbonate ($HOOCO_2^-$), peroxynitrate (O_2NOO^-), peroxynitrite ($ONOO^-$), peroxynitrous acid (ONOOH), and organic peroxides (ROOH) (Pantopoulos and Schipper, 2012; Wu et al., 2014).

H_2O_2 was instantly reduced to $O_2^{\cdot -}$ by the catalyst superoxide dismutases (SOD) (Johnson and Giuliyi, 2005). Nicotinamide adenine dinucleotide phosphate (NADPH), oxidase (NOX), and mitochondria produce superoxide, which inactivates various enzymes and helps in the lipid peroxidation of cellular membranes (Brand, 2010). The production of ROS causes oxidative stress, leading to the degradation of DNA (Fu et al., 2014), causing single- and double-strand breaks, base and sugar lesions, and DNA–protein crosslinks, and generating basic sites (Valko et al., 2006). Silver ions bind to proteins with cysteine moieties present on the plasma membrane, causing physiological and biochemical damage. Silver nanoparticles penetrate the cytoplasm, inactivating the functioning of potential enzymes (respiratory enzymes) and leading to cell death (Ocsoy et al., 2013).

Copper nanoparticles interact with the cell membrane, releasing sodium (Na^+) ions, which disorganize the membrane and produce cellular lysates (Diaz-Visurraga et al., 2010). Metallic nanoparticles interrupt activity, causing cell death by binding thiol moieties of bacterial proteins. This alters the cell's electrical potential and respiratory process, causing a disturbance in the permeability (Radzig et al., 2013). Respiratory enzymes were also inhibited by the production of ROS (Park et al., 2009) (Figure 4.2).

4.4 ANTIVIRAL MECHANISM AGAINST VIRUSES

Nanoparticles attach to the glycoprotein envelope of the virus and bind to the disulfide bond region of CD^{4-}, the binding domain preventing the entry of viruses (Elbeshehy et al., 2015). This inhibits the virus and causes injury to coat protein and its aggregation. Viral proteins are decreased as NPs interfere with the biosynthesis of protein and the post-translational modification process activating defense mechanisms (Kochkina et al., 1994). They also interact with the nucleic acid of the viruses present in the cell and compete with them. Chitosan nanoparticles showed an antimicrobial effect,

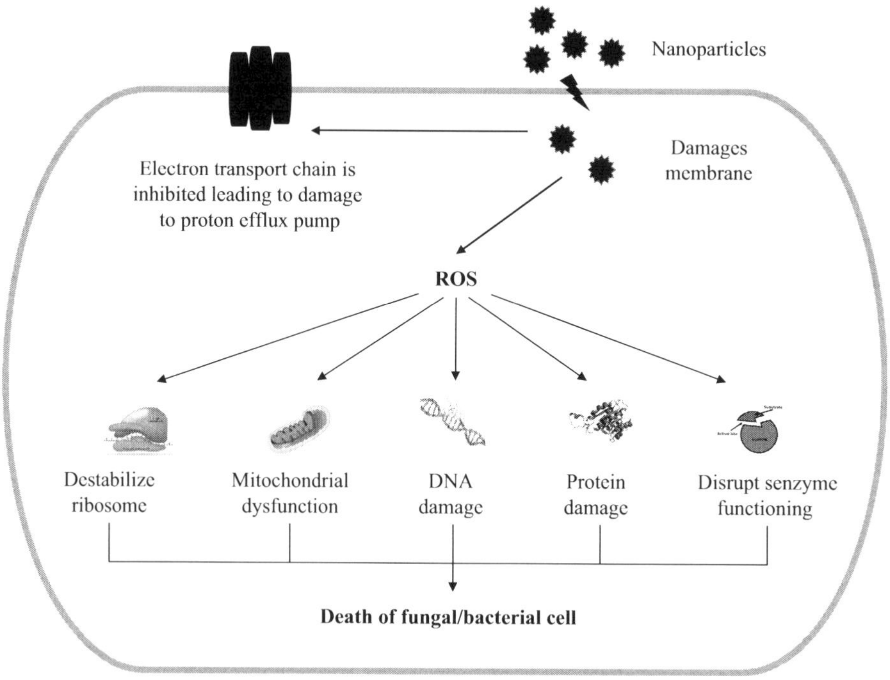

FIGURE 4.2 Antimicrobial mechanisms of various nanofungicides on bacterial/fungal pathogens.

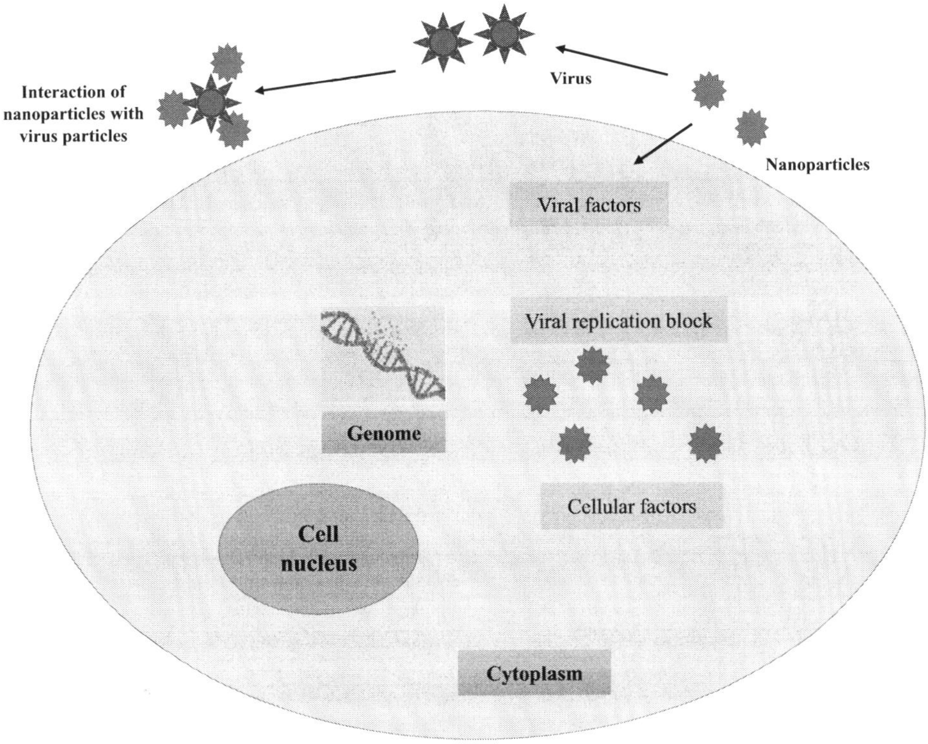

FIGURE 4.3 Antiviral mechanism of nanopesticides.

causing cell membrane disruption, toxin production, and microbial growth inhibition. They also inhibited the activity of H^+-ATPase and prevented the synthesis of proteins and mRNA (Malerba and Cerana, 2016). The production of plant hormones such as salicylic acid (SA), jasmonic acid (JA), and ethylene (Derbalah and Elsharkawy, 2019), plant secondary metabolites, pathogenesis-related genes (PRs), and ROS (Cai et al., 2019) are also activated by nanoparticles (Figure 4.3).

4.5 ANTINEMATIC MECHANISMS AGAINST NEMATODES

The ingestion of nanoparticles by nematodes causes adverse effects on their digestive system, leading to the impairment of the intestine, neurons, and reproductive organs. Extended exposure to nanoparticles causes hyperpermeability of the intestines, inhibiting defecation behavior (Wu et al., 2013; Yin et al., 2014). Accumulation of nanoparticles in the lumen of the intestines causes dysregulation of innate immunity (Wu et al., 2014). The immune system is damaged by exposure to nanoparticles, changing the level of expression of genes that are involved in antimicrobial peptide formation, leading to immunotoxicity (Shakoor et al., 2016). Nanoparticles directly cause immune cell damage by necrosis and apoptosis or indirectly cause an imbalance of immune-specific signaling pathways. The movement, development, fecundity, and reproduction of nematodes are affected by the toxicity of nanoparticles (Zhang et al., 2020). The toxicity causes germ cell apoptosis, cell cycle arrest, and DNA damage, affecting gonad development (Zhao et al., 2016) (Figure 4.4).

The nanotoxicity of nanoparticles is caused by activation of receptors, alteration in signaling, destabilization of lysosomes, toxicity to mitochondria, and damaged proteins. Nanoparticles interact with the cell surface, causing activation of receptors that trigger cascades of intercellular signaling such as MAPK, which alters the expression of stress response genes, affecting the production of ROS. The expression of p38 MAPK PMK-1 at both protein and gene levels, causing oxidative stress, is increased

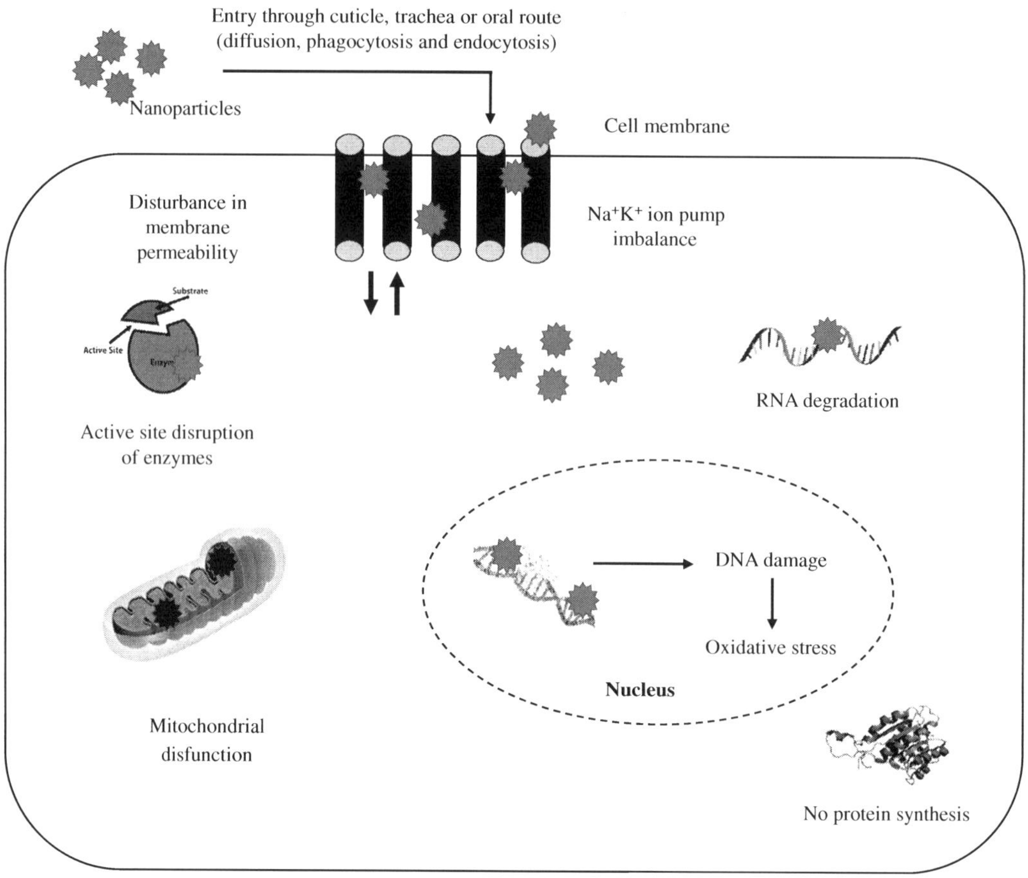

FIGURE 4.4 Antinematicidal mechanism of nanopesticides.

significantly (Lim et al., 2012; Tsyusko et al., 2012). Nanoparticles destabilize the outer membrane of mitochondria, which alters their membrane potential and disrupts the electron transport chain and oxidative phosphorylation (Xia et al., 2006), thereby increasing ROS production (Meyer et al., 2013). Protein denaturation causes misfolded protein accumulation and a change in Ca^{2+} homeostasis (Lindholm et al., 2006). High levels of Ca levels induce oxidative stress (Soenen et al., 2011). Oxidative stress, DNA damage, and endoplasmic reticulum stress lead to apoptosis and necrosis.

4.6 INSECTICIDAL MECHANISM AGAINST INSECTS

Nanoparticles activate insect-target alteration pathways and trigger the release of stomach poisons. The nanopesticides enter the insects through physical contact, ingestion, and inhalation. The nanoparticles enhance the insecticidal action, as they effectively increase pest coverage, adherence, and permeability, and penetrate the exoskeleton of insects, causing cellular disruption and cell death (Abd-Elsalam et al., 2018). Nano specks and dust enter the exoskeleton of stored grain pests and enhance their permeability and dispersal, increasing the entry of pesticides into insects (Yang et al., 2017; Zhao et al., 2017). Transport and conductivity are also increased by accelerating the dose effect, pesticide efficacy, pesticide poisoning, and bioactivity. Figure 4.4 represents the possible mechanism of action of nanopesticides on insects.

The body of insects is dehydrated by attachment to the cuticular wax layer, which also affects the head, thorax, siphon, and abdominal region. Nanoalumina, nanoclay, and nanosilica become attached to the cuticle and absorb water from the insects. Silver nanoparticles split the insect body, alter the membrane, change permeability and respiration, damage DNA, and interfere with melanin synthesis. Nanoparticle inhalation causes internalization of cells by phagocytosis, resulting in deterioration of the midgut. It also alters the activity of metabolic genes, denatures cellular organelles, and reduces protein, glucose, and lipid levels. It reduces food intake, leading to the death of the insect (Gustafson et al., 2015). It also affects membrane potential and the activity of nervous system enzymes by binding to acetylcholine esterase and β carboxyl esterase. Silver nanoparticles affect the release of gonadotrophin and the activity of acetylcholine esterase, causing developmental damage and reproductive failure. Silver nanoparticles act as inhibitors of amylase and glutathione S transferase enzymes (Vinutha et al., 2013). Nanoparticles affect proteins and nucleic acids, causing mutation and leading to cell death (Benali, 2018). ROS leading to oxidative stress are generated, along with an impact on antioxidants and detoxifying enzymes (Saini et al., 2020) (Figure 4.5).

4.7 DRAWBACKS AND FUTURE OUTLOOK

These new-age nanopesticides interact with the ecosystem, disturbing the food chain equilibrium (Kaphle et al., 2018). Nanoparticles used in plant disease management can cause threats to humans and other organisms if they are released inadvertently into the environment. The incredibly minute-sized particles with high surface area cause a bigger risk, as they can easily pass through cellular barriers (Zoomi et al., 2020). This may cause harmful effects on muscle activity, disturbance in the brain, developmental abnormalities, infertility, and endocrine-related problems in mammals (Kumar et al., 2017). This is important, as plants and animals

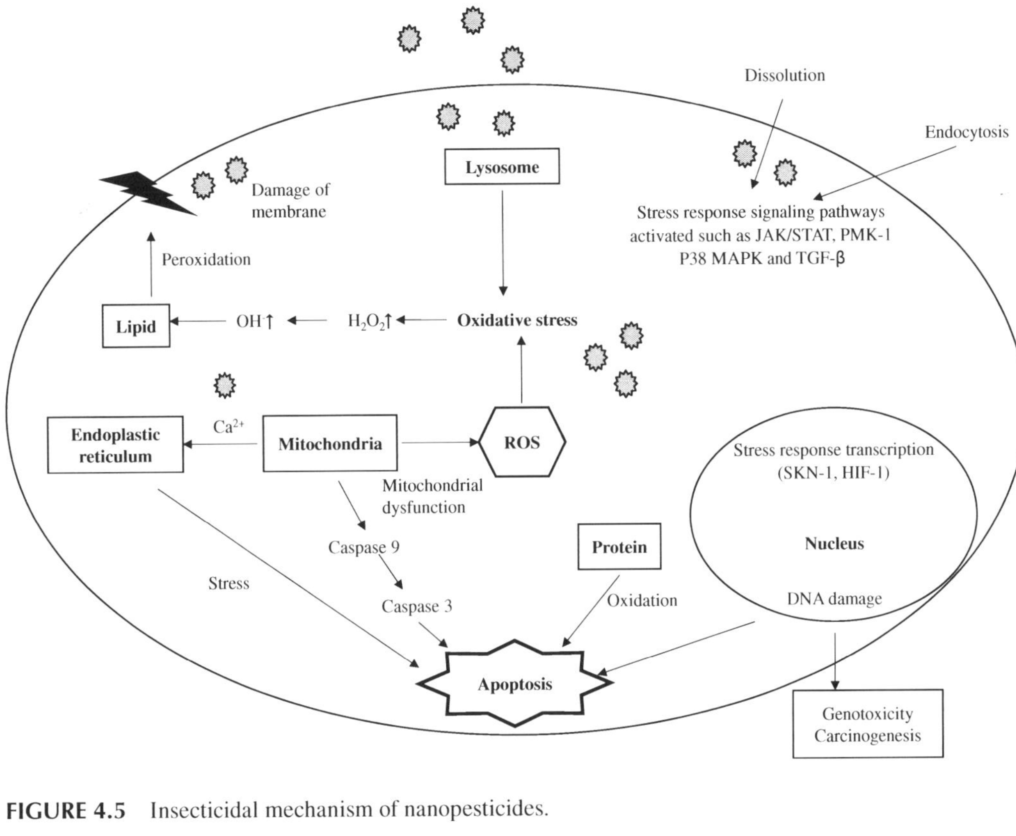

FIGURE 4.5 Insecticidal mechanism of nanopesticides.

treated with nanoparticles during agricultural practices are consumed by humans, comprising an important source of transmission. Insufficient information has been obtained related to the environmental toxicity, adsorption capacity, and permissible limit of metal nanoparticles. Metal nanoparticles exposed to soil change the microbial biomass and affect the growth of plants as well as biochemical, physiological, and molecular parameters (Castillo-Henriquez et al., 2020; Courtois et al., 2019). Nanoparticles are considered environmentally friendly, but highly toxic metal nanoparticles needed to be removed and cleared from the agricultural soil, which causes harmful effects on humans and the environment. Proper testing and safe manufacturing practices with physiological characterization before use should be done to avoid their side effect on human health as well as the environment (Sardoiwala et al., 2018; Hochella et al., 2019). Considering the adverse effects of metallic nanopesticides on the environment and public health (Savić-Zdravković et al., 2021), toxicological evaluation should be done before they are commercially available, and their long-term effects on plant physiology and the ecosystem should be further studied.

The production of metal nanoparticles of precise size, shape, and monodispersity is difficult (Khandel and Shahi, 2018). The mechanism of action of metallic nanoparticles has been studied in detail, but more study is required on nanoparticles such as zinc, nickel, iron, and titanium. The mechanisms of nanoparticles according to size, shape, composition, and arrangement are poorly understood (Van Aken, 2015). Engineered nanoparticles undergo chemical, physical, and biological modifications due to agglomeration and dispersal when they enter the soil. The target selectivity, stability, reactivity, and toxicity change at various levels due to their transformation. Therefore, effective studies and detailed analysis should be done on their fate in soil for sustainable agricultural practices (Boros and Ostafe, 2020). Improper use of nanopesticides by untrained employees adds risk when they are dissolved and dispersed in soil and water. Non-target organisms such as mineral solubilizers and symbionts are also affected by the direct soil and foliar application of nanoparticles. Therefore, before applying these nanoparticles, critical analysis should be done of their effect on non-target organisms.

However, the application of nanotechnology in plant pathology is still in the initial phases. Further studies related to nanophytopathology on disease diagnosis, the physiology of pathogen and host, interaction, and the infection process will benefit the development of novel nanopesticides that are less harmful than conventional pesticides. In the near future, a huge quantity of nanopesticide formulations will be used commercially in fields, leading to sustainable agriculture. Nanoparticle size, stability, and monodispersity are the three main factors for the assessment of their synthesis. Hence, effective control of these factors should be studied extensively. The synthesis of nanoparticles is time-consuming, so reducing the time taken for their synthesis is an intriguing approach. Less phytotoxic metal nanoparticles should be developed, and their effect on plant growth and soil microflora should be examined in future studies. Furthermore, for the commercialization and effective usage of nanomaterials, various experts should work in collaboration to design and evaluate biomimetic nanomaterials in the agricultural sector.

4.8 CONCLUSIONS

Nanotechnology serves as a potential source of plant disease management, as it enhances the plant's resistance against pests and pathogens. This method is considered a safer method as compared with conventional pesticides for the control of plant diseases and pests. The use of nanopesticides in agriculture has advantages over crop protection (crop rotation), transgenic/genetically engineered plants, chemical pesticides, biological control agents, and botanical extracts. Nanoparticles integrated with crop protection and several pest suppression methods can be used to develop sustainable, long-lasting crop protection management. Thus, the use of nanopesticides in plant protection is considered as the start of a new era in the field of plant protection.

REFERENCES

Abbas, A., Naz, S.S., & Syed, S.A. (2019). Antimicrobial activity of silver nanoparticles (AgNPs) against *Erwinia carotovora* subsp. *atroseptica* and *Alternaria alternata*. *Pak J Agric Sci*, *56*(1), 113–117.

Abdallah, Y., Liu, M., Ogunyemi, S.O., Ahmed, T., Fouad, H., Abdelazez, A., et al. (2020). Bioinspired green synthesis of chitosan and zinc oxide nanoparticles with strong antibacterial activity against rice pathogen Xanthomonas oryzae pv. oryzae. *Molecules*, *25*(20), 4795.

Abdellatif, K.F., Abdelfattah, R.H., El-Ansary, M.S.M. (2016). Green nanoparticles engineering on root-knot nematode infecting eggplants and their effect on plant DNA modification. *Iran J Biotechnol*, *14*(4), 250–259.

Abd-Elsalam, K., & Prasad, R. (2018). *Nanobiotechnology Applications in Plant Protection*. Springer International Publishing (ISBN 978-3-319-91161-8).

Adeel, M., Farooq, T., White, J.C., Hao, Y., He, Z., & Rui, Y. (2021). Carbonbased nano-materials suppress Tobacco Mosaic Virus (TMV) infection and induce resistance in Nicotiana benthamiana. *J Hazard Mater*, *404*(A), 124167.

Ahmed, T., Shahid, M., Noman, M., Niazi, M.B.K., Mahmood, F., Manzoor, I., et al. (2020). Silver nanoparticles synthesized by using Bacillus cereus SZT1 ameliorated the damage of bacterial leaf blight pathogen in rice. *Pathogens*, *9*(3), 160.

Akther, T., & Hemalatha, S. (2019). Mycosilver nanoparticles: Synthesis, characterization and its efficacy against plant pathogenic fungi. *Bionanosci*, *9*(2), 296–301.

Alam, T., Khan, R.A.A., Ali, A., Sher, H., Ullah, Z., & Ali, M. (2019). Biogenic synthesis of iron oxide nanoparticles via Skimmia laureola and their antibacterial efficacy against bacterial wilt pathogen Ralstonia solanacearum. *Mater Sci Eng C Mater Biol Appl*, *98*, 101–108.

Alkubaisi, N.A., & Aref, N.M.A. (2016). Dispersed gold nanoparticles potentially ruin gold barley yellow dwarf virus and eliminate virus infectivity hazards. *Appl Nanosci*. https://doi.org/10.1007/s13204-016-0540-0.

Almaary, K.S., Sayed, S.R.M., Abd-Elkader, O.H., Dawoud, T.M., El Orabi, N.F., & Elgorban, A.M. (2020). Complete green synthesis of silver-nanoparticles applying seed-borne Penicillium duclauxii. *Saudi J Biol Sci*, *27*(5), 1333–1339.

Almoneafy, A.A., Kakar, K.U., Nawaz, Z., Li, B., Saand, M.A., Chun-lan, Y., et al. (2014). Tomato plant growth promotion and antibacterial related-mechanisms of four rhizobacterial Bacillus strains against *Ralstonia solanacearum*. *Symbiosis*, *63*(2), 59–70.

Al-Zubaidi, S., Al-Ayafi, A., & Abdelkader, H. (2019). Biosynthesis, characterization and antifungal activity of silver nanoparticles by *Aspergillus niger* isolate. *J Nanopart Res*, *1*(1), 23–36.

Anderson, P.K., Cunningham, A.A., Patel, N.G., Morales, F.J., Epstein, P.R., & Daszak, P. (2004). Emerging infectious diseases of plants: Pathogen pollution, climate change and agrotechnology drivers. *Trends Ecol Evol*, *19*(10), 535–544.

Arunthirumeni, M., Veerammal, V., & Shivakumar, M.S. (2022). Biocontrol efficacy of mycosynthesized selenium nanoparticle using Trichoderma sp. on insect Pest Spodoptera litura. *J Clust Sci*, *33*, 1645–1653.

Badawy, A.A., Abdelfattah, N.A.H., Salem, S.S., Awad, M.F., & Fouda, A. (2021). Efficacy assessment of biosynthesized copper oxide nanoparticles (CuO-NPs) on stored grain insects and their impacts on morphological and physiological traits of wheat (Triticum aestivum L.) plant. *Biology*, *10*(3), 233.

Banik, S., & Sharma, P. (2011). Plant pathology in the era of nanotechnology. *Indian Phytopath*, *64*(2), 120–127.

Benelli, G. (2018). Mode of action of nanoparticles against insects. *Environ Sci Pollut Res Int*, *25*(13), 12329–12341.

Bhuyar, P., Rahim, M.H.A., Sundararaju, S., Ramaraj, R., Maniam, G.P., & Govindan, N. (2020). Synthesis of silver nanoparticles using marine macroalgae Padina sp. and its antibacterial activity towards pathogenic bacteria. And its antibacterial activity towards pathogenic bacteria. *Beni Suef Univ J Basic Appl Sci*, *9*(1), 3.

Boros, B.-V., & Ostafe, V. (2020). Evaluation of ecotoxicology assessment methods of nanomaterials and their effects. *Nanomaterials (Basel)*, *10*(4), 610.

Boruah, S., & Dutta, P. (2021). Fungus mediated biogenic synthesis and characterization of chitosan nanoparticles and its combine effect with Trichoderma asperellum against Fusarium oxysporum, Sclerotium rolfsii and Rhizoctonia solani. *Indian Phytopathol*, *74*(1), 81–93.

Bouwmeester, H., Deekkers, S., Noordam, M.Y., Hagens, W.I., Bulder, A.S., de Heer, C. et al. (2009). Review of health safety aspects of nanotechnogies in food production. *Regul Toxicol Pharmacol*, *53*, 52–62.

Brand, M.D. (2010). The sites and topology of mitochondrial superoxide production. *Exp Gerontol*, *45*(7–8), 466–472.

Bukhari, S.I., Hamed, M.M., Al-Agamy, M.H., Gazwi, H.S., Radwan, H.H., & Youssif, A.M. (2021). Biosynthesis of copper oxide nanoparticles using streptomyces mhm38 and its biological applications. *J Nanomater*, 6693302, 16 pages, 2021. https://doi.org/10.1155/2021/6693302.

Cai, L., Liu, C., Fan, G., Liu, C., & Sun, X. (2019). Preventing viral disease by ZnONPs through directly deactivating TMV and activating plant immunity in Nicotiana benthamiana. *Environ Sci Nano*, *6*(12), 3653–3669.

Cai, L., Liu, C., Fan, G., Liu, C., & Sun, X. (2019). Preventing viral disease by ZnONPs through directly deactivating TMV and activating plant immunity in Nicotiana benthamiana. *Environ Sci Nano*, *6*(12), 3653–3669.

Castillo-Henriquez, L., Alfaro-Aguilar, K., Ugalde-Alvarez, J., Vega-Fernandez, L., Montes de Oca-Vasquez, G., & Vega-Baudrit, J.R. (2020). Green synthesis of gold and silver nanoparticles from plant extracts and their possible applications as antimicrobial agents in the agricultural area. *Nanomaterials (Basel)*, *10*(9), 1763.

Chen, J., Mao, S., Xu, Z., & Ding, W. (2019). Various antibacterial mechanisms of biosynthesized copper oxide nanoparticles against soil borne *Ralstonia solanacearum*. *RSC Adv*, *9*(7), 3788–3799.

Chen, J., Wu, L., Lu, M., Lu, S., Li, Z., & Ding, W. (2020). Comparative study on the fungicidal activity of metallic MgO nanoparticles and macroscale MgO against soilborne fungal phytopathogens. *Front Microbiol*, *11*, 365.

Chhipa, H. (2017). Nanofertilizers and nanopesticides for agriculture. *Environ Chem Lett*, *15*(1), 15–22.

Chikte, R.G., Paknikar, K.M., Rajwade, J.M., & Sharma, J. (2019). Nanomaterials for the control of bacterial blight disease in pomegranate: Quo vadis? *Appl Microbiol Biotechnol*, *103*(11), 4605–4621.

Chinnaperumal, K., Govindasamy, B., Paramasivam, D., Dilipkumar, A., Dhayalan, A., Vadivel, A., et al. (2018). Bio-pesticidal effects of Trichoderma viride formulated titanium dioxide nanoparticle and their physiological and biochemical changes on Helicoverpa armigera (hub.). *Pestic Biochem Physiol*, *149*, 26–36.

Costa, M.I., Alvarez-Cerimedo, M.S., Urquiza, D., Ayude, M.A., Hoppe, C.E., Fasce, D.P., et al. (2020). Synthesis, characterization and kinetic study of silver and gold nanoparticles produced by the archaeon Haloferax volcanii. *J Appl Microbiol*, *129*(5), 1297–1308.

Courtois, P., Rorat, A., Lemiere, S., Guyoneaud, R., Attard, E., Levard, C., et al. (2019). Ecotoxicology of silver nanoparticles and their derivatives introduced in soil with or without sewage sludge: A review of effects on microorganisms, plants and animals. *Environ Pollut*, *253*, 578–598.

Derbalah, A., & Elsharkawy, M. (2019). A new strategy to control Cucumber mosaic virus using fabricated NiO-nanostructures. *J Biotechnol*, *306*, 134–141.

Derbalah, A.S.H., & Elsharkawy, M.M. (2019). A new strategy to control Cucumber mosaic virus using fabricated NiO-nanostructures. *J Biotechnol*, *306*, 134–141.

Diaz-Visurraga, J., Cardenas, G., & Garcia, A. (2010). Morphological changes induced in bacteria as evaluated by electron microscopy. In: Mendez-Vilas, A., Diaz, J. (Eds.) *Microscopy Science, Technology, Applications*. Formatex, Madrid, Spain, pp. 307–315.

Elansary, M., Hamouda, R., & Elshamy, M. (2022). Using biosynthesized zinc oxide nanoparticles to alleviate the toxicity on banana parasitic-nematode. *Waste Biomass Valor*, *13*, 405–415.

Elazzazy, A.M., Elbeshehy, E.K.F., & Betiha, Md.A. (2017). In vitro assessment of activity of graphene silver composite sheets against multidrugresistant bacteria and Tomato Bushy Stunt Virus. *Trop J Pharmaceut Res*, *16*, 2705–2711.

El-Belely, E.F., Farag, M., Said, H.A., Amin, A.S., Azab, E., Gobouri, A.A., et al. (2021). Green synthesis of zinc oxide nanoparticles (ZnO-NPs) using Arthrospira platensis (class: Cyanophyceae) and evaluation of their biomedical activities. *Nanomaterials (Basel)*, *11*(1), 95.

Elbeshehy, E.K., Elazzazy, A.M., & Aggelis, G. (2015). Silver nanoparticles synthesis mediated by new isolates of Bacillus spp., nanoparticle characterization and their activity against bean yellow mosaic virus and human pathogens. *Front Microbiol*, *6*, 453.

El-Dougdoug, N.K., Bondok, A.M., & El-Dougdoug, K.A. (2018). Evaluation of silver nanoparticles as antiviral agent against ToMV and PVY in tomato plants. *Sciences*, *8*, 100–111.

Elgorban, A.M., El-Samawaty, A.E.M., Abd-Elkader, O.H., Yassin, M.A., Sayed, S.R.M., Khan, M., et al. (2017). Bioengineered silver nanoparticles using Curvularia pallescens and its fungicidal activity against Cladosporium fulvum. *Saudi J Biol Sci*, *24*(7), 1522–1528.

Elmer, W.H., Zuverza-Mena, N., Triplett, L.R., Roberts, E.L., Silady, R.A., & White, J.C. (2021). Foliar application of copper oxide nanoparticles suppresses Fusarium wilt development on chrysanthemum. *Environ Sci Technol*, *55*(15), 10805–10810.

El-Saadony, M.T., Abd El-Hack, M.E., Taha, A.E., Fouda, M.M., Ajarem, J.S., Maodaa, S.N., et al. (2020). Ecofriendly synthesis and insecticidal application of copper nanoparticles against the storage pest Tribolium castaneum. *Nanomaterials (Basel)*, *10*(3), 587.

Elsharkaway, M., & Derbalah, A. (2018). Antiviral activity of titanium dioxide nanostructures as a control strategy for broad bean strain virus in faba bean: Control of broad bean stain virus by TiO2 nanostructures. *Pest Manag Sci*, *75*(3), 828–834.

El-Shazly, M.A., Attia, Y.A., Kabil, F.F., Anis, E., & Hazman, Md. (2017). Inhibitory effects of salicylic acid and silver nanoparticles on potato virus Y infected potato plants in Egypt. *Middle East J. Agricult Res*, *6*, 835–848.

Eugene, K., & Zholobak, N. (2016). Antiviral activity of cerium dioxide nanoparticles on tobacco mosaic virus model. In: *Proceedings of the Topical Issues of New Drugs Development*. Kharkiv, Ukraine.

Fabiyi, O., Olatunji, G., Atolani, O., & Olawuyi, O. (2020). Preparation of bio-nematicidal nanoparticles of Eucalyptus officinalis for the control of cyst nematode (Heterodera sacchari). *J Anim Plant. Sci*, *30*, 1172–1177.

Fan, Q., Liao, Y.-Y., Kunwar, S., Da Silva, S., Young, M., Santra, S., et al. (2021). Antibacterial effect of copper composites against Xanthomonas euvesicatoria. *Crop Prot*, *139*, 105366.

Fu, P.P., Xia, Q., Hwang, H.M., Ray, P.C., & Yu, H. (2014). Mechanisms of nanotoxicity: Generation of reactive oxygen species. *J Food Drug Anal*, *22*(1), 64–75.

Ghareeb, R.Y., Alfy, H., Fahmy, A.A., Ali, H.M., & Abdelsalam, N.R. (2020). Utilization of Cladophora glomerata extract nanoparticles as eco-nematicide and enhancing the defense responses of tomato plants infected by meloidogyne javanica. *Sci Rep*, *10*(1), 1–15.

Ghormade, V., Deshpande, M.V., & Paknikar, K.M. (2011). Perspectives for nano-biotechnology enabled protection and nutrition of plants. *Biotechnol Adv*, *29*(6), 792–803.

Ginsburg, I. (1989). Bacteriolysis is inhibited by hydrogen peroxide and by proteases. *Agents Actions*, *28*(3–4), 238–242.

Gkanatsiou, Ch., Ntalli, N., Menkissoglu-Spiroudi, U., & Dendrinou-Samara, C. (2019). Essential metal-based nanoparticles (copper/iron NPs) as potent nematicidal agents against meloidogyne spp. *J Nanotechnol Res*, *1*, 044–058.

Gupta, A., & Kumar, A. (2018). Climate resilient agro-technologies for enhanced crop and water productivity under water deficit agro-ecologies. In: Meena, R.S. (Ed.) *Sustainable Agriculture*. Scientific Publisher, Jodhpur, pp. 339–356.

Gustafson, H.H., Holt-Casper, D., Grainger, D.W., & Ghandehari, H. (2015). Nanoparticle uptake: The phagocyte problem. *Nano Today*, *10*(4), 487–510.

Hao, Y., Cao, X., Ma, C., Zhang, Z., Zhao, N., Ali, A. et al. (2017). Potential applications and antifungal activities of engineered nanomaterials against gray mold disease agent Botrytis cinerea on rose petals. *Frontiers in Plant Science*, *8*, 1332.

Hao, Y., Yuan, W., Ma, C., White, J.C., Zhang, Z., Adeel, M., et al. (2018). Engineered nanomaterials suppress turnip mosaic virus infection in tobacco (Nicotiana benthamiana). *Environ Sci Nano*, *5*(7), 1685–1693.

Hasan, K.A., Soliman, H., Baka, Z., & Shabana, Y.M. (2020). Efficacy of nano-silicon in the control of chocolate spot disease of Vicia faba L. caused by Botrytis fabae. *Egypt J Basic Appl Sci*, *7*(1), 53–66.

Hassan, S.E., Fouda, A., Radwan, A.A., Salem, S.S., Barghoth, M.G., Awad, M.A. et al. (2019). Endophytic actinomycetes Streptomyces spp. mediated biosynthesis of copper oxide nanoparticles as a promising tool for biotechnological applications. *J Biol Inorg Chem JBIC Publ Soc Biol Inorg Chem*, *24*(3), 377–393.

Hochella, M.F., Mogk, D.W., Ranville, J., Allen, I.C., Luther, G.W., Marr, L.C. et al. (2019). Natural, incidental, and engineered nanomaterials and their impacts on the Earth system. *Science*, *363*(6434), 1414.

Hossain, A., Hong, X., Ibrahim, E., Li, B., Sun, G., Meng, Y., et al. (2019). Green synthesis of silver nanoparticles with culture supernatant of a bacterium Pseudomonas rhodesiae and their antibacterial activity against soft rot pathogen Dickeya dadantii. *Molecules*, *24*(12), 2303.

Ibrahim, E., Luo, J., Ahmed, T., Wu, W., Yan, C., & Li, B. (2020). Biosynthesis of silver nanoparticles using onion endophytic bacterium and its antifungal activity against rice pathogen Magnaporthe oryzae. *J Fungi (Basel)*, *6*(4), 294.

Iqtedar, M., Mirza, N., Aihetasham, A., Iftikhar, S., Kaleem, A., & Abdullah, R. (2020). Termiticidal Activity of Mycosynthesized Silver Nanoparticles from Aspergillus fumigatus BTCB15. *Rev. Mex. Ing. Qui´M*, *19*(3), 1201–1211.

Islam, M.S., Akter, N., Rahman, M.M., Shi, C., Islam, M.T., Zeng, H., & Azam, M.S. (2018). Mussel-inspired immobilization of silver nanoparticles toward antimicrobial cellulose paper. *ACS Sustain Chem Eng*, *6*(7), 9178–9188.

Jain, D., & Kothari, S.L. (2014). Green synthesis of silver nanoparticles and their application in plant virus inhibition. *J Mycol Plant Pathol*, *44*, 21–24.

Jiang, H., Lv, L., Ahmed, T., Jin, S., Shahid, M., Noman, M., et al. (2021). Effect of the nanoparticle exposures on the tomato bacterial wilt disease control by modulating the rhizosphere bacterial community. *Int J Mol Sci*, *23*(1), 414.

Joginder, S.D., Ravinder, K., Naresh, K., Pawan, K., Kiran, N., & Duhan, S. (2017). Nanotechnology: The new perspective in precision agriculture. *Biotechnol Rep*, *15*, 11–23.

Johnson, F., & Giulivi, C. (2005). Superoxide dismutases and their impact upon human health. *Mol Asp Med*, *26*(4–5), 340–352.

Johnson, M., Gaffney, C., White, V., Bechelli, J., Balaraman, R., & Trad, T. (2020). Non-hydrolytic synthesis of caprylate capped cobalt ferrite nanoparticles and their application against *Erwinia carotovora* and *Stenotrophomonas maltophilia*. *J Mater Chem B*, *8*(47), 10845–10853.

Juan-ni, C., Lin-tong, W., Kun, S., Yun-song, Z., & Wei, D. (2022). Nonphytotoxic copper oxide nanoparticles are powerful "nanoweapons" that trigger resistance in tobacco against the soil-borne fungal pathogen Phytophthora nicotianae. *J Integr Agric*, *21*(11), 3245–3262.

Kaman, P.K., & Dutta, P. (2019). Synthesis, characterization and antifungal activity of biosynthesized silver nanoparticle. *Indian Phytopathol*, *72*(1), 79–88.

Kaphle, A., Navya, P.N., Umapathi, A., & Daima, H.K. (2018). Nanomaterials for agriculture, food and environment: Applications, toxicity and regulation. *Environ Chem Lett*, *16*(1), 43–58.

Khan, M., Khan, A.U., Bogdanchikova, N., & Garibo, D. (2021). Antibacterial and antifungal studies of biosynthesized silver nanoparticles against plant parasitic nematode meloidogyne incognita, plant pathogens Ralstonia solanacearum and Fusarium oxysporum. *Molecules*, *26*(9), 2462.

Khan, M.R., & Rizvi, T.F. (2014). Nanotechnology: Scope and application in plant disease management. *Plant Pathol J*, *13*(3), 214–231.

Khandel, P., & Shahi, S.K. (2018). Mycogenic nanoparticles and their bioprospective applications: Current status and future challenges. *J Nanostructure Chem*, *8*(4), 369–391.

Kim, J.S., Kuk, E., Yu, K.N., Kim, J.H., Park, S.J., Lee, H.J., et al. (2007). Antimicrobial effects of silver nanoparticles. *Nanomedicine*, *3*(1), 95–101.

Kim, K.J., Sung, W.S., Suh, B.K., Moon, S.K., Choi, J.S., Kim, J.G., & Lee, D.G. (2009). Antifungal activity and mode of action of silver nano-particles on *Candida albicans*. *Biometals*, *22*(2), 235–242.

Kitherian, S. (2017). Nano and bio-nanoparticles for insect control. *Res J Nanosci Nanotechnol*, *7*(1), 1–9.

Kochkina, Z., Pospeshny, G., & Chirkov, S. (1994). Inhibition by chitosan of productive infection of T-Series bacteriophages in the Escherichia coli culture. *Mikrobiologiia*, *64*(2), 211–215.

Kotb, O.M., Abd El-Latif, F.M., Atawia, A.R., Saleh, S.S., & El-Gioushy, S.F. (2020). Green synthesis of chromium nanoparticles by aqueous extract of *Melia azedarach*, Artemisia herbaalba and bacteria fragments against *Erwinia amylovora*. *Asian J Biotechnol Bioresour*, *6*(2), 22–30.

Kumar, S., Bhanjana, G., Sharma, A., Dilbaghi, N., Sidhu, M.C., & Kim, K. (2017). Development of nanoformulation approaches for the control of weeds. *Sci Total Environ*, *586*, 1272–1278.

Kumaravel, J., Lalitha, K., Arunthirumeni, M., & Shivakumar, M.S. (2021). Mycosynthesis of bimetallic zinc oxide and titanium dioxide nanoparticles for control of Spodoptera frugiperda. *Pestic Biochem Physiol*, *178*, 104910.

Lamberth, C., Jeanmart, S., Luksch, T., & Plant, A. (2013). Current challenges and trends in the discovery of agrochemicals. *Science*, *341*(6147), 742–746.

Lasso-Robledo, J.L., Torres, B., & Peralta-Videa, J.R. (2022). Do all cu nanoparticles have similar applications in nano-enabled agriculture? *Plant Nano Biol*, *1*, 100006.

Lim, D., Roh, J.-Y., Eom, H.-J., Choi, J.-Y., Hyun, J., & Choi, J. (2012). Oxidative stress-related *pmk-1* P38 MAPK activation as a mechanism for toxicity of silver nanoparticles to reproduction in the nematode *Caenorhabditis elegans*. *Environ Toxicol Chem*, *31*(3), 585.

Lindholm, D., Wootz, H., & Korhonen, L. (2006). ER stress and neurodegenerative diseases. *Cell Death Differ*, *13*(3), 385–392.

Lok, C.N., Ho, C.M., Chen, R., He, Q.Y., Yu, W.Y., Sun, H. et al. (2007). Silver nanoparticles: Partial oxidation and antibacterial activities. *J Biol Inorg Chem JBIC Publ Soc Biol Inorg Chem*, *12*(4), 527–534.

Malandrakis, A.A., Kavroulakis, N., Avramidou, M., Papadopoulou, K.K., Tsaniklidis, G., & Chrysikopoulos, C.V. (2021). Metal nanoparticles: Phytotoxicity on tomato and effect on symbiosis with the Fusarium solani FsK strain. *Sci Total Environ*, *787*, 147606.

Malerba, M., & Cerana, R. (2016). Chitosan effects on plant systems. *Int J Mol Sci*, *17*(7), 996.

Mallmann, E.J.J., Cunha, F.A., Castra, B.N.M.F., Maciel, A.M., Menezes, E.A., Fechine, P.B.A. (2015). Antifungal activity of silver nanoparticles obtained by green synthesis. *Rev. do Instituto Medicina Trop. São Paulo,* 57(2), 165–167.

Manimegalai, T., Raguvaran, K., Kalpana, M., & Maheswaran, R. (2022). Facile synthesis of silver nanoparticles using Vernonia anthelmintica (L.) Willd. And their toxicity against Spodoptera litura (Fab.), Helicoverpa armigera (H€ub.), Aedes aegypti Linn. and Culex quinquefasciatus Say. *J Clust Sci, 33,* 2287–2303.

Marathe, K., Naik, J., & Maheshwari, V. (2020). Biogenic synthesis of silver nanoparticles using *Streptomyces* spp. and their antifungal activity against *Fusarium verticillioides. J Clust Sci.* https://doi.org/10.1007/s10876-020-01894-5.

Masum, M.M.I., Siddiqa, M.M., Ali, K.A., Zhang, Y., Abdallah, Y., Ibrahim, E., et al. (2019). Biogenic synthesis of silver nanoparticles using Phyllanthus emblica fruit extract and its inhibitory action against the pathogen Acidovorax oryzae strain RS-2 of rice bacterial brown stripe. *Front Microbiol, 10,* 820.

McQuillan, J. (2010). Bacterial-nanoparticle interactions. Thesis for the degree of Doctor of Philosophy in Biological Sciences. University of Exeter.

Memarizadeh, N., Ghadamyari, M., Adeli, M., & Talebi, K. (2014). Preparation, characterization and efficiency of nanoencapsulated Imidacloprid under laboratory conditions. *Ecotoxicol Environ Saf, 107,* 77–83.

Meyer, J.N., Leung, M.C.K., Rooney, J.P., Sendoel, A., Hengartner, M.O., Kisby, G.E., & Bess, A.S. (2013). Mitochondria as a target of environmental toxicants. *Toxicol Sci, 134*(1), 1.

Mogul, M.G., Akin, H., Hasirci, N., Trantolo, D.J., Gresser, J.D., & Wise, D.L. (1996). Controlled release of biologically active agents for purposes of agricultural crop management. *Resour. Conserv. Rec, 16,* 289–320.

Mohamed, A.A., Abu-Elghait, M., Ahmed, N.E., & Salem, S.S. (2020). Eco-friendly mycogenic synthesis of zno and cuo nanoparticles for in vitro antibacterial, antibiofilm, and antifungal applications. *Biol Trace Elem Res, 199,* 2788–2799.

Mohammed, T.G., Gomah, A.A., & Abd El-Rahman, A.F. (2020). Biogenic synthesis of silver nanoparticles using *Pimpinella anisum* L seed aqueous extract and its inhibitory action against some phytopathogens. *J Mater Sci Res Rev, 6*(2), 30–39.

Moore, D., Robson, G.D., & Trinci, A.P. (2011). *21st Century Guidebook to Fungi with CD.* Cambridge University Press, Cambridge, 368–370.

Moradian, F., Ghorbani, R., & Biparva, P. (2018). Assessment of different antibacterial effects of Fe and Cu nanoparticles on Xanthomonas campestris growth and expression of its pathogenic gene hrpE. *J Agric Sci Technol, 20*(5), 1059–1070.

Mueller, C.F., Laude, K., McNally, J.S., & Harrison, D.G. (2005). ATVB in focus: Redox mechanisms in blood vessels. *Arterioscler Thromb Vasc Biol, 25*(2), 274–278.

Narware, J., Yadav, R., Keswani, C., Singh, S., & Singh, H. (2019). "Silver nanoparticle-based biopesticides for phytopathogens: Scope and potential in agriculture". In: Opender, K. (Ed.) *Nano-biopesticides Today and Future Perspectives.* Elsevier, Amsterdam, pp. 303–314.

Nuruzzaman, M., Rahman, M.M., Liu, Y., & Naidu, R. (2016). Nanoencapsulation, nano-guard for pesticides: A new window for safe application. *J Agri Food Chem, 64*(7), 1447–1483.

Ocsoy, I., Paret, M.L., Ocsoy, M.A., Kunwar, S., Chen, T., You, M., & Tan, W. (2013). Nanotechnology in plant disease management: DNA-directed silver nanoparticles on graphene oxide as an antibacterial against *Xanthomonas perforans. ACS Nano, 7*(10), 8972–8980.

Ogunyemi, S.O., Zhang, M., Abdallah, Y., Ahmed, T., Qiu, W., Ali, M. et al. (2020). The Bio-Synthesis of three metal oxide nanoparticles (ZnO, MnO_2, and MgO) and their antibacterial activity against the bacterial leaf blight pathogen. *Front Microbiol, 11,* 3099.

Omanovi´c-Mikli˘canina, E., & Maksimovi´c., M. (2016). Nanosensors applications in agriculture and food industry. *Bull. Chem. Technol. Soc. Bosnia Herzegovina, 47,* 59–70.

Pandey, P., Irulappan, V.B.M.V., Kumar, M.S., & Senthil-Kumar, M. (2017). Impact of combined abiotic and biotic stresses on plant growth and avenues for crop improvement by exploiting physio-morphological traits. *Front Plant Sci, 8,* 537.

Pantopoulos, K., & Schipper, H.M. (2012). *Principles of Free Radical Biomedicine.* Nova Science Publications, Hauppauge.

Park, J., Kim, J.Y., Kim, J., Lee, J.H., Hahn, J.S., Gu, M.B., & Yoon, J. (2009). Silver-ion mediated reactive oxygen species generation affecting bactericidal activity. *Water Res, 43*(4), 1027–1032.

Parveen, A., & Siddiqui, Z.A. (2021). Impact of silicon dioxide nanoparticles on growth, photosynthetic pigments, proline, activities of defense enzymes and some bacterial and fungal pathogens of tomato. *Vegetos*, *35*(1), 83–93.

Poddar, K., Vijayan, J., Ray, S., & Adak, T. (2018).Nanotechnology for Sustainable Agriculture. In: R.L. Singh, & S. Mondal (Eds.) *Biotechnology for Sustainable Agriculture*. Elsevier, pp. 281–303.

Prasad, R. (2016). *Advances and Applications through Fungal Nanobiotechnology*. Springer, International Publishing.

Puebla, R.A., Dos Santos, D.S. Jr., & Aroca, R.F. (2004). Surface-enhanced Raman scattering for ultrasensitive chemical analysis of 1 and 2-naphthalenethiols. *Analyst*, *129*(12), 1251–1256.

Pusztahelyi, T., Holb, I.J., & Pócsi, I. (2015). Secondary metabolites in fungus-plant interactions. *Front Plant Sci*, *6*, 1–23.

Radzig, M.A., Nadtochenko, V.A., Koksharova, V.A., Kiwi, J., Lipasova, V.A., & Khmel, I.A. (2013). Antibacterial effects of silver nanoparticles on gram-negative bacteria: Influence on the growth and biofilms formation, mechanisms of action. *Colloids Surf B Biointerfaces*, *102*, 300–306.

Rajan, M.S. (2004). *Nano: The Next Revolution*. National Book Trust, 81-237-7961-5, New Delhi, India, pp. 1–179.

Roseline, T.A., Murugan, M., Sudhakar, M., & Arunkumar, K. (2019). Nanopesticidal potential of silver nanocomposites synthesized from the aqueous extracts of red seaweeds. *Environ Technol Innov*, *13*, 82–93.

Saini, R.K., Patel, S., Bajpai, J., & Bajpai, A.K. (2020). Advanced controlled nanopesticide delivery systems for managing insect pests. In: K.R.K. Thomas, & T.K.J. Volva (Eds.) *Controlled Release of Pesticides for Sustainable Agriculture*. Springer, Cham, pp. 155–184.

Sardoiwala, M.N., Kaundal, B., & Choudhury, S.R. (2018). Toxic impact of nanomaterials on microbes, plants and animals. *Environ Chem Lett*, *16*(1), 147–160.

Savić-Zdravković, D., Milošević, D., Conić, J., Marković, K., Ščančar, J., Mililša, M., & Jovanović, B. (2021). Revealing the effects of ceriumdioxide nanoparticles through the analysis of morphological changes in Chironomus RIPARIUS. *Sci Total Environ*, *786*, 147439.

Schaad, N.W., Jones, J., Chun, W. (2001). Laboratory Guide for Identification of Plant Pathogenic Bacteria. *Biologica Plantarum*, *44*, 546.

Shafie, R.M., Salama, A.M., & Farroh, K.Y. (2018). Silver nanoparticles activity against tomato spotted wilt virus. *Middle East J Agric Res*, *7*, 1251–1267.

Shahryari, F., Rabiei, Z., & Sadighian, S. (2020). Antibacterial activity of copper nanoparticles synthesized with sumac extract and copper-chitosan nanocomposite against some plant pathogenic bacteria in laboratory. https://doi.org/10.22084/ab.2020.19130.1410.

Shakoor, S., Sun, L., & Wang, D. (2016). Multi-walled carbon nanotubes enhanced fungal colonization and suppressed innate immune response to fungal infection in nematodes. *Toxicol Res (Camb)*, *5*(2), 492–499.

Shobha, G., Moses, V., & Ananda, S. (2014). Biological synthesis of copper nanoparticles and its impact. *Int J Pharm Sci Invent*, *3*(8), 6–28.

Shukla, G., Gaurav, S.S., Rani, V., Singh, A., Rani, P., Verma, P., & Kumar, B. (2020). Evaluation of larvicidal effect of mycogenic silver nanoparticles against the white grubs (Holotrichia sp.). *J Adv Sci Res*, *11*(1), 296–304.

Soenen, S.J., Rivera-Gil, P., Montenegro, J.M., Parak, W.J., De Smedt, S.C., & Braeckmans, K. (2011). Cellular toxicity of inorganic nanoparticles: Common aspects and guidelines for improved nanotoxicity evaluation. *Nano Today*, *6*(5), 446.

Sumanth, B., Lakshmeesha, T.R., Ansari, M.A., Alzohairy, M.A., Udayashankar, A.C., Shobha, B., et al. (2020). Mycogenic synthesis of extracellular zinc oxide nanoparticles from Xylaria acuta and its nano-antibiotic potential. *Int J Nanomedicine*, *15*, 8519–8536.

Sundin, G.W., Castiblanco, L.F., Yuan, X., Zeng, Q., & Yang, C.-H. (2016). Bacterial disease management: Challenges, experience, innovation and future prospects. *Mol Plant Pathol*, *17*(9), 1506–1518.

Taghizadeh, R., & Mohammadkhani, N. (2017). Antifeedant activity of *Descura Iniasophia* and *Thuja orientalis* extracts against *Tribolium castaneum* (Coleoptera: Tenebrionidae). *Crop Prot*, *6*(4), 487–495.

Taghizadeh, S.-M., Lal, N., Ebrahiminezhad, A., Moeini, F., Seifan, M., Ghasemi, Y., & Berenjian, A. (2020). Green and economic fabrication of zinc oxide (ZnO) nanorods as a broadband uv blocker and antimicrobial agent. *Nanomaterials (Basel)*, *10*(3), 530.

Tan, B.L., Norhaizan, M.E., Liew, W.P.P., & Sulaiman, R.H. (2018). Antioxidant and oxidative stress: A mutual interplay in age-related diseases. *Front Pharmacol*, *9*, 1162.

Tauseef, A., Hisamuddin,G. J., Rehman, A., & Uddin, I. (2021). Differential response of cowpea towards theCuOnanoparticles under meloidogyne incognita stress. *S Afr J Bot*, *139*, 175–182.

Tomah, A.A., Alamer, I.S.A., Li, B., & Zhang, J.Z. (2020). Mycosynthesis of silver nanoparticles using screened Trichoderma isolates and their antifungal activity against Sclerotinia sclerotiorum. *Nanomaterials (Basel)*, *10*(10), 1955.

Tsyusko, O., Unrine, J.M., Spurgeon, D.J., Blalock, E.M., Starnes, D., Tseng, M.T. et al. (2012). Toxicogenomic responses of the model organism *Caenorhabditis elegans* to gold nanoparticles. *Environ Sci Technol*, *46*(7), 4115.

Valko, M., Rhodes, C., Moncol, J., Izakovic, M.M., & Mazur, M. (2006). Free radicals, metals and antioxidants in oxidative stress-induced cancer. *Chem Biol Interact*, *160*(1), 1–40.

Van Aken, B. (2015). Gene expression changes in plants and microorganisms exposed to nanomaterials. *Curr Opin Biotechnol*, *33*, 206–219.

Vargas-Hernandez, M., Macias-Bobadilla, I., Guevara-Gonzalez, R.G., Rico-Garcia, E., Ocampo-Velazquez, R.V., Avila-Juarez, L., & Torres-Pacheco, I. (2020). Nanoparticles as potential antivirals in agriculture. *Agriculture*, *10*(10), 444.

Vinutha, J.S., Bhagat, D., & Bakthavatsalam, N. (2013). Nanotechnology in the management of polyphagous pest *Helicoverpa armigera*. *J Acadamic Ind Res*, *1*(10), 606–608.

Wang, Y., Sun, C., Xu, C., et al. (2016). Preliminary experiments on nano-silver against tobacco mosaic virus and its mechanism. *Tob Sci Technol*, *49*, 22–30.

Wang, Z.L., Zhang, X., Fan, G.J., Que, Y., Xue, F., & Liu, Y.H. (2022). Toxicity effects and mechanisms of MgO nanoparticles on the oomycete pathogen Phytophthora infestans and its host Solanum tuberosum. *Toxics*, *10*(10), 553.

Wu, H., Yin, J.J., Wamer, W.G., Zeng, M., & Lo, Y.M. (2014). Reactive oxygen species-related activities of nano-iron metal and nano-iron oxides. *J Food Drug Anal*, *22*(1), 86–94.

Wu, Q., Zhao, Y., Fang, J., & Wang, D. (2014). Immune response is required for the control of in vivo translocation and chronic toxicity of graphene oxide. *Nanoscale*, *6*(11), 5894–5906.

Wu, Q.L., Yin, L., Li, X., Tang, M., Zhang, T., & Wang, D.Y. (2013). Contributions of altered permeability of intestinal barrier and defecation behavior to toxicity formation from graphene oxide in nematode Caenorhabditis elegans. *Nanoscale*, *5*(20), 9934–9943.

Xia, T., Kovochich, M., Brant, J., Hotze, M., Sempf, J., Oberley, T. et al. (2006). Comparison of the abilities of ambient and manufactured nanoparticles to induce cellular toxicity according to an oxidative stress paradigm. *Nano Lett*, *6*(8), 1794.

Yang, D., Cui, B., Wang, C., Zhao, X., Zeng, Z., Wang, Y. et al. (2017). Preparation and characterization of emamectin benzoate solid nanodispersion. *J Nanomater*, *2017*(2), 1–9.

Yin, J.A., Liu, X.J., Yuan, J., Jiang, J., & Cai, S.Q. (2014). Longevity manipulations differentially affect serotonin/dopamine level and behavioral deterioration in aging Caenorhabditis elegans. *J Neurosci*, *34*(11), 3947–3958.

Yosri, M., Abdel-Aziz, M.M., & Sayed, R.M. (2018). Larvicidal potential of irradiated mycoinsecticide from Metarhizium anisopliae and larvicidal synergistic effect with its mycosynthesized titanium nanoparticles (TiNPs). *J Radiat Res Appl Sci*, *11*, 328–334.

Zaki, S.A., Ouf, S.A., Abd-Elsalam, K.A., Asran, A.A., Hassan, M.M., Kalia, A., et al. (2022). Trichogenic silver-based nanoparticles for suppression of fungi involved in damping-off of cotton seedlings. *Microorganisms*, *10*(2), 344.

Zhang, F., You, X., Zhu, T., Gao, S., Wang, Y., Wang, R. et al. (2020). Silica nanoparticles enhance germ cell apoptosis by inducing reactive oxygen species (ROS) formation in Caenorhabditis elegans. *J Toxicol Sci*, *45*(3), 117–129.

Zhao, X., Cui, H., Wang, Y., Sun, C., Cui, B., & Zeng, Z. (2017). Development strategies and prospects of nano-based smart pesticide formulation. *J Agric Food Chem*, *66*(26), 6504–6512.

Zhao, X., Cui, H., Wang, Y., Sun, C., Cui, B., & Zeng, Z. (2017). Development strategies and prospects of nano-based smart pesticide formulation. *J Agric Food Chem*, *66*(26), 6504–6512.

Zhao, Y., Yang, R., Rui, Q., & Wang, D. (2016). Intestinal insulin signaling encodes two different molecular mechanisms for the shortened longevity induced by graphene oxide in Caenorhabditis elegans. *Sci Rep*, *6*, 24024.

Zoomi, I., Kehri, H.K., Akhtar, O., Pandey, D., Srivastava, P., & Narayan, R.P. (2020). *Nanomaterials and Environmental Biotechnology*. Switzerland: Springer, pp. 41–53.

5 Synthesis and Application of Nano-biofertilizers in Plants

Battana Swapna, Srinivasan Kameswaran, and Mandala Ramakrishna

5.1 INTRODUCTION

The increasing population has raised the demand for the development of high-yielding food and forage crops to support global food security, and agricultural production plays a pivotal role in achieving this. Across the world, agricultural production is enormously affected by climate change, and extreme temperatures can reduce the yields of major crops and in the meantime enhance the development of diseases and weeds, eventually affecting food production. Changes in precipitation intensify the possibility of a decline in crop production. Though the impact of climate change is optimistic for a small number of crops in a few locations of the world, to a larger extent, the effects on agriculture are predicted to be negative, instigating a risk to global food security. Populations of developing countries are perhaps the most severely affected.

Extensive agriculture faces several problems, including reduced soil health, excessive fertilizer usage, developing multiple deficiencies of mineral nutrients, and diminishing productivity. Soil offers stability for plants and holds water and nutrients, essential for plant growth. Growing plants rely on the blend along with the concentration of mineral nutrients available in the soil. Mineral nutrient management is one of the key steps in raising a successful crop.

Fertilizer is such an important input material, which provides available mineral nutrients for improving sustainable crop production. Global experiences in agriculture have demonstrated that balanced fertilization is the most efficient measure for enhancing crop production. The intention is to obtain a higher economic yield with the lowest inputs. Among the various inputs, fertilizer is the most expensive. This costly input must be efficiently utilized. Fertilizers are regarded as efficient when maximal yield is obtained with the lowest possible amount of fertilizer utilization. It is certainly tough to quantify the efficiency of a specific fertilizer, as it depends on loss due to leaching, immobilization, losses in gaseous forms, adsorption on microbial cells, chemical reactions occurring among different constituents during mixing of fertilizers before application to soil, soil physical properties, soil chemical properties, and characteristic features of fertilizers.

5.1.1 Chemical Fertilizers

Chemical fertilizers are extensively used to compensate for a shortage of nutrients like boron, nitrogen, phosphorus, zinc, potassium, and iron in the soil substratum. However, a large amount of these fertilizers is lost, as they move or evaporate and as a result, cannot be reached by the plant they are applied to. In addition, this has been inferred to cause monetary loss together with environmental pollution. Recent studies on the disproportionate use of fertilizers revealed adverse effects such as soil salinity, eutrophication of aquatic bodies, and accumulation of nitrates and heavy metals, leading to air, water, and soil pollution (Savci, 2012; Duhan et al., 2017; Pahalvi et al., 2021). Biofertilizers came into existence when the development of a low-priced, safe, and biodegradable alternative to chemical fertilizers became necessary to maximize agricultural productivity (Wu et al., 2005; Sagar et al., 2022).

 DOI: 10.1201/b23308-5

5.1.2 Biofertilizers

Biofertilizers are biological fertilizers that comprise living biomass or inactive cells of microorganisms. Biofertilizers are stimulated in the course of soil interaction with the plant roots, thus increasing the accessibility of nutrients to the plants. Biofertilizer is constituted of helpful microorganisms, namely, blue-green algae, plant growth–promoting rhizobacteria, mycorrhizae, and plant extracts (Vafa et al., 2021). The use of microorganisms as biofertilizers was initiated at the end of the nineteenth century (Malusa et al., 2016). Biofertilizers are sustainable, environmentally safe, and economical, and have long-lasting effects (Du et al., 2018). They inhabit the plant's rhizosphere and increase nourishment through mineral nutrient accessibility (Enrico et al., 2020).

Biofertilizers discharge substances that promote plant growth, and carry out solubilization of phosphate using *Mycobacterium*, *Pseudomonas*, *Flavobacterium*, and *Micrococcus* among bacteria and *Aspergillus* and *Penicillium* among fungal organisms (Sharma et al., 2013; Ahmed et al., 2021). Phosphorus is considered the most crucial mineral nutrient after nitrogen, and is essential for plant development. It is essential for improving nodulation by nitrogen fixers. Phosphorus-solubilizing bacteria make phosphorus available to plants. This is one of the widely used biofertilizers. Biofertilizers improve the solubility of potassium using potassium-solubilizing bacteria such as *Acidothiobacillus ferrooxidans*, *Bacillus edaphicus*, *B. mucilaginosus*, *B. circulans*, and *Paenibacillus* spp. (Etesami et al., 2017; Baba et al., 2021). Potassium-solubilizing bacteria (KSB) can transform potassium from an insoluble to a soluble form, making it accessible to plants.

Biofertilizers fix the atmospheric nitrogen in the soil using bacteria like *Rhizobium* (Ketema and Tefera, 2022; Kusale et al., 2021a; Ramoneda et al., 2020; Riah et al., 2021). *Rhizobium* species, free-living soil bacteria, are among the very important mutualistic nitrogen-fixing bacteria that take food from plants and in turn, supply nitrate to the plants.

Azospirillum is another rhizospheric genus involved in nitrogen fixation that exhibits associative mutualism and takes plant exudates as food material. Cyanobacteria like *Nostoc* and *Anabaena* form a symbiotic association with liverworts, ferns, cycad roots, and lichens, and then have the ability to fix nitrogen, which is available to the plants.

Zinc is considered a key micronutrient for plant nutrition, which has many functions in the life cycle of plants. The unavailability of zinc in the soil causes deficiency, leading to reduced crop yield. Zinc-solubilizing bacteria (Zn-SB) are favorable microorganisms to utilize for agricultural sustainability. Zn-SB solubilizes Zn, P, and K, fixes nitrogen, and synthesizes growth regulators such as auxins, gibberellic acid, and kinetin in addition to siderophore generation. Zn-SB augments growth, the yield of plants, and soil fertility by emitting various compounds that transform the zinc from an unavailable to an available form (Naz et al., 2016; Kumar et al., 2019; Sindhu et al., 2019; Saboor et al., 2021; Khan et al., 2022).

Some of the biofertilizers generate siderophore, which can sequester iron (Kusale et al., 2021b; Nithyapriya et al., 2021), causing unavailability of iron for pathogens and thus impeding the development of pathogenic organisms. They release different growth regulators, such as auxins, gibberellins, cytokinin, and abscisic acid, etc. (Hamid et al.,2021), which help in promoting plant growth (Mahanty et al., 2017; Fasusi et al., 2021; Aloo et al., 2022).

Biofertilizers that show good performance in laboratory and controlled greenhouse conditions often do not perform similarly in the field, and they may tend to be susceptible to biotic and abiotic stress. Normally, crops are grown under varied ecological conditions, comprising various ranges of rainfall, temperature, soil microbial diversity, and soil type. Therefore, such differences contribute to the discrepancy in the efficacy of the biofertilizers (Ali et al., 2021; Zambrano-Mendoza et al., 2021). In addition, the concerns about biofertilizers include inferior quality, extreme dose requirement, insignificant performance in a changing environment, and poor shelf life (Kalam et al., 2020; Basu et al., 2021).

To avoid the limitations of chemical fertilizers as well as biofertilizers, nanobiotechnology focused on the development of nano-biofertilizer to boost the nutrition use efficiency of plant systems for sustainable agriculture. The utilization of nano-biofertilizers brought several problems under control, and nanotechnology is considered one of the tools to perpetuate agriculture and the ecosystem sustainably. Nano-biofertilizers improve the sustained and slow discharge of fertilizer and retain its quality, and this chapter, therefore, highlights the synthesis of nano-biofertilizers and their application to plants.

5.2 NANO-BIOFERTILIZERS: AN EMERGING ECO-FRIENDLY APPROACH

Recently, nanobiotechnology has sprung up as a valuable implement within the agricultural field to improve crop production by the design and development of tiny particles possessing exclusive physicochemical characteristics in comparison to traditional methods. Nanotechnology has acquired swift pace in the modern agricultural field and is anticipated to become crucial in times to come by providing huge applications. Agri-nanobiotechnology is an eco-friendly technology that can contribute to the avoidance of worldwide food insecurity and the consequences of climate change (Mishra et al., 2014; Sangeetha et al., 2017a, b).

Biogenic nanofertilizers are nanotechnology-based environment-friendly fertilizers that serve as efficient plant growth and yield enhancers, soil health maintainers, and nutrient use managers. Microorganism-driven biocatalysis is a green process for the production of nanoparticles. The application of nano-biofertilizer leads to a threefold rise in nutrient use capability, and approximately 80–100 times less is required than chemical fertilizer (Tarafdar, 2021). Well-established microbial technologies and the processes used in biotechnology for scaling up biomass are utilized for the synthesis of nanomaterials (Prasad et al., 2016). This leads to microbial growth on large surface areas, and financial viability, which is advantageous in agriculture, facilitating the easy synthesis of nano-biofertilizers. The sustainable production of nano-biofertilizer includes the usage of enzymes from microbes for salts to be broken down into particles of nano size (Duhan et al., 2017). Moreover, biotic polymers, polysaccharides, and extracellular polymeric substances (EPS) act as suitable candidates for the synthesis of nano-biofertilizers. Additionally, the organic polymers retain moisture content in the soil, playing a major part in ecosystem balance (Ding et al., 2018). EPS plays a significant role in assemblage and adherence to cells and biofilm formation, which consequently guards cells against threatening environmental conditions (Ding et al., 2018). Also, some active microbe-derived polysaccharides are utilized as metal-chelating compounds in the environmental domain (Raliya et al., 2013). Intense attempts are being made to reduce the production cost of nano-biofertilizer globally.

The formulation of nano-biofertilizer should involve the integration of all desired characteristics, like higher solubility, extended discharge, stability, rigorous activity with the minimum application, and low eco-toxicity with proper delivery (Torney et al., 2007; Rana and Kalaichelvan, 2013; Prasad et al., 2014, 2017; Akhtar et al., 2022). Of late, fungal nanotechnology has developed into an impressive area, in which fungal organisms are utilized in the production of nano-sized products that have enormous applications in agriculture. Fungi, due to their high enzyme synthesis efficiency extracellularly as well as intracellularly in comparison to bacteria and actinomycetes, are playing a prominent role in biosynthesizing nanoparticles (Rai et al., 2009; Alghuthaymi et al., 2015; Prasad, 2016, 2017; Prasad et al., 2018; Sonawane et al., 2022; Anjum et al., 2023).

Inadequate growth and lower yield in several crops are caused by utilizing some fertilizers that have problems such as instability in field conditions due to variations in climate, reduced storage, low durability, and the necessity of plentiful fertilizer to spread over an extensive field area (Çakmakçi, 2019). The utilization of nano-biofertilizers offers vital advantages to farmers regarding performance in the field, low price, and contributing to higher yield (Mala et al., 2017). Nano-biofertilizers, besides being environment-friendly, generate good-quality crops through the improvement of microbial population efficiency by way of utilizing essential mineral nutrients like N, P, and K. This will also stimulate the enzyme activity of microorganisms that contribute to soil fertility.

5.2.1 Synthesis of Nano-biofertilizer

Nanoparticles and biofertilizers can be combined to form nano-biofertilizer. Nano-biofertilizer releases the nutrients slowly to increase plant nutrient absorption. It also enhances soil fertility by having extended effects on the properties of soil.

Allium cepa has been utilized as a natural reducing agent for synthesizing silver nanoparticles. A homogeneous slurry was prepared by blending 10 g of onion with 100 ml of distilled water, adding 70% alcohol, boiling, filtering, and then storing at 40 °C for 24 h. The onion extract was mixed with 1.0 mM $AgNO_3$ using continuous movement at 60–65 °C for 5 min and kept at room temperature for 5 days. Onion silver nanoparticles were collected via centrifugation and stored in the form of lyophilized powder (Lekshmi et al., 2012). This powder is used as a nano-biofertilizer (Gosavi et al., 2020). Corradini et al. (2010) assessed chitosan nanoparticles comprising nitrogen, phosphorus, and potassium fertilizers for use in agriculture. Encapsulation and lyophilization techniques are used for the synthesis of nano-biofertilizers. Encapsulation is one of the recently evolving techniques that help in the sustained release of biogenic fertilizer into the soil. This technique is usually used to grow microbial species having adaptable structural characteristics (Schoebitz et al., 2013). Microencapsulation is the promising subfield of encapsulation technology that constructs a defensive shield over the active molecule, forming a capsule or a shell, to enhance crop yield and the longevity of the biogenic fertilizer. This kind of nano-biofertilizer generation is advantageous in terms of time-controlled release into the soil as well as plant stress tolerance (Cakmakci, 2019).

The formulation of nano-biofertilizer includes encapsulation of nanoparticles with biofertilizer. It protects bacterial strains from mechanical stress, and the slow release of nutrients further enhances the efficiency of this product. The synthesis of nano-biofertilizers consists of three critical steps: (1) culture of microorganisms; (2) encapsulating biofertilizer with nano-sized particles; (3) assessment of purity, nature, effectiveness, and life span (Panichikkal et al., 2021b).

Encapsulation is a method of placing microbial cells into a nanoparticulate shell (Figure 5.1). It includes the utilization of non-hazardous, degradable substances such as calcium alginate and starch. It has been reported that starch can enhance the growth of microorganisms (Vafa et al.,2021).

Nanoencapsulation is the covering of biofertilizer at the nanoscale with the support of a nanomaterial. Plant growth–promoting rhizobacteria (PGPR) containing biofertilizers can be synthesized

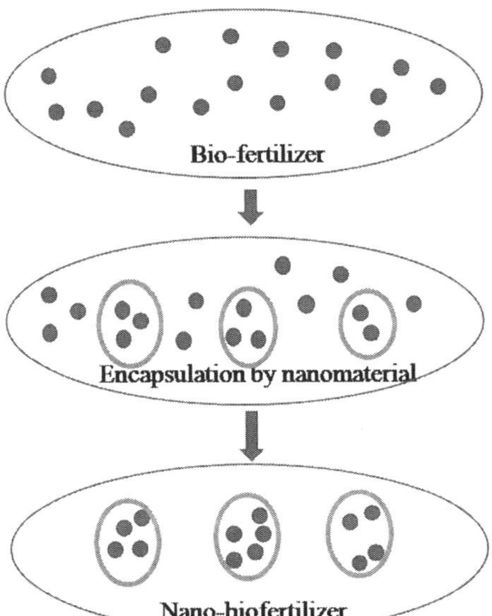

FIGURE 5.1 Synthesis of nano-biofertilizers using encapsulation.

using nanoencapsulation technology (Golbashy et al.,2017). This is an adaptable technology that enables increased durability and time-controlled release of PGPR. Gradual and controlled bioavailability of mineral nutrients to plants can be achieved through nanoencapsulation of several mineral nutrients and organic compounds such as zeolite, polymers, and chitosan (Qureshi et al., 2018). Nano-biofertilizer acts synergistically with the microbes and arbuscular mycorrhizae fungi (AMF) inoculants by enriching the soil through various mechanisms, including producing phytohormones, solubilizing phosphorus, and producing chelating agents (Mala et al., 2017).

The freeze-drying technique is used in lyophilization for preserving and storing microorganisms. It is a commonly used technique that aids in the laboratory formulation of microorganisms. Lyophilization is an effective method to increase the rate of bacterial survival and storage for an extended period. Hence, the lyophilized microorganisms are applied directly along with an appropriate carrier in the crop field (Hernández et al., 2006).

Biofilm biofertilizer is a recent advancement made in the synthesis of biofertilizers by using biofilms. Biofilms are communities of microbes that are attached to surfaces that can be either biotic or abiotic (Junaid and Khan, 2018). Microorganisms go hand in hand, forming an association with each other via quorum sensing. The biofilm is synthesized either on the plant surface or inside the plant. It is advantageous for the fixation of nitrogen by PGPR and enhances the productivity of crops (Parween et al., 2017). Biofilm also helps in Zn solubilization and production of chelating agents (Igiehon and Babalola, 2017). Biofilms can also provide a suitable environment for synthesizing and entrapping nano-sized particles (Bhatia et al., 2021). Nutrient uptake in addition to abiotic stress tolerance is considerably improved by bacterial and fungal biofilms (Velmourougane et al., 2017). Biofilm biofertilizer has a lot of applications to augment ecosystem resilience. These comprise improving soil microbial diversity, reviving soil quality, and developing environmental stress resistance (Mitter et al., 2021).

Nano-biofertilizer preparation with microcapsular structure includes integration of PGPR with starch, bentonite, and sodium alginate mix in a 2:1 ratio. The blend is covered with $CaCl_2$ solution to form tiny capsule structures (Saberi-Rise and Moradi-Pour, 2020).

A blend of salicylic acid and nanoparticles has been used to prepare nano-biofertilizer. In this method, a mixture of 2% sodium alginate, ZnONPs, and salicylic acid is combined with the biofertilizer. The resultant mixture is covered with 3% $CaCl_2$ to form 1-mm beads of nano-biofertilizer (Panichikkal et al., 2021a). In another approach, nano-biofertilizer was synthesized using degradable waste including flowers, cow dung, and kitchen waste united with nano-sized particles to enhance soil fertility. The bio-waste was cleaned using water, cut, and subjected to thermal decomposition (pyrolysis). The pyrolyzed waste was combined with nanoparticles for synthesizing nano-biofertilizer (Singh et al., 2019; Beig et al., 2022).

Rios et al. (2019) reported the synthesis of Fe and B nano-biofertilizer using proteoliposomes extracted from *Brassica* plants. Vesicles were extracted from *Brassica oleracea* leaves using 0.5 g of polyvinylpyrrolidone (PVP) and buffer comprising sucrose (0.5 M), dithiothreitol (DTT) (1 mM), 4-(2-hydroxyethyl)-1-Piperazineethanesulfonic acid (HEPES) (50 mM), and ascorbic acid (1.37 mM) at pH 7.5. Then, the sample was homogenized, filtered, and centrifuged for 30 min at 10,000 × g. $FeSO_4$ and H_3BO_3 solutions were vortexed with the vesicles to obtain Fe and B nano-biofertilizers.

Carbon dots are advanced carbon nanomaterials. New Se-doped carbon quantum dots (Se–CQDs) were produced using biomass of algae through different doping processes. The performance of Se–CQDs was evaluated in tomatoes to assess their role in the growth and development of plant life. This research proposes the potential role of alga as a natural resource for synthesizing selenium-supplemented nano-biofertilizers that improve crop quality (Li et al., 2023).

Nano-biofertilizer was made utilizing certain PGPR and gold nanoparticles. Precipitates from *Pseudomonas fluorescens* and *Bacillus subtilis* (PGPR) culture were mixed with gold nanoparticles according to 0.5 McFarland turbidity (McFarland,1907) and centrifuged. The samples were gradually dehydrated, accompanied by coating with gold (Shukla et al., 2015).

5.2.2 Applications

5.2.2.1 Nano-biofertilizer as Soil Fertility Enhancer

Productiveness and fertility of the soil are the capabilities of the soil to maintain plant growth and maximize crop yield by furnishing nutrients required for growing plants. Soil mineral nutrients must be accurately provided according to the crop's requirements without causing any negative effects on the ecosystem (Janmohammadi et al., 2016; Mahapatra et al., 2022).

Nano-biofertilizer boosts soil fertility through several approaches that include restoration of mineral nutrients through fixing nitrogen and solubilizing phosphorus; increasing nutrient availability; producing siderophores that sequestrate heavy metals, in turn chelating iron; production of phytohormones; retention of soil moisture; and developing resistance against plant stresses (Kumari and Singh, 2020; Sharma et al.,2021).

5.2.2.2 Nano-biofertilizer for Maintaining Soil Health

At present, the unrestricted usage of chemical fertilizers disturbs the structure of the soil, pollutes soil as well as water resources, and causes the draining of nutrients, showing adverse impacts on human health and the ecosystem. Extensive nutrient deficiencies and reduced soil health give rise to low nutrient use efficiency and productivity. It is essential to work on managing soil health for fruitful and sustainable agriculture. A recent study found that inoculating plants with AMF coupled with Fe-NPs significantly improved the root zone of plants and remediated heavy metals (Mokarram-Kashtiban et al., 2019; Sumbul et al., 2020; El-Saadony et al., 2021).

5.2.2.3 Nano-biofertilizer for Enhancing Plant Growth

The impact of nano Fe integrated with PGPR was studied in maize crop growth and physiological responses and harvest. It was found that spraying *Azospirillum brasilens* along with nanoparticles enhanced maize crop growth and yield (Heidari et al., 2018). The application of iron and boron on *Prunus dulcis* leaves using nano-biofertilization technology showed augmented mineral nutrient entry and movement in the plant (Rios et al., 2020).

Nano-biofertilizer synthesized from *Allium cepa* extract was an effective fertilizer for brinjal and tomato plants, but this will need to be evaluated in field conditions and also other crops (Gosavi et al., 2020).

5.2.2.4 Nano-biofertilizer for Enhancing Crop Yield

Soybean yield, oil quantity, and quality increased with the application of zinc nanoparticles combined with the inoculum of *Bradyrhizobium japonica* on the leaves (Sharifi, 2016). Under water stress, the yield of maize was enhanced by spraying nano-biofertilizer (Farnia and Omidi, 2015)

5.2.2.5 Nano-biofertilizer for Tolerance against Environmental Stresses

Moreover, the application of bio-agents along with nano Zn-Fe oxide to wheat plants under salt stress remarkably enhanced the photosynthetic rate, osmolytes like proline, soluble sugars, antioxidative enzymes, and seed development (Babaei et al., 2017).

5.3 ADVANTAGES OF NANO-BIOFERTILIZERS

In conclusion, the slow release of nutrients, providing better soil health, an increase in nutrient use efficiency, an increase in yield stability, improving crop characteristics, and reduction of environmental pollution are the advantages of nano-biofertilizers (Figure 5.2). Certainly, nanotechnology has contributed to the possibility of taking advantage of nanoscale materials as carriers of fertilizer or extended-release vectors for creating "smart fertilizer" as a novel substitution for augmenting nutrient use efficiency and minimizing expenses while safeguarding the environment.

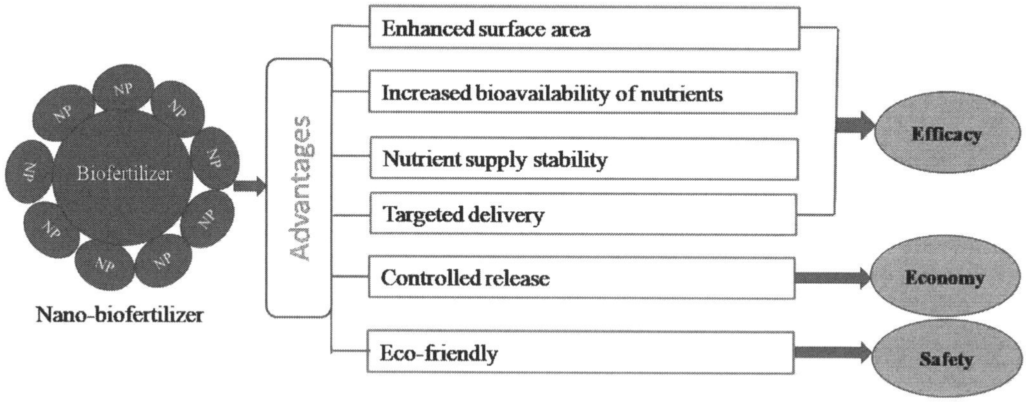

FIGURE 5.2 Advantages of nano-biofertilizers.

5.4 LIMITATIONS

As nano-biofertilizer technology is in an early stage of development, the standardization of formulation and nano-toxicity assessment of these fertilizers have not yet been well established in studies (Babu et al., 2022). The adverse effects of nano-biofertilizers should be assessed before their application in open-field conditions, while some nanomaterials may have toxic effects. To ensure a sustainable approach, preference must be given to non-toxic, biocompatible, and biodegradable nanoparticles in the future. Indeed, the high cost of nano-biofertilizers compared with conventional biofertilizers at present and the lack of technical expertise are some of the limitations. To allay the doubts of the general public, the optimum dose of various nano-biofertilizers, and their effect on plants, soil, and ecosystem on long-term exposure at optimal and supra-optimal levels, has to be evaluated intensively before application.

5.5 CONCLUSION AND FUTURE PERSPECTIVES

Biofertilizers are an emerging kind of agent that may become a vital component in modern agricultural production, a part of sustainable farming practices and alternatives to chemical fertilizers that cause various environmental hazards. Biofertilizers can fix atmospheric nitrogen in the soil, making it accessible to plants, solubilize phosphate into available forms, obtain phosphates from soil layers, produce hormones to support root development, and decompose organic matter. Their application may lead to improved crop yields, enhanced water retention in soil, and induced stress tolerance in plants.

The development of nano-biofertilizers affords a distinctive chance to progress in generating food with the least harmful effect on the surroundings. These nano-sized products will have a good influence on agriculture as well as providing a sustainable environment. Nano-biofertilizer synthesis and application in agriculture is still in its infancy. However, day after day, nanotechnology is showing potential effects, and many researchers and agriculturalists are moving toward applying it widely in modern agriculture. This confronts them with the effective dosage, toxic effect, and eco-footprint of nano-biofertilizers in the soils over an extended time. Therefore, further work is essential to fill the research gaps on the cost-effective large-scale production approaches, dosage, shelf-life period, effect on various crops grown at diverse climatic conditions, and eco-safety of nano-biofertilizers, and the safety of human consumption of these crops, before they are marketed on a large scale. The most suitable nano-biofertilizers with known efficacy and safety must be explored so that they can be commercialized and available to farmers at affordable prices.

Furthermore, there is a need for the development of more microbial nanotechnologies that are well-matched and appropriate to a specific crop. These support emerging crop-specific nano-biofertilizers, consequently increasing the yield. In conclusion, nano-biofertilizer technology has high potential and can quickly be upgraded to a promising way to support global food security.

REFERENCES

Ahmed, S., Choudhury, A.R., Roy, S.K., Choi, J., Sayyed, R.Z., & Sa, T. (2021). Biomolecular painstaking utilization and assimilation of phosphorus under indigent stage in agricultural crops. In: Harikesh Bahadur Singh, Anukool Vaishnav, R.Z. Sayyed (Ed.) *Antioxidants in Plant-Microbe Interaction*. Springer, Singapore, 565–588.

Akhtar, N., Ilyas, N., Meraj, T.A., Pour-Aboughadareh, A., Sayyed, R.Z., Mashwani, Z.U.R., & Poczai, P. (2022). Improvement of plant responses by nanobiofertilizer: A step towards sustainable agriculture. *Nanomaterials*, 12(6), 965.

Alghuthaymi, M.A., Almoammar, H., Rai, M., Said-Galiev, E., & Abd-Elsalam, K.A. (2015). Myconanoparticles: Synthesis and their role in phytopathogens management. *Biotechnology and Biotechnological Equipment*, 29(2), 221–236.

Ali, S.S., Kornaros, M., Manni, A., Al-Tohamy, R., El-Shanshoury, A.E.R.R., Matter, I.M., Elsamahy, T., Sobhy, M., & Sun, J. (2021). Advances in microorganisms-based biofertilizers: Major mechanisms and applications. In: Amitava Rakshit, Vijay Singh Meena, Manoj Parihar, H.B. Singh, A.K. Singh *Biofertilizers*. Woodhead Publishing (Elsevier), 371–385.

Aloo, B.N., Tripathi, V., Makumba, B.A., & Mbega, E.R. (2022). Plant growth-promoting rhizobacterial biofertilizers for crop production: The past, present, and future. *Frontiers in Plant Science*, 13, 1002448.

Anjum, S., Vyas, A., & Sofi, T.A. (2023). Fungi-mediated synthesis of nanoparticles: Characterization process and agricultural applications. *Journal of the Science of Food and Agriculture*.https://doi.org/10.1002/jsfa.12496.

Baba, Z.A., Hamid, B., Sheikh, T.A., Alotaibi, S.H., El Enshasy, H.A., Ansari, M.J., Zuan, A.T.K., & Sayyed, R.Z. (2021). Psychrotolerant *Mesorhizobium* sp. isolated from temperate and cold desert regions solubilizes potassium and produces multiple plant growth promoting metabolites. *Molecules*, 26(19), 5758.

Babaei, K., Seyed Sharifi, R., Pirzad, A., & Khalilzadeh, R. (2017). Effects of bio fertilizer and Nano Zn-Fe oxide on physiological traits, antioxidant enzymes activity and yield of wheat (*Triticum aestivum* L.) under salinity stress. *Journal of Plant Interactions*, 12(1), 381–389.

Babu, S., Singh, R., Yadav, D., Rathore, S.S., Raj, R., Avasthe, R., Yadav, S.K., Das, A., Yadav, V., Yadav, B. et al. (2022). Nanofertilizers for agricultural and environmental sustainability. *Chemosphere*, 292, 133451.

Basu, A., Prasad, P., Das, S.N., Kalam, S., Sayyed, R.Z., Reddy, M.S., & El Enshasy, H. (2021). Plant growth promoting rhizobacteria (PGPR) as green bioinoculants: Recent developments, constraints, and prospects. *Sustainability*, 13(3), 1140.

Beig, B., Niazi, M.B.K., Sher, F., Jahan, Z., Malik, U.S., Khan, M.D., Américo-Pinheiro, J.H.P., & Vo, D.V.N. (2022). Nanotechnology-based controlled release of sustainable fertilizers. A review. *Environmental Chemistry Letters*, 20(4), 2709–2726.

Bhatia, R., Gulati, D., & Sethi, G. (2021). Biofilms and nanoparticles: Applications in agriculture. *Folia Microbiologica*, 66(2), 159–170.

Cakmakci, R.A. (2019). Review of biological fertilizers current use, new approaches and future perspectives. *International Journal of Innovative Studies in Sciences and Engineering Technology*, 5, 83–92.

Corradini, E., De Moura, M.R., & Mattoso, L.H.C. (2010). A preliminary study of the incorporation of NPK fertilizer into chitosan nanoparticles. *Express Polymer Letters*, 4(8), 509–515.

Ding, P., Song, W., Yang, Z., & Jian, J. (2018). Influence of Zn(II) stress-induction on component variation and sorption performance of extracellular polymeric substances (EPS) from *Bacillus vallismortis*. *Bioprocess and Biosystems Engineering*, 41(6), 781–791.

Du, C., Abdullah, J.J., Greetham, D., Fu, D., Yu, M., Ren, L., Li, S., & Lu, D. (2018). Valorization of food waste into biofertilizer and its field application. *Journal of Cleaner Production*, 187, 273–284.

Duhan, J.S., Kumar, R., Kumar, N., Kaur, P., Nehra, K., & Duhan, S. (2017). Nanotechnology: The new perspective in precision agriculture. *Biotechnology Reports*, 15, 11–23.

El-Saadony, M.T., ALmoshadak, A.S., Shafi, M.E. Albaqami, N.M., Saad, A.M., El-Tahan, A.M., Desoky, E.S.M., Elnahal, A.S., Almakas, A., Abd El-Mageed, T.A. et al. (2021). Vital roles of sustainable nanofertilizers in improving plant quality and quantity-an updated review. *Saudi Journal of Biological Sciences*, 28(12), 7349–7359.

El-Saadony, M.T., Desoky, E.S.M., Saad, A.M., Eid, R.S., Selem, E., & Elrys, A.S. (2021). Biological silicon nanoparticles improve *Phaseolus vulgaris* L. yield and minimize its contaminant contents on a heavy metals-contaminated saline soil. *Journal of Environmental Sciences*, 106, 1–14.

Enrico, J.M., Piccinetti, C.F., Barraco, M.R., Agosti, M.B., Eclesia, R.P., & Salvagiotti, F. (2020). Biological nitrogen fixation in field pea and vetch: Response to inoculation and residual effect on maize in the Pampean region. *European Journal of Agronomy*, 115, 126016.

Etesami, H., Emami, S., & Alikhani, H.A. (2017). Potassium solubilizing bacteria (KSB): Mechanisms, promotion of plant growth, and future prospects: A review. *Journal of Soil Science and Plant Nutrition*, 17(4), 897–911.

Farnia, A., Omidi, M.M., & Farnia, A. (2015). Effect of nano-zinc chelate and nano-biofertilizer on yield and yield components of maize (*Zea mays* L.), under water stress condition. *Indian Journal of Natural Sciences*, 5(29), 4614–4624.

Fasusi, O., Cruz, C., & Babalola, O. (2021). Agricultural sustainability: Microbial biofertilizers in rhizosphere management. *Agriculture*, 11(2), 163.

Golbashy, M., Sabahi, H., Allahdadi, I., Nazokdast, H., & Hosseini, M. (2017). Synthesis of highly intercalated urea-clay nanocomposite via domestic montmorillonite as eco-friendly slow-release fertilizer. *Archives of Agronomy and Soil Science*, 63(1), 84–95.

Gosavi, V.C., Daspute, A.A., Patil, A., Gangurde, A., Wagh, S.G., Sherkhane, A., & AnandraoDeshmukh, V. (2020). Synthesis of green nanobiofertilizer using silver nanoparticles of *Allium cepa* extract. *IJCS*, 8(4), 1690–1694.

Hamid, B., Zaman, M., Farooq, S., Fatima, S., Sayyed, R., Baba, Z., Sheikh, T., Reddy, M., El Enshasy, H., Gafur, A. et al. (2021). Bacterial plant biostimulants: A sustainable way towards improving growth, productivity, and health of crops. *Sustainability*, 13(5), 2856.

Heidari, M., Salmanpour, I., Ghorbani, H., & Asghari, H.R. (2018). Iron chelate and rhizobacteria changed growth, grain yield, and physiological characteristics in maize. *Scientia Agriculturae Biochemical*, 49(4), 245–254.

Hernandez, A., Weekers, F., Mena, J., Borroto, C., & Thonart, P. (2006). Freeze-drying of the biocontrol agent *Tsukamurella paurometabola* C-924 Predicted stability of formulated powders. *Industrial Biotechnology*, 2(3), 209–212.

Igiehon, N.O., & Babalola, O.O. (2017). Biofertilizers and sustainable agriculture: Exploring arbuscular mycorrhizal fungi. *Applied Microbiology and Biotechnology*, 101(12), 4871–4881.

Janmohammadi, M., Navid, A., Segherloo, A.E., & Sabaghnia, N. (2016). Impact of nano-chelated micronutrients and biological fertilizerson growth performance and grain yield of maize under deficit irrigation condition. *Biologia*, 62, 134–147.

Junaid, P.M., & Khan, F. (2018). Biofilm: A future generation biofertilizer. *Indian Journal of Applied & Pure Biology*, 33(2), 131–134.

Kalam, S., Basu, A., Ahmad, I., Sayyed, R.Z., El-Enshasy, H.A., Dailin, D.J., & Suriani, N.L. (2020). Recent understanding of soil Acidobacteria and their ecological significance: A critical review. *Frontiers in Microbiology*, 11, 580024.

Ketema, P., & Tefera, T. (2022). Effectiveness of *rhizobium* strains on faba bean (*Vicia fabae* L.) at Gumer district, highland area of Southern Ethiopia. *Ukraine J. Ecol*, 12, 13–18.

Khan, M.A., Yasmin, H., Shah, Z.A., Rinklebe, J., Alyemeni, M.N., & Ahmad, P. (2022). Co application of biofertilizer and zinc oxide nanoparticles upregulate protective mechanism culminating improved arsenic resistance in maize. *Chemosphere*, 294, 133796.

Kumar, A., Dewangan, S., Lawate, P., Bahadur, I., & Prajapati, S. (2019). Zinc-solubilizing bacteria: A boon for sustainable agriculture. In: R.Z. Sayyed, Naveen Kumar Arora, M.S. Reddy (eds.) *Plant Growth Promoting Rhizobacteria for Sustainable Stress Management Volume 1: Rhizobacteria in Abiotic Stress Management*, Springer, Singapore, 139–155.

Kumari, R., & Singh, D.P. (2020). Nano-biofertilizer: An emerging eco-friendly approach for sustainable agriculture. In: *Proceedings of the National Academy of Sciences, India Section B: (Biological Sciences)*, 733–741.

Kusale, S., Attar, Y., Sayyed, R., Malek, R., Ilyas, N., Suriani, N., Khan, N., & El Enshasy, H. (2021a). Production of plant beneficial and antioxidants metabolites by *Klebsiellavariicola* under salinity stress. *Molecules*, 26(7), 1894.

Kusale, S., Attar, Y., Sayyed, R., El Enshasy, H., Hanapi, S., Ilya, S.N., Elgorban, A., Bahkali, A., & Marraiki, N. (2021b). Inoculation of *Klebsiella variicola* alleviated salt stress and improved growth and nutrients in wheat and maize. *Agronomy*, 11(5), 927.

Lekshmi, N.C., Sumi, S.B., Viveka, S., Jeeva, S., & Brindha, J.R. (2012). Antibacterial activity of nanoparticles from *Allium* sp. *Journal of Microbiology and Biotechnology Research*, 2(2), 115–119.

Li, J., Wang, Z., Zhang, Y., Cao, X., Lian, F., & Gu, S. (2023). Novel selenium-doped carbon quantum dots derived from algae effectively enhanced the delivery and accumulation of selenium in tomato plants (*Lycopersicum esculentum*) via foliar application. *Environmental Science: Nano*, 10(3), 866–878.

Mahanty, T., Bhattacharjee, S., Goswami, M., Bhattacharyya, P., Das, B., Ghosh, A., & Tribedi, P. (2017). Biofertilizers: A potential approach for sustainable agriculture development. *Environmental Science and Pollution Research International*, 24(4), 3315–3335.

Mahapatra, D.M., Satapathy, K.C., & Panda, B. (2022). Biofertilizers and nanofertilizers for sustainable agriculture: Phycoprospects andchallenges. *Science of the Total Environment*, 803, 149990.

Mala, R., Selvaraj, C.A., Sundaram, B.V., Rajan, B.S.S., & Gurusamy, M.U. (2017). Evaluation of Nano structured slow-release fertilizer on the soil fertility, yield and nutritional profile of *Vigna radiata*. *Recent Patents on Nanotechnology*, 11(1), 50–62.

Malusà, E., Pinzari, F., & Canfora, L. (2016). Efficacy of biofertilizers: Challenges to improve crop production. In: Singh, D., Singh, H., & Prabha, R. (eds.) *Microbial Inoculants in Sustainable Agricultural Productivity*. Springer, New Delhi, 17–40.

McFarland, J. (1907). The nephelometer, an instrument for estimating the number of bacteria in suspension used for calculating the opsonic index and for vaccines. *Journal of the American Medical Association*, 49, 1176–1178.

Mishra, S., Singh, A., Keswani, C., & Singh, H.B. (2014). Nanotechnology: Exploring potential application in agriculture and its opportunities and constraints. *Biotech. Today*, 4(1), 9–14.

Mitter, E.K., Tosi, M., Obregon, D., Dunfield, K.E., & Germida, J.J. (2021). Rethinking crop nutrition in times of modern microbiology: Innovative biofertilizer technologies. *Frontiers in Sustainable Food Systems*, 5, 29.

Mokarram-Kashtiban, S., Hosseini, S.M., Kouchaksaraei Tabari, M., & Younesi, H. (2019). The impact of nanoparticles zero-valent iron (nZVI) and rhizosphere microorganisms on the phytoremediation ability of white willow and its response. *Environmental Science and Pollution Research International*, 26(11), 10776–10789.

Naz, I., Ahmad, H., Khokhar, S.N., Khan, K., & Shah, A.H. (2016). Impact of zinc solubilizing bacteria on zinc contents of wheat. American-Eurasian Journal of Agricultural & Environmental Sciences, 16, 449–454.

Nithyapriya, S., Lalitha, S., Sayyed, R.Z., Reddy, M., Dailin, D., El Enshasy, H., Suriani, N.L., & Herlambang, S. (2021). Production, purification, and characterization of bacillibactin siderophore of *Bacillus subtilis* and its application for improvement in plant growth and oil content in sesame. *Sustainability*, 13(10), 5394.

Pahalvi, H.N., Rafiya, L., Rashid, S., Nisar, B., & Kamili, A.N. (2021). Chemical fertilizers and their impact on soil health. Microbiota and biofertilizers. *Ecofriendly Tools for Reclamation of Degraded Soil Environs*, 2, 1–20.

Panichikkal, J., Prathap, G., Nair, R.A., & Krishnankutty, R.E. (2021a). Evaluation of plant probiotic performance of *Pseudomonas* sp. encapsulated in alginate supplemented with salicylic acid and zinc oxide nano-particles. *International Journal of Biological Macromolecules*, 166, 138–143.

Panichikkal, J., Puthiyattil, N., Raveendran, A., Nair, R.A., & Krishnankutty, R.E. (2021b). Application of encapsulated *Bacillus licheniformis* supplemented with chitosan nanoparticles and rice starch for the control of *Sclerotium rolfsii* in Capsicum annuum (L.) seedlings. *Current Microbiology*, 78(3), 911–919.

Parween, T., Bhandari, P., Siddiqui, Z.H., Jan, S., Fatma, T., & Patanjali, P.K. (2017). Biofilm: A next-generation biofertilizer. In: Ram Prasad (ed.) *Mycoremediation and Environmental Sustainability*, Springer, Cham. 39–51.

Prasad, R., Bhattacharyya, A., & Nguyen, Q.D. (2017). Nanotechnology in sustainable agriculture: Recent developments, challenges, and perspectives. *Frontiers in Microbiology*, 8, 1014.

Prasad, R., Kumar, V., Kumar, M., & Wang, S. (eds.) (2018). *Fungal Nanobionics: Principles and Applications*, 316. Springer, Singapore.

Prasad, R., Kumar, V., & Prasad, K.S. (2014). Nanotechnology in sustainable agriculture: Present concerns and future aspects. *African Journal of Biotechnology*, 13(6), 705–713.

Prasad, R., Pandey, R., & Barman, I. (2016). Engineering tailored nanoparticles with microbes: Quo vadis? *WIREs Nanomedicine and Nanobiotechnology*, 8(2), 316–330.

Qureshi, A., Singh, D.K., & Dwivedi, S. (2018). Nano-fertilizers: A novel way for enhancing nutrient use efficiency and crop productivity. *International Journal of Current Microbiology and Applied Sciences*, 7(2), 3325–3335.

Rai, M., Yadav, A., & Gade, A. (2009). Silver nanoparticles as a new generation of antimicrobials. *Biotechnology Advances*, 27(1), 76–83.

Raliya, R., Tarafdar, J.C., Gulecha, K., Choudhary, K., Ram, R., Mal, P., & Saran, R.P. (2013). Scope of nanoscience and nanotechnology in agriculture. *Journal of Applied Biology and Biotechnology*, 1(3), 041–044.

Ramoneda, J., Roux, J.J.L., Frossard, E., Frey, B., & Gamper, H.A. (2020). Geographical patterns of root nodule bacterial diversity in cultivated and wild populations of a woody legume crop. *FEMS Microbiology Ecology*, 96(10), 145.

Rana, S., & Kalaichelvan, P. (2013). Ecotoxicity of nanoparticles. *International Scholarly Research Notices*.

Riah, N., de Lajudie, P., Béna, G., Heulin, K., & Djekoun, A. (2021). Variability in symbiotic efficiency with respect to the growth of pea and lentil inoculated with various rhizobial genotypes originating from sub-humid and semi-arid regions of eastern Algeria. *Symbiosis*, 85(3), 371–384.

Rios, J.J., Garcia-Ibañez, P., & Carvajal, M. (2019). Theuse of biovesicles to improve the efficiency ofZnfoliar fertilization. *Colloids Surfaces Biointerfaces*, 173, 899–905.

Rios, J.J., Yepes-Molina, L., Martinez-Alonso, A., & Carvajal, M. (2020). Nanobiofertilization as a novel technology for highly efficient foliar application of Fe and B in almond trees. *Royal Society Open Science*, 7(11), 200905.

Saberi-Rise, R., & Moradi-Pour, M. (2020). The effect of *Bacillus subtilis* Vru1 encapsulated in alginate-bentonite coating enriched with titanium nanoparticles against *Rhizoctonia solani* on bean. *International Journal of Biological Macromolecules*, 152, 1089–1097.

Saboor, A., Ali, M.A., Ahmed, N., Skalicky, M., Danish, S., Fahad, S., Hassan, F., Hassan, M.M., Brestic, M., El Sabagh, A., & Datta, R. (2021). Biofertilizer-based zinc application enhances maize growth, gas exchange attributes, and yield in zinc-deficient soil. *Agriculture*, 11(4), 310.

Sagar, A., Rai, S., Ilyas, N., Sayyed, R.Z., Al-Turki, A.I., El Enshasy, H.A., & Simarmata, T. (2022). Halotolerant rhizobacteria for salinity-stress mitigation: Diversity, mechanisms and molecular approaches. *Sustainability*, 14(1), 490.

Sangeetha, J., Thangadurai, D., Hospet, R., Purushotham, P., Karekalammanavar, G., Mundaragi, A.C., David, M., Shinge, M.R., Thimmappa, S.C., Prasad, R. et al. (2017a). Agricultural nanotechnology: Concepts, benefits, and risks. In: Prasad, R., Kumar, M., & Kumar, V. (eds.) *Nanotechnology*. Springer Nature Singapore Pte Ltd, Singapore, 1–17.

Sangeetha, J., Thangadurai, D., Hospet, R., Harish, E.R., Purushotham, P., Mujeeb, M.A., Shrinivas, J., David, M., Mundaragi, A.C., Thimmappa, A.C. et al. (2017b). Nanoagrotechnology for soil quality, crop performance and environmental management. In: Prasad, R., Kumar, M., & Kumar, V. (eds.) *Nanotechnology*. Springer Nature Singapore Pte Ltd, Singapore, 73–97.

Savci, S. (2012). Investigation of effect of chemical fertilizers on environment. *APCBEE Procedia*, 1, 287–292.

Schoebitz, M., López, M.D., & Roldán, A. (2013). Bioencapsulation of microbial inoculants for better soil-plant fertilization. A review. *Agronomy for Sustainable Development*, 33(4), 751–765.

Seyed Sharifi, R. (2016). Application of biofertilizers and zinc increases yield, nodulation and unsaturated fatty acids of soybean. *Zemdirbyste-Agriculture*, 103(3), 251–258.

Sharma, S.B., Sayyed, R.Z., Trivedi, M.H., & Gobi, T.A. (2013). Phosphate solubilizing microbes: Sustainable approach for managing phosphorus deficiency in agricultural soils. *SpringerPlus*, 2, 587.

Sharma, A., Patel, S., & Menghani, E. (2021). Synthesis, application and prospects of nanobiofertilizers: A reappraisal. *Journal of Phytological Research*, 34, 79–85.

Shukla, S.K., Kumar, R., Mishra, R.K., Pandey, A., Pathak, A., Zaidi, M.G.H., Srivastava, S.K., & Dikshit, A. (2015). Prediction and validation of gold nanoparticles (GNPs) on plant growth promoting rhizobacteria (PGPR): A step toward development of nano-biofertilizers. *Nanotechnology Reviews*, 4(5), 439–448.

Sindhu, S.S., Sharma, R., Sindhu, S., & Phour, M. (2019). Plant nutrient management through inoculation of zinc-solubilizing bacteria for sustainable agriculture. In: Giri, B., Prasad, R., Wu, Q.S., & Varma, A. (eds.) *Biofertilizers for Sustainable Agriculture and Environment*. Springer, Cham, *Soil Biology, 55*, 173–201.

Singh, P., Ghosh, D., Manyapu, V., Yadav, M., & Majumder, S. (2019). Synergistic impact of iron (iii) oxide nano-particles and organic waste on growth and development of *Solanum lycopersicum* plants: New paradigm in nanobiofertilizer. *Plant Archives*, 19, 339–344.

Sonawane, H., Shelke, D., Chambhare, M., Dixit, N., Math, S., Sen, S., Borah, S.N., Islam, N.F., Joshi, S.J., Yousaf, B. et al. (2022). Fungi-derived agriculturally important nanoparticles and their application in crop stress management–Prospects and environmental risks. *Environmental Research*, 212(D), 113543.

Sumbul, A., Ansari, R.A., Rizvi, R., & Mahmood, I. (2020). Azotobacter: A potential bio-fertilizer for soil and plant health management. *Saudi Journal of Biological Sciences*, 27(12), 3634–3640.

Tarafdar, J.C. (2021). Bionanofertilizer: A new invention for global farming. In: Amitava Rakshit, Vijay Singh Meena, Manoj Parihar, H.B. Singh, A.K. Singh (eds.) *Biofertilizers*. Woodhead Publishing, Elsevier, 347–358.

Torney, F., Trewyn, B., Lin, V.Y., & Wang, K. (2007). Mesoporous silica nanoparticles deliver DNA and chemicals into plants. *Nature Nanotechnology*, 2(5), 295–300.

Vafa, Z., Sohrabi, Y., Sayyed, R., Suriani, N.L., & Datta, R. (2021). Effects of the combinations of rhizobacteria, mycorrhizae, and seaweed, and supplementary irrigation on growth and yield in wheat cultivars. *Plants*, 10(4), 811.

Velmourougane, K., Prasanna, R., & Saxena, A.K. (2017). Agriculturally important microbial biofilms: Present status and future prospects. *Journal of Basic Microbiology*, 57(7), 548–573.

Wu, S.C., Cao, Z.H., Li, Z.G., Cheung, K.C., & Wong, M.H. (2005). Effects of biofertilizer containing N-fixer, P and K solubilizers and AM fungi on maize growth: A greenhouse trial. *Geoderma*, 125(1–2), 155–166.

Zambrano-Mendoza, J.L., Sangoquiza-Caiza, C.A., Campaña-Cruz, D.F., & Yánez-Guzmán, C.F. (2021). Use of biofertilizers in agricultural production. *Technology in Agriculture*, 193–210.

6 Biofortification with Zinc
The Use of Nanoparticles as an Alternative to Improve the Nutritional Quality of Crops

Alejandro Palacio-Márquez, Carlos A. Ramírez-Estrada, Esteban Sánchez, and Celia Chávez-Mendoza

6.1 INTRODUCTION

About 12% of the world's population was affected by severe food insecurity in 2020, and according to United Nations estimates, the world population will increase by approximately 2.2 billion by 2050 (FAO et al., 2021). In addition, the problem of malnutrition affects one in three people, and the main problems they suffer are wasting, stunting, being underweight, and micronutrient deficiencies (WHO, 2021).

Micronutrients are essential elements for the formation of healthy bones, brains, and bodies, which are required in small amounts. These nutrients should be obtained from food, ideally from a balanced diet. Deficiencies in iodine, iron, and zinc are the most important problems and pose a threat to the health and development of populations worldwide, particularly to children and pregnant women in underdeveloped countries (Dahdouh et al., 2019; WHO, 2021).

In the case of zinc (Zn), it is estimated that 33% of the world's population is affected by or at risk of suffering from a deficiency of this microelement. It has been reported that levels of Zn intake below the recommended levels can cause problems in the development of pregnancy, growth in children, diseases related to the immune system, as well as poor neurobehavioral development. In turn, severe Zn deficiency causes pustular dermatitis, alopecia, chronic diarrhea, emotional disorder, weight loss, recurrent infections, hypogonadism in men, neuro-sensory disorders, and difficulty in ulcer healing; problems that could become fatal if not recognized and treated in time (Dahdouh et al., 2019; Sida-Arreola et al., 2017; Prasad, 2012; Wessells et al., 2012).

An innovative strategy for the problem of micronutrient deficiency in the food diet is "biofortification", which has been defined as the process of increasing the bioavailable concentrations of essential elements in edible portions of cultivated plants through agronomic management (Pachón, 2010; White and Broadley, 2005). The exogenous application of such nutrients through fertilization is necessary to carry out the process of increasing the concentration of minerals in plant tissues (Blasco, 2015).

Currently, the main objective of agricultural production systems is to increase the profits of farmers and agricultural industries, leaving product quality and human health issues in the background (Bulgari et al., 2014). In addition, resources such as water and soil are becoming increasingly scarce, and the effects of climate change are progressively more severe and frequent, aggravating the situation (Singh and Prasad, 2014; Rouphael and Colla, 2018). The use of nanotechnology, in the form of nanofertilizers, emerges as an option to achieve optimal plant development in a sustainable and precise manner; given their size, plants can absorb them with a different dynamic, which presents an additional advantage. Several studies suggest that the use of nanofertilizers increases by the

DOI: 10.1201/b23308-6

yield and nutritional quality of plants by 20–30% in relation to conventional fertilizers (Raliya et al., 2017; Kah et al., 2018).

Applications of Zn nanoparticles, at low concentrations, have been shown to have positive effects on germination, vegetative growth, and photosynthetic pigment content, and significant increases in yield (Rossi et al., 2019; Siddiqui et al., 2019). It has been proven that the use of this technology by foliar application improves fruit quality, increasing the absorption, translocation, and accumulation of Zn in the fruit, which makes it a novel tool for biofortification programs (Subbaiah et al., 2016; Davarpanah et al., 2016; El-Ramady et al., 2018). Based on all this, the general objective of this review is to describe the action of nanoparticles in biofortification, emphasizing the use of Zn nanoparticles and their effects on plants.

6.2 MALNUTRITION

The word *malnutrition* refers to nutritional imbalances and deficiencies or excesses in a person's caloric intake, causing decreased physical and mental activity that in the long term results in the clinical condition of the individual deteriorating (WHO, 2021; Cederholm et al., 2015).

The concept of malnutrition encompasses both undernutrition and overweight and obesity disorders. Despite manifesting as distinct physical problems, various studies indicate that they share symptoms, such as altered metabolism, chronic inflammation, and intestinal dysfunction. There is also a correlation between excessive energy and macronutrient intake and micronutrient deficiencies in overweight individuals. In addition, early undernutrition increases the risk of obesity in adulthood, leading to clinical problems such as a high percentage of adipose tissue, high circulating levels of triglycerides, and high blood pressure (WHO, 2021; Bourke et al., 2016).

Globally, malnutrition affects more than 2 billion people around the world, with women and children being the sector of the population with the most alarming data. It has been estimated that in 2020, 29.9% of women aged 15 to 49 years suffered from anemia, while 22% of children under five years of age were stunted, 6.7% were wasted, and 5.7% were overweight (FAO et al., 2021). UNICEF (2021) reports that 2 out of every 3 children between 6 and 23 months do not have a sufficiently balanced diet for healthy growth, and only 52% of children under 2 years of age have access to the recommended frequency of meals. Potential causes of malnutrition include poor nutrition, mental health problems, mobility problems, digestive disorders, stomach disorders, and alcoholism (Ahmed et al., 2022).

Despite global efforts to address food insecurity, malnutrition continues to be a growing problem and is a key factor in increased infectious mortality, neurological disability, and limited mental and physical growth. In turn, it has been linked to an increase in cardiovascular disease, cancer, stroke, and diabetes (FAO et al., 2021; Wakeel et al., 2018). More recently, malnutrition has been associated with an increased risk of SARS-CoV-2 infection, and people with this condition present a more severe clinical picture, with a higher number of hospitalizations requiring mechanical ventilation. In addition, the side effects of SARS-CoV-2 could be leading to malnutrition or impairments in related physiological processes, as a higher incidence of malnutrition was found in older adults with severe SARS-CoV-2 infection (Kurtz et al., 2021; Huizar et al., 2021).

6.2.1 Micronutrient Deficiency

The concept of malnutrition refers to three major conditions, the most studied being undernutrition and obesity; however, there is also malnutrition related to micronutrient imbalances, which generally has invisible health implications and is therefore called "hidden hunger", since its development is progressive over time and its effects are not noticeable until it has caused irreversible damage (Praharaj et al., 2021; WHO, 2021; UNICEF, 2021b).

Although humans require micronutrients in minimal amounts, approximately 3 billion people are deficient in micronutrients, making it a problem with a negative global socioeconomic impact.

The problem mainly affects women, children, and the elderly, especially those in low-income countries whose diet is cereal-based or poorly balanced (Kiran et al., 2022; Praharaj et al., 2021; Peña-Rosas et al., 2019). However, micronutrient deficiency is caused not only by incorrectly chosen diets or limited access to nutrient-rich foods, but also by altered metabolism leading to changes in nutrient absorption, distribution, or excretion because of systemic inflammation caused by obesity. This problem occurs mainly in developed countries where people obtain more than 30% of their daily calories from nutrient-poor foods (Kobylińska et al., 2022).

Among the characteristic symptoms of micronutrient deficiency are anemia, blindness, birth defects, growth retardation, cognitive delay, immune weakness, and constant infectious diseases, which are the main causes of morbidity and mortality in low-developed countries (Peña-Rosas et al., 2019; Tulchinsky, 2010). In turn, Song et al. (2015) relate micronutrient deficiency to a higher rate of suffering from depression, finding that people with this problem suffer from anxiety and chronic sadness.

Deficiencies in nutrients such as iron and vitamin B12 generate important problems such as anemia, which in addition to affecting a significant number of women of childbearing age, is estimated to reduce the mental and work capacity of people by 20% and increase the risk of maternal mortality, premature births, and low birth weight. Also, iron deficiency during the first years of life has been linked to problems in growth, neurological development, and cognitive performance, and an increased risk of contracting infections (Peña-Rosas et al., 2019; Wakeel et al., 2018; Ritchie and Roser, 2017).

Vitamin A deficiency is the main cause of preventable blindness in children but progresses to permanent blindness in more severe cases. Also, in lactating women and infants, it potentiates serious infections, increasing mortality rates. Furthermore, this deficiency has been linked to epithelial defects in the respiratory, urinary, and gastrointestinal tracts and glandular ducts, causing alterations in organ growth and development and thyroid dysfunction (Timoneda et al., 2018; Ritchie and Roser, 2017). Whereas iodine deficiency is the main cause of brain damage in children, it is also related to thyroid problems and can cause a loss of about 13 IQ points, generating mental retardation (Yadav and Pandav, 2018). The lack of other essential nutrients, such as folic acid and zinc, causes various problems in human health (Ritchie and Roser, 2017).

6.2.1.1 Zinc Deficiency in Humans

Zn is an essential nutrient for almost all organisms and has a fundamental role as a metal cofactor in more than 300 proteins within the human body. In addition, it plays an essential role as a stabilizer and receptor for steroid hormones, thyroid hormones, retinoids, and cell membranes. It also acts as a regulator, is essential for the synthesis of biomolecules such as DNA, and binds to nuclear proteins, forming complexes called "zinc fingers". These functions of Zn are related to its ability to bind to histidine and cysteine, stabilize active sites, and intervene in maintaining the integrity of the immune system (Taboada-Lugo, 2017; Caro et al., 2016).

The adult requirement for Zn is 11 mg day^{-1}. It is present in the human body in an amount of 2 to 3 g and is second only to iron as the most abundant micronutrient. Zinc deficiency is associated with impaired immune responses and leads to an increased risk of respiratory viral infections, particularly in the elderly. In addition, it generates a decrease in antibody production through altered immune cell function and number (Gorji and Ghadiri, 2020). This problem is very common in countries where the diet is unbalanced, mainly in places with cereal-based diets and low protein intake (Bilski et al., 2012). Zinc deficiency, unlike anemia, can occur equally in men, women, and children. In 2014, approximately 33% of the world's population was at risk for such a deficiency, accounting for 5% of deaths in children under 5 years of age in developing countries, representing the fifth leading cause of disease and mortality (Reed et al., 2015; Singh and Prasad, 2014).

This deficiency can affect the central nervous, gastrointestinal, immune, epidermal, reproductive, and skeletal systems. It has been reported that Zn levels below the recommended levels can cause problems in the development of pregnancy and growth in children, difficulties with the immune system, as well as poor neurobehavioral development. In turn, a severe deficiency of this microelement causes pustular dermatitis, alopecia, chronic diarrhea, emotional disorder, weight loss, recurrent infections, hypogonadism in men, neuro-sensory disorders, and ulcer healing problems, conditions that could become fatal if not recognized and treated in time (Dahdouh et al., 2019; Sida-Arreola et al., 2017; Prasad, 2012; Wessells et al., 2012).

Zinc is considered essential in recovery after any illness or injury, so deficiencies of this element can increase the time of this process, resulting in increased susceptibility to diseases and infections, in addition to the fact that it can reduce mental capacity and increase the prevalence of maternal, neonatal, and infant complications (Ritchie and Roser, 2017). Recent studies suggest Zn supplementation as an alternative for decreasing the effects caused by SARS-CoV-2 infections, finding that Zn enhances the antiviral potential of mammalian cells by improving the immune system (Sharma et al., 2021; Joachimiak, 2021).

6.3 BIOFORTIFICATION: HISTORY AND ADVANCES

An innovative strategy for the problem of micronutrient deficiency in the diet is biofortification. Although there is still no consensus on the term, biofortification has been defined as the process of increasing the bioavailable concentrations of essential elements in edible portions of cultivated plants through agronomic management (fertilization) or genetic improvement (Gupta et al., 2021; Velu et al.,2014; White and Broadley, 2005). Biofortification of staple crops is a recent strategy, in addition to others such as food fortification through industrial processes, the addition of supplements at the time of consuming food, or a balanced diet (Pachón, 2010; FAO, 2014). Through its different alternatives, biofortification attempts to be a tool that seeks to improve the nutritional content of staple foods consumed mainly by low-income people, providing a medium- and long-term solution to the serious problem of malnutrition suffered by these populations, and at the same time, it is presented as a relatively inexpensive, profitable, and sustainable solution (Bouis, 2018a; Bouis et al., 2011).

Since 2004, the HarvestPlus project, coordinated by the International Food Policy Research Institute (IFPRI) and the International Center for Tropical Agriculture (CIAT), has been conducting programs aimed at increasing the nutrient content of staple crops in developing countries (FAO and HarvestPlus, 2019; Pachón, 2010). In 2008, expert economists ranked biofortification among the most cost-effective and sustainable solutions to address the global issue of poor nutrition and the problems it causes. Also in 2016, the World Food Prize was awarded to Howarth Bouis of the HarvestPlus organization and Maria Andrade, Robert Mwanga, and Jan Low of IPC (International Potato Center) for their work related to food biofortification (Andersson et al., 2017; Bouis, 2018b).

According to the World Bank, biofortification programs have proven to have excellent cost-benefit. They found that for every $1 invested in biofortification, up to $17 in benefits can be realized, since for most crops, only a one-time investment is required for the propagation of varieties with the elevated nutrient levels for this to become self-sustaining. However, a concern for markets could be a change in the coloration and appearance of biofortified fruits, so awareness campaigns among the population are required (Bouis, 2018b; De Steur et al., 2017).

For biofortification to have an impact on society, it is necessary to thoroughly evaluate the crop to be used, because the amount of nutrients that can be increased depends on the crop as well as the population that will benefit and access to the product. A biofortifiable crop has high yields, tastes good, is easy to cook, and is considered a good source of nutrients to be biofortified. Crops such as rice (*Oryza sativa* L.), sweet potato (*Ipomoea batatas* L.), beans (*Phaseolus vulgaris* L.), maize (*Zea mays* L), wheat (*Triticum* spp.), or cassava (*Manihot esculenta* L.) are of great importance in the diet of most developing countries, so the focus of biofortification has centered on studies on these crops (Table 6.1) (Ghosh et al., 2019; De Steur et al., 2017; Pachón, 2010).

TABLE 6.1
Biofortification with Micronutrients in Staple Crops

Crops	Biofortification forms	Key results	References
Rice	Genetic modification with overexpression of ferritin, NAS, and OsYSL2.	The Fe concentration of the modified seeds increased up to 6 times in the greenhouse crop and 4.4 times in the field crop.	Masuda et al., 2012
	Genetic modification by *OsVIT* gene silencing.	Increased iron translocation and a 1.8-fold increase in grain Fe concentration.	Bashir et al., 2013
	Foliar and soil application of Zn.	Yield increases by 7% and grain Zn content increases by 58 to 176%.	Gómez-Coronado et al., 2016
	Foliar application of selenium (Se) in the form of selenite and selenate.	Se content in the fruit increased up to 884% without yield change in different rice cultivars.	Lidon et al., 2019
Bean	Genetic improvement through the crossing of varieties.	Obtaining varieties with greater storage of Fe in grain, reaching up to 99 ppm per gram.	Blair and Izquierdo, 2012
	Application of silicon to the nutritive solution through fertirrigation.	Significant increases in the content of silicon in the fruit in both raw and cooked form.	Montesano et al., 2016
	Foliar application of Zn sulfate in 31 different production locations.	14.7% increase in Zn content in grains.	Ram et al., 2016
	The addition of iodine (I) in the nutritive solution is applied in the form of fertigation.	Increase of 0.6 ppm I in the fruit without affecting the development of the plant.	Dobosy et al., 2020
	Edaphic and foliar application of Zn.	The results indicate that the foliar application increases the concentration of Zn in a better way than the edaphic one, while the latter helps in the increase of amino acids and proteins.	Kachinski et al., 2022
Wheat	Foliar application of Fe citrate, Fe sulfate, and Zn sulfate.	Increases in the content of Fe and Zn in the treatments applied both in grain and in flour. The increase in Zn is more promising.	Zhang et al., 2010
	Inoculation with *Providencia* sp., *Anabaena* sp., and *Calothrix* sp.	Increase in the content of protein, Fe, Mn, and Cu in grain.	Rana et al., 2012
	Foliar and edaphic application of different types of fertilizers with Se.	Increases in the concentration of Se in grains, especially when it was applied in an edaphic way and without another nutrient.	Ramkissoon et al., 2019
	Mn application in the form of seed coverage, priming, soil application, and foliar application.	The content of Mn increased by 29% when it was applied in a foliar way in relation to the control without application.	Zulfiqar et al., 2021
Maize	They used a partial diallele mating design to develop 36 maize lines with provitamin A production.	They found that the development of hybrid varieties may be the appropriate approach for biofortification programs related to carotenoid content.	Halilu et al., 2016
	Foliar application of Zn in different forms.	Increase in growth, yield, and zinc content in grain.	Subbaiah et al., 2016
	Edaphic application of N, Fe, Zn, and Se.	Increase in the content of the four elements applied without finding significant effects on biomass production.	Grujcic et al., 2021

6.3.1 Types of Biofortification

Several studies indicate that the main forms of biofortification are genetic improvement and agronomic management. The former includes conventional plant breeding, the application of molecular tools such as gene editing, and even the use of genetically modified organisms. Generally, the institutions in charge of carrying out biofortification programs prefer the use of traditional breeding techniques, using the crossing of varieties until the expected results are obtained, as they consider that this is the simplest way to bring biofortified products to consumers, since it does not have the same regulations as the use of more modern techniques (Cakmak and Kutman, 2018; Bouis and Saltzman, 2017).

One of the most promising techniques is genetic modification, which is an alternative when conventional breeding does not achieve the proposed objectives. This technique focuses on achieving greater absorption of the desired nutrient, increased translocation from the leaf or root to the grain, specialization of storage in the endosperm, reduction of compounds that interfere with the absorption of the nutrient, and increased bioavailability at the time of consumption (Ludwig and Slamet-Loedin, 2019). Several studies report the efficiency of this technique, reaching the desired biofortification levels for nutrients such as Zn, Fe, vitamin A, protein, and selenium in crops such as rice and wheat, among others. However, its use presents many questions, as it faces numerous regulatory processes and biosafety standards (Saini et al., 2020; Ludwig and Slamet-Loedin, 2019; Bouis and Saltzman, 2017).

For its part, agronomic improvement consists of applying various sources of fertilization or beneficial compounds through a foliar or edaphic application (Figure 6.1), which is mainly used when dealing with micronutrients (Bouis and Saltzman, 2017); likewise, it is a simple tool to apply and is very cost-effective, besides being the most effective when there are problems in the soil that limit the absorption and mobility of nutrients. Agronomic management potentiates its effects depending on the method and timing of application, agroclimatic conditions, and crop variety. Also, it has the advantage that in addition to increasing the content of a nutrient, it can increase crop yields, so it is considered a promising technique and a short-term solution to the problem of poor food quality of staple crops (Praharaj et al., 2021; Ludwig and Slamet-Loedin, 2019).

In recent years, micronutrient deficiency has become very important worldwide, and various strategies to improve the nutritional quality of crops have focused on it. Key micronutrients for human nutrition, such as Fe, Zn, I, and Se, have been the subject of study because they are the main deficiency problems in the diet, especially in resource-poor populations (Jha and Warkentin, 2020; Wakeel et al., 2018). Biofortification, because of its advantages and benefits, has gained momentum in the last decade, finding a balanced way to overcome deficiencies of these micronutrients (Jha and Warkentin, 2020).

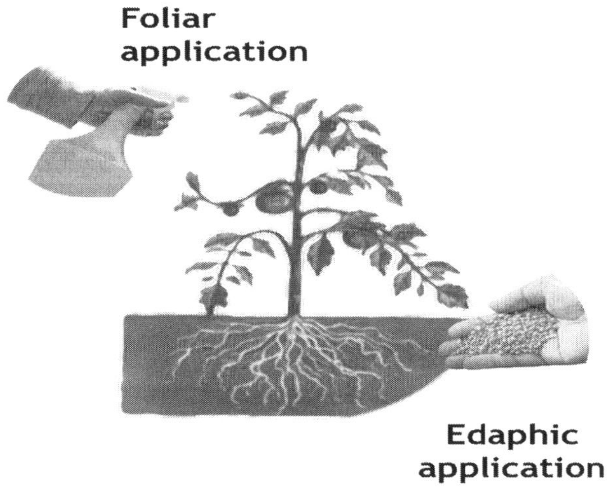

FIGURE 6.1 Forms of fertilization used in agronomic biofortification.

6.3.2 Biofortification with Zinc

A quarter of the world's population relies on diets containing predominantly cereals, legumes, and seeds, which do not provide the necessary amount of Zn for proper physiological development (Ramesh et al., 2014; Bilski et al., 2012). Zinc accumulation in edible portions of plants is highly variable depending on the species and is frequently reduced by the environmental factors to which the crops are exposed. In addition, it may depend on the interactions that may occur with other nutrients; for example, negative correlations have been found between Fe and Zn content in sorghum grains, and negative correlations between P and Zn content were also found in wheat (White and Broadley, 2009; McDonald et al., 2008).

During the last 10 years, organizations worldwide have carried out work in more than 12 countries to increase the concentration of zinc in edible portions of important staple crops in each region. The greatest response was found in wheat grains, with an 83% increase in this nutrient, followed by rice with 27%, and corn with the lowest response, with a 9% increase. In the case of beans, in plants to which Zn was applied, a substantial increase in its content in the fruit was obtained, in addition to the discovery that proteins are responsible for storing zinc in the grain (Cakmak and Kutman, 2018; Ram et al., 2016).

6.4 NANOTECHNOLOGY: HISTORY AND CONCEPTS

Nanotechnology is defined as "the design, synthesis, characterization, and application of materials, devices, and systems while controlling size and shape at the nanoscale". It can also be defined as the practical application of the theory of nanoscience, which is defined as the study of materials and/or substances with a dimension between 1 and 100 nm (Rafique et al., 2020). In consensus, nanotechnology is an emerging engineering discipline that applies nanoscience methods to create usable, marketable, and economically viable products by manipulating and characterizing materials between molecular and micrometer sizes. The characteristics offered by nanoparticles are attributed to the principles of quantum mechanics and the combination of multiple scientific disciplines, including biology, physics, chemistry, medicine, and engineering (Bhagyaraj and Oluwafemi, 2018; Sudha et al., 2018).

The history of nanotechnology dates to Mesopotamian and Roman civilizations, which used gold, silver, and copper nanoparticles in the vessels and cups they manufactured to give a shiny effect on the surface and make the color change when exposed to light. This fact was confirmed in 1990, when scientists subjected these vessels to analysis with a transmission electron microscope (Bayda et al., 2019; Sudha et al., 2018). Englishman John Utynam, 1449, patented glass based on gold nanoparticles. In the 16th century, the Swiss physician Theophrastus von Hohenheim used gold nanoparticles to treat patients suffering from various ailments. The Austrian chemist Richard Zsigmondy, the winner of the Nobel Prize for chemistry in 1925, was the first to introduce the term *nanometer* in his research on the measurement of particles such as colloidal gold (Rafique et al., 2020).

The concept of nanotechnology was imagined by the American physicist and Nobel laureate Richard Feynman in 1959. Considered the father of nanotechnology, he described a vision of using machines to build smaller machines and so on down to the molecular level. Fifteen years later, Japanese scientist Norio Taniguchi was the first to define the concept of nanotechnology to refer to the separation, consolidation, and deformation processing of materials by an atom or molecule. Another important point in the history of nanotechnology was the introduction of the concepts of molecular nanotechnology by the American scientist Eric Drexler in 1977, in addition to the invention of the tunneling microscope in 1981 by physicists Gerd Binnig and Heinrich Rorher at the IBM Zurich research laboratory. From this, the growth of this emerging science was exponential (Bayda et al., 2019; Bhagyaraj and Oluwafemi, 2018).

The advantages offered using nanotechnology compared with other technologies are typical, since nanoparticles have a larger surface area by weight than larger particles, which makes them more reactive with other molecules. These properties have allowed the use of this technology in various fields such as energy storage, water and contaminated soil treatment, food-protective films, and nanocarriers that enable greater efficiency in drug intake, among many other functions that allow its application in the fields of medicine, mining, electronics, computing, and robotics, in resource conservation processes, in energy generation, and in the pharmaceutical and food industries, among others (Chakraborty et al., 2020; Pathakoti et al., 2018).

6.4.1 Nanotechnology in Agriculture

In the past, agriculture was the center of attention for the application of new technologies, being the predecessor of industrial revolutions for more than 90 centuries. However, in the case of nanotechnology, its benefits have been used in the industrial field and the health sciences for more than half a century, whereas in agriculture, research on its potential benefits is still minuscule, since less than 5% of the total research related to this topic pertains to the agricultural area. Despite that, the upward trend suggests that in the coming years, the application of nanoparticle-based products in agricultural fields will be a regular practice (Mukhopadhyay, 2014; Ditta, 2012).

Nanotechnology emerges as an alternative because as an emerging science, it is aimed at understanding and creating materials, devices, and systems that exploit the nanoscale properties of various compounds. In agriculture, nanotechnology has begun to develop products that help improve soil fertility, reduce diseases, and increase crop quality and production, moving from conventional agriculture to precision agriculture. Examples such as carbon nanotubes, Cu, Ag, Mn, Mo, Zn, Fe, Si, and Ti, or in their oxide forms, in addition to nanoformulations of phosphorus, urea, and sulfur, are considered as future sources of agricultural fertilization (Chhipa, 2017; Solanki et al., 2017).

6.4.2 Nanofertilizers

Crop fertilization is of utmost importance when producing food. However, the low effectiveness at which fertilizers are used by crops, reaching less than 25%, in addition to their excessive application, has a negative environmental impact, causing soil deterioration, contamination in aquifers, and air pollution. Recent studies focus on finding less harmful alternatives for the environment, without affecting the quality and production of crops (Raliya et al., 2017; Naderi and Danesh-Shahraki, 2013). In the last decade, research on nanofertilizers has grown exponentially, being the target of various communities such as the American Chemical Society (ACS), the Society of Environmental Toxicology and Chemistry (SETAC), and Crop Chemical Europe, as well as regulatory institutions such as the United States Department of Agriculture (USDA), causing great impact on the scientific community; however, in the industrial branch, there is very little information on nanotechnology-based products. Nanoparticle-based pesticides are the most common products on the market, but nanotechnology is still considered as a dream, not too far away (Kah et al., 2018).

Nanotechnology offers different aspects with potential applications in different fields of agriculture and biotechnology to solve environmental problems and biological demands with greater precision. The use of these nanofertilizers causes an increase in fertilizer efficiency, reduces soil toxicity, and minimizes over-application and its potentially toxic effects. Thus, there is an opportunity for nanotechnology to have a significant influence on energy, the economy, and the environment (Naderi and Danesh-Shahraki, 2013).

Nanofertilizers are all those particles that contain nutrients and whose dimension is less than 100 nm, preferably between 30 and 40 nm. In addition, a very high contact surface of the nutrient ion and a slower and more precise release than traditional fertilizers stand out within their characteristics. This is due to the reduction of the particle size, which causes the specific surface to increase and as a consequence, the contact area with the plants as well. Due to this, the plants can absorb fertilizers with a different dynamic, which presents an additional advantage compared with traditional sources of fertilization (Subbaiah et al., 2016; Subramanian et al., 2015).

Nanofertilizers are classified into three categories: 1) nanofertilizers based on macronutrients, 2) nanofertilizers based on micronutrients, and 3) nanomaterials that act as nutrient carriers. Several studies compare the effects of different categories of nanomaterials against traditional fertilizers, obtaining an average efficacy of nanofertilizers over conventional fertilizers of 19, 18, and 29% for categories 1, 2, and 3, respectively (Kah et al., 2018). Because of this, nanoformulations of active ingredient agricultural inputs could cause a significant reduction in environmental problems caused by over-fertilization, such as eutrophication. Nanofertilizers such as Ca nanoparticles, P-hydroxyapatite, Fe^{+2}, ZnO, TiO_2, Ag, and carbon nanotubes can be used as an alternative to conventional agricultural inputs (Kah et al., 2018; Chhipa, 2017).

Despite the promising outlook for the use of nanofertilizers, the studies that have been presented so far are not yet sufficient to determine that this new technology is superior to traditional fertilizers, in addition to the fact that studies should also focus on establishing the effects generated by these materials on plant development (Kah et al., 2018; Subramanian et al., 2015).

6.4.2.1 Zinc Nanoparticles

There are several forms of Zn-containing nanoparticles, such as ZnS and ZnSe, or CdSe/ZnS quantum dots, which are currently used in various fields of science. However, the most widely used form of zinc as nanoparticles continues to be zinc oxide (ZnO) due to its wide range of positive properties, such as high electrical and thermal conductivity, high-temperature stability, and neutral pH, as well as mild antimicrobial effects. Also, its use is related to easy availability and low procurement price (Sturikova et al., 2018). The production of ZnO is estimated at close to 1000 tons per year. It is commonly used in plastic, glass, ceramic, cement, and rubber materials, as well as in pigments, paints, and sunscreens (Piccinno et al., 2012). In agriculture, applications of Zn nanoparticles, at low concentrations, have been shown to have positive effects on germination, vegetative growth, and chlorophyll content, and significant increases in yield (Rossi et al., 2019; Siddiqui et al., 2019).

It has been proven that the use of this technology by foliar application improves fruit quality, increasing the absorption, translocation, and accumulation of Zn in fruits, making it a novel tool for biofortification programs (Subbaiah et al., 2016; Davarpanah et al., 2016; El-Ramady et al., 2018). On the other hand, applications at high concentrations have stressful effects that can inhibit root growth, delay crop development, and affect protein and DNA synthesis (Ma et al., 2015; Sturikova et al., 2018).

6.5 BIOFORTIFICATION WITH NANOFERTILIZERS

Despite the positive results of biofortification work, there is a serious problem caused by the excess fertilization required to reach the levels of nutrients to be biofortified. Among the most common problems caused by excess fertilization are over-accumulation of heavy metals in soil and water, eutrophication, and soil degradation. In addition, the success of biofortification depends on the form, uptake, mobilization, and accumulation of the nutrient, so there is a need for alternative, sustainable, and accurate strategies to deliver the nutrient to the plant (Khan et al., 2021; Kulkarni and Goswami, 2019).

Modern technologies, such as nanotechnology, emerge as an efficient alternative for increasing production and improving food quality. Nanotechnology is a technique that can be progressively applied to biofortification programs (Solanki and Laura, 2018). Nanotechnology in the form of nanofertilizers, as mentioned earlier, presents advantages over the application of traditional fertilizers; in addition to increasing yield, it increases the activity of nitrogen metabolism, helps to mitigate stress situations, and increases the photosynthetic rate, and above all, several studies have found greater translocation of the nutrient from the roots and leaves to the grain (Khan et al., 2021; Lv et al., 2019; Zulfiqar et al., 2019).

Biofortification projects are successfully using three main forms of nanofertilizers: I) nanoscale coating fertilizers, in which conventional fertilizers are enclosed in nanoparticles or intercalated in nanopores of zeolites and/or nanometer clays to increase nutrient availability, postpone release, or supplement with an extra nanoscale element; II) fertilizers with nanoscale additions, in which traditional fertilizers are supplemented with nutrient nanoparticles; and III) nanoscale fertilizers, which are nanoparticles less than 100 nm in size that contain nutrients and are used directly as fertilizers (Khan et al., 2021).

In addition to the forms of nanofertilizers, the manner of application greatly influences the achievement of biofortification. There are three main methods of application: seed priming, soil application, and foliar application. Also, the chemical composition, stability, concentration, and size of the nanoparticles must be considered for nutrient absorption, translocation, and accumulation to take place. Numerous studies indicate that, like traditional biofortification, biofortification with nanoparticles focuses mainly on the application of micronutrients, such as Zn, Fe, I, Cu, and Se. Among these micronutrients, zinc stands out and represents the largest number of studies in biofortification programs (Dhaliwal et al., 2022; El-Ramady et al., 2021; Szerement et al., 2021).

6.5.1 Biofortification with Zinc Nanofertilizers

In recent years, the use of Zn nanoparticles to improve plant uptake has become more popular. This is because decreasing the particle size results in a greater number of particles per unit of Zn applied, in addition to increasing the specific surface area of a fertilizer, which will increase the dissolution rate of fertilizers with low water solubility, such as Zn oxide (ZnO), so gradual increases in Zn uptake can be observed as particle size decreases (Solanki and Laura, 2018).

In relation to this, Du et al. (2019), examined the effects of ZnO nanoparticles and zinc sulfate foliar on wheat (*Triticum aestivum* L.) growth at the same doses. The highest Zn content in grain was found with 100 ppm ZnO. It was almost 30% higher compared with the application of 2000 ppm zinc sulfate. Similarly, Bautista-Diaz et al. (2021) reported significant differences in Zn content in bean fruits when ZnO nanoparticles were foliar applied compared with Zn sulfate application, also finding increases in leaves and roots. Similar results were published by Mahdieh et al. (2018), who applied zinc nanoparticles in the form of ZnO on two occasions in a foliar manner in pinto bean plants and compared this with Zn sulfate and chelate at twice the concentration, obtaining an increase of 28.5% in the zinc content in grains compared with the application of Zn sulfate and 11.3% compared with chelate.

In recent years, studies with ZnO nanoparticles have grown exponentially, because their physicochemical properties make them a potential solution to counteract Zn deficiency in the human diet, and these same properties allow them to dissolve in the soil, thus avoiding soil contamination from excess fertilization that could be present in biofortification programs. In addition, ZnO nanoparticles can produce a longer-lasting release of Zn^{2+}, which is the form of Zn that can be more easily absorbed by plants (Doolette et al., 2020; Du et al., 2019). Almendros et al. (2022) found a higher translocation factor of Zn within the plant when applying ZnO nanoparticles and pointed out that the translocation of Zn to the fruits is influenced by its concentration; with high doses of Zn, the roots avoid translocating it to prevent toxicity in other organs of the plant, so that low applications of ZnO in the form of nanoparticles could stimulate to a greater extent the movement of Zn within the plant.

Umar et al. (2021) found the maximum Zn concentration in corn grains with foliar application of ZnO nanoparticles and reported that Zn absorption and translocation are comparable to when Zn sulfate is applied. Similarly, Subbaiah et al. (2016) achieved a 37% increase in Zn content in corn grains with foliar application of 100 ppm ZnO. Velázquez-Gamboa et al. (2021) found a 240.31% increase in Zn content in *Stevia* plants when ZnO nanoparticles were applied, showing that this form of Zn is more assimilable than Zn sulfate. Meanwhile, de Souza et al. (2021) found that the application of ZnO in the particulate form increased the Zn content in cowpea beans; however, they found no differences with the application of ZnO in the conventional form. Several studies indicate the effects of the use of ZnO in biofortification (Table 6.2).

TABLE 6.2
Application of ZnO in the Form of Nanoparticles in Biofortification Work

References	Crops	Doses	Methods	Key results
Subbaiah et al., 2016	Maize	50, 100, 200, 400, 600, 800, 1000, 1500, and 2000 ppm	Foliar application	It increased the content of Zn in grain by 37% at a dose of 100 ppm in relation to the control without application.
Mahdieh et al., 2018	Pinto beans	0.05%, 0.1%, and 0.15% p/v of ZnO	Seed application and foliar application in two stages of crop development	Increase of 28.5% in the content of zinc in grains at a dose of 0.15% compared with the application of Zn sulfate and 11.3% compared with the chelate at twice the concentration.
Ponce-García et al., 2019	Green beans	25, 50, and 100 ppm	Edaphic application in nutrient solution	The concentration of Zn in the fruit increased in the three treatments in relation to the control.
Du et al., 2019	Wheat	0, 10, 20, 50, 100, 200, and 1000 ppm	Edaphic application in nutrient solution	Increases of 48, 138, 175, 186, and 230% in the doses from 20 to 1000 ppm, respectively.
Velázquez-Gamboa et al., 2020	Stevia	50, 75, and 100 ppm	Edaphic application in nutrient solution	They found a 240.31% increase in Zn content in *Stevia* plants by applying ZnO nanoparticles compared with Zn sulfate.
Umar et al., 2021	Maize	8 kg of Zn.ha^{-1} in an edaphic way and a 2% solution in a foliar way	Edaphic application in nutrient solution and foliar application	A relative increase of 82 and 59% in grain Zn concentration was recorded with soil and foliar application, respectively, in relation to the control.
de Souza et al., 2021	Cowpea bean	0, 100, 300, 500, and 800 mg.dm^{-3}	Soil application	At the dose of 100 mg dm^{-3}, it represented an increase of 14% of Zn in grain compared with the control.
Aslam et al., 2021	Mung bean	0.1% ZnO and 0.1% ZnO plus lysine	Foliar application	Increases of 33.81 and 40.87% in the content of Zn in grain, respectively.
Yang et al., 2021	Rice	25 and 100 mg.kg^{-1} at different stages of development: basal, tillering, and panicle formation	Soil application	The application of ZnO increased the Zn concentration of the rice by 13.5, 19.8, and 39.4% for each one of the stages compared with the traditional fertilization of the crop.
Dhaliwal et al., 2021	Chickpea	0.5% ZnO and 0.5% ZnO + 0.5% Fe nanoparticles	Foliar application	12.7% increase in Zn concentration in the grain for both treatments compared with the control with the standard fertilization dose.
Palacio-Márquez et al., 2022	Green bean	25, 50, and 100 ppm	Foliar application	Increases of 32.96, 97.77, and 103.35% in the content of Zn in the fruit in relation to the control without foliar application.

6.6 CONCLUSIONS

Advances in the use of Zn nanoparticles in biofortification have been exponential in recent years, positioning them as a unique, efficient, and sustainable product, capable of increasing the Zn content in various fruits. Zn nanoparticles, thanks to their specific properties, can be absorbed with a different dynamic, which presents an additional advantage compared with traditional sources of fertilization, improving efficiency and reducing the doses to be used, making them an alternative to the problem of environmental pollution caused by excessive fertilization. In addition, although their physiological effects are not completely proven, the benefits on growth and yield allow their use in biofortification without affecting agricultural production. However, despite these positive effects, there are still many differences in terms of their synthesis, doses, and forms of application, which generates discrepancies in the results, so more studies on the use of Zn nanoparticles in biofortification are needed to determine that the use of this new technology is superior to the use of traditional fertilizers.

REFERENCES

Ahmed, M. H., Vasas, D., Hassan, A., & Molnár, J. (2022). The impact of functional food in prevention of malnutrition. *PharmaNutrition*, 100288. https://doi.org/10.1016/j.phanu.2022.100288

Almendros, P., González, D., Fernández, M. D., García-Gomez, C., & Obrador, A. (2022). Both Zn biofortification and nutrient distribution pattern in cherry tomato plants are influenced by the application of ZnO nanofertilizer. *Heliyon*, 8(3), e09130. https://doi.org/10.1016/j.heliyon.2022.e09130

Andersson, M. S., Saltzman, A., Virk, P. S., & Pfeiffer, W. H. (2017). Progress update: Crop development of biofortified staple food crops under HarvestPlus. *African Journal of Food, Agriculture, Nutrition and Development*, 17(2), 11905–11935. https://doi.org/10.18697/ajfand.78.HarvestPlus05

Aslam, Z., Bashir, S., Shahzad, M., Ahmad, J. N., Bashir, S., Ahmad, A., ... & El-Shehawi, A. M. (2021). Comparative efficacy of zinc sources for zinc-biofortification of mung bean (*Vigna radiata* L.). *Fresenius Environmental Bulletin*, 30(8), 9903–9912.

Bashir, K., Takahashi, R., Akhtar, S., Ishimaru, Y., Nakanishi, H., & Nishizawa, N. K. (2013). The knockdown of OsVIT2 and MIT affects iron localization in rice seed. *Rice*, 6(1), 1–7. https://doi.org/10.1186/1939-8433-6-31

Bautista-Diaz, J., Cruz-Alvarez, O., Hernández-Rodríguez, O. A., Sánchez-Chávez, E., Jacobo-Cuellar, J. L., Preciado-Rangel, P., ... & Ojeda-Barrios, D. L. (2021). Zinc sulphate or zinc nanoparticle applications to leaves of green beans. *Folia Horticulturae*, 33(2), 365–375. https://doi.org/10.2478/fhort-2021-0028

Bayda, S., Adeel, M., Tuccinardi, T., Cordani, M., & Rizzolio, F. (2019). The history of nanoscience and nanotechnology: From chemical–physical applications to nanomedicine. *Molecules*, 25(1), 112. https://doi.org/10.3390/molecules25010112

Bhagyaraj, S. M., & Oluwafemi, O. S. (2018). Nanotechnology: The science of the invisible. In: *Synthesis of Inorganic Nanomaterials*. Eds: Bhagyaraj, S.M., Oluwafemi, O. S., Kalarikkal, N., & Thomas, S. (pp. 1–18). Woodhead Publishing, Sawston.

Bilski, J., Jacob, D., Soumaila, F., Kraft, C., & Farnsworth, A. (2012). Agronomic biofortification of cereal crop plants with Fe, Zn, and Se, by the utilization of coal fly ash as plant growth media. *Advances in Biologicalresearch*, 3(4), 130. https://pubmed.ncbi.nlm.nih.gov/30288015

Blair, M. W., & Izquierdo, P. (2012). Use of the advanced backcross-QTL method to transfer seed mineral accumulation nutrition traits from wild to Andean cultivated common beans. *Theoretical and Applied Genetics*, 125(5), 1015–1031. https://doi.org/10.1007/s00122-012-1891-x

Blasco, B. (2015). Biofortificación: Beneficios potenciales para los cultivos y la salud humana. Disponible en. http://aefa-agronutrientes.org/la-biofortificacion (*Fecha de Consulta* Febr. 10, 2021).

Bouis, H. E. (2018b). Reducing mineral and vitamin deficiencies through biofortification: Progress under HarvestPlus. In: *Hidden Hunger: Strategies to Improve Nutrition Quality*. Eds: Biesalski, H. K., & Birner, R. (Vol. 118, pp. 112–122). Karger Publishers, Basilea, Switzerland.

Bouis, H. E. (2018a). Biofortification: An agricultural tool to address mineral and vitamin deficiencies. In: *Food Fortification in a Globalized World*. Eds: Venkatesh-Mannar, M. G., & Hurrell, R. F. (pp. 69–81). Academic Press. Cambridge, MA.

Bouis, H. E., & Saltzman, A. (2017). Improving nutrition through biofortification: A review of evidence from HarvestPlus, 2003 through 2016. *Global Food Security*, 12, 49–58. https://doi.org/10.1016/j.gfs.2017.01.009

Bouis, H. E., Hotz, C., McClafferty, B., Meenakshi, J. V., & Pfeiffer, W. H. (2011). Biofortification: A new tool to reduce micronutrient malnutrition. *Food and Nutrition Bulletin*, 32(1_suppl1), S31–S40. https://doi.org/10.1177/15648265110321S105

Bourke, C. D., Berkley, J. A., & Prendergast, A. J. (2016). Immune dysfunction as a cause and consequence of malnutrition. *Trends in Immunology*, 37(6), 386–398. https://doi.org/10.1016/j.it.2016.04.003

Bulgari, R., Cocetta, G., Trivellini, A., Vernieri, P., & Ferrante, A. (2014). Biostimulants and crop responses: A review. *Biological Agriculture and Horticulture*, 31(1), 1–17. https://doi.org/10.1080/01448765.2014.964649

Cakmak, I., & Kutman, U. B. (2018). Agronomic biofortification of cereals with zinc: A review. *European Journal of Soil Science*, 69(1), 172–180. https://doi.org/10.1111/ejss.12437

Caro, C. R., Coronell, M. D. C., Arrollo, J., Martínez, G., Majana, L. S., & Sarmiento-Rubiano, L. A. (2016). La deficiencia de zinc: Un problema global que afecta la salud y el desarrollo cognitivo. *Archivos Latinoamericanos de Nutrición*, 66(3), 165–175.

Cederholm, T., Bosaeus, I., Barazzoni, R., Bauer, J., Van Gossum, A., Klek, S., … & de van der Schueren, M. A. (2015). Diagnostic criteria for malnutrition–an ESPEN consensus statement. *Clinical Nutrition*, 34(3), 335–340. https://doi.org/10.1016/j.clnu.2015.03.001

Chakraborty, A., Guha, S., & Patra, P. (2020). Nanotechnology and global applications: Bench to community. *Journal of Biology and Life Science*, 11(2), 181–189.

Chhipa, H. (2017). Nanofertilizers and nanopesticides for agriculture. *Environmental Chemistry Letters*, 15(1), 15–22. https://doi.org/10.1007/s10311-016-0600-4

Dahdouh, S., Grande, F., Espinosa, S. N., Vincent, A., Gibson, R., Bailey, K., … & Charrondière, U. R. (2019). Development of the FAO/INFOODS/IZINCG global food composition database for phytate. *Journal of Food Composition and Analysis*, 78, 42–48.. https://doi.org/10.1016/j.jfca.2019.01.023

Davarpanah, S., Tehranifar, A., Davarynejad, G., Abadía, J., & Khorasani, R. (2016). Effects of foliar applications of zinc and boron nano-fertilizers on pomegranate (*Punica granatum* cv. Ardestani) fruit yield and quality. *Scientia Horticulturae*, 210, 57–64. https://doi.org/10.1016/j.scienta.2016.07.003

de Souza, A. A. B., do Nascimento, C. W. A., & de Souza, E. R. (2021). Mineral composition, chlorophyll fluorescence and zinc biofortification in Vigna unguiculata fertilized with bulk and nanoparticulate zinc oxides. *Acta Physiologiae Plantarum*, 43(12), 1–10. https://doi.org/10.1007/s11738-021-03333-y

De Steur, H., Demont, M., Gellynck, X., & Stein, A. J. (2017). The social and economic impact of biofortification through genetic modification. *Current Opinion in Biotechnology*, 44, 161–168. https://doi.org/10.1016/j.copbio.2017.01.012

Dhaliwal, S. S., Sharma, V., Shukla, A. K., Verma, V., Behera, S. K., Singh, P., … & Hossain, A. (2021). Comparative efficiency of mineral, chelated and Nano forms of zinc and iron for improvement of zinc and iron in chickpea (*Cicer arietinum* L.) through biofortification. *Agronomy*, 11(12), 2436. https://doi.org/10.3390/agronomy11122436

Dhaliwal, S. S., Sharma, V., Shukla, A. K., Verma, V., Kaur, M., Shivay, Y. S., … & Hossain, A. (2022). Biofortification—A frontier novel approach to enrich micronutrients in field crops to encounter the nutritional security. *Molecules*, 27(4), 1340. https://doi.org/10.3390/molecules27041340

Ditta, A. (2012). How helpful is nanotechnology in agriculture? *Advances in Natural Sciences: Nanoscience and Nanotechnology*, 3(3), 033002. http://doi.org/10.1088/2043-6262/3/3/033002

Dobosy, P., Kröpfl, K., Óvári, M., Sandil, S., Németh, K., Engloner, A., … & Záray, G. (2020). Biofortification of green bean (*Phaseolus vulgaris* L.) and lettuce (*Lactuca sativa* L.) with iodine in a plant-calcareous sandy soil system irrigated with water containing KI. *Journal of Food Composition and Analysis*, 88, 103434. https://doi.org/10.1016/j.jfca.2020.103434

Doolette, C. L., Read, T. L., Howell, N. R., Cresswell, T., & Lombi, E. (2020). Zinc from foliar-applied nanoparticle fertiliser is translocated to wheat grain: A 65Zn radiolabelled translocation study comparing conventional and novel foliar fertilisers. *Science of the Total Environment*, 749, 142369. https://doi.org/10.1016/j.scitotenv.2020.142369

Du, W., Yang, J., Peng, Q., Liang, X., & Mao, H. (2019). Comparison study of zinc nanoparticles and zinc sulphate on wheat growth: From toxicity and zinc biofortification. *Chemosphere*, 227, 109–116. https://doi.org/10.1016/j.chemosphere.2019.03.168

El-Ramady, H., Abdalla, N., Alshaal, T., El-Henawy, A., Elmahrouk, M., Bayoumi, Y., & Domokos-Szabolcsy, É. (2018). Plant nano-nutrition: Perspectives and challenges. In: *Nanotechnology, Food Security and Water Treatment* (pp. 129–161). Springer, Cham.

El-Ramady, H., Abdalla, N., Elbasiouny, H., Elbehiry, F., Elsakhawy, T., Omara, A. E. D., … & Zia-ur-Rehman, M. (2021). Nano-biofortification of different crops to immune against COVID-19: A review. *Ecotoxicology and Environmental Safety*, 222, 112500. https://doi.org/10.1016/j.ecoenv.2021.112500

FAO (2014). *Programa Conjunto FAO/OMS sobre Normas Alimentarias Comité del Codex sobre Nutrición y Alimentos para Regímenes Especiales. Documento de Debate sobre la Biofortificación con Nutrientes Esenciales*. Pags 1–9.

FAO, & HarvestPlus (2019). *Biofortification: A food-systems solution to help end hidden hunger*. Organización de Las Naciones Unidas Para La Alimentación Y, La Agricultura. https://www.fao.org/publications/card/es/c/CA8711EN/

FAO, FIDA, O. M. S., PMA, & UNICEF (2021). *El Estado de la Seguridad alimentaria y la Nutrición en el Mundo 2021. Transformación de los Sistemas Alimentarios en Aras de la Seguridad alimentaria, una Mejor Nutrición y Dietas Asequibles y Saludables para Todos*. FAO, Roma. https://doi.org/10.4060/cb5409es

Ghosh, A., Hasim Reja, M., Nalia, A., Kanthal, S., Maji, S., Venugopalan, V., & Nath, R. (2019). Micronutrient biofortification in pulses: An agricultural approach. *Current Journal of Applied Science and Technology*, 35(4), 1–12.

Gomez-Coronado, F., Poblaciones, M. J., Almeida, A. S., & Cakmak, I. (2016). Zinc (Zn) concentration of bread wheat grown under Mediterranean conditions as affected by genotype and soil/foliar Zn application. *Plant and Soil*, 401(1), 331–346. https://doi.org/10.1007/s11104-015-2758-0

Gorji, A., & Ghadiri, M. K. (2020). The potential roles of micronutrient deficiency and immune system dysfunction in COVID-19 pandemic. *Nutrition*, 111047. https://doi.org/10.1016/j.nut.2020.111047

Grujcic, D., Yazici, A. M., Tutus, Y., Cakmak, I., & Singh, B. R. (2021). Biofortification of silage maize with zinc, iron and selenium as affected by nitrogen fertilization. *Plants*, 10(2), 391. https://doi.org/10.3390/plants10020391

Gupta, P. K., Balyan, H. S., Sharma, S., & Kumar, R. (2021). Biofortification and bioavailability of Zn, Fe and Se in wheat: Present status and future prospects. *Theoretical and Applied Genetics*, 134(1), 1–35. https://doi.org/10.1007/s00122-020-03709-7

Halilu, A. D., Ado, S. G., Aba, D. A., & Usman, I. S. (2016). Genetics of carotenoids for provitamin A biofortification in tropical-adapted maize. *The Crop Journal*, 4(4), 313–322. https://doi.org/10.1016/j.cj.2016.05.002

Huizar, M. I., Arena, R., & Laddu, D. R. (2021). The global food syndemic: The impact of food insecurity, Malnutrition and obesity on the healthspan amid the COVID-19 pandemic. *Progress in Cardiovascular Diseases*, 64, 105. https://doi.org/10.1016%2Fj.pcad.2020.07.002

Jha, A. B., & Warkentin, T. D. (2020). Biofortification of pulse crops: Status and future perspectives. *Plants*, 9(1), 73. https://doi.org/10.3390/plants9010073

Joachimiak, M. P. (2021). Zinc against COVID-19? Symptom surveillance and deficiency risk groups. *PLOS Neglected Tropical Diseases*, 15(1), e0008895. https://doi.org/10.1371/journal.pntd.0008895

Kachinski, W. D., Ávila, F. W., dos Reis, A. R., Muller, M. M. L., Mendes, M. C., & Petranski, P. H. (2022). Agronomic biofortification increases concentrations of zinc and storage proteins in common bean (*Phaseolus vulgaris* L.) grains. *Food Research International*, 111105. https://doi.org/10.1016/j.foodres.2022.111105

Kah, M., Kookana, R. S., Gogos, A., & Bucheli, T. D. (2018). A critical evaluation of nanopesticides and nanofertilizers against their conventional analogues. *Nature Nanotechnology*, 13(8), 677. https://doi.org/10.1038/s41565-018-0131-1

Khan, M., Pandey, A., Hamurcu, M., Gezgin, S., Athar, T., Rajput, V. D., ... & Minkina, T. (2021). Insight into the prospects for nanotechnology in wheat biofortification. *Biology*, 10(11), 1123. https://doi.org/10.3390/biology10111123

Kiran, A., Wakeel, A., Mahmood, K., Mubaraka, R., Haefele, S. M. (2022). Biofortification of staple crops to alleviate human malnutrition: Contributions and potential in developing countries. *Agronomy*, 12(2), 452. https://doi.org/10.3390/agronomy12020452

Kobylińska, M., Antosik, K., Decyk, A., & Kurowska, K. (2022). Malnutrition in obesity: Is it possible? *Obesity Facts*, 15(1), 19–25. https://doi.org/10.1159/000519503

Kulkarni, S., & Goswami, A. (2019). Effect of excess fertilizers and nutrients: A review on impact on plants and human population. In: *Proceedings of International Conference on Sustainable Computing in Science, Technology and Management (SUSCOM)*. Amity University, Rajasthan, Jaipur-India.

Kurtz, A., Grant, K., Marano, R., Arrieta, A., Feaster, W., Steele, C., & Ehwerhemuepha, L. (2021). Long-term effects of malnutrition on severity of COVID-19. *Scientific Reports*, 11(1), 1–8. https://doi.org/10.1038/s41598-021-94138-z

Lidon, F. C., Oliveira, K., Galhano, C., Guerra, M., Ribeiro, M. M., Pelica, J., ... & Reboredo, F. H. (2019). Selenium biofortification of rice through foliar application with selenite and selenate. *Experimental Agriculture*, 55(4), 528–542. https://doi.org/10.1017/S0014479718000157

Ludwig, Y., & Slamet-Loedin, I. H. (2019). Genetic biofortification to enrich rice and wheat grain iron: from genes to product. *Frontiers in Plant Science*, 10, 833. https://doi.org/10.3389/fpls.2019.00833

Lv, J., Christie, P., & Zhang, S. (2019). Uptake, translocation, and transformation of metal-based nanoparticles in plants: Recent advances and methodological challenges. *Environmental Science: Nano*, 6(1), 41–59. https://doi.org/10.1039/C8EN00645H

Ma, C., White, J. C., Dhankher, O. P., & Xing, B. (2015). Metal-based nanotoxicity and detoxification pathways in higher plants. *Environmental Science and Technology*, 49(12), 7109–7122. https://doi.org/10.1021/acs.est.5b00685

Mahdieh, M., Sangi, M. R., Bamdad, F., & Ghanem, A. (2018). Effect of seed and foliar application of nano-zinc oxide, zinc chelate, and zinc sulphate rates on yield and growth of pinto bean (*Phaseolus vulgaris*) cultivars. *Journal of Plant Nutrition*, 41(18), 2401–2412. https://doi.org/10.1080/01904167.2018.1510517

Masuda, H., Ishimaru, Y., Aung, M. S., Kobayashi, T., Kakei, Y., Takahashi, M., ... & Nishizawa, N. K. (2012). Iron biofortification in rice by the introduction of multiple genes involved in iron nutrition. *Scientific Reports*, 2(1), 1–7. https://doi.org/10.1038/srep00543

McDonald, G. K., Genc, Y., & Graham, R. D. (2008). A simple method to evaluate genetic variation in grain zinc concentration by correcting for differences in grain yield. *Plant and Soil*, 306(1–2), 49–55. https://doi.org/10.1007/s11104-008-9555-y

Montesano, F. F., D'Imperio, M., Parente, A., Cardinali, A., Renna, M., & Serio, F. (2016). Green bean biofortification for Si through soilless cultivation: Plant response and Si bioaccessibility in pods. *Scientific Reports*, 6(1), 1–9. https://doi.org/10.1038/srep31662

Mukhopadhyay, S. S. (2014). Nanotechnology in agriculture: Prospects and constraints. *Nanotechnology, Science and Applications*, 7, 63. https://doi.org/10.2147%2FNSA.S39409

Naderi, M. R., & Danesh-Shahraki, A. (2013). Nanofertilizers and their roles in sustainable agriculture. *International Journal of Agriculture and Crop Sciences*, 5(19), 2229.

Pachón, H. (2010). *El Impacto Nutricional de Cultivos Biofortificados o Cultivos con Mayor Calidad Nutricional*. Colombia: Centro Internacional de Agricultura Tropical (CIAT). Agrosalud. 14 Pags.

Palacio-Márquez, A., Ramírez-Estrada, C. A., Sánchez, E., Ojeda-Barrios, D. L., Chávez-Mendoza, C., & Sida-Arreola, J. P. (2022). Biofortification with nanoparticles and zinc nitrate plus chitosan in green beans: Effects on yield and mineral content. *Notulae Botanicae Horti Agrobotanici Cluj-Napoca*, 50(2), 12672–12672. https://doi.org/10.15835/NBHA50212672

Pathakoti, K., Manubolu, M., & Hwang, H. M. (2018). Nanotechnology applications for environmental industry. In: *Handbook of Nanomaterials for Industrial Applications* (pp. 894–907). Elsevier.

Peña-Rosas, J. P., Mithra, P., Unnikrishnan, B., Kumar, N., De-Regil, L. M., Nair, N. S., ... & Solon, J. A. (2019). Fortification of rice with vitamins and minerals for addressing micronutrient malnutrition. *Cochrane Database of Systematic Reviews*, 10(10). https://doi.org/10.1002/14651858.CD009902.pub2

Piccinno, F., Gottschalk, F., Seeger, S., & Nowack, B. (2012). Industrial production quantities and uses of ten engineered nanomaterials in Europe and the world. *Journal of Nanoparticle Research*, 14(9), 1109. https://doi.org/10.1007/s11051-012-1109-9

Ponce-García, C. O., Soto-Parra, J. M., Sánchez, E., Muñoz-Márquez, E., Piña-Ramírez, F. J., Flores-Córdova, M. A., ... & Yáñez Muñoz, R. M. (2019). Efficiency of nanoparticle, sulfate, and zinc-chelate use on biomass, yield, and nitrogen assimilation in green beans. *Agronomy*, 9(3), 128. https://doi.org/10.3390/agronomy9030128

Praharaj, S., Skalicky, M., Maitra, S., Bhadra, P., Shankar, T., Brestic, M., ... & Hossain, A. (2021). Zinc biofortification in food crops could alleviate the zinc malnutrition in human health. *Molecules*, 26(12), 3509. https://doi.org/10.3390/molecules26123509

Prasad, A. S. (2012). Discovery of human zinc deficiency: 50 years later. *Journal of Trace Elements in Medicine and Biology*, 26(2–3), 66–69. https://doi.org/10.1016/j.jtemb.2012.04.004

Rafique, M., Tahir, M. B., Rafique, M. S., & Hamza, M. (2020). History and fundamentals of nanoscience and nanotechnology. In: *Nanotechnology and Photocatalysis for Environmental Applications* (pp. 1–25). Elsevier.

Raliya, R., Saharan, V., Dimkpa, C., & Biswas, P. (2017). Nanofertilizer for precision and sustainable agriculture: Current state and future perspectives. *Journal of Agricultural and Food Chemistry*, 66(26), 6487–6503. https://doi.org/10.1021/acs.jafc.7b02178

Ram, H., Rashid, A., Zhang, W., Duarte, A. P., Phattarakul, N., Simunji, S., ... & Mahmood, K. (2016). Biofortification of wheat, rice and common bean by applying foliar zinc fertilizer along with pesticides in seven countries. *Plant and Soil*, 403(1–2), 389–401. https://doi.org/10.1007/s11104-016-2815-3

Ramesh, P., Rajendran, A., & Meenakshisundaram, M. (2014). Green syntheis of zinc oxide nanoparticles using flower extract Cassia auriculata. *Journal of Nanoscience and Nanotechnology*, 2(1), 41–45

Ramkissoon, C., Degryse, F., da Silva, R. C., Baird, R., Young, S. D., Bailey, E. H., & McLaughlin, M. J. (2019). Improving the efficacy of selenium fertilizers for wheat biofortification. *Scientific Reports*, 9(1), 1–9. https://doi.org/10.1038/s41598-019-55914-0

Rana, A., Joshi, M., Prasanna, R., Shivay, Y. S., & Nain, L. (2012). Biofortification of wheat through inoculation of plant growth promoting rhizobacteria and cyanobacteria. *European Journal of Soil Biology*, 50, 118–126. https://doi.org/10.1016/j.ejsobi.2012.01.005

Reed, S., Neuman, H., Moscovich, S., Glahn, R. P., Koren, O., & Tako, E. (2015). Chronic zinc deficiency alters chick gut microbiota composition and function. *Nutrients*, 7(12), 9768–9784. https://doi.org/10.3390/nu7125497

Ritchie, H., & Roser, M. (2017). Micronutrient deficiency. Our World in data. https://ourworldindata.org/micronutrient-deficiency?utm_medium=syndication [Accessed 16 Mar 2022].

Rossi, L., Fedenia, L. N., Sharifan, H., Ma, X., & Lombardini, L. (2019). Effects of foliar application of zinc sulfate and zinc nanoparticles in coffee (*Coffea arabica* L.) plants. *Plant Physiology and Biochemistry*, 135, 160–166. https://doi.org/10.1016/j.plaphy.2018.12.005

Rouphael, Y., & Colla, G. (2018). Synergistic biostimulatory action: Designing the next generation of plant biostimulants for sustainable agriculture. *Frontiers in Plant Science*, 9, 1655. https://doi.org/10.3389/fpls.2018.01655

Saini, D. K., Devi, P., & Kaushik, P. (2020). Advances in genomic interventions for wheat biofortification: A review. *Agronomy*, 10(1), 62. https://doi.org/10.3390/agronomy10010062

Sharma, P., Reddy, P. K., & Kumar, B. (2021). Trace element zinc, a nature's gift to fight unprecedented global pandemic COVID-19. *Biological Trace Element Research*, 199(9), 3213–3221. https://doi.org/10.1007/s12011-020-02462-8

Sida-Arreola, J. P., Sánchez, E., Ojeda-Barrios, D. L., Ávila-Quezada, G. D., Flores-Córdova, M. A., Márquez-Quiroz, C., & Preciado-Rangel, P. (2017). Can biofortification of zinc improve the antioxidant capacity and nutritional quality of beans? *Emirates Journal of Food and Agriculture*, 29(3), 237–241. https://doi.org/10.9755/ejfa.2016-04-367

Siddiqui, Z. A., Parveen, A., Ahmad, L., & Hashem, A. (2019). Effects of graphene oxide and zinc oxide nanoparticles on growth, chlorophyll, carotenoids, proline contents and diseases of carrot. *Scientia Horticulturae*, 249, 374–382. https://doi.org/10.1016/j.scienta.2019.01.054

Singh, M. K., & Prasad, S. K. (2014). Agronomic aspects of zinc biofortification in rice (Oryza sativa L.). *Proceedings of the National Academy of Sciences India Section B: biological Sciences*, 84(3), 613–623. https://doi.org/10.1007/s40011-014-0329-4

Solanki, P., & Laura, J. S. (2018). Biofortification of crops using nanoparticles to alleviate plant and human Zn deficiency: A review. *Research Journal of Life Sciences, Bioinformatics, Pharmaceutical and Chemical Sciences*, 4(5), 364–385.

Solanki, R. G., & Rajaram, P. (2017). Structural, optical and morphological properties of CdS nanoparticles synthesized using hydrazine hydrate as a complexing agent. *Nano-Structures and Nano-Objects*, 12, 157–165. https://doi.org/10.1016/j.nanoso.2017.10.003

Song, E. K., Moser, D. K., Kang, S. M., & Lennie, T. A. (2015). Association of depressive symptoms and micronutrient deficiency with cardiac event–free survival in patients with heart failure. *Journal of Cardiac Failure*, 21(12), 945–951. https://doi.org/10.1016/j.cardfail.2015.10.007

Sturikova, H., Krystofova, O., Huska, D., & Adam, V. (2018). Zinc, zinc nanoparticles and plants. *Journal of Hazardous Materials*, 349, 101–110. https://doi.org/10.1016/j.jhazmat.2018.01.040

Subbaiah, L. V., Prasad, T. N. V. K. V., Krishna, T. G., Sudhakar, P., Reddy, B. R., & Pradeep, T. (2016). Novel effects of nanoparticulate delivery of zinc on growth, productivity, and zinc biofortification in maize (*Zea mays* L.). *Journal of Agricultural and Food Chemistry*, 64(19), 3778–3788. https://doi.org/10.1021/acs.jafc.6b00838

Subramanian, K. S., Manikandan, A., Thirunavukkarasu, M., & Rahale, C. S. (2015). Nano-fertilizers for balanced crop nutrition. In: *Nanotechnologies in Food and Agriculture* (pp. 69–80). Springer, Cham.

Sudha, P. N., Sangeetha, K., Vijayalakshmi, K., & Barhoum, A. (2018). Nanomaterials history, classification, unique properties, production and market. In: *Emerging Applications of Nanoparticles and Architecture Nanostructures*. Eds: Barhoum, A., & Hamdy-Makhlouf, A. S. (pp. 341–384). Elsevier, Amsterdam.

Szerement, J., Szatanik-Kloc, A., Mokrzycki, J., & Mierzwa-Hersztek, M. (2021). Agronomic biofortification with Se, Zn, and Fe: An effective strategy to enhance crop nutritional quality and stress defense—A review. *Journal of Soil Science and Plant Nutrition*, 1–31. https://doi.org/10.1007/s42729-021-00719-2

Taboada-Lugo, N. (2017). El Zinc y el cobre: Micronutrientes esenciales para la salud humana. *Acta Médica del Centro*, 11(2), 79–89

Timoneda, J., Rodríguez-Fernández, L., Zaragozá, R., Marín, M. P., Cabezuelo, M. T., Torres, L., Viña, J. R., & Barber, T. (2018). Vitamin A deficiency and the lung. *Nutrients*, 10(9), 1132. https://doi.org/10.3390/nu10091132

Tulchinsky, T. H. (2010). Micronutrient deficiency conditions: Global health issues. *Public Health Reviews*, 32(1), 243. https://doi.org/10.1007/BF03391600

Umar, W., Hameed, M. K., Aziz, T., Maqsood, M. A., Bilal, H. M., & Rasheed, N. (2021). Synthesis, characterization and application of ZnO nanoparticles for improved growth and Zn biofortification in maize. *Archives of Agronomy and Soil Science*, 67(9), 1164–1176. https://doi.org/10.1080/03650340.2020.1782893

UNICEF (2021a). Feed to fall. *Fondo de las Naciones Unidas para la Infancia (online)*. https://www.unicef.org/reports/fed-to-fail-child-nutrition [Accessed 15 Mar 2022].

UNICEF (2021b). Nutrition. *Fondo de las Naciones Unidas para la Infancia (online)*. https://www.unicef.org/nutrition [Accessed 16 Mar. 2022].

Velázquez-Gamboa, M. C., Rodríguez-Hernández, L., Abud-Archila, M., Gutiérrez-Miceli, F. A., González-Mendoza, D., Valdez-Salas, B., González-Terreros, E., & Luján-Hidalgo, M. C. (2021). Agronomic biofortification of *Stevia rebaudiana* with zinc oxide (ZnO) phytonanoparticles and antioxidant compounds. *Sugar Tech*, 23(2), 453–460. https://doi.org/10.1007/s12355-020-00897-w

Velu, G., Ortiz-Monasterio, I., Cakmak, I., Hao, Y., & Singh, R. Á. (2014). Biofortification strategies to increase grain zinc and iron concentrations in wheat. *Journal of Cereal Science*, 59(3), 365–372. https://doi.org/10.1016/j.jcs.2013.09.001

Wakeel, A., Farooq, M., Bashir, K., & Ozturk, L. (2018). Micronutrient malnutrition and biofortification: Recent advances and future perspectives. *Plant Micronutrient Use Efficiency*, 225–243. https://doi.org/10.1016/B978-0-12-812104-7.00017-4

Wessells, K. R., & Brown, K. H. (2012). Estimating the global prevalence of zinc deficiency: Results based on zinc availability in national food supplies and the prevalence of stunting. *PLOS ONE*, 7(11), e50568. https://doi.org/10.1371/journal.pone.0050568

White, P. J., & Broadley, M. R. (2005). Biofortifying crops with essential mineral elements. *Trends in Plant Science*, 10(12), 586–593. https://doi.org/10.1016/j.tplants.2005.10.001

White, P. J., & Broadley, M. R. (2009). Biofortification of crops with seven mineral elements often lacking in human diets–iron, zinc, copper, calcium, magnesium, selenium and iodine. *New Phytologist*, 182(1), 49–84. https://doi.org/10.1111/j.1469-8137.2008.02738.x

WHO. (2021). Malnutrición. Organización Mundial de la Salud [online]. https://www.who.int/es/news-room/fact-sheets/detail/malnutrition [Accessed 20 Feb 2022].

Yadav, K., & Pandav, C. S. (2018). National iodine deficiency disorders control programme: Current status & future strategy. *The Indian Journal of Medical Research*, 148(5), 503. https://doi.org/10.4103%2Fijmr.IJMR_1717_18

Yang, G., Yuan, H., Ji, H., Liu, H., Zhang, Y., Wang, G., Chen, L., & Guo, Z. (2021). Effect of ZnO nanoparticles on the productivity, Zn biofortification, and nutritional quality of rice in a life cycle study. *Plant Physiology and Biochemistry*, 163, 87–94. https://doi.org/10.1016/j.plaphy.2021.03.053

Zhang, Y., Shi, R., Rezaul, K. M., Zhang, F., & Zou, C. (2010). Iron and zinc concentrations in grain and flour of winter wheat as affected by foliar application. *Journal of Agricultural and Food Chemistry*, 58(23), 12268–12274. https://doi.org/10.1021/jf103039k

Zulfiqar, F., Navarro, M., Ashraf, M., Akram, N. A., & Munné-Bosch, S. (2019). Nanofertilizer use for sustainable agriculture: Advantages and limitations. *Plant Science*, 289, 110270. https://doi.org/10.1016/j.plantsci.2019.110270

Zulfiqar, U., Hussain, S., Ishfaq, M., Ali, N., Ahmad, M., Ihsan, F., Sheteiwy, M. S., Rauf, A., Hano, C., & El-Esawi, M. A. (2021). Manganese supply improves bread wheat productivity, economic returns and grain biofortification under conventional and no tillage systems. *Agriculture*, 11, 142. https://doi.org/10.3390/agriculture11020142

7 Green Synthesis of Nanoparticles and Applications

Arshi Anees, Rishil Gupta, Phanindra P V, Oluwatoyin Adenike Fabiyi, Uday Kumar Thera, Tesleem Taye Bello, and Faheem Ahmad

7.1 INTRODUCTION

Due to the increasing importance and wide range of applications of nanoparticles (NPs), their production has been on the increase in recent years, and workers have employed different approaches and ways of achieving the synthesis of these nanoparticles (Fabiyi et al., 2018). Chemical synthesis of metal nanoparticles has been the most commonly utilized method, but due to the fact that it is usually costly, consumes high amounts of energy, and involves the use of toxic reagents as stabilizing agents during synthesis, it has given rise to a series of environmental concerns. Furthermore, harmful effects of chemically synthesized nanoparticles have been reported from biomedical applications as a result of the accumulation of chemical residues (Shanker et al., 2003; Noruzi et al., 2011). Thus, there arose the need to develop easier, ecologically sound, and cost-effective alternatives to chemical methods. Recently, the process referred to as green synthesis, which basically involves the use of naturally occurring reducing agents such as plant tissues, microbes (fungi, bacteria, and actinomyces), and enzymes, has provided a cost-effective and ecologically friendly alternative in this regard (Abdel-Aziz et al., 2018; Marooufpour et al., 2019).

7.2 SYNTHESIS OF NANOPARTICLES VIA THE GREEN ROUTE

Preparation of nanoparticles via the green route refers to the chemical-free, environmentally benign, and costless method of utilizing natural resources such as plants and microorganisms as reducing and stabilizing agents to produce nanoparticles (Marooufpour et al., 2019; Stone et al., 2020; Fabiyi et al., 2020a). This method, due to its non-toxic and highly stable nature, has attracted a lot of attention, and numerous global studies have been conducted to demonstrate the exceptional efficacy of several plants and microbes involved in producing nanoparticles from metals (Devi et al., 2019). The green synthesis method for the production of nanoparticles is highly diversified and employs plants, bacteria, fungi, viruses, and their products to model the process of synthesis and characterization of well-defined and structured nanoparticles (Gahlawat and Choudhury, 2019; Stone et al., 2020). These natural organisms and their products are introduced into a metallic salt, leading to a biological reduction that converts the metallic materials to nanoparticles (Mittal et al., 2013). When compared with the use of microbes, plant-mediated synthesis of nanoparticles offers a more convenient approach, since reports have shown that plant extracts possess the capacity to diminish metallic ions faster than microbes and also produce more stable metal nanoparticles (Korbekandi et al., 2016; Prasad et al., 2018; Fabiyi et al., 2020b). The biogenic fusion of nanoparticles using microbes is achieved through a green chemistry approach, which operates at low ambient temperatures and pressures and favors simpler downstream processing at a reduced cost (Abdel-Aziz et al., 2018; Prasad et al., 2018). The microbial blend of metal nanoparticles is achieved in two ways: intracellularly and extracellularly (Jain et al., 2011). The former method utilizes suitable detergents and

DOI: 10.1201/b23308-7

special ultrasound treatment to produce synthesized nanoparticles (Kalimuthu et al., 2008), while the latter method is costless, requiring easier downstream reactions, which explains the reason why more researchers now concentrate more on the use of extracellular methods in achieving a microbial synthesis of nanoparticles (Prasad et al., 2018).

7.2.1 Protocols for Green Synthesis of Nanoparticles

The green synthesis of nanoparticles refers to a method of synthesizing nanoparticles using environmentally friendly and sustainable approaches, typically utilizing natural sources such as plant extracts, microorganisms, or other biomaterials. Here is a general protocol for the green synthesis of nanoparticles:

1. *Selection of the biological source:* Choose a suitable biological source that contains phytochemicals or biomolecules capable of reducing and stabilizing nanoparticles. This can include plant extracts, microbial cultures, or other natural sources.
2. *Preparation of the biological extract:* Collect the biological material and process it to obtain an extract. For plant extracts, this typically involves washing, drying, and grinding the plant material, followed by extraction using a suitable solvent (e.g., water, ethanol) or a combination of solvents. For microorganisms, culture them under appropriate conditions to obtain biomass or filtrate.
3. *Nanoparticle synthesis:* Mix the biological extract with a precursor solution containing metal salts (e.g., silver nitrate, gold chloride) or other compounds that will serve as the source for nanoparticle formation. The phytochemicals or biomolecules present in the biological extract act as reducing and stabilizing agents for the nanoparticles.
4. *Reaction conditions:* The reaction conditions, such as temperature and pH, can influence the nanoparticle synthesis. Optimize these parameters based on the specific requirements of the nanoparticles being synthesized and the characteristics of the biological source.
5. *Characterization:* After the reaction, characterize the synthesized nanoparticles using appropriate techniques such as UV-Vis spectroscopy, X-ray diffraction (XRD), transmission electron microscopy (TEM), dynamic light scattering (DLS), or Fourier transform infrared spectroscopy (FTIR). These techniques will help determine the size, shape, structure, and stability of the nanoparticles.
6. *Purification and stabilization:* Depending on the synthesis method and desired application, it may be necessary to purify the synthesized nanoparticles to remove impurities. This can be achieved through centrifugation, filtration, or other purification techniques. Stabilization of the nanoparticles can be achieved by adding capping agents or dispersing agents to prevent agglomeration.
7. *Application-specific modifications:* The synthesized nanoparticles can be further functionalized or modified for specific applications by adding surface coatings or functional groups or conjugating them with biomolecules.

It's important to note that the exact protocol may vary depending on the specific nanoparticles being synthesized, the biological source used, and the desired application.

7.2.2 Synthesis of Nanoparticles via Plants

Extracts from green plants extracts generally are known to contain several compounds, such as proteins, organic acids, polysaccharides, and phytochemicals like flavonoids, terpenoids, alkaloids, and polyphenols, as well as hydrocarbons, that can serve as reducing and stabilizing agents for nanoparticle production (Can, 2020; Fabiyi, 2021). A wide variety of plant species have proved effective in their abilities to accumulate metallic ions on the surface and in different organs, which they subsequently reduce and transform into nanoparticles through a series of complex synthesis steps (Makarov et al., 2014). In the last few decades, several in vitro methods have been adopted, which involve utilizing plant materials as reducing agents for the synthesis of nanoparticles. Extracts of various plant species obtained and prepared from dissimilar parts of the plants, like stem, roots, leaves, barks, flowers, seeds, and fruits, in combination with acids and metal salts like gold, silver, iron, copper, platinum, and copper, have been deployed for these purposes (Gao et al., 2016; Kumar et al., 2017; Fabiyi et al., 2021). A number of green composited nanoparticles obtained from dissimilar plant species have been documented lately, for example, extracts from plant parts such as leaves of different common plant species like green tea (*Camellia sinensis*) (Nune et al., 2009), neem (*Azadirachta indica*) (Shukla et al., 2012), acalypha (*Acalypha indica*), *Cnidoscolus aconitifolius* (Fabiyi, 2021), aloe species (Chandran et al., 2006), lemon grass (Shankar et al., 2005), Jatropha (Jogleka et al., 2011), and *Citrus* spp. (Veeraputhiran, 2013) have been used in the green synthesis of several metal nanoparticles. Plant-mediated synthesis of nanoparticles is achieved through the reduction of metallic ions in combination with biomolecules like terpenoids, carboxylic acids, ketones, aldehydes, enzymes, flavonoids, and amides (Masum et al., 2019; Abdallah et al., 2019; Ogunyemi et al., 2019), with plant extracts acting as reducing and stabilizing agents in the process of blending the nanoparticles (Hossain et al., 2019). The plant-mediated process for the synthesis of nanoparticles is regarded as being more beneficial than the microbe-based or chemical methods, since it is devoid of any hazardous effects of microbes and chemical contamination (Devi et al., 2019). Also, the production of smaller-sized particles has been attributed to products of green synthesis where, for example, iron nanoparticles of 2–80 nm were produced via green synthesis compared with 87–400 nm nanoparticles obtained from the same metal when prepared through the wet chemical method (Gokila et al., 2021). Furthermore, the process of green synthesis requires less energy, thereby making it easier to accomplish the task of reducing metallic ions to well-structured nanoparticles (Ren, 2019; Gokila et al., 2021).

7.2.3 Synthesis of Nanoparticles via Fungi

Fungi include microorganisms such as molds, mushrooms, and yeasts. They are eukaryotic organisms that are found abundantly worldwide. They perform important roles such as organic matter decomposition, nutrient recycling, and exchange within the environment, since they exist mostly as symbionts of animals, plants, and other fungi as well as parasites. Several species of fungi have been engaged in the microbial blending of metal nanoparticles, which is ascribed to their ability to secrete large amounts of the proteins required for this purpose (Manjunath et al., 2017). Various species of fungi have been successfully deployed in the green blending of several metal nanoparticles, resulting in the production of differently sized and shaped nanoparticles. For example, fungi-mediated biosynthesis of gold and silver nanoparticles by species of *Penicillium* and *Aspergillus* has been reported (Honary et al., 2013). *Penicillium aculeatum* was found to produce spherical nanoparticles of gold with a size of just about 60 nm (Barabadi et al., 2017; Shen et al., 2017). Also, a mitosporic species of the *Aspergillus* strain produced nanoparticles with a size of 4–29 nm. Furthermore, common mushrooms such as *Ganoderma lucidum*, *Lentinus edodes*, and *Pleurotus ostreatus* have been used to synthesize some metal nanoparticles (Vetchinkina et al., 2017). Yeasts belong to single-celled eukaryotic organisms, which are grouped as fungi. The biosynthesis of metal nanoparticles by *Saccharomyces cerevisiae* (Lim et al., 2011) and *Candida guilliermondii* (Mishra et al., 2011) has been reported. Another example is the production of 15-nm triangular and hexagonal nanoparticles by the intracellular activity of *Yarrowia lipolytica* (Agnihotri et al., 2009). A more recent report is the process of producing gold nanoparticles through biological means by the non-conventional yeast *Magnusiomyces ingens* which produced spherical, triangular, and hexagonal nanoparticles in the region of 80.1 nm (Zhang et al., 2016). Generally, the green synthesis of nanoparticles using fungi usually involves culturing the fungi to obtain a mycelia-free filtrate, which would be mixed and incubated with the metallic salts to serve as reducing agents that produce the nanoparticles. Although the presence of certain proteins and polyphenols in most fungi is largely responsible for their bio-reduction and stabilizing potential, many authors believe that certain electrostatic forces play a vital role in determining the stability of most nanoparticles produced by microbial biosynthesis. The different shapes produced on biosynthesized nanoparticles were found to be largely due to the capping agents employed, as determined by their individual binding properties (Graily-Moradi et al., 2020).

7.2.4 Synthesis of Nanoparticles via Bacteria

Bacteria are grouped into Prokaryotes, which are unicellular organisms with a simple structure that lacks a nucleus. Green synthesis of several metallic ions to produce different shapes and sizes of nanoparticles has been attained through the utilization of different bacterial strains: *Pseudomonas* species (Husseiny et al., 2007; Rajasree and Suman, 2012), *Bacillus* (Guo et al., 2013), *Lactobacillus* (Kanmani and Lim, 2013), *Trichoderma* (Ponmurugan, 2017), *Salmonella* (Mortazavi et al., 2017), and *Streptomyces* (Wypij et al., 2018). Fusion of nanoparticles using bacteria is achieved through two pathways: intracellular and extracellular, based on the site of the nanoparticles' synthesis. The intercellular method involves the transfer of the metallic ions into the bacterial cell to achieve nanoparticles in combination with enzymes, while in the extracellular method, nanoparticles are formed by trapping the metallic ions on the bacterial cell membrane by an enzyme (Kasthuri et al., 2009; Hulkoti and Taranath, 2014). However, intracellular methods have produced smaller-sized nanoparticles when compared with those produced using extracellular methods (Narayanan and Sakthivel, 2010). The bacterial cell wall is composed of peptidoglycan, which supplies polyanions for the required stoichiometric action between the metallic ions and the reactive chemical groups of the wall, to be followed by inorganic deposition of metal. The bacterial cell wall has innumerable latent metal binding points, which play a significant role during the green synthesis process. Metal ions must penetrate the cell wall into the cytoplasm before being transferred back through the cell wall meshwork to achieve extracellular liberation (Beveridge and Murray, 1980). These reports have demonstrated that successful bacterial-mediated nanoparticle synthesis largely depends on pH and temperature (Marooufpour et al., 2019). However, unlike fungi, culturing bacteria is tedious and requires extra precautions. Also, the reaction time required for the reduction process is long, ranging from hours to days, and this has been a major constraint limiting the bacterial-mediated synthesis of nanoparticles on a commercial scale.

7.2.5 Synthesis of Nanoparticles via Algae

Algae are a group of photosynthetic eukaryotic microorganisms. They are known to possess membranous organelles and a nuclear envelope. They exist in almost all habitats that have light and moisture but are mostly found on the surface of fresh or marine water as well as on moist rock surfaces. They are broadly grouped into macroalgae and microalgae (Oscar et al., 2014). They possess a cell wall that contains proteins, enzymes, and polysaccharides with characteristics of reducing agents, which has made them ideal for use in the biosynthesis of metal nanoparticles. Furthermore, they are easier to handle due to their capacity to synthesize nanoparticles at low temperatures with very low toxicity, which makes them easy to use (Lewis Oscar et al., 2016). The successful synthesis of metals into nanoparticles has been recorded using different groups of algae, like *Chlorophyceae*, *Rhodophyceae*, *Cyanophyceae*, and *Phaeophyceae* (Aziz et al., 2015; Sharma et al., 2015). Some algae have been utilized in the green synthesis of nanoparticles from various metal salts by the action of their functional groups and enzymes present in the cell wall. Also, several algae strains, such as *Spirullina platensis*, *Fucus vesiculosus*, and *Pitthophora oedogonia*, have been utilized as reducing agents in the processing of metal oxide nanoparticles (Siddiqi and Husen, 2016). Some other specific examples include the green synthesis of silver and gold nanoparticles by *Sargassum longifolium* (Rajeshkumar et al., 2014) and *Sargassum wightii* (Singaravelu et al., 2007), respectively. Also, the use of *Plectonema boryanum* to extracellularly and intracellularly produce cubic-shaped gold nanoparticles of sizes ranging between 9 and 25 nm has been reported (Lengke et al., 2006). A more recent report by Abdel-Raouf et al. (2017) detailed a successful biosynthesis of spherical gold nanoparticles, 3.9–77 nm in size, using *Galaxaura elongata*. Furthermore, the alga *Laurencia papillosa* was also successfully used to synthesize silver nanoparticles (Omar et al., 2017). The process of green synthesis of nanoparticles by algae is grouped into four possible pathways: the first method involves the use of biomolecules obtained from ruptured microalgae cells; the second method involves removal of supernatant after centrifugation and filtration of culture media from which cells have been previously removed; the third method involves the suspension of the whole microalgae in distilled water; the fourth involves the exploitation of live cells maintained in their normal culture (Dahoumane et al., 2016).

7.3 CHARACTERIZATION TECHNIQUES FOR NANOPARTICLES

The efficiency and applicability of any nanoparticle determine its application for various purposes. The utilization of a nanoparticle for a specific purpose is entirely based on its physical and chemical properties, which regulate biological effects (Reddy et al., 2012; Mughal et al., 2021). In this context, characterization is a necessary process that helps determine whether particles are at the nanoscale. It refers to the processes in which the different properties of nanoparticles are studied. Properties of nanoparticles related to morphology, physiochemical, optical, structural, and elemental characteristics, and size are determined by various technologies (Table 7.1). Newer and more efficient methods are becoming available with the advent of cutting-edge technologies (Titus et al., 2019). They help in studying the qualitative and quantitative parameters of nanoparticles. The upcoming sections briefly describe some of the most used methods, namely, microscopy-based, spectroscopy-based, X-ray-based, and dynamic light scattering-based techniques and zeta potential, for determining various properties of nanoparticles.

7.3.1 Microscopy-based Techniques

Optical microscopy, with limited resolution, is only helpful for micron-level observations. The nanoscale resolution is obtained by other technologically advanced imaging techniques like transmission electron microscopy (TEM), scanning tunnel microscopy (STM), atomic force microscopy (AFM), and scanning electron microscopy (SEM), which produce highly magnified images. The utilization of SEM enables the acquisition of surface-related data like the morphology and dispersion of nanoparticles (Leppard, 2008). SEM uses high-energy electron beams, emitted from either

TABLE 7.1
Properties of Nanoparticles

Properties	Studied parameters	Used techniques
Morphological	Shape, size, surface morphology, texture	Scanning electron microscopy, transmission electron microscopy, atomic force microscopy, scanning tunnel microscopy, dynamic light scattering (DLS)
Optical	Absorbance, reflectance, transmittance, photochemical features	UV-visible spectrophotometer, Raman spectroscopy
Structural	Phase, particle size, crystalline nature, surface area	X-ray diffractometry, Brunauer–Emmett–Teller (BET) analysis
Elemental	Composition of elements and their bonding nature	Fourier transform infrared spectroscopy (FT-IR), Raman spectroscopy, X-ray photoelectron spectroscopy
Electrical	Charge estimation	Zeta potential

field emission guns or thermionic guns, to scan the surface of a sample. It reveals the surfaces by producing high-resolution images of around 1–5 nm. During the scanning, X-rays are also emitted, which can be analyzed through a technique known as energy dispersive X-ray (EDX) analysis to identify the essential characteristics of the nanoparticles (Mohan and Renjanadevi, 2016).

In transmission electron microscopy (TEM), the sample is subjected to electron beam irradiation, resulting in an interaction between the beam and the sample. The resulting transmitted electrons are magnified and focused using different objective lenses to form an image. It provides information about the size and shape of nanoparticles. However, sample preparation requires high precision, as it requires a fragile foil of the sample so that electrons can penetrate through it (Werner et al.,1997). In addition, high-resolution transmission electron microscopy produces a crystallographic structure image of a sample at an atomic level. It can identify the defects in crystals as well as individual atoms.

Atomic force microscopy uses a laser beam to generate three-dimensional images at atomic or angstrom-scale resolution. The measurement of surface topography, roughness, texture, and size of various materials can be advantageous (Hoo et al.,2008). On the other hand, scanning tunnel microscopy produces surface images of conducting samples at atomic-scale resolution with the help of a piezoelectric crystal. Contact between the surface and conducting tip results in the tunneling of electrons through the vacuum between them.

7.3.2 Spectroscopy-based Techniques

Spectroscopic techniques use the interaction of nanoparticles with different electromagnetic radiations to determine their shape, size, and concentration (Nurmi et al., 2005). In addition, they reveal specific optical properties like reflectance, transmittance, photochemical features, and elemental composition of nanoparticles. The predominant spectroscopic methodologies employed in scientific research include ultraviolet-visible, Raman, and Fourier transform infrared spectroscopy. Brief descriptions of these techniques are provided in the following.

In a UV-visible spectrophotometer, samples are exposed to a fixed wavelength in the range of UV to visible light based on the maximum absorption level of the sample. When a light beam is passed through the solution, particles absorb a part of the light and transmit the rest. Absorbance increases with an increase in the concentration of the sample. Moreover, metal nanoparticles with a diameter of 100 nm scatter the light with great efficiency due to the plasmon resonance of their surface. The magnitude and bandwidth of the spectrum and wavelength peak of plasmon resonance depend on the nanoparticles' size, shape, and composition.

Raman spectroscopy uses inelastic or Raman scattering of monochromatic laser light to study vibrational, rotational, and other system modes (Slavov et al., 2010). The laser light interacts with phonons, elastic atoms in the nanoparticles, resulting in an energy shift. This energy shift provides typical information about the mode and arrangement of phonons in the system. As the laser beam illuminates the sample, molecules become excited and then relax to a vibrationally excited state. This change creates Raman scattering, and the electromagnetic radiation from the molecules/laser hit spots is analyzed. Advanced versions include surface-enhanced Raman spectroscopy, stimulated Raman, and resonance Raman spectroscopy.

Fourier transform infrared spectroscopy, as the name suggests, utilizes infrared light to scan the samples and reveal the details of material composition. The aforementioned methodology is additionally employed to characterize materials of unknown origin, detect the presence of impurities within a given material, ascertain the existence of additives, and identify instances of oxidation and decomposition within the sample. Infrared radiation of about 100–10000 cm^{-1} is sent through the sample. The sample absorbs some of the radiation, and the rest passes through. The radiation that is absorbed undergoes a conversion process into either rotational or vibrational energy, which is detected, and the resulting signal spectrum generally ranges from 400 to 4000 cm^{-1}. Each molecule has a specific spectrum representing the molecular fingerprint of the sample. Therefore, FTIR is an invaluable tool for identifying the chemical composition of particles (Taha et al., 2013).

7.3.3 X-Ray-based Techniques

Frequently used techniques are X-ray diffraction and X-ray photoelectron spectroscopy. X-ray diffraction defines the particle size, phase, crystallinity of the sample, and elemental proportions in mixture samples. In this technique, the sample is illuminated with X-rays that interact with the sample's atomic plane and result in the partial transmission of the beam. The rest of the beam is absorbed, scattered, refracted, and diffracted by the sample. The diffraction of X-rays is unique for each sample based on its type and arrangement of atoms.

X-ray photoelectron spectroscopy is a sensitive and quantitative spectroscopic method used to estimate the composition of particles and the bonding nature of their elements (Korin et al., 2017). The sample is irradiated with X-rays under an ultra-high vacuum, which leads to the emission of electrons from the elements at the surface. Measurement of the kinetic energy and the number of electrons escaped from the material provides information about the state of oxidation of the surface elements and the proportion of elements, respectively. The techniques can also be used to detect the contaminants present in the sample.

7.3.4 Dynamic Light Scattering (DLS)

The dynamic light scattering (DLS) technique, also called photon correlation spectroscopy, identifies nanoparticles and analyzes the size distribution of nanoparticles. This technique uses a laser beam to illuminate the sample, and the scattered light is analyzed. The light scattering is based on the Brownian motion of particles, random motion in every direction. Smaller particles will move faster and scatter less light than larger particles. The intensity modulation of scattered light as a function of time determines the particle size. Particles in the size range of 10 μm to 3 nm can be identified and measured. The particle size analysis by DLS in the liquid phase gives information about the hydrodynamic particle size of the nanoparticles (Statefeld et al., 2016).

7.3.5 Zeta Potential

Zeta potential measures electrostatic attraction or repulsion between particles in a suspension. The surface of a charged particle attracts oppositely charged ions, forming a thin layer around it, creating an electric double layer. Zeta potential is the electric potential of this double layer, which is determined by measuring the velocity of charged particles moving toward the electrode in the presence of an electric field (Xu, 2008). Zeta potential is analyzed by electrophoretic light scattering (ELS) and electroacoustic determination techniques. The magnitude of zeta potential is the best estimator of the physical stability of nanoparticles. Particles aggregate and form complexes at a lower magnitude of 0–5 mV. As the magnitude increases, the stability of the particles increases due to electrostatic repulsion. Magnitudes of 5–20 mV, 20–40 mV, and 40+ mV indicate particles with minimal, moderate, and higher stability, respectively.

7.4 MECHANISMS THAT FACILITATE NANOPARTICLE SYNTHESIS

Green synthesis of nanoparticles is a bottom-up approach where simpler atoms or molecules are arranged into complexes by various biological agents (Rana et al., 2020). These agents use their enzymatic machinery to reduce the metallic ions and stabilize them by adding carbohydrates, proteins, or other organic molecules. These agents can be microorganisms, including bacteria, fungi, algae, yeast, or plants acting as bio-nano factories.

7.4.1 Microorganism-based

Microorganisms employ different mechanisms for the synthesis of nanoparticles. Microorganisms adopt these mechanisms to survive under the heavy stress of different metals. Nanoparticles are by-products of these survival mechanisms, synthesized either intracellularly or extracellularly (Duran et al., 2011). Cells maintain homeostasis of metal ions, leading to the generation of nanoparticles through functions like energy production by chemolithotrophs, integrating metal ions into different cellular functions, and detoxification by efflux systems or transformation using alkylation/oxidation, reduction/methylation, bioaccumulation, and biosorption. Microorganisms' reduction of metals into nanoparticles includes various enzymes such as reductases, flavonoids, naphthoquinones, and anthraquinones. Specific biomolecules involving amino acids, peptides or proteins, enzymes, and aliphatic and aromatic compounds help reduce metal ions through the shuttle of electrons or charge-capping mechanisms (Ovais et al., 2018). Synthesis of Ag, Au, Au, and Ag alloy nanoparticles, Pt, Pb, Se, SiO_2, TiO_2, ZrO_2, and quantum dots using various biological agents has been reported so far (reviewed by Hussain et al., 2016).

7.4.2 Plant-Based

Plant extracts are mixed with solutions of precursor metal ions to synthesize nanoparticles mediated by plants. Plant leaf extracts contain various phytochemicals with great potential to reduce metal ions, which act as reducing and stabilizing agents to facilitate the synthesis (Jadoun et al., 2021). The main phytochemicals include terpenoids, flavones, sugars, ketones, carboxylic acids, aldehydes, and amides. The synthesis mechanism involves three phases: activation, growth phase, and process termination. In the activation phase, metal ions are reduced to form new structures through self-organization. In the next phase, these newly formed structures grow, and the reduction of metal ions further increases with the increased stability of the formed nanoparticles. In the last phase, the final shape of the nanoparticle is achieved.

7.5 NANOPARTICLES AND THEIR APPLICATIONS IN SUSTAINABLE AGRICULTURAL PRACTICES

The major foundation of the developing economy, agriculture supplies people with food and a higher quality of life. The field of agriculture is currently confronting a number of difficulties, including unexpected climate change, soil pollution by dangerous environmental pollutants like fertilizers and pesticides, and significantly increased food demand due to a growing worldwide population (Pouratashi and Iravani, 2012). In conventional agriculture, a lot of pesticides and fertilizers are used, which is bad for ecosystems and other living creatures. In order to preserve multiple species and protect the ecosystem, sustainable agriculture should prioritize minimizing the usage of agrochemicals. In this regard, agrochemical encapsulated nanocarrier systems, nano-plant growth promoters, nanopesticides, nanofertilizers, nanoherbicides, etc., have been created for possible applications in agriculture (Figure 7.1). These can offer significant advantages for agriculture, such as increased crop yield, prevention of plant diseases, and elimination of undesirable weeds and insects with reduced expense, energy use, and waste creation. The issues involved are methods for synthesis, how they penetrate applied surfaces, and the dangers they pose (Singh et al., 2021). Science and technological advances help the agricultural industry by supplying fresh approaches and answers to challenging issues. The advent of nanotechnology has facilitated the frequent generation of nanoformulations, which are formulations designed for sustainable agriculture. These nanoformulations exhibit enhanced efficacy and reduced levels of pollutants (Fraceto et al., 2016). Sustainable nanotechnology can be defined as the "[j]udicious use of nanotechnology that leads to the development of present without affecting the future generation needs" (World Commission on Environment and Development, Our Common Future, Oxford, 1987). In 2011, the Sustainable Nanotechnology Organization came into being. The phrase "triple bottom line," or TBL for short, was first used by John Elkington in 1999. It refers to the three fundamental pillars of sustainability, which are the environment, society, and economy (Elkington, 1999). An alternative to conventional, inefficient, cumbersome, and detrimental methods of plant production

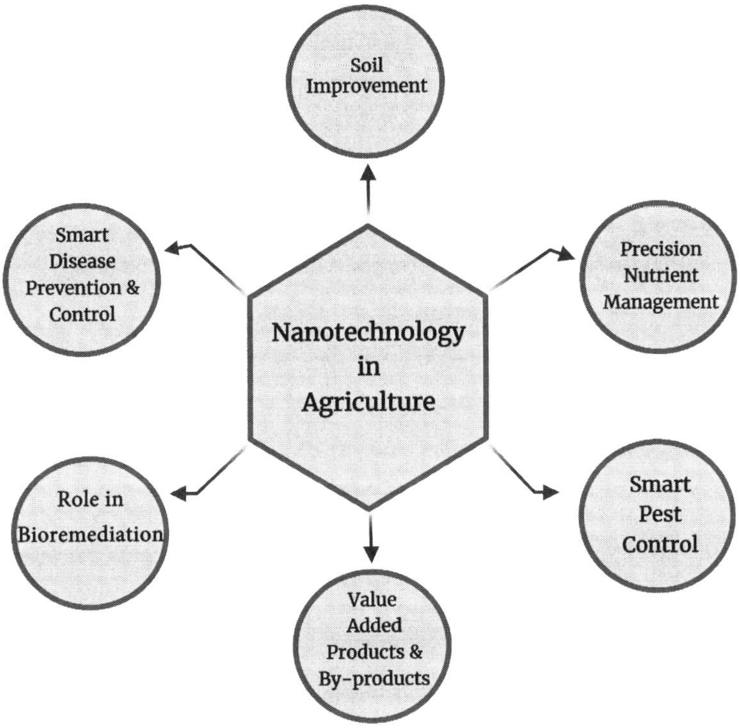

FIGURE 7.1 Various uses of nanotechnology in agriculture.

and protection could be provided with the use of nanotechnology-mediated agri-application for sustainable crop production (phytonanotechnology). This is due to the target-specificity and controlled-release systems of nano-derived fertilizers, insecticides, and herbicides, which deliver the biomolecules to key places (Chen, 2018). The interaction of NPs with the substrate on which they have been administered determines their precise function. The uptake and transport of NPs in crops directly affect their real, practical overall influence and impact (Sabo-Attwood et al., 2012; Taylor et al., 2014). Nanoparticles are essentially minute metal-based matter particles with the potential to function as an integrated array of various active principle–consisting particles to carry out functioning inside an investigational organism in order to make our chosen organism more efficient based on functioning across every key area (Baker et al., 2015). A crucial role in boosting plant development and crop productivity is played by nano-derived products for agricultural purposes. They are a notable alternative to conventional chemicals and practices for farmers because of a variety of advantages, including their small size, ease of handling, long-term storage, high effectiveness, and environmental friendliness. The commercialization of nanoparticles is now significantly expanding globally. For advancing sustainable nanotechnology applications, a novel nanoparticle has been created. It is referred to as "chitosan," and it is the most effective polymer for creating nanoparticles that can be employed as delivery systems. Several characteristics of chitosan nanoparticles include high permeability, biocompatibility, cost-effectiveness, biodegradability, etc. (Shukla et al., 2013). Nanotechnology has significantly benefited the agricultural sector by creating nanoscale transporters, biosensors, and manufactured xylem vessels. These are examples of some of the most significant advances in nanotechnology that have influenced the modernization of the agricultural sector. The utilisation of nanoparticles (NPs) in order to enhance several aspects of plant growth and development, such as germination rate, growth parameters, chlorophyll contents, cell elongation, nutrient uptake, the release of necessary chemicals, NPs solubility, microbial activities, crop yield, and the evaluation of phytotoxicity under both controlled and excessive doses, has been investigated (Arif et al.,2016).

7.5.1 Synthesis of Nanoparticles

Synthesizing nanomaterials or nanoparticles for use in agriculture can be done in a few different ways, including top-down, bottom-up, or through the use of biological techniques (Figure 7.2). The top-down methodology involves breaking down large, unorganized materials into smaller, more manageable pieces using nanotechnology assembly techniques. Milling the materials is the foundation of the top-down physical technique. These nanomaterials can only be controlled to a limited extent, and there is a higher concentration of impurities, which is a major limitation of this method. In the bottom-up strategy, the synthesis of nanoparticles through chemical reactions starts at the atomic or molecular scale. Since it is a synthetic process under chemical control, the particulate size can be better managed, and impurities can be minimized using this technique (Singh et al., 2014; Pradhan et al., 2017). Furthermore, in addition to chemical and physical methods, nanoparticles can also be created using the widely used biosynthesis technique. Plants, fungi, and microbes are just some of the many natural options available. This method is preferable because it allows more precise regulation of both particulate toxicity and size (El-Ramady et al., 2018; Yadav et al.,2012). The most effective approach for each application will need a synthesis that can produce large-scale particles with precise control over their physicochemical characteristics, resulting in a uniform and target-specific nanoformulation. For this reason, the bottom-up method is typically the most successful one used today for the synthesis of nanofertilizers (Raliya et al., 2017; Singh et al., 2014). To achieve precision agriculture, it is essential to maximize the efficacy with which these nanofertilizers are used. As such, nano-based smart technology, such as nanofertilizers, is anticipated to pave the way toward a more sustainable agricultural industry (Khot et al., 2012; Davarpanah et al., 2016).

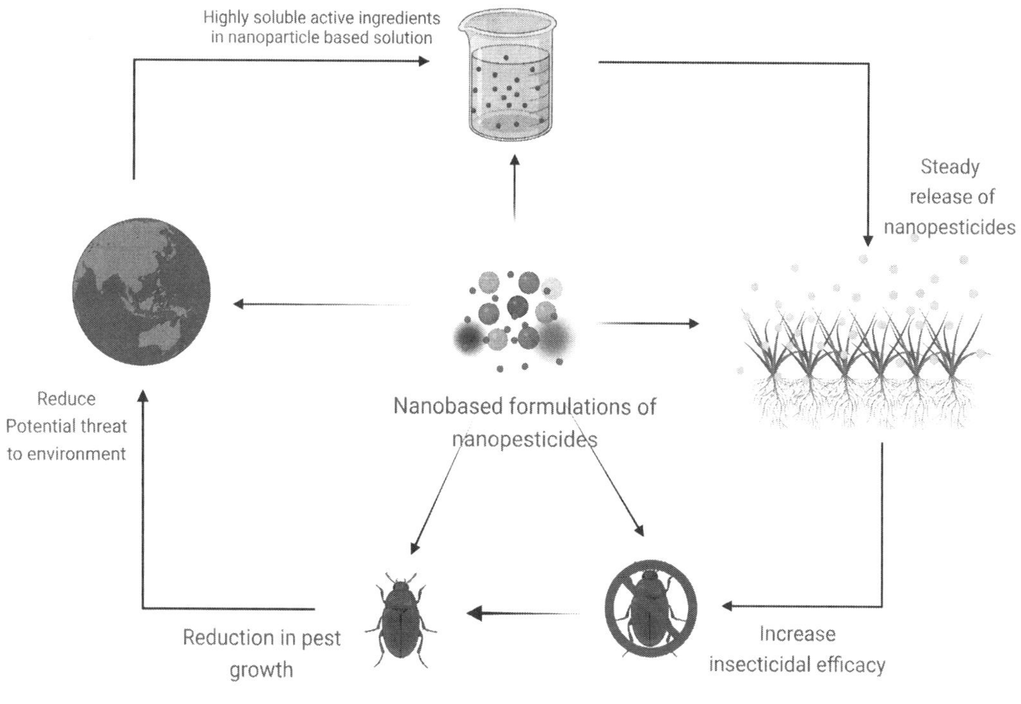

FIGURE 7.2 Nano-based formulations of nanopesticides and their mechanism of action.

These fertilizers have the potential to reduce nutrient pollution in the land and water and serve as a viable alternative to the large quantities of traditional fertilizers (Dimkpa et al., 2017). Distribution of controlled and efficient discharge of the contents for pinpointed agricultural applications can be achieved through the use of a variety of nanofertilizers, such as nitrogen, phosphorus, potassium, copper, manganese, zinc, and carbon nanotubes (Solanki et al., 2015; Liu et al., 2015). Organic and inorganic nanomaterials are used to synthesize nanoparticles. Furthermore, there is also a wide range of physical and chemical processes that can be used to synthesize them. Consequently, the two primary categories of nanofertilizers are those that focus on micronutrients and those that focus on macronutrients. Additionally, nanobiofertilizers are also on the rise as a complementary technique. Scientists are working hard to reduce the disparity between food supply and demand by implementing cutting-edge technological solutions. Through the development of nanofertilizers, the application of more effective herbicides and pesticides, the control of soil characteristics, the management of wastewater, and the detection of pathogens, nanotechnology is one of the potential technologies that could increase agricultural output (Neme et al., 2021).

7.5.2 Nanofertilizers – Smart Nanoformulations

The application of fertilizers is central to increasing crop yields and overall agricultural output. However, only 30–40% of nutrients actually make it to the crops, with the rest being lost to evaporation, leaching, and degradation. As a result, fertilizer effectiveness will decline, and farming costs will rise. Moreover, the wasteful application of fertilizers endangers all forms of life in the ecosystem by polluting its land, water, and air. Inefficient farming practices, such as over-fertilization, have led to a decline in soil organic matter and a corresponding drop in agricultural yields. In order to combat contamination over the long term and improve crop yields in the meantime, nanofertilizers may prove to be the most effective solution. Nanofertilizers are currently being synthesized in order to regulate the release of nutrients based on the needs of crops and are thus expected to be more effective than traditional fertilizers (Liu et al., 2006). Poor use efficacy is a major problem of typical fertilizers (Tilman et al.,2002). When applied traditionally, nitrogenous fertilizer has an effective range of only 30–60% in terms of its role as a fertilizer. Losses of 8–90% of phosphatic nutrients occur as a result of chemical bonding in the soil (Abdel-aziz et al., 2018). Statistics show that a staggering 40–70% of the nitrogen gas, 80–90% of the phosphorus, and 50–90% of the potassium in the fertilizers used are lost to the atmosphere, which results in a significant decline in economic activity (Trenkel, 2010; Solanki et al., 2015). These fertilizers have numerous negative consequences for the environment, including eutrophication (Tilman et al., 2002).

Nanofertilizers are nanoformulations of micro- and macronutrients like iron, zinc, nitrogen, magnesium, potassium, phosphorus, manganese, and others that improve plant growth (Dimkpa et al., 2017). As a result of their nanoscale, nanomaterials with a large surface area can potentially solve this challenge. These could be used as nanocoatings, such as sulfur nanocoatings (with a layer thickness of ≤100–nm), which would guarantee a controlled release of the substance, surface protection, and eventually an increased use efficiency (Brady and Weil, 1996). Another potential benefit of controlled-release fertilizers is that they mitigate the toxic effects caused by using excessive amounts of conventional chemical fertilizer on a given area of soil (Suman et al., 2015). By encapsulating individual fertilizer fragments within nanoparticle membranes, a slow and consistent release of nutrients from the fertilizers can be achieved. Nanocomposites containing macro- (N, P, and K) and micronutrients increase the uptake of nutrients by cereal crops.

When compared with conventional fertilizers, nanocomposite fertilizers made from hydroxyapatite and urea are found to have sustained phosphorus availability for up to four weeks after application, with optimum nitrogen release and ammonia volatilization (Mejías et al., 2021). This is made possible by the hydroxyapatite crystal structure having a large available surface area, which in combination with the strong bonding that exists between the crystals, results in a gradual release of nitrogen from the urea (Broda et al., 2022; Shenoy et al., 2021). Kah et al. (2018) demonstrated that compared with traditional fertilizers, nanofertilizers had been shown to improve median efficacy by 18–29%. Liu and Lal found that compared with conventional fertilizers, phosphatic nanofertilizers boost development rates by 32%. According to a study by Syu et al. (2014), nanoparticles alter impulses transmitted within the roots. Roots' internal signals for nutrient release in reaction to P and N deficiency would represent a promising new avenue for achieving controlled release of nanofertilizers. With nanofertilizers, controlled release can be connected to factors like time or weather. With slow-release fertilizers, nutrients are released gradually throughout a crop's development, allowing plants to absorb as much as possible without waste. In addition, nanofertilizers boost soil quality by reducing the harmful impacts of excessive fertilizer application (Suman et al., 2015). Increased photosynthesis rate and translocation to various parts of the plant are just two of the many ways in which nanofertilizers outperform traditional fertilizers. Other ways in which nanofertilizers benefit plants include an increase in plant height and leaf area, as well as a decrease in dry mass and the rate at which chlorophyll is produced (Ali and Al-Juthery, 2017; Singh, 2017; Salama et al., 2021).

Materials with nanoscale structures are utilized in the production of nanofertilizers. These include chitin, zeolite, polyacrylic acid, clay minerals, and hydroxyapatite. They can be spread on the surface of the soil as fertilizer or sprinkled directly onto the leaves of the plants in a process known as a foliar application (Figure 7.3) (Iqbal et al., 2019)

Typically, fertilizers are encased in nanoparticles to create nanofertilizers. The other methods include:

1. Incorporating the nutrient within nanoporous substances
2. Coating with a covering of thin polymer layer
3. Providing as nanoparticles or nanoemulsions (Rai et al., 2012)

Up until now, the majority of the proposed nanofertilizers or nanopesticides have involved the reformulation of registered bioactive ingredients. This is done in order to achieve enhanced performance in comparison to the active constituents that are currently in use, as well as to surmount the major drawbacks associated with the use of agrochemicals in actual agricultural settings (Zhang et al., 2021).

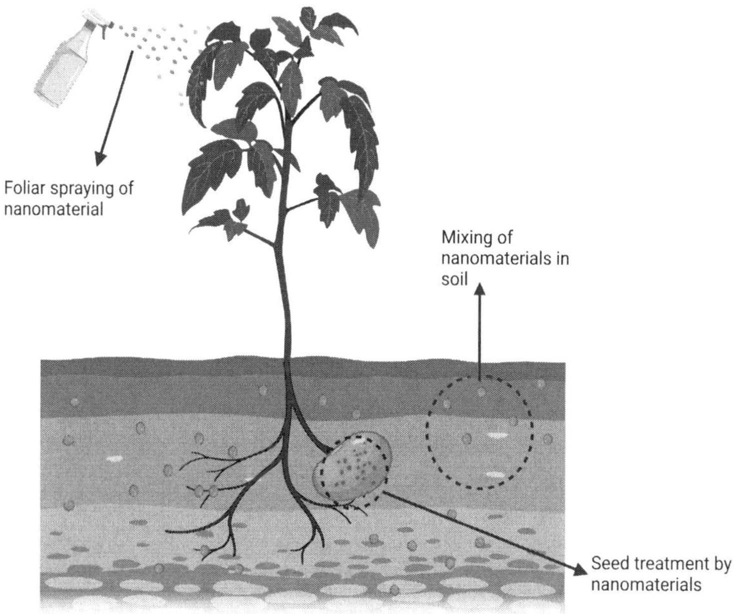

FIGURE 7.3 Uptake of nanomaterial by seed, plant, and soil.

7.5.3 Nanoparticles in Plant Protection

Nanotechnology has also benefited plant security measures, making them more effective against insects, other pests, and weeds. Possible applications for nanoparticles include novel forms of pesticides, insecticides, and insect repellents (Barik et al., 2008; Gajbhiye et al., 2009). Nanotechnology has a broad range of applications in making the agricultural sector more sustainable, including nanofungicides, nanoherbicides, nanoinsecticides, nanofertilizers, nanomicronutrients, nanobiopesticides, nanobiosensors, etc.

7.5.3.1 Nanoparticles as Nanopesticides

Annually, pests and diseases affecting plants cause a global harvest loss of 20–40% (Flood, 2010). Agriculture commonly incorporates pesticide use into its routine in order to increase crop yields. There is a constant effort to create new, original pesticides with a higher rate of success against certain types of pests. Therefore, year after year, a wide variety of pesticides are being developed (Natrayan et al., 2019). When pesticides are applied to crops, only a small amount (0.1%) of the pesticides actually reach the pests they are intended to kill. The rest of the pesticides (99.99%) end up contaminating the environment. This contamination can have serious consequences for human health and the food chain. Additionally, the widespread use of pesticides can lead to the development of pesticide-resistant pathogens, weeds, and insects (Gamal et al., 2018). Using biopesticides instead of conventional pesticides can mitigate these negative outcomes. Although they are making some headway, the efficacy of biopesticides is still limited, as biopesticides' effectiveness against pests is primarily based on the environment. Since the nanopesticides release their active ingredients slowly and in a controlled fashion, they can provide effective insect control for an extended period (Hashem et al.,2018). To effectively and sustainably manage various pests while reducing the use of synthetic pesticides and the associated environmental dangers, nanopesticides are crucial.

The use of pesticides, such as insecticides, fungicides, and herbicides, is integral to contemporary pest management strategies in agriculture. Effective, low-cost, environmentally friendly pesticides must be developed. Pesticides may benefit from the application of novel ideas like nanotechnology in ways that are good for the environment. These include lowering their toxicity, extending their storage life, and making them more water-soluble (Mali et al., 2020; Worrall et al., 2018). Insect pests and host diseases may also be thwarted with the help of nanoparticles (Khot et al., 2012). For the production of nanoinsecticides, polysaccharides like chitosan, alginates, starch, and polyesters have been explored (Mali et al., 2020). Generally speaking, there are two ways in which nanomaterials can be used to safeguard plants: either using nanoparticles as a form of self-protecting field armor or spraying existing herbicides that have been modified to use nanoparticles as carriers is a viable option (Worrall et al., 2018). Nanomaterials have many potential applications, but they are only beginning to be studied for use in crop protection and food supply (Prasad et al., 2017). When compared with traditional pesticides, nanopesticides exhibit remarkably novel behaviors (Seeniappan et al., 2021). Nanoparticles are small enough to be carried in a colloidal or dissolved form during transport. Due to their diminutive size, the soluble state of the active ingredients will be enhanced; they can dissolve more readily in a liquid medium, which can potentially increase their effectiveness. As a result, the use of nanopesticides has a far less harmful effect on the ecosystem than typical pesticides. On the one hand, using smaller amounts of pesticides can reduce the overall environmental impact of agriculture, as less water and energy are required for their production and application (Dubey et al., 2016). There is a direct correlation between the use of nanopesticides and increased agricultural yields with decreased input costs, particularly in the areas of labor and water. However, the Environmental Protection Agency has warned that certain health problems may be caused by nanopesticides.

7.5.3.1.1 Nanoemulsions

The term "nanoemulsions" describes an oil-in-water emulsion in which a pesticide is dissolved in water as nanoscale particles. The primary benefit of manufacturing nanoemulsions is an increase in the absorption of active ingredients in pesticides.

7.5.3.2 Nanoparticles as Nanoherbicides

Weeds are the most pressing problem facing contemporary farming. Nanoherbicides rely heavily on biodegradable polymers for their effectiveness. Biopolymer encapsulation of atrazine, for instance, improves the chemical's biocompatibility, bioavailability, and physiochemical characteristics (T. Hofmann et al., 2020). Polymer-encapsulated atrazine is superior to free atrazine in terms of stability and herbicidal activity, as well as decreased mobility in soil (Mughal et al., 2021). Typically, atrazine causes contamination in agricultural areas and water supplies. It was found that the nanoformulated atrazine was more successful than standard atrazine against the specified weeds. It was seen from the foliar interaction of atrazine nanoformulation with mustard (India) that the vascular tissues of leaves are where the medicinal activity is directly mediated (Meng et al., 2021). This demonstrates that it is feasible for a low herbal activity content to be preserved through nanoformulation of the active ingredients. Nanocapsules containing atrazine are highly successful against both hairy beggarticks and narrow amaranth. For this reason, nanoherbicides are also an effective tool for weed control (Mitter and Hussey, 2019)

7.5.3.3 Nanoparticles as Nanofungicides

The presence of fungal growth in crops and produce is a major cause of concern in the agricultural industry. Although conventional fungicides are functioning effectively, their risk to human and environmental health makes them unacceptable. Nanoparticles have also been explored for their potential use as nanofungicides, which are agents designed to target and kill fungi at the nanoscale level. As a result, nanofungicides have appeared as promising new antifungal remedies to combat this problem (Singh et al., 2021).

Silver nanoparticles have been shown to have broad-spectrum antifungal activity against a wide range of fungi, including both yeasts and molds. The mechanism of action of silver nanoparticles is believed to involve the release of silver ions, which can damage fungal cell membranes and interfere with cellular processes.

Researchers found that at lower quantities, Cu NPs were more effective than commercial fungicides against *Phytophthora infestans* in tomatoes. The cereal-infecting fungus *Fusarium graminearum* can be stopped in its tracks when Cu(II) is complexed with chitosan nanogels (Brunel et al., 2013).

However, it's important to note that while nanoparticles may have promising antifungal activity, more research is needed to evaluate their safety and efficacy in various applications, as well as their potential impact on the environment. It is also important to consider the potential for nanoparticles to accumulate in living organisms and the potential impact on human health and the environment.

7.5.3.4 Nanoparticles as Nanobactericides

Nanoparticles have been explored for their potential use as nanobactericides, which are agents designed to target and kill bacteria at the nanoscale level. Several types of nanoparticles have been studied for this purpose, including silver, gold, copper, and zinc oxide nanoparticles.

Silver nanoparticles have received the most attention for their antimicrobial properties. They have been shown to have broad-spectrum antimicrobial activity against a wide range of bacteria, including both Gram-positive and Gram-negative bacteria, as well as some viruses and fungi. The mechanism of action of silver nanoparticles is believed to involve the release of silver ions, which can disrupt bacterial cell membranes and interfere with cellular processes.

Gold nanoparticles have also shown promising results as nanobactericides. They are effective against both Gram-positive and Gram-negative bacteria, as well as some viruses. The mechanism of action of gold nanoparticles is not well understood, but it is believed to involve the disruption of bacterial cell membranes and interference with cellular processes.

Copper nanoparticles have been shown to have potent antibacterial activity, particularly against Gram-negative bacteria. The mechanism of action of copper nanoparticles is believed to involve the release of copper ions, which can damage bacterial cell membranes and interfere with cellular processes.

Zinc oxide nanoparticles have also shown antibacterial activity against a range of bacteria, including both Gram-positive and Gram-negative bacteria. The mechanism of action of zinc oxide nanoparticles is believed to involve the disruption of bacterial cell membranes and interference with cellular processes.

Overall, nanoparticles have the potential to be effective nanobactericides, but more research is needed to fully understand their mechanisms of action and to evaluate their safety and efficacy in various applications.

7.5.4 Nanoparticles in Mitigating Abiotic Stress Response

The utilization of nanoparticles has a significant impact on several crucial aspects of plant growth and development, such as photosynthesis, seedling vigor, and the growth of roots and shoots. Plants are ubiquitous and consequently, are subjected to a range of environmental stressors, including but not limited to drought, salt stress, high temperature, and UV radiation. Plants exhibit a variety of stress responses, including modifications in molecular mechanisms, expression of stress-responsive genes, and synthesis of antioxidative enzymes, that aid in the mitigation of harsh environmental conditions. These responses are crucial in enabling plants to survive in adverse conditions (Rejeb et al., 2014). Plants have developed mechanisms to counteract the effects of osmotic stress, including the synthesis of organic osmolytes such as polyols and trehalose, as well as amino acids like proline and glycine. Nanoparticles offer an effective means of supporting plants in mitigating their defense mechanisms.

Apart from facilitating the growth of plants, NPs also provide a safeguard against abiotic stressors. The binding of toxic metals to the surface of nanoparticles is attributed to their significant surface area and diminutive size, leading to a decrease in their accessibility. Abiotic stressors encompass a range of environmental factors such as drought, salinity, alkalinity, temperature variability, and mineral and metal toxicity. Nanoparticles can imitate the function of antioxidant enzymes, taking the form of nano-enzymes. These nano-enzymes are capable of scavenging oxidative stress (Sharifi et al., 2020). Photosynthesis is a crucial metabolic process in plants and is considered to be highly vulnerable. Therefore, in order to sustain its normal functioning, it is essential to reduce oxidative and osmotic stress. Under stress conditions such as nutrient deficiency, salinity, water scarcity, drought, and heat, the primary targets in the photosynthetic machinery are Photosystem II (PS II), Rubisco, and ATP. The study investigated the defense response of plants to abiotic stress, specifically examining the effects of SiO_2 NPs on the *Cucurbita pepo* plant. The results indicated that SiO_2 NPs enhanced plant transpiration rate, water use efficiency, total chlorophyll, and carbonic anhydrase activity as part of a defense mechanism against salt stress (Siddiqui et al., 2014).

In a similar vein, researchers discovered that the photoreduction activity undergoes alterations, and linolenic acid is inhibited in the electron transport chain due to the presence of TiO_2 (anatase). Additionally, it resulted in a reduction in the rate at which chloroplasts produce oxygen (Mingyu et al., 2008). Likewise, under conditions of stress, reactive oxygen species (ROS) are generated by cellular organelles, serving as a distinctive marker of abiotic stress. Plants possess the necessary enzymatic mechanisms to cope with the oxidative stress exerted on them by their surroundings. Nevertheless, plants are subject to the ramifications of such circumstances when stress surpasses their defense mechanisms. Nanomaterials help in mitigating such stress by activating specific genes, accumulating osmolytes, and providing free nutrients and amino acids.

7.5.5 Nanoclays

In their nanoclay form, clays are mineral materials that play an exceptionally important part in agriculture (Merino et al., 2020). It has been learned that nanoclays, which are sourced from nature, are inexpensive and non-toxic (Yendluri et al.,2019). Using these materials has a number of benefits for farmers, including increased effectiveness with inputs like fertilizer and water, reduced nutrient deficiency (as a nanofertilizer), increased water-holding capacity, encapsulation of bioagent materials, and better water absorption (Merino et al., 2020).

Among the most prominent nanofertilizers are nanoclays. These are depicted as multilayered, bidimensional silicate particles with a nanometer thickness (1 nm) and a length in micrometers (Othman, 2014). Nanoclays have many uses beyond their widespread use as nanofertilizers, including food and beverage packaging, medicinal applications, and the effective delivery of nanopesticides (Garcia et al.,2012).

Nanoclays have the ability to preserve nutrients for extended periods of time, bolster the development of plants, maintain a stable supply of nutrients, improve nitrogen utilization efficiency, and reduce environmental contamination to a minimum level (Songkhum et al.,2018; Pereira et al.,2012).

7.5.6 Nanoparticles in Seed Science

Seed is nature's nano-gift to man. It is a self-perpetuating biological entity that is able to survive in the harsh environment on its own. Nanotechnology can be used to harness the full potential of seeds. Nanoparticles have shown potential in the field of seed science due to their unique properties and potential applications. They can be used to improve seed germination, growth, and development, as well as enhance plant tolerance to various environmental stressors.

One way nanoparticles can be used in seed science is to coat seeds with them. This can improve the seeds' performance and protect them from environmental stressors such as drought, salinity, and pathogens. For example, nanoparticles such as titanium dioxide and silver have been shown to improve seed germination and growth, as well as to protect against fungal and bacterial infections.

Nanoparticles can also be used to deliver nutrients or other beneficial compounds to the seeds. This can enhance seedling growth and development, and improve plant productivity. For example, nanoparticles can be used to deliver micronutrients such as iron or zinc to seeds, which can improve plant growth and prevent nutrient deficiencies.

Carbon nanotubes (CNTs) have been identified as potential conduits for water permeation, facilitating the movement of water from the substrate into seeds by penetrating the seed coat. These processes aid in the initiation of seed growth, which can be utilized in agricultural systems that rely on rainfall for irrigation.

Another application of nanoparticles in seed science is in genetic modification. Nanoparticles can be used to deliver genetic material such as DNA or RNA to seeds, which can then be incorporated into the plant's genome. This can be used to create plants with desirable traits such as increased disease resistance, drought tolerance, or yield.

The process of seed production can be arduous, particularly in crops that rely on wind pollination. A reliable approach to ensuring genetic purity is by identifying the pollen load that may result in contamination. The dispersion of pollen is influenced by various factors such as air temperature, humidity, wind velocity, and the crop's pollen production. The implementation of nanobiosensors that are tailored to detect contaminating pollen has the potential to provide early warning of contamination, thereby mitigating the extent of the contamination. The aforementioned technique can also be employed to avert the infiltration of genetically modified crop pollen into non-genetically modified field crops.

The utilization of nanomembrane-coated seeds that possess the ability to detect the presence of water and subsequently facilitate water absorption during optimal germination conditions has the potential to enhance seed germination efficacy and minimize water consumption. The nanomembrane possesses the capability to be tailored to react to distinct environmental stimuli such as humidity, temperature, or illumination, thereby enabling the seeds to initiate the germination process solely under favorable circumstances.

Aerial broadcasting of seeds embedded with magnetic particles can also help improve seed distribution and reduce waste. The magnetic particles can be used to guide the seeds to specific locations, such as areas with optimal soil conditions, and improve the efficiency of seed sowing.

The application of nanotechnology and nanoparticles has been found to reduce seed damage and improve seed viability by detecting the moisture content of stored seeds. The utilization of nanosensors can facilitate the monitoring of seed moisture levels during storage, thereby enabling the implementation of suitable measures such as the modulation of temperature or humidity levels to avert moisture-induced harm.

The incorporation of bioanalytical nanosensors can facilitate the assessment of seed aging and enhance the efficacy of seed quality control measures. Nanosensors can be designed to detect specific biomarkers associated with seed aging, such as changes in enzyme activity or gene expression. This can help identify seed batches with reduced viability and prevent the planting of low-quality seeds.

Overall, the use of nanotechnology in seed science research shows great potential for improving seed performance and productivity, as well as reducing waste and improving sustainability. However, more research is needed to fully understand the potential risks and benefits of using nanotechnology in agriculture.

7.5.7 Clay Nanotubes

Clay nanotubes (Halloysite) are an additional, recent accomplishment in the field of plant protection. Clay nanotubes have potential applications in agriculture due to their unique properties, such as their high surface area, porosity, and adsorption capacity. Some possible applications of clay nanotubes in agriculture are as follows:

1. *Soil amendment:* Clay nanotubes can be used as a soil amendment to improve soil quality and nutrient retention. The high surface area and porosity of clay nanotubes can increase the cation exchange capacity of soils, which can enhance nutrient uptake by plants.
2. *Slow-release fertilizers:* Clay nanotubes can be loaded with fertilizers and used as a slow-release fertilizer. The porous structure of clay nanotubes can control the release of nutrients over time, which can improve the efficiency of fertilization and reduce nutrient loss through leaching.
3. *Pesticide delivery:* Clay nanotubes can be used as carriers for pesticides, herbicides, and other agrochemicals. The high surface area and adsorption capacity of clay nanotubes can enhance the retention and efficacy of agrochemicals while reducing their environmental impact.
4. *Seed coating:* Clay nanotubes can be used as a coating material for seeds to improve seed germination, seedling growth, and plant development. The porous structure of clay nanotubes can provide a protective environment for seeds, enhance water retention, and facilitate nutrient uptake.

Overall, the potential applications of clay nanotubes in agriculture are still being explored, and there is ongoing research to optimize their synthesis and application methods. However, the unique properties of clay nanotubes make them a promising material for sustainable agriculture.

7.5.8 Nanosensors

Nanotechnology-based nanosensors are evolving into potent instruments for monitoring plant pathogens and preventing their spread. Recent years have seen an increase in the use of nanosensors in combination with Global Positioning System (GPS) systems for real-time monitoring of cultivated regions. It is possible to create a network of wireless nanosensors in agricultural areas, allowing farmers to track soil health and crop development in real time. Nanosensors are ideal for precision farming, which seeks to maximize output with a minimum of resources expended, and the data they generate can aid farmers in making more informed choices (Joseph and Marrison, 2006). Nanosensors have the potential to identify a wide variety of threats to plants, including toxins, pests, nutrient content, drought, temperature, and pressure. They can help farmers answer concerns about food safety, such as whether or not their crops have been contaminated by microbes or chemicals, and when the optimal time for harvesting is. When placed within the cultivated fields, nanosensors will collect and transmit crucial data that will aid in the targeted application of agricultural inputs for maximum crop production (Scott and Chen, 2012). Thereby, they also have the potential to assist producers in increasing their efficiency by ensuring that inputs are only used when they are absolutely necessary. Nanosensors with an improved ability to identify crop health and harvest time have the potential to be a key component of controlled environmental agriculture (CEA).

7.5.8.1 Nanobiosensors

The term "nanobiosensor" refers to nanosensors that use immobilized bioreceptors specific for target analyte molecules. Nanobiosensors can be used to identify a wide variety of substances (such as urea, glucose, pesticides, etc.) and microorganisms/pathogens (Rai et al., 2012). These tools shorten the duration of traditionally laborious microbial tests and immunoassays. Nanobiosensors play an essential role in the detection and quantification of low concentrations of contaminants such as viruses, bacteria, toxins, pollutants, and more in the water supply, agricultural, and food industries. In addition, disease outbreaks can be predicted before their occurrence, demonstrating that biosensors offer a novel diagnostic technology for semi-quantitatively detecting soil conditions. Nanosensors can be used to overcome the insufficiency of current sensors for precise quantitative monitoring.

7.5.8.2 Enzymatic Biosensors

Due to their specificity in binding to biomolecules, enzymes can serve as a detecting element (Le Goff et al., 2011; Sassolas et al., 2012). There are four categories of immobilization-based enzyme biosensors: (1) controlled-pore glass beads with optical transducer element, (2) polyurethane foam with photothermal transducer element, (3) ion-selective membrane with either potentiometric or amperometric transducer element, and (4) screen-printed electrode with amperometric transducer element (Su and Li, 2004).

7.5.9 Micro-fabricated Xylem Vessels

Xylem vessels are structures in plants that transport water and nutrients from the roots to the rest of the plant. Micro-fabricated xylem vessels are a type of microfluidic device that mimics the function of natural xylem vessels in plants. One potential application of micro-fabricated xylem vessels is in the development of new technologies for water transport in agriculture. By studying the transport of water and nutrients in micro-fabricated xylem vessels and the effects of environmental stresses, such as drought or salinity, on plant physiology, researchers can gain a better understanding of how plants regulate water and nutrient uptake and develop new technologies for improving water use efficiency in crops.

As a result of advances in nanofabrication and characterization tools, we are now able to investigate the physicochemical and biological interactions between plant cell bodies and the pathogens responsible for causing disease. With the aid of these instruments, we have gained insight into the underlying mechanisms and developed more effective methods for treating these ailments (Cursino et al., 2009). Micro-fabricated xylem vessels with nano-sized features have allowed us to research the aforementioned mechanisms, which were previously impossible to investigate using more conventional approaches (Zaini et al., 2009).

7.5.10 Smart Dust

The quantity of airborne dust and pollutants may be measured with the help of high-tech dust sensors (Scott and Chen, 2012). By deploying "smart dust" sensors throughout vines and orchards, we can keep tabs on environmental factors like temperature and humidity as well as potentially harmful pests and diseases.

7.5.11 ZigBee – A Mesh Networking Standard

ZigBee is a standard for wireless mesh networks that are both inexpensive and energy efficient. The terms "Smart Fields" and "Soil Net" are both derivatives of this idea. They can be used to keep an eye out for and record the presence of pests, dry conditions, or excess moisture, all of which can have a negative impact on agricultural yields (Nath and Chilkoti, 2004).

7.5.12 Zeolites

Natural minerals, including nanoclays and zeolites, are also used to improve fertilizer use efficiency (Chinnamuthu and Boopathi, 2009). These minerals can have their crystalline layers filled with macro- and micronutrients to allow a controlled discharge of those nutrients when needed. In the agricultural sector, zeolites are typically utilized to capture, store, and facilitate the gradual release of nitrogen (Leggo, 2000). As nutrients are released from the absorbed form (zeolite) more slowly than they are from the commonly used ionic form of fertilizers, these kinds of nano-coated fertilizers also aid in reducing soil, water, and air pollution. Nitrate can be effectively absorbed by surfactant-modified zeolites, which can subsequently be employed as a fertilizer vehicle for the timed discharge of nitrate and other anions. In addition, the ammonium-containing zeolite particles can boost phosphate uptake and crop yield by boosting phosphate solubilization capacity, which ultimately results in a higher yield from the crop.

7.5.13 Zeolites for Water Retention

Zeolites are a group of porous minerals that are commonly used as adsorbents, catalysts, and ion exchangers in a variety of applications. They are composed of aluminosilicates with a crystalline structure that contains uniform pores and channels, allowing them to selectively adsorb or catalyze specific molecules.

Using zeolites would help control water availability for crops in times of extreme drought by decreasing overland flow (surface runoff) and protecting the soil from erosion. Therefore, the zeolite may be used for dry land farming, though it would require some sort of technological breakthrough, which needs to be investigated (Ramesh et al., 2010). It possesses a notably high cation exchange capacity (CEC). A surface area of approximately 1150.5 m^2 g^{-1}, a high CEC (ten times that of soil), and a large quantity of free water in the structural channels all contribute to its exceptional adsorption capacity (Sand and Mumpton, 1978). Zeolite has the potential to increase infiltration by between 7 and 30% on land with a gentle slope and by more than 50% on land with a steep slope. In addition, the treated soil has the potential to increase its moisture content by a range of 0.4–1.8% in intense drought conditions and by a range of 5–15% in general scenarios (Xiubin and Zhanbin, 2001).

It is able to form compounds with other cations, including Na^+, K^+, Ca^{2+}, and so on (Dana, 1977; Navrotsky et al., 1995).

7.5.14 Purification of Irrigation Water

In place of conventional purification processes, such as those involving UV radiation or chemicals, nanofiltration could be used to disinfect irrigation water (Hillie and Hlophe, 2007). Nanofiltration employs nanofilters with nanopores, which can eliminate water-borne pathogens and heavy metals like lead, uranium, and arsenic (Gao et al., 2014; Zhu et al., 2014). Fused mesh carbon nanotubes have been used effectively for this purpose and are the most cost-effective option. Nanoceram filters with a positive charge on their surface are effective at filtering out microbe-produced toxins, DNA, and pathogenic viruses, as well as tiny particulates (Foley et al., 2012).

7.5.15 Nanobarcode Technology

Nanobarcode technology is a type of nanotechnology that involves the use of uniquely coded metallic nanoparticles to tag molecules or cells for identification and tracking. The technology is based on the principle of surface-enhanced Raman scattering (SERS). One of the main advantages of nanobarcode technology is its high sensitivity and specificity, which allow the detection of very low concentrations of molecules or cells. The technology also allows multiplexed detection, meaning that multiple barcodes can be used simultaneously to detect multiple targets in a single sample. Nanobarcodes can potentially have several applications in sustainable agriculture, including identifying and tracking plant growth and health, monitoring soil quality, and tracking agricultural products through the supply chain.

One potential use of nanobarcodes in sustainable agriculture is in the monitoring of soil quality. Nanobarcodes can be used to tag soil samples with unique barcodes, identifying and tracking different soil types and their properties. This information can be used to optimize soil management practices, such as the application of fertilizers and other soil amendments, to improve soil health and productivity. Since this method allows the simultaneous tagging and detection of numerous pathogens in a farm using any fluorescence-based tools, it has been shown to be an economically efficient, rapid, and effortless approach to the decoding and recognition of diseases (Li et al., 2005).

7.6 THE ROLE OF NANOPARTICLES IN PLANT HEALTH AND DISEASE CONTROL

The agricultural sector and farmers have faced numerous challenges due to the rising demand for enhanced food production. The confluence of climate variations that are anticipated to prolong periods of drought and raise yearly temperatures will render crop production in numerous agriculturally susceptible areas more susceptible to diseases and other forms of stress. According to the majority of agricultural experts, it is possible to attain the required augmentation in food and fiber production. However, this can only be accomplished at a cost that is both unsustainable and ecologically detrimental (Tilman et al., 2011). The emergence of nanotechnology presents a potent tool for combating various factors that pose a risk to the well-being of plants. Nanotechnology offers numerous benefits. A compelling factor is a significant decrease in the number of active chemicals that enter the agroecosystem. A significant portion of the bactericides, fungicides, and fertilizers that are applied fail to reach their intended targets, instead infiltrating groundwater and posing a threat to delicate ecosystems. Due to the significant active surface area of nanoparticles, smaller quantities can effectively administer equivalent or greater amounts of active ingredients to pest species or tissue sites while reducing exposure to non-target regions as compared with larger bulk or non-nano forms. In addition, nanotechnology presents a diverse range of distinctive characteristics that can be utilized in the management of plant health. These include enhanced solubility, delivery, and extended residual activity of agrichemicals, resulting in prolonged effectiveness. Nanoparticles have shown great potential in improving plant health and controlling diseases. They are tiny particles, typically less than 100 nm in size, which allows them to penetrate plant tissues and cell walls, enhancing their efficacy in various applications.

Engineered NPs within the size range of 1–100 nm have demonstrated activity in suppressing plant diseases, and the types of NPs that have shown such activity include metalloids, metallic oxides, nonmetals, and carbon nanomaterials. Studies have shown that these nanoparticles can have various mechanisms of action against plant pathogens, including disruption of cell membranes, induction of ROS, and activation of plant defense responses. NPs have been increasingly used in agriculture to improve plant health and increase crop yields. Nanoparticles have unique physical and chemical properties that make them useful in various applications, including as bactericides and fungicides. Some NPs act directly as antimicrobial agents, while others act by altering the host plant's nutritional status and activating its defense mechanisms. NPs of Ag and Cu, for instance, have the potential to cause direct toxicity in microorganisms. As nutrients, it seems that NPs of B, Cu, Mn, Si, and Zn contribute to host defense. Nanoparticles will help mitigate the new challenges in disease management caused by active metals and other chemical inputs, allowing the food supply to keep up in the face of a warming environment (Elmer et al., 2018).

Viruses are responsible for about 20% of crop failures globally.

It has been demonstrated that nanomaterials such as nanosilica can be used effectively for the delivery of specific genes into cells (Torney, 2009). In addition, nanosilica has been used as a standalone nanoinsecticide, with its insecticidal characteristics proposed to be on account of the material's ability to be absorbed into the cuticular layer of insect pests, which normally serves as a barrier for protection against pesticides (Barik et al., 2008). In addition to surface charge, their diminutive size (~3–5 nm) and hydrophobicity make them a promising tool for controlling a wide range of ectoparasites of livestock and farming insect pests (Ulrichs et al., 2005). One advantage of using nanosilica as an insecticide is that it is a natural substance considered safe for the environment. It is also biodegradable, which means it breaks down over time and does not accumulate in the environment. It has also been hypothesized that nanosilver can alter the biochemistry of the plasma membrane (Pal et al., 2007) and suppresses the transcription of ATP-making proteins (Yamanka et al., 2005).

Furthermore, it has been found that nanoemulsions, which are mixtures of oil and water, could potentially be used in the production of insecticides to combat a wide variety of insect pests (Wang et al., 2007). For instance, polyethylene glycol–coated nanoparticles filled with garlic essential oil have shown promising results in tests conducted against store-product pests such as the insect *Tribolium castaneum* (Yang et al., 2009).

Some ways in which nanoparticles are being used in plant health and disease control include the following.

7.6.1 Fertilizers

Nanoparticles such as zinc oxide and iron oxide have been shown to increase nutrient uptake and improve plant growth, leading to increased crop yields.

7.6.2 Pesticides

Nanoparticles can be used as carriers for pesticides, allowing more targeted delivery and reducing the amount of pesticide needed to achieve the same level of control.

7.6.3 Disease Control

Nanoparticles have antimicrobial properties that can be used to control plant diseases. For example, silver nanoparticles have been shown to be effective against a range of plant pathogens.

7.6.4 Stress Tolerance

Nanoparticles can improve plant tolerance to various stresses, such as drought, salinity, and heavy metal toxicity.

However, it is important to note that there are still many questions surrounding the long-term effects of nanoparticles on plants and the environment. Further research is needed to fully understand the potential benefits and risks of using nanoparticles in agriculture.

7.6.5 Potential of Foliar NP Application for Disease Suppression

Studies of plant physiology have indicated that particles have the ability to penetrate plant leaves via the stomata and cuticle structures (Tocha et al.,, 2006). It is hypothesized that the process of entry or absorption may exhibit greater speed and efficacy for nanoparticles with smaller diameters. Nanoparticles were observed to infiltrate the cells of pumpkin leaves through the stomatal openings (Corredor et al., 2009). Furthermore, it is widely acknowledged that NP metals/metal oxides exhibit higher translocation than their corresponding bulk elements. This enhanced transportability is observed in both xylem and, more significantly, phloem-based vasculature (Wang et al., 2012). The utilization of NP TiO_2 in the field of agriculture, which encompasses safeguarding against pathogenic ailments, is predicated on the photocatalytic surface characteristics of the material (Ahmad and Rasool, 2014). The application of NP TiO_2 on crops has the potential to enhance yield by up to 30% while concurrently mitigating pathogenic disease to a significant degree (Chao and Choi, 2005).

7.6.6 Potential of Soil NP Application for Disease Suppression

The use of soil NPs for disease suppression is an emerging area of research. NPs are particles that range in size from 1 to 100 nm and have unique physicochemical properties that make them potentially useful for a wide range of applications, including disease suppression in agriculture.

Soil nutrient management is an important aspect of sustainable agriculture. Nutrient deficiencies can lead to reduced crop yields and in some cases, can increase susceptibility to certain diseases. Regarding disease suppression, soil nutrient management strategies can have a positive impact, including applying soil nitrogen-phosphorus fertilizers. Studies have shown that the application of nitrogen-phosphorus fertilizers can enhance plant growth and increase the resistance of crops to certain diseases. For example, nitrogen fertilization can improve plant vigor and increase the plant's ability to defend itself against fungal pathogens. Phosphorus fertilization, on the other hand, can improve root development and increase the plant's ability to take up nutrients, which can enhance its overall health and resistance to disease. Nitrogen-phosphorus can interact with plant cells in a variety of ways, including penetration of plant tissues and modification of gene expression, leading to changes in plant physiology and defense mechanisms. Nitrogen-phosphorus can also interact with soil microorganisms, which can have a positive impact on soil health and disease suppression.

Several studies have shown promising results for the use of nitrogen-phosphorus in disease suppression. For example, silver NPs have been shown to inhibit the growth of bacterial and fungal pathogens in tomato plants. Similarly, zinc oxide NPs have been shown to enhance the resistance of cucumber plants to powdery mildew. However, the effects of NP application on disease suppression can vary depending on the specific pathogen and crop involved. In some cases, excessive NP application can actually increase the incidence of certain diseases. For example, excessive nitrogen can increase foliar diseases such as rust, while excessive phosphorus can promote root rots.

Therefore, it is important to use soil nutrient management strategies, including the application of nitrogen-phosphorus fertilizers, in a balanced and targeted manner, taking into account the specific needs of the crop and the soil. Integrated pest management approaches, which combine nutrient management with other disease prevention strategies such as crop rotation, disease-resistant varieties, and cultural practices, can effectively reduce disease incidence and maintain healthy crops.

Specific nanoparticles can play various roles in soil depending on their properties and functions. Here are some examples of the roles of specific NPs in soil:

7.6.6.1 Carbon-based NPs

Carbon-based NPs, such as graphene and carbon nanotubes, have high surface area and unique mechanical and electrical properties, making them useful for soil remediation, water purification, and soil amendment to improve soil quality.

7.6.6.2 Metal-based NPs

Metal-based NPs, such as silver, copper, and iron oxide, have antimicrobial properties and can be used for the control of soil-borne pathogens. They can also improve soil fertility and plant growth by supplying essential nutrients such as iron and zinc.

7.6.6.3 Metal oxide NPs

Metal oxide NPs, such as titanium dioxide and zinc oxide, have unique optical and electronic properties, making them useful in photocatalytic reactions for the degradation of pollutants in soil and water. They can also enhance plant growth and protect plants from environmental stresses such as UV radiation and heavy metal toxicity.

7.6.6.4 Clay-based NPs

Clay-based NPs, such as montmorillonite and halloysite, can improve soil structure, water-holding capacity, and nutrient retention. They can also enhance plant growth and protect against plant pathogens by adsorbing toxins and reducing the availability of nutrients to pathogens.

Overall, the specific roles of NPs in soil depend on their physicochemical properties, the soil environment, and the intended application. It is important to carefully consider the potential risks and benefits of NP use in soil and to conduct further research to better understand their behavior and potential impacts on the soil environment.

Therefore, before NPs can be widely used for disease suppression in agriculture, further research is needed to better understand their mechanisms of action, potential risks, and optimal application rates and methods. It is also important to consider the potential for unintended consequences and to carefully evaluate the environmental impacts of NP use in agriculture.

7.7 THE ROLE OF NANOPARTICLES IN BIOREMEDIATION

In addition to mitigating the effects of climate change, removing pollutants from the environment is currently regarded as one of the most difficult challenges worldwide. Increasing human activities result in the release of a flood of environmentally harmful compounds that have not been treated, some of which have the capacity for long-term persistence and migration across food chains. Although a great number of technologies for cleanup have been researched and put into practice up to this point, there is still a need for the development of a method that is economical, efficient, and kind to the environment. This method should be able to deal with residues of low concentration while still ensuring maximum results in highly polluted environments. The term "bioremediation" refers to the process of enhancing naturally occurring remedial processes that rely on living organisms to degrade, transform, or remove dangerous organic contaminants. The catabolic processes of microorganisms and their potential to contribute to the breakdown of organic contaminants are essential to the success of this biological technique. Nanobioremediation is a relatively new technique that has demonstrated exceptional potential for use in the elimination of contaminants in the natural environment. Nanoparticles, which are characterized by their small size, strong reactivity, and catalytic activities, have proven to be an excellent tool for cleaning up contaminated soil and water. Different species of bacteria, fungi, algae, and other types of microorganisms are involved in the remediation process (Figure 7.4). The chemical and physical features of nanoparticles, such as their high ratio of surface area to volume, are responsible for this phenomenon. On the other hand, microorganisms have been the subject of a significant amount of research and are known to play an important role in the maintenance of many processes that occur within an ecosystem, including the

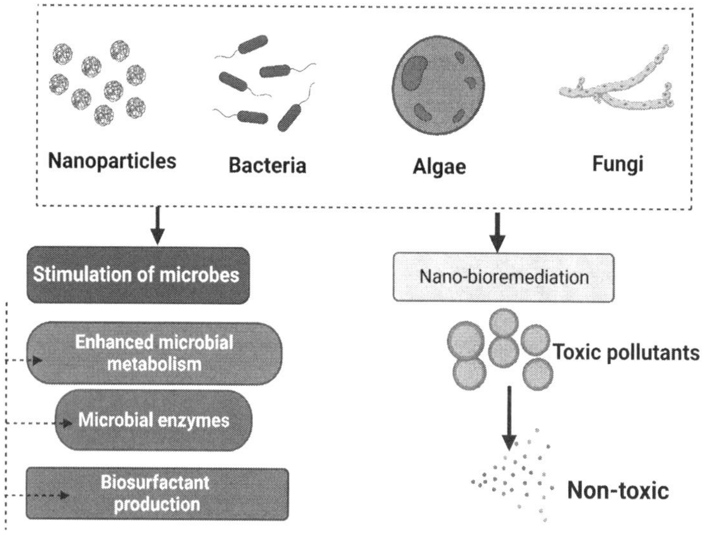

FIGURE 7.4 Nanobioremediation – removal of toxic pollutants.

intake of nutrients, the rise of food production, and the biodegradation process. They are extremely useful in a variety of fields, including agriculture, food processing, biotechnology, medicine, and aquaculture (Díaz-Rodríguez et al., 2017).

The use of nanoparticles in bioremediation has given a sustainable and efficient means of accelerating the breakdown of contaminants. Various technologies, including adsorption, reaction of chemicals, photocatalysis, and filtration, are employed to eliminate toxins from the environment. The attenuation of environmental damage is dependent on these technologies. Conventional methods have a number of drawbacks, including their high cost, high demand for energy, generation of potentially harmful toxic chemicals, the necessity of elevated temperature and pressure, and non-economical attributes due to their incapacity to eliminate impurities and absence of material reusability (Sharma et al., 2015).It is possible that naturally occurring hyperaccumulators and indigenous soil bacteria, with the assistance of nanoparticles, could enhance biodegradation processes. This would increase the potential extent of remediation. Both nano-phytoremediation and nano-remediation mediated by microbes could be used to describe this process. Due to their tiny size (1–100 nm) and high surface area and reactivity, the incorporation of NPs into bioremediation techniques may result in numerous advantages (Alazaiza et al., 2021). Iron oxide nanoparticles (IO-NPs) generated from *Aspergillus tubingensis* were investigated by Mahanty et al. (2020), who found that they may be used sustainably to remove lead(II) 98%, nickel(II) 96.45%, copper(II) 92.19%, and zinc(II) 92.19% from wastewater (Table 7.2). Pesticides get into water sources when runoff from farms carries them there. This is a very serious problem for both water ecosystems and people. Galdames et al. (2017) developed innovative strategies for cleaning contaminated soils by basing them on zero-valent iron nanoparticles (nZVI), bioremediation using compost made from organic waste, and a process that combines both. In situ immobilization of As and Cr in two soils (Soil I from a contaminated industrial site and Soil II, which was artificially contaminated) using nZVI decreased the concentration of As and Cr in both soils and their leachates. Baragaño et al. (2020) studied the ability of graphene oxide nanoparticles (nGox) to decrease or increase the availability of As and metals in polluted soils compared with that of nZVi. The application of nZVI and nGOx to polluted soils significantly impacted the availability of As and metals. Iron nanoparticles were also found to have a positive impact (nZVi). According to the findings of this research, both NPs may act as approaches for the immobilization and stabilization of contaminants, which can then be used for phytoremediation. Gil-Díaz et al. (2022) evaluated the effectiveness of different

TABLE 7.2
Nanobioremediation of Heavy Metals

Nanoparticles	Heavy metal	Synthesis source	References
Fe	Removal of chromium (VI)	*Eichhornia crassipes*	Wei et al. (2017)
Pd	Adsorption of arsenic	*Chlorella vulgaris*	Arsiya et al. (2017)
Iron oxide nanoparticles (IO-NPs)	Pb(II), Ni(II), Cu(II), and Zn(II)	*Aspergillus tubingensis*	Mahanty et al. (2020)
Ag	Adsorption of Cu^{2+}	*Convolvulus arvensis*	Al-Senani et al. (2020)
AgNPs and CuO-NPs	Cadmium	*Catharanthus roseus*	Verma et al. (2022)
Nickel-bentonite nanoparticles (NBNPs)	Heavy metal ions rejection	*Scrophularia striata*	Dadari et al. (2022)
FeO	Cr-VI, Cd, Pb	*Eucalyptus globulus*	Andrade-Zavaleta et al. (2022)
TiO_2	Cd	*Trianthema portulacastrum*, *Chenopodium quinoa*	Irshad et al. (2022)
Fe_3O_4 NPs	Hg(II) and Cr(VI)	*Thymus schimperi*	Geneti et al. (2022)

types of commercial iron nanoparticles (nZVI, bimetallic nZVI-Pd, and nano-magnetite [nFe_3O_4]) for the remediation of an industrial soil co-contaminated with Cr and polychlorinated biphenyls (PCBs). The immobilization of the chromium was stable for at least 70 days under the experimental conditions after adding nZVI, nZVI-Pd, or nFe_3O_4 to soil that was co-contaminated with chromium and PCBs. This significantly reduced the leachability of chromium in the soil. According to the findings, the recovery of soils contaminated with lead and PCBs might be possible with the help of pseudo-anaerobic conditions and the addition of either nZVI or nZVI-Pd. Nanoparticles have gained popularity as potentially useful materials for soil remediation. Vu et al. (2022) presented evidence that soil contaminated with oil may be cleaned up using suspensions of Fe-Cu nanoparticles and biosurfactants. Based on these findings, a suspension of biosurfactants and Fe-Cu nanoparticles showed promise as a potential method for the cleaning up of oil-contaminated soil. Integrated nanoparticle-bioremediation strategies, also known as nano-bioremediation, are another potentially useful approach that boosts the functions of the microbiome to remove hazardous pollutants from polluted areas. Improved bioremediation of Pb (lead) and Cd (cadmium) contaminated soil by *Halomonas* sp., Gram-negative bacteria, was achieved by adding polyvinylpyrrolidone (PVP)-coated FeO nanoparticles. In comparison to the removal performed just by NPs or bacteria, this method was substantially more effective in removing approximately 100% of the lead after 24 h and the cadmium after 48 h (Alabresm et al.,2018). The elimination of heavy metals through the application of a variety of NPs, including biogenic ones, has recently attracted considerable attention. Biogenic nanoparticles are those that are produced by the use of biological organisms in the synthesis process. In order to dispose of cadmium (Cd), Laslo et al. (2022) created a new experimental design involving two bacteria, *L. fermentum* or *L. casei,* and exogenous selenium nanoparticles (SeNPs). By reducing selenite using an *L. casei* strain, they were able to create SeNPs that were both biologically produced and spherical, with a size distribution peaking at around 200 nm. Using two probiotic bacteria (*L. fermentum* and *L. casei*) to establish Cd's binding affinity was a demonstration of the protective relevance of SeNPs in this process. According to their research findings, *L. fermentum* and *L. casei* can remove Cd from an aqueous solution, and the consumption of SeNPs boosts the efficiency with which Cd is remediated. Marine bacteria can better adapt to harsh environments than terrestrial microorganisms, making their employment in noble metal recovery an intriguing research area. In the first study of its kind, Zhang and Hu (2018) showed that a marine *Bacillus* sp. GP was capable of synthesizing lead and gold nanoparticles when the media included sodium lactate, and they showed that Pd/Au nanoparticles have catalytic activity for 4-nitrophenol reduction. Toxic pollutant loads in the soil can be reduced or eliminated with the application of NPs, which enhance the microbial community in the soil. The rhizospheric bacteria that are beneficial to soil health have been shown to increase colonization and biomass when exposed to Si NPs (Srivastava et al., 2021).

7.8 CONCLUSIONS

Nanoparticles are incredibly small particles with unique physical, chemical, and biological properties that make them suitable for a wide range of applications. They have been extensively studied and utilized in various fields, including agriculture, medicine, electronics, energy, and environmental remediation. Despite their numerous benefits, nanoparticles also pose potential risks to human health and the environment. Therefore, it is essential to continue researching their properties and potential impacts to ensure their safe and effective use. Overall, nanoparticles represent a promising avenue for innovation and advancement in various fields, but their use must be balanced with caution and responsibility to minimize any negative consequences. The use of nanoparticles in agriculture significantly contributes to sustainable agricultural practices and has the potential to revolutionize the industry by increasing crop yields, enhancing nutrient uptake, improving plant growth and stress tolerance, and reducing the harmful impact of agrochemicals on the environment. Nanoparticles have shown promise in improving soil quality, controlling pests and diseases, and increasing the efficiency of fertilizers and water usage. However, more research is needed to fully understand their effects on plant growth, soil microorganisms, and potential risks to human health and the environment.

By using nanoparticles, farmers can reduce the negative impact of conventional agricultural practices on the environment and human health. Nanoparticles can improve the efficiency of inputs such as water, fertilizers, and pesticides, leading to lower environmental contamination and increased cost-effectiveness. Moreover, using nanoparticles can help enhance crops' resilience to abiotic and biotic stresses, contributing to food security and mitigating the impacts of climate change on agriculture. While the use of nanoparticles in agriculture offers significant benefits, it is important to approach their implementation with caution and responsibility. Regulations must be established to ensure their safe use, and the potential risks to the environment and human health must be carefully evaluated.

REFERENCES

Abdallah, Y., Ogunyemi, S. O., Abdelazez, A., Zhang, M., Hong, X., Ibrahim, E., Hossain, A., Fouad, H., Li, B., & Chen, J. (2019). The green synthesis of MgO nano-flowers using *Rosmarinus officinalis* L.(Rosemary) and the antibacterial activities against *Xanthomonas oryzae pv. oryzae*. *BioMed Research International*, 5620989.

Abdel-Aziz, S. M., Prasad, R., Hamed, A. A., & Abdelraof, M. (2018). Fungal nanoparticles: A novel tool for a green biotechnology? In: R. Prasad, V. Kumar, M. Kumar, S. Wang (eds) *Fungal Nanobionics: Principles and Applications*, Springer, Singapore. pp. 61–87.

Abdel-Raouf, N., Al-Enazi, N. M., & Ibraheem, I. B. (2017). Green biosynthesis of gold nanoparticles using *Galaxaura elongata* and characterization of their antibacterial activity. *Arabian Journal of Chemistry*, *10*, S3029–S3039.

Agnihotri, M., Joshi, S., Kumar, A. R., Zinjarde, S., & Kulkarni, S. (2009). Biosynthesis of gold nanoparticles by the tropical marine yeast *Yarrowia lipolytica* NCIM 3589. *Materials Letters*, *63*(15), 1231–1234.

Ahmad, P., & Rasool, S. (Eds.). (2014). *Emerging Technologies and Management of Crop Stress Tolerance: Biological Techniques*. Academic Press, Elsevier, New York.

Alabresm, A., Chen, Y. P., Decho, A. W., & Lead, J. (2018). A novel method for the synergistic remediation of oil-water mixtures using nanoparticles and oil-degrading bacteria. *Science of the Total Environment*, *630*, 1292–1297.

Alazaiza, M. Y., Albahnasawi, A., Ali, G. A., Bashir, M. J., Copty, N. K., Amr, S. S. A., … & Al Maskari, T. (2021). Recent advances of nanoremediation technologies for soil and groundwater remediation: A review. *Water*, *13*(16), 2186.

Ali, N. S., & Al-Juthery, H. W. A. (2017). The application of nanotechnology for micronutrient in agricultureal production. *The Iraqi Journal of Agricultural Science*, *48*(4), 984.

Al-Senani, G. M., & Al-Kadhi, N. (2020). The synthesis and effect of silver nanoparticles on the adsorption of Cu2+ from aqueous solutions. *Applied Sciences*, *10*(14), 4840.

Andrade-Zavaleta, K., Chacon-Laiza, Y., Asmat-Campos, D., & Raquel-Checca, N. (2022). Green synthesis of superparamagnetic iron oxide nanoparticles with *Eucalyptus globulus* extract and their application in the removal of heavy metals from agricultural soil. *Molecules*, *27*(4), 1367.

Arif, N., Yadav, V., Singh, S., Singh, S., & Mishra, R. K. (2016). Current trends of engineered 745 nanoparticles (ENPs) in sustainable agriculture: An overview. *Journal of Environmental and Analytical Toxicology*, *6*(746), 397. https://doi.org/10.4172/2161-0525.1000397.

Arsiya, F., Sayadi, M., & Sobhani, S. (2017). Arsenic (III) adsorption using palladium nanoparticles from aqueous solution. *Journal of Water and Environmental Nanotechnology*, *2*(3), 166–173.

Aziz, N., Faraz, M., Pandey, R., Shakir, M., Fatma, T., Varma, A., … & Prasad, R. (2015). Facile algae-derived route to biogenic silver nanoparticles: Synthesis, antibacterial, and photocatalytic properties. *Langmuir*, *31*(42), 11605–11612.

Baker, S., Kumar, K. M., Santosh, P., Rakshith, D., & Satish, S. (2015). Extracellular synthesis of silver nanoparticles by novel *Pseudomonas veronii* AS41G inhabiting *Annona squamosa* L. and their bactericidal activity. *Spectrochimica Acta. Part A: Molecular and Biomolecular Spectroscopy*, *136*(C), 1434–1440.

Barabadi, H., Honary, S., Ali Mohammadi, M., Ahmadpour, E., Rahimi, M. T., Alizadeh, A., … & Saravanan, M. (2017). Green chemical synthesis of gold nanoparticles by using *Penicillium aculeatum* and their scolicidal activity against hydatid cyst protoscolices of *Echinococcus granulosus*. *Environmental Science and Pollution Research International*, *24*(6), 5800–5810.

Baragaño, D., Forján, R., Welte, L., & Gallego, J. L. R. (2020). Nanoremediation of As and metals polluted soils by means of graphene oxide nanoparticles. *Scientific Reports*, *10*(1), 1–10.

Barik, T. K., Sahu, B., & Swain, V. (2008). Nanosilica—from medicine to pest control. *Parasitology Research*, *103*(2), 253–258.

Ben Rejeb, I., Pastor, V., & Mauch-Mani, B. (2014). Plant responses to simultaneous biotic and abiotic stress: Molecular mechanisms. *Plants*, *3*(4), 458–475.

Beveridge, T. J., & Murray, R. G. (1980). Sites of metal deposition in the cell wall of *Bacillus subtilis*. *Journal of Bacteriology*, *141*(2), 876–887.

Brady, N. C., & Weil, R. R. (1996). *The Nature and Properties of Soils*. Pearson Education, Incl., New York, USA.

Broda, M., Yelle, D. J., & Serwańska, K. (2022). Bioethanol production from lignocellulosic biomass—Challenges and solutions. *Molecules*, *27*(24), 8717.

Brundtland, G. H. (1987). World Commission on Environment and Development. Our Common Future. New York, UN.

Brunel, F., El Gueddari, N. E., & Moerschbacher, B. M. (2013). Complexation of copper (II) with chitosan nanogels: toward control of microbial growth. *Carbohydrate Polymers*, *92*(2), 1348–1356.

Can, M. (2020). Green gold nanoparticles from plant-derived materials: An overview of the reaction synthesis types, conditions, and applications. *Reviews in Chemical Engineering*, *36*(7), 859–877.

Chandran, S. P., Chaudhary, M., Pasricha, R., Ahmad, A., & Sastry, M. (2006). Synthesis of gold nanotriangles and silver nanoparticles using Aloevera plant extract. *Biotechnology Progress*, *22*(2), 577–583.

Chen, H. (2018). Metal based nanoparticles in agricultural system: Behavior, transport, and interaction with plants. *Chemical Speciation and Bioavailability*, *30*(1), 123–134.

Chinnamuthu, C. R., & Boopathi, P. M. (2009). Nanotechnology and agroecosystem. *Madras Agricultural Journal*, *96*, 1.

Corredor, E., Testillano, P. S., Coronado, M. J., González-Melendi, P., Fernández-Pacheco, Ibarra, M.R., de la Fuente, J.M., Rubiales, D., Pérez-de-Luque, A., & Risueño, M. C. (2009). Nanoparticle penetration and transport in living pumpkin plants: In situ subcellular identification. *BMC Plant Biology*, *9*, 1–11.

Cursino, L., Li, Y., Zaini, P. A., De La Fuente, L., Hoch, H. C., & Burr, T. J. (2009). Twitching motility and biofilm formation are associated with tonB1 in *Xylella fastidiosa*. *FEMS Microbiology Letters*, *299*(2), 193–199.

Dadari, S., Rahimi, M., & Zinadini, S. (2022). Novel antibacterial and antifouling PES nanofiltration membrane incorporated with green synthesized nickel-bentonite nanoparticles for heavy metal ions removal. *Chemical Engineering Journal*, *431*, 134116.

Dahoumane, S. A., Mechouet, M., Alvarez, F. J., Agathos, S. N., & Jeffryes, C. (2016). Microalgae: An outstanding tool in nanotechnology. *Bionatura*, *1*(4), 196–201.

Dana, J. D. (1977). *Manual of Mineralogy*. Wiley, New York.

Davarpanah, S., Tehranifar, A., Davarynejad, G., Abadía, J., & Khorasani, R. (2016). Effects of foliar applications of zinc and boron nano-fertilizers on pomegranate (Punica granatum cv. Ardestani) fruit yield and quality. *Scientia Horticulturae*, *210*, 57–64.

Devi, H. S., Boda, M. A., Shah, M. A., Parveen, S., & Wani, A. H. (2019). Green synthesis of iron oxide nanoparticles using *Platanus orientalis* leaf extract for antifungal activity. *Green Processing and Synthesis*, *8*(1), 38–45.

Díaz-Rodríguez, A., Ibarra-Villareal, A., de los Santos-Villalobos, S., Parra-Cota, F. I., Herrera-Sepúlveda, A., & Soto-Padilla, M. Y. (2017). Potencial biotecnológico y ambiental de los microorganismos. *La Soc Acad*, *1*, 52–63.

Dimkpa, C. O., White, J. C., Elmer, W. H., & Gardea-Torresdey, J. (2017). Nanoparticle and ionic Zn promote nutrient loading of sorghum grain under low NPK fertilization. *Journal of Agricultural and Food Chemistry*, *65*(39), 8552–8559.

Dubey, A., & Mailapalli, D. R. (2016). Nanofertilisers, nanopesticides, nanosensors of pest and nanotoxicity in agriculture. In: E. Lichtfouse (eds) *Sustainable Agriculture Reviews*, Springer, Switzerland, 307–330.

Durán, N., Marcato, P. D., Durán, M., Yadav, A., Gade, A., & Rai, M. (2011). Mechanistic aspects in the biogenic synthesis of extracellular metal nanoparticles by peptides, bacteria, fungi, and plants. *Applied Microbiology and Biotechnology*, *90*(5), 1609–1624.

Eichert, T., Kurtz, A., Steiner, U., & Goldbach, H. E. (2008). Size exclusion limits and lateral heterogeneity of the stomatal foliar uptake pathway for aqueous solutes and water suspended nanoparticles. *Physiologia Plantarum*, *134*(1), 151–160.

Elkington, J., & Rowlands, I. H. (1999). Cannibals with forks: The triple bottom line of 21st century business. *Alternatives Journal*, *25*(4), 42.

Elmer, W., Ma, C., & White, J. (2018). Nanoparticles for plant disease management. *Current Opinion in Environmental Science & Health*, *6*, 66–70.

El-Ramady, H., Abdalla, N., Alshaal, T., El-Henawy, A., Elmahrouk, M., Bayoumi, Y., … & Schnug, E. (2018). Plant nano-nutrition: Perspectives and challenges. In: In: K. Gothandam, S. Ranjan, N. Dasgupta, C. Ramalingam, E. Lichtfouse (eds) *Nanotechnology, food security and water treatment. Environmental Chemistry for a Sustainable World*. Springer, Cham. pp. 129–161.

Fabiyi, O. A. (2021). Sustainable management of meloidogyne incognita infecting carrot (*Daucus carota*): Green synthesis of silver nanoparticles with *Cnidoscolus aconitifolius*. *Vegetos*, *34*(2), 277–285.

Fabiyi, O. A., Alabi, R. O., & Ansari, R. A. (2020). Nanoparticles' synthesis and their application in the management of phytonematodes: An overview. In: R. Ansari, R. Rizvi, I. Mahmood (eds) *Management of Phytonematodes: Recent Advances and Future Challenges*. Springer, Singapore, pp. 125–140.

Fabiyi, O. A., Claudius-Cole, A. O., Olatunji, G. A., Abubakar, D. O., & Adejumo, O. A. (2021). Response of *Meloidogyne javanica* to silver nanoparticle liquid from agricultural wastes. *AGRIVITA Journal of Agricultural Science*, *43*(3), 507–517.

Fabiyi, O. A., Olatunji, G. A., & Saadu, A. O. (2018). Suppression of *Heterodera sacchari* in rice with agricultural waste-silver Nano particles. *The Journal of Solid Waste Technology and Management*, *44*(2), 87–91.

Fabiyi, O., Olatunji, G., Atolani, O., & Olawuyi, O. (2020). Preparation of bio-nematicidal nanoparticles of *Eucalyptus officinalis* for the control of cyst nematode (*Heterodera sacchari*). *Journal of Animal and Plant Sciences*, *30*(5), 1172–1177.

Flood, J. (2010). The importance of plant health to food security. *Food Security*, *2*(3), 215–231.

Foley, C. P., Nishimura, N., Neeves, K. B., Schaffer, C. B., & Olbricht, W. L. (2012). Real-time imaging of perivascular transport of nanoparticles during convection-enhanced delivery in the rat cortex. *Annals of Biomedical Engineering*, *40*, 292–303.

Fraceto, L. F., Grillo, R., de Medeiros, G. A., Scognamiglio, V., Rea, G., & Bartolucci, C. (2016). Nanotechnology in agriculture: Which innovation potential does it have? *Frontiers in Environmental Science*, *4*, 20. https://doi.org/10.3389/fenvs.2016. 00020.

Gahlawat, G., & Choudhury, A. R. (2019). A review on the biosynthesis of metal and metal salt nanoparticles by microbes. *RSC Advances*, *9*(23), 12944–12967.

Gajbhiye, M., Kesharwani, J., Ingle, A., Gade, A., & Rai, M. (2009). Fungus-mediated synthesis of silver nanoparticles and their activity against pathogenic fungi in combination with fluconazole. *Nanomedicine: Nanotechnology, Biology and Medicine*, *5*(4), 382–386.

Galdames, A., Mendoza, A., Orueta, M., de Soto García, I. S., Sánchez, M., Virto, I., & Vilas, J. L. (2017). Development of new remediation technologies for contaminated soils based on the application of zero-valent iron nanoparticles and bioremediation with compost. *Resource-Efficient Technologies*, *3*(2), 166–176.

Gamal, M. M. Z. (2018). Nano-particles: A recent approach for controlling stored grain insect pests. *Academia Journal of Agricultural Research*, *6*(5 Conference Proceedings), 88–94.

Gao, A., Hang, R., Huang, X., Zhao, L., Zhang, X., Wang, L., Tang, B., Ma, S., & Chu, P. K. (2014). The effects of titania nanotubes with embedded silver oxide nanoparticles on bacteria and osteoblasts. *Biomaterials*, *35*(13), 4223–4235.

Gao, J. F., Li, H. Y., Pan, K. L., & Si, C. Y. (2016). Green synthesis of nanoscale zero-valent iron using a grape seed extract as a stabilizing agent and the application for quick decolorization of azo and anthraquinone dyes. *RSC Advances*, *6*(27), 22526–22537.

Garcia, M. S., & Lagaron, J. M. (2012). Nanocomposites for food and beverage packaging materials. In : Qingrong Huang (eds.) *Nanotechnology in the Food, Beverage and Nutraceutical Industries*, Springer. pp 335–361. .

Geneti, S. T., Mekonnen, G. A., Murthy, H. C., Mohammed, E. T., Ravikumar, C. R., Gonfa, B. A., & Sabir, F. K. (2022). Biogenic synthesis of magnetite nanoparticles using leaf extract of *Thymus schimperi* and their application for monocomponent removal of chromium and mercury ions from aqueous solution. *Journal of Nanomaterials*,Springer, 473–505.

Gil-Díaz, M., Pérez, R. A., Alonso, J., Miguel, E., Diez-Pascual, S., & Lobo, M. C. (2022). Iron nanoparticles to recover a co-contaminated soil with Cr and PCbs. *Scientific Reports*, *12*(1), 3541.

Gokila, V., Perarasu, V. T., & Rufina, R. (2021). Qualitative comparison of chemical and green synthesized Fe 3 O 4 nanoparticles. *Advances in Nano Research*, *10*(1), 71–76.

Graily-Moradi, F., Mallak, A. M., & Ghorbanpour, M. (2020). Biogenic synthesis of gold nanoparticles and their potential application in agriculture. In *Biogenic Nano-particles and Their Use in Agro-Ecosystems*, 187–204. Springer.

Guo, D., Xie, G., & Luo, J. (2013). Mechanical properties of nanoparticles: Basics and applications. *Journal of Physics D: Applied Physics*, *47*(1), 013001.

Hashem, A. S., Awadalla, S. S., Zayed, G. M., Maggi, F., & Benelli, G. (2018). Pimpinella anisum essential oil nanoemulsions against *Tribolium castaneum*—Insecticidal activity and mode of action. *Environmental Science and Pollution Research International*, *25*(19), 18802–18812.

Hillie, T., & Hlophe, M. (2007). Nanotechnology and the challenge of clean water. *Nature nanotechnology*, *2*(11), 663–664.

Hofmann, T., Lowry, G. V., Ghoshal, S., Tufenkji, N., Brambilla, D., Dutcher, J. R., ... & Wilkinson, K. J. (2020). Technology readiness and overcoming barriers to sustainably implement nanotechnology-enabled plant agriculture. *Nature Food*, *1*(7), 416–425.

Honary, S., Gharaei-Fathabad, E., Barabadi, H., & Naghibi, F. (2013). Fungus-mediated synthesis of gold nanoparticles: A novel biological approach to nanoparticle synthesis. *Journal of Nanoscience and Nanotechnology*, *13*(2), 1427–1430.

Hoo, C. M., Starostin, N., West, P., & Mecartney, M. L. (2008). A comparison of atomic force microscopy (AFM) and dynamic light scattering (DLS) methods to characterize nanoparticle size distributions. *Journal of Nanoparticle Research*, *10*(S1), 89–96.

Hossain, A., Abdallah, Y., Ali, M. A., Masum, M. M. I., Li, B., Sun, G., ... & An, Q. (2019). Lemon-fruit-based green synthesis of zinc oxide nanoparticles and titanium dioxide nanoparticles against soft rot bacterial pathogen *Dickeya dadantii*. *Biomolecules*, *9*(12), 863.

Hulkoti, N. I., & Taranath, T. C. (2014). Biosynthesis of nanoparticles using microbes—A review. *Colloids and Surfaces, Part B: Biointerfaces*, *121*, 474–483. https://doi.org/10.1016/j.colsurfb.2014.05.027.

Hussain, I., Singh, N. B., Singh, A., Singh, H., & Singh, S. C. (2016). Green synthesis of nanoparticles and its potential application. *Biotechnology Letters*, *38*(4), 545–560.

Husseiny, M. I., Abd El-Aziz, M., Badr, Y., & Mahmoud, M. A. (2007). Biosynthesis of gold nanoparticles using *Pseudomonas aeruginosa*. *Spectrochimica Acta. Part A: Molecular and Biomolecular Spectroscopy*, *67*(3–4), 1003–1006.

Iqbal, M., & Umar, S. (2019). Nano-fertilization to enhance nutrient use efficiency and productivity of crop plants. In: A. Husen, M. Iqbal (eds) *Nanomaterials and Plant Potential*. Springer, Cham. pp 473–505.

Irshad, M. A., Shakoor, M. B., Nawaz, R., Yasmeen, T., Arif, M. S., Rizwan, M., ... & Ali, S. (2022). Green and eco-friendly synthesis of TiO2 nanoparticles and their application for removal of cadmium from wastewater: Reaction kinetics study. *Zeitschrift für Physikalische Chemie*, *236*(5), 637–657.

Jadoun, S., Arif, R., Jangid, N. K., & Meena, R. K. (2021). Green synthesis of nanoparticles using plant extracts: A review. *Environmental Chemistry Letters*, *19*(1), 355–374. https://doi.org/10.1007/s10311-020-01074-x.

Jain, N., Bhargava, A., Majumdar, S., Tarafdar, J. C., & Panwar, J. (2011). Extracellular biosynthesis and characterization of silver nanoparticles using *Aspergillus flavus* NJP08: A mechanism perspective. *Nanoscale*, *3*(2), 635–641.

Joglekar, S., Kodam, K., Dhaygude, M., & Hudlikar, M. (2011). Novel route for rapid biosynthesis of lead nanoparticles using aqueous extract of *Jatropha curcas* L. latex. *Materials Letters*, *65*(19–20), 3170–3172.

Joseph, T., & Morrison, M. (2006). Nanotechnology in agriculture and food. *Nanoforum Report*, *2*(2), 3.

Kah, M., Kookana, R. S., Gogos, A., & Bucheli, T. D. (2018). A critical evaluation of nanopesticides and nanofertilizers against their conventional analogues. *Nature Nanotechnology*, *13*(8), 677–684.

Kalimuthu, K., Babu, R. S., Venkataraman, D., Bilal, M., & Gurunathan, S. (2008). Biosynthesis of silver nanocrystals by *Bacillus licheniformis*. *Colloids and Surfaces, Part B: Biointerfaces*, *65*(1), 150–153.

Kanmani, P., & Lim, S. T. (2013). Synthesis and structural characterization of silver nanoparticles using bacterial exopolysaccharide and its antimicrobial activity against food and multidrug resistant pathogens. *Process Biochemistry*, *48*(7), 1099–1106.

Kasthuri, J., Kathiravan, K., & Rajendiran, N. J. N. R. (2009). Phyllanthin-assisted biosynthesis of silver and gold nanoparticles: A novel biological approach. *Journal of Nanoparticle Research*, *11*(5), 1075–1085.

Khot, L. R., Sankaran, S., Maja, J. M., Ehsani, R., & Schuster, E. W. (2012). Applications of nanomaterials in agricultural production and crop protection: A review. *Crop Protection*, *35*, 64–70.

Korbekandi, H., Mohseni, S., Mardani Jouneghani, R., Pourhossein, M., & Iravani, S. (2016). Biosynthesis of silver nanoparticles using Saccharomyces cerevisiae. *Artificial Cells, Nanomedicine and Biotechnology*, *44*(1), 235–239.

Korin, E., Froumin, N., & Cohen, S. (2017). Surface analysis of nanocomplexes by X-ray photoelectron spectroscopy (XPS). *ACS Biomaterials Science and Engineering*, *3*(6), 882–889.

Kumar, B., Smita, K., Cumbal, L., Debut, A., & Angulo, Y. (2017). Biofabrication of copper oxide nanoparticles using Andean blackberry (*Rubus glaucus* Benth.) fruit and leaf. *Journal of Saudi Chemical Society*, *21*, S475–S480.

Laslo, V., Pinzaru, S. C., Zaguła, G., Kluz, M., Vicas, S. I., & Cavalu, S. (2022). Synergic effect of selenium nanoparticles and lactic acid bacteria in reduction cadmium toxicity. *Journal of Molecular Structure*, *1247*, 131325.

Le Goff, A., Holzinger, M., & Cosnier, S. (2011). Enzymatic biosensors based on SWCNT-conducting polymer electrodes. *Analyst*, *136*(7), 1279–1287.

Leggo, P. J. (2000). An investigation of plant growth in an organo-zeolitic substrate and its ecological significance. *Plant and Soil*, *219*(1–2), 135–146.

Lengke, M. F., Fleet, M. E., & Southam, G. (2006). Morphology of gold nanoparticles synthesized by filamentous cyanobacteria from gold (I)–orphology of gold nanoparticles synthesized by . *Langmuir*, *22*(6), 2780–2787.

Leppard, G. G. (2008). Nanoparticles in the environment as revealed by transmission electron microscopy: Detection, characterisation and activities. *Current Nanoscience*, *4*(3), 278–301.

LewisOscar, F., Vismaya, S., Arunkumar, M., Thajuddin, N., Dhanasekaran, D., & Nithya, C. (2016). Algal nanoparticles: Synthesis and biotechnological potentials. *Algae-Organisms for Imminent Biotechnology*, *7*, 157–182.

Li, Y., Cu, Y. T. H., & Luo, D. (2005). Multiplexed detection of pathogen DNA with DNA-based fluorescence nanobarcodes. *Nature Biotechnology*, *23*(7), 885–889.

Lim, H. A., Mishra, A., & Yun, S. I. (2011). Effect of pH on the extra cellular synthesis of gold and silver nanoparticles by *Saccharomyces cerevisae*. *Journal of Nanoscience and Nanotechnology*, *11*(1), 518–522.

Liu, X. M., Feng, Z. B., Zhang, F. D., Zhang, S. Q., & He, X. S. (2006). Preparation and testing of cementing and coating nano-subnanocomposites of slow/controlled-release fertilizer. *Agricultural Sciences in China*, *5*(9), 700–706.

Mahanty, S., Chatterjee, S., Ghosh, S., Tudu, P., Gaine, T., Bakshi, M., … & Chaudhuri, P. (2020). Synergistic approach towards the sustainable management of heavy metals in wastewater using mycosynthesized iron oxide nanoparticles: Biofabrication, adsorptive dynamics and chemometric modeling study. *Journal of Water Process Engineering*, *37*, 101426.

Makarov, V. V., Love, A. J., Sinitsyna, O. V., Makarova, S. S., Yaminsky, I. V., Taliansky, M. E., & Kalinina, N. O. (2014). "Green" nanotechnologies: Synthesis of metal nanoparticles using plants. *Acta Naturae (а н г л о я з ы ч н а я в е р с и я)*, *6*(1(20)), 35–44.

Mali, S. C., Raj, S., & Trivedi, R. (2020). Nanotechnology a novel approach to enhance crop productivity. *Biochemistry and Biophysics Reports*, *24*, 100821.

Manjunath, H. M., Joshi, C. G., & Raju, N. G. (2017). Biofabrication of gold nanoparticles using marine endophytic fungus–*Penicillium citrinum*. *IET Nanobiotechnology*, *11*(1), 40–44.

Marooufpour, N., Alizadeh, M., Hatami, M., & Asgari Lajayer, B. (2019). Biological synthesis of nanoparticles by different groups of bacteria. In: R. Prasad, (eds) *Microbial Nanobionics. Nanotechnology in the Life Sciences*. Springer, Cham, . pp 63–85.

Masum, M. M. I., Siddiqa, M. M., Ali, K. A., Zhang, Y., Abdallah, Y., Ibrahim, E., … & Li, B. (2019). Biogenic synthesis of silver nanoparticles using *Phyllanthus emblica* fruit extract and its inhibitory action against the pathogen *Acidovorax oryzae* strain RS-2 of rice bacterial brown stripe. *Frontiers in Microbiology*, *10*, 820.

Mejias, J. H., Salazar, F., Pérez Amaro, L., Hube, S., Rodriguez, M., & Alfaro, M. (2021). Nanofertilizers: A cutting-edge approach to increase nitrogen use efficiency in grasslands. *Frontiers in Environmental Science*,*9*, 52.

Meng, R., Wu, Z., Xie, Q. T., Cheng, J. S., & Zhang, B. (2021). Preparation and characterization of zein/carboxymethyl dextrin nanoparticles to encapsulate curcumin: Physicochemical stability, antioxidant activity and controlled release properties. *Food Chemistry*, *340*, 127893.

Merino, D., Tomadoni, B., Salcedo, M. F., Mansilla, A. Y., Casalongué, C. A., & Alvarez, V. A. (2020). Nanoclay as carriers of bioactive molecules applied to agriculture. In: O. V. Kharissova et al. (eds.), *Handbook of Nanomaterials and Nanocomposites for Energy and Environmental Applications*. Springer Nature, Switzerland. pp 1–22.

Mingyu, S., Chao, L., Chunxiang, Q., Lei, Z., Liang, C., Hao, H., … & Fashui, H. (2008). Nano-anatase relieves the inhibition of electron transport caused by linolenic acid in chloroplasts of spinach. *Biological Trace Element Research*, *122*(1), 73–81.

Mishra, A., Tripathy, S. K., & Yun, S. I. (2011). Bio-Synthesis of gold and silver nanoparticles from *Candida guilliermondii* and their antimicrobial effect against pathogenic bacteria. *Journal of Nanoscience and Nanotechnology*, *11*(1), 243–248. https://doi.org/10.1166/jnn.2011.3265.

Mittal, A. K., Chisti, Y., & Banerjee, U. C. (2013). Synthesis of metallic nanoparticles using plant extracts. *Biotechnology Advances*, *31*(2), 346–356.

Mitter, N., & Hussey, K. (2019). Moving policy and regulation forward for nanotechnology applications in agriculture. *Nature Nanotechnology*, *14*(6), 508–510.

Mohan, A. C., & Renjanadevi, B. J. P. T. (2016). Preparation of zinc oxide nanoparticles and its characterization using scanning electron microscopy (SEM) and X-ray diffraction (XRD). *Procedia Technology*, *24*, 761–766.

Mortazavi, S. M., Khatami, M., Sharifi, I., Heli, H., Kaykavousi, K., Sobhani Poor, M.H., Kharazi, S., & Nobre, M. A. L. Bacterial biosynthesis of gold nanoparticles using *Salmonella enterica* subsp. enterica serovar Typhi isolated from blood and stool specimens of patients. *Journal of Cluster Science*, *28*, pp.2997–3007.

Mughal, B., Zaidi, S. Z. J., Zhang, X., & Hassan, S. U. (2021). Biogenic nanoparticles: Synthesis, characterisation and applications. *Applied Sciences*, *11*(6), 2598.

Narayanan, K. B., & Sakthivel, N. (2010). Biological synthesis of metal nanoparticles by microbes. *Advances in Colloid and Interface Science*, *156*(1–2), 1–13.

Nath, N., & Chilkoti, A. (2004). Label free colorimetric biosensing using nanoparticles. *Journal of Fluorescence*, *14*(4), 377–389.

Natrayan, L., Yogeshwaran, S., Yuvaraj, L., & Kumar, M. S. (2019, October). Effect of graphene reinforcement on mechanical and microstructure behavior of AA8030/graphene composites fabricated by stir casting technique. In *AIP Conference Proceedings* (Vol. 2166, No. 1, 020012). AIP Publishing LLC.

Navrotsky, A., Petrovic, I., Hu, Y., Chen, C. Y., & Davis, M. E. (1995). Energetics of microporous materials. *Journal of Non-Crystalline Solids*, *192*, 474–477.

Neme, K., Nafady, A., Uddin, S., & Tola, Y. B. (2021). Application of nanotechnology in agriculture, postharvest loss reduction and food processing: Food security implication and challenges. *Heliyon*, *7*(12), e08539.

Noruzi, M., Zare, D., Khoshnevisan, K., & Davoodi, D. (2011). Rapid green synthesis of gold nanoparticles using *Rosa hybrida* petal extract at room temperature. *Spectrochimica Acta. Part A: Molecular and Biomolecular Spectroscopy*, *79*(5), 1461–1465.

Nune, S. K., Chanda, N., Shukla, R., Katti, K., Kulkarni, R. R., Thilakavathy, S., … & Katti, K. V. (2009). Green nanotechnology from tea: Phytochemicals in tea as building blocks for production of biocompatible gold nanoparticles. *Journal of Materials Chemistry*, *19*(19), 2912–2920.

Nurmi, J. T., Tratnyek, P. G., Sarathy, V., Baer, D. R., Amonette, J. E., Pecher, K., … & Driessen, M. D. (2005). Characterization and properties of metallic iron nanoparticles: Spectroscopy, electrochemistry, and kinetics. *Environmental Science and Technology*, *39*(5), 1221–1230.

Ogunyemi, S. O., Zhang, F., Abdallah, Y., Zhang, M., Wang, Y., Sun, G., … & Li, B. (2019). Biosynthesis and characterization of magnesium oxide and manganese dioxide nanoparticles using *Matricaria Chamomilla* L. extract and its inhibitory effect on *Acidovorax oryzae* strain RS-2. *Artificial Cells, Nanomedicine and Biotechnology*, *47*(1), 2230–2239.

Omar, H. H., Bahabri, F. S., & El-Gendy, A. M. (2017). Biopotential application of synthesis nanoparticles as antimicrobial agents by using *Laurencia papillosa*. *International Journal of Pharmacology*, *13*(3), 303–312.

Oscar, F. L., Bakkiyaraj, D., Nithya, C., & Thajuddin, N. (2014). Deciphering the diversity of microalgal bloom in wastewater-an attempt to construct potential consortia for bioremediation. *International Journal of Current Microbiology and Applied Sciences*, *2278*, 92.

Othman, S. H. (2014). Bio-nanocomposite materials for food packaging applications: Types of biopolymer and nano-sized filler. *Agriculture and Agricultural Science Procedia*, *2*, 296–303.

Ovais, M., Khalil, A. T., Islam, N. U., Ahmad, I., Ayaz, M., Saravanan, M., … & Mukherjee, S. (2018). Role of plant phytochemicals and microbial enzymes in biosynthesis of metallic nanoparticles. *Applied Microbiology and Biotechnology*, *102*(16), 6799–6814.

Yadav, P., Yadav, R. M., & Singh, D. P. (2012). Mechanical milling: A top down approach for the synthesis of nanomaterials and nanocomposites. *Nanoscience and Nanotechnology*, *2*(3), 22–48.

Pal, S., Tak, Y. K., & Song, J. M. (2007). Does the antibacterial activity of silver nanoparticles depend on the shape of the nanoparticle? A study of the gram-negative bacterium *Escherichia coli*. *Applied and Environmental Microbiology*, *73*(6), 1712–1720.

Pereira, E. I., Minussi, F. B., da Cruz, C. C., Bernardi, A. C., & Ribeiro, C. (2012). Urea–montmorillonite-extruded nanocomposites: A novel slow-release material. *Journal of Agricultural and Food Chemistry*, *60*(21), 5267–5272.

Ponmurugan, P. (2017). Biosynthesis of silver and gold nanoparticles using *Trichoderma atroviride* for the biological control of Phomopsis canker disease in tea plants. *IET Nanobiotechnology*, *11*(3), 261–267.

Pouratashi, M., & Iravani, H. (2012). Farmers' knowledge of integrated pest management and learning style preferences: Implications for information delivery. *International Journal of Pest Management*, *58*(4), 347–353.

Pradhan, S., & Mailapalli, D. R. (2017). Interaction of engineered nanoparticles with the agri-environment. *Journal of Agricultural and Food Chemistry*, *65*(38), 8279–8294.

Prasad, R., Bhattacharyya, A., & Nguyen, Q. D. (2017). Nanotechnology in sustainable agriculture: Recent developments, challenges, and perspectives. *Frontiers in Microbiology*, *8*, 1014.

Prasad, R., Jha, A. K., & Prasad, K. (Eds.). (2018). *Exploring the Realms of Nature for Nanosynthesis*. Springer International Publishing, Cham.

Rai, V., Acharya, S., & Dey, N. (2012). Implications of nanobiosensors in agriculture. *Journal of Biomaterials and Nanobiotechnology*, *3*(2), 315–324.

Rajasree, S. R., & Suman, T. Y. (2012). Extracellular biosynthesis of gold nanoparticles using a gram negative bacterium *Pseudomonas fluorescens*. *Asian Pacific Journal of Tropical Disease*, *2*, S796–S799.

Rajeshkumar, S., Malarkodi, C., Paulkumar, K., Vanaja, M., Gnanajobitha, G., & Annadurai, G. (2014). Algae mediated green fabrication of silver nanoparticles and examination of its antifungal activity against clinical pathogens. *International Journal of Metals*, *2014*, 1–8.

Raliya, R., Saharan, V., Dimkpa, C., & Biswas, P. (2017). Nanofertilizer for precision and sustainable agriculture: Current state and future perspectives. *Journal of Agricultural and Food Chemistry*, *66*(26), 6487–6503.

Ramesh, K., Biswas, A. K., Somasundaram, J., & Rao, A. S. (2010). Nanoporous zeolites in farming: Current status and issues ahead. *Current Science*, 760–764.

Rana, A., Yadav, K., & Jagadevan, S. (2020). A comprehensive review on green synthesis of nature-inspired metal nanoparticles: Mechanism, application and toxicity. *Journal of Cleaner Production*, *272*, 122880. https://doi.org/10.1016/j.jclepro.2020.122880.

Reddy, L. H., Arias, J. L., Nicolas, J., & Couvreur, P. (2012). Magnetic nanoparticles: Design and characterization, toxicity and biocompatibility, pharmaceutical and biomedical applications. *Chemical Reviews*, *112*(11), 5818–5878. http://doi.org/10.1021/cr300068p.

Ren, J. (2019, October). Research on green synthetic iron nanoparticles. In *IOP Conference Series: Materials Science and Engineering* (Vol. 612, No. 2, 022025). IOP Publishing.

Sabo-Attwood, T., Unrine, J. M., Stone, J. W., Murphy, C. J., Ghoshroy, S., Blom, D., ... & Newman, L. A. (2012). Uptake, distribution and toxicity of gold nanoparticles in tobacco (*Nicotiana xanthi*) seedlings. *Nanotoxicology*, *6*(4), 353–360.

Salama, D. M., Abd El-Aziz, M. E., Rizk, F. A., & Abd Elwahed, M. S. A. (2021). Applications of nanotechnology on vegetable crops. *Chemosphere*, *266*, 129026.

Sand, L. B., & Mumpton, F. A. (1978). *Natural Zeolites: Occurrence, Properties, and Use (No. CONF-760626-(Exc.))*. Pergamon Press, Inc., Elmsford, NY.

Sassolas, A., Blum, L. J., & Leca-Bouvier, B. D. (2012). Immobilization strategies to develop enzymatic biosensors. *Biotechnology Advances*, *30*(3), 489–511.

Scott, N., & Chen, H. (2012). Nanoscale science and engineering for agriculture and food systems. *Industrial Biotechnology*, *8*(6), 340–343.

Seeniappan, H., Venkatesan, B., & Krishnan, N. N. (2021). A comparative assessment of performance and emission characteristics of a DI diesel engine fuelled with ternary blends of two higher alcohols with lemongrass oil biodiesel and diesel fuel. *Energy and Environment*, *33*(6), 1134–1159.

Shankar, S. S., Ahmad, A., Pasricha, R., & Sastry, M. (2003). Bioreduction of chloroaurate ions by geranium leaves and its endophytic fungus yields gold nanoparticles of different shapes. *Journal of Materials Chemistry*, *13*(7), 1822–1826.

Shankar, S. S., Rai, A., Ahmad, A., & Sastry, M. (2005). Controlling the optical properties of lemongrass extract synthesized gold nanotriangles and potential application in infrared-absorbing optical coatings. *Chemistry of Materials*, *17*(3), 566–572.

Sharifi, M., Faryabi, K., Talaei, A.J., Shekha, M.S., Ale-Ebrahim, M., Salihi, A., Nanakali, N.M.Q., Aziz, F.M., Rasti, B., Hasan, A., & Falahati, M., 2020. Antioxidant properties of gold nanozyme: A review. *Journal of Molecular Liquids*, 297, p.112004.

Sharma, A., Sharma, S., Sharma, K., Chetri, S.P., Vashishtha, A., Singh, P., Kumar, R., Rathi, B.,& Agrawal, V., 2016. Algae as crucial organisms in advancing nanotechnology: a systematic review. *Journal of applied phycology*, 28, pp.1759-1774.

Sharma, G., Jasuja, N. D., Kumar, M., & Ali, M. I. (2015). Biological synthesis of silver nanoparticles by cell-free extract of *Spirulina platensis*. *Journal of nanotechnology*, *2015*, 1–6.

Shen, W., Qu, Y., Pei, X., Li, S., You, S., Wang, J., Zhang, Z., & Zhou, J., 2017. Catalytic reduction of 4-nitrophenol using gold nanoparticles biosynthesized by cell-free extracts of Aspergillus sp. WL-Au. *Journal of Hazardous Materials*, 321, 299–306.

Shenoy, S., Tarafder, K., & Sridharan, K. (2021). Bimetallic nanoparticles grafted ZnO hierarchical structures as efficient visible light driven photocatalyst: An experimental and theoretical study. *Journal of Molecular Structure, 1236*, 130355.

Shukla, S. K., Mishra, A. K., Arotiba, O. A., & Mamba, B. B. (2013). Chitosan-based nanomaterials: A state-of-the-art review. *International Journal of Biological Macromolecules, 59*, 46–58.

Shukla, V. K., Yadav, R. S., Yadav, P., & Pandey, A. C. (2012). Green synthesis of nanosilver as a sensor for detection of hydrogen peroxide in water. *Journal of Hazardous Materials, 213*, 161–166.

Siddiqi, K. S., & Husen, A. (2016). Fabrication of metal and metal oxide nanoparticles by algae and their toxic effects. *Nanoscale Research Letters, 363*, 1–11.

Siddiqui, M. H., & Al-Whaibi, M. H. (2014). Role of nano-SiO2 in germination of tomato (*Lycopersicum esculentum* seeds Mill.). *Saudi Journal of Biological Sciences, 21*(1), 13–17.

Singaravelu, G., Arockiamary, J. S., Kumar, V. G., & Govindaraju, K. (2007). A novel extracellular synthesis of monodisperse gold nanoparticles using marine alga, *Sargassum wightii* Greville. *Colloids and Surfaces, Part B: Biointerfaces, 57*(1), 97–101.

Singh, M. D. (2017). Nano-fertilizers is a new way to increase nutrients use efficiency in crop production. *International Journal of Agriculture Sciences, ISSN, 9*(7), 0975–3710.

Singh, R. P., Handa, R., & Manchanda, G. (2021). Nanoparticles in sustainable agriculture: An emerging opportunity. *Journal of Controlled Release, 329*, 1234–1248.

Slavov, L., Abrashev, M. V., Merodiiska, T., Gelev, C., Vandenberghe, R. E., Markova-Deneva, I., &Nedkov, I. (2010). Raman spectroscopy investigation of magnetite nanoparticles in ferrofluids. *Journal of Magnetism and Magnetic Materials, 322*(14), 1904–1911. https://doi.org/10.1016/j.jmmm.2010.01.005.

Solanki, P., Bhargava, A., Chhipa, H., Jain, N., & Panwar, J. (2015). Nano-fertilizers and their smart delivery system. In: Mahendra Rai, Caue Ribeiro, Luiz Mattoso, Nelson Duran (eds.),*Nanotechnologies in Food and Agriculture*, Springer,pp. 81–101.

Sone, B. T., Diallo, A., Fuku, X. G., Gurib-Fakim, A., & Maaza, M. (2020). Biosynthesized CuO nano-platelets: Physical properties & enhanced thermal conductivity nanofluidics. *Arabian Journal of Chemistry, 13*(1), 160–170.

Songkhum, P., Wuttikhun, T., Chanlek, N., Khemthong, P., & Laohhasurayotin, K. (2018). Controlled release studies of boron and zinc from layered double hydroxides as the micronutrient hosts for agricultural application. *Applied Clay Science, 152*, 311–322.

Srivastava, S., Shukla, A., Rajput, V.D., Kumar, K., Minkina, T., Mandzhieva, S., Shmaraeva, A., & Suprasanna, P., 2021. Arsenic remediation through sustainable phytoremediation approaches. *Minerals*, 11(9),936.

Stetefeld, J., McKenna, S. A., & Patel, T. R. (2016). Dynamic light scattering: A practical guide and applications in biomedical sciences. *Biophysical Reviews, 8*(4), 409–427. https://doi.org/10.1007/s12551-016-0218-6.

Su, X. L., & Li, Y. (2004). Quantum dot biolabeling coupled with immunomagnetic separation for detection of escherichia coli O157: H7. *Analytical Chemistry, 76*(16), 4806–4810.

Suman, T. Y., Rajasree, S. R., & Kirubagaran, R. (2015). Evaluation of zinc oxide nanoparticles toxicity on marine algae *Chlorella vulgaris* through flow cytometric, cytotoxicity and oxidative stress analysis. *Ecotoxicology and Environmental Safety, 113*, 23–30.

Syu, Y. Y., Hung, J. H., Chen, J. C., & Chuang, H. W. (2014). Impacts of size and shape of silver nanoparticles on *Arabidopsis* plant growth and gene expression. *Plant physiology and biochemistry, 83*, 57–64.

Taha, M., Hassan, M., Essa, S., & Tartor, Y. (2013). Use of Fourier transform infrared spectroscopy (FTIR) spectroscopy for rapid and accurate identification of Yeasts isolated from human and animals. *International Journal of Veterinary Science and Medicine, 1*(1), 15–20.

Taylor,A. F., Rylott, E. L., Anderson, C. W., & Bruce, N. C. (2014). Investigating the toxicity, uptake, nanoparticle formation and genetic response of plants to gold. *PLOS One, 9*(4), e93793. https://doi.org/10.1371/journal.pone.0093793.

Tilman, D., Balzer, C., Hill, J., & Befort, B. L. (2011). Global food demand and the sustainable intensification of agriculture. *Proceedings of the National Academy of Sciences of the United States of America, 108*(50), 20260–20264.

Tilman, D., Cassman, K. G., Matson, P. A., Naylor, R., & Polasky, S. (2002). Agricultural sustainability and intensive production practices. *Nature, 418*(6898), 671–677.

Titus, D., Samuel, E. J. J., & Roopan, S. M. (2019). Green synthesis, characterization and applications of nanoparticles-chapter 12. *Micro and Nano Technologies*, 303–319.

Tocha, E., Schönherr, H., & Vancso, G. J. (2006). Quantitative nanotribology by AFM: A novel universal calibration platform. *Langmuir, 22*(5), 2340–2350.

Torney, F. (2009). Nanoparticle mediated plant transformation. *Emerging technologies in plant science research. Interdepartmental plant physiology major fall seminar series Physics* (Vol. 696).

Trenkel, M. E. (2010). *Slow-and Controlled-Release and Stabilized Fertilizers: An Option for Enhancing Nutrient Use Efficiency in Agriculture.* IFA, International Fertilizer Industry Association.

Ulrichs, C., Mewis, I., & Goswami, A. (2005). Crop diversification aiming nutritional security in West Bengal: Biotechnology of stinging capsules in nature's water-blooms. *Ann Tech Issue of State Agri Technologists Service Assoc*, 1–18.

Veeraputhiran, V. (2013). Bio-catalytic synthesis of silver nanoparticles. *International Journal of ChemTech Research*, *5*(5), 255–2562.

Verma, A., & Bharadvaja, N. (2022). Plant-mediated synthesis and characterization of silver and copper oxide nanoparticles: Antibacterial and heavy metal removal activity. *Journal of Cluster Science*, *33*(4), 1697–1712.

Vetchinkina, E. P., Loshchinina, E. A., Vodolazov, I. R., Kursky, V. F., Dykman, L. A., & Nikitina, V. E. (2017). Biosynthesis of nanoparticles of metals and metalloids by basidiomycetes. Preparation of gold nanoparticles by using purified fungal phenol oxidases. *Applied Microbiology and Biotechnology*, *101*(3), 1047–1062.

Vu, K. A., & Mulligan, C. N. (2022). Remediation of oil-contaminated soil using Fe/Cu nanoparticles and biosurfactants. *Environmental Technology*, Taylor and Francis, *44*, 3446–3458.

Wang, L., Li, X., Zhang, G., Dong, J., & Eastoe, J. (2007). Oil-in-water nanoemulsions for pesticide formulations. *Journal of Colloid and Interface Science*, *314*(1), 230–235.

Wang, A. Z., Langer, R., & Farokhzad, O. C. (2012). Nanoparticle delivery of cancer drugs. *Annual Review of Medicine*, *63*, 185–198.

Wei, Y., Fang, Z., Zheng, L., & Tsang, E. P. (2017). Biosynthesized iron nanoparticles in aqueous extracts of *Eichhornia crassipes* and its mechanism in the hexavalent chromium removal. *Applied Surface Science*, *399*, 322–329.

Werner, P., Eichler, S., Mariani, G., Kögler, R., & Skorupa, W. (1997). Investigation of C x Si defects in C implanted silicon by transmission electron microscopy. *Applied Physics Letters*, *70*(2), 252–254.

Worrall, E. A., Hamid, A., Mody, K. T., Mitter, N., & Pappu, H. R. (2018). Nanotechnology for plant disease management. *Agronomy*, *8*(12), 285.

Wypij, M., Czarnecka, J., Świecimska, M., Dahm, H., Rai, M., & Golinska, P. (2018). Synthesis, characterization and evaluation of antimicrobial and cytotoxic activities of biogenic silver nanoparticles synthesized from *Streptomyces xinghaiensis* of1 strain. *World Journal of Microbiology and Biotechnology*, *34*(2), 1–13.

Xiubin, H., & Zhanbin, H. (2001). Zeolite application for enhancing water infiltration and retention in loess soil. *Resources, Conservation and Recycling*, *34*(1), 45–52.

Xu, R. (2008). Progress in nanoparticles characterization: Sizing and zeta potential measurement. *Particuology*, *6*(2), 112–115.

Yamanaka, M., Hara, K., & Kudo, J. (2005). Bactericidal actions of a silver ion solution on Escherichia coli, studied by energy-filtering transmission electron microscopy and proteomic analysis. *Applied and Environmental Microbiology*, *71*(11), 7589–7593.

Yang, F. L., Li, X. G., Zhu, F., & Lei, C. L. (2009). Structural characterization of nanoparticles loaded with garlic essential oil and their insecticidal activity against *Tribolium castaneum* (Herbst)(Coleoptera: Tenebrionidae). *Journal of Agricultural and Food Chemistry*, *57*(21), 10156–10162.

Yendluri, R. (2019). Nanoclays: A new avenue for drug delivery. *EC Pharmacology Toxicology ECO*, *2*, 20–22.

Zaini, P. A., De La Fuente, L., Hoch, H. C., & Burr, T. J. (2009). Grapevine xylem sap enhances biofilm development by *Xylella fastidiosa*. *FEMS Microbiology Letters*, *295*(1), 129–134.

Zhang, H., & Hu, X. (2018). Biosynthesis of Pd and Au as nanoparticles by a marine bacterium *Bacillus sp.* GP and their enhanced catalytic performance using metal oxides for 4-nitrophenol reduction. *Enzyme and Microbial Technology*, *113*, 59–66.

Zhang, X., Qu, Y., Shen, W., Wang, J., Li, H., Zhang, Z., Li, S., & Zhou, J., 2016. Biogenic synthesis of gold nanoparticles by yeast Magnusiomyces ingens LH-F1 for catalytic reduction of nitrophenols. *Colloids and Surfaces A: Physicochemical and Engineering Aspects*, 497, 280–285.

Zhang, Q., Jeppesen, D. K., Higginbotham, J. N., Graves-Deal, R., Trinh, V. Q., Ramirez, M.A., Sohn, Y., Neininger, A.C., Taneja, N., McKinley, E.T., & Coffey, R. J. (2021). Supermeres are functional extracellular nanoparticles replete with disease biomarkers and therapeutic targets. *Nature Cell Biology*, *23*(12), 1240–1254.

Zhu, L., Cao, Y., & Cao, G. (2014). Electrochemical sensor based on magnetic molecularly imprinted nanoparticles at surfactant modified magnetic electrode for determination of bisphenol A. *Biosensors and Bioelectronics*, *54*, 258–261.

8 Green Nanotechnology for Remediation of Heavy Metal–Contaminated Soils

Current Status and Future Scope

Kanika Khanna, Nandni Sharma, Jaspreet Kour, Arun Dev Singh, Sandeep Kour, Tamanna Bhardwaj, Neerja Sharma, Puja Ohri, and Renu Bhardwaj

8.1 INTRODUCTION

Soil is considered to be the central player of the ecosystem that supports an enormous variety of life, and it is also a pre-requisite for overall plant development as well as crop production. It is a fundamental part of the ecosystem, which regulates numerous planetary processes and protects crop yield and production. However, in the past few years, we have witnessed soil deterioration due to several factors, such as industrialization, urbanization, and abundant usage of chemical fertilizers in agriculture and many other anthropogenic activities (Lal, 2015). Pollutants linked with public health interests include many toxic metals, metalloids, pesticides, hydrocarbons, greenhouse gases, and nuclear residues. All of these are non-biodegradable as well as having high toxicity and cause major environmental pollution (Amin et al., 2021). Heavy metals are notably the most significant contaminants, which are rising exponentially, causing negative effects on human health. Hence, it is quite challenging to confer food safety and security by enhancing crop productivity in nations with a skyrocketing population (Rajput et al., 2022). They originate from weathering of rocks and are continuously increasing in soils (Bakshi and Abhilash, 2020) due to anthropogenic activities, namely, mining, textile mills, metal industries, tanneries, effluent, and paper and pulp industries, etc., from which they are discharged as effluent into water bodies and further accumulate within soils (Rai et al., 2019).

The existence of heavy metals within soils has caused negative effects on plants and soil microbiomes. The adverse effects they cause lead to the generation of reactive oxygen species (ROS), which further disrupt the macro-molecular structures of living organisms, thereby affecting their ecology as well as nutritional quality. For instance, Cu within soil disturbs the growth and developmental processes of plants and affects certain biochemical and physiological attributes (Saleem et al., 2020). The heavy metal–mediated alterations within plants vary with plant type, species, and heavy metal type, as well as physicochemical properties of soil (Chaplygin et al., 2021). Several studies have also reported that metal-polluted soils impair many biological processes of plants as well as microbiomes and their diversity (Rajput et al., 2022). The level of antioxidants, enzymatic activities, and biomass as well as crop yield is affected, while free radical production is considerably increased (Rajput et al., 2022; Huang et al., 2019). Interestingly, heavy metal stabilization at a particular site limits its mobility, availability, and leaching toward other sites, thereby affecting its translocation (Suman et al., 2018).

DOI: 10.1201/b23308-8

The use of nanotechnology is exponentially gaining importance due to its expanding role in multidisciplinary fields and broad potential (Tripathi et al., 2017). The swift expansion of this novel technology has reached a stable position with a positive impact on living organisms as well as the environment. Due to characteristic features possessed by nanoparticles (NPs) in terms of small size, shape, surface area, and reactivity, they are being explored for extensive applications (Brunner et al., 2006). The smaller size allows them to penetrate various plant barriers and help in targeting a specific cellular location, while the additional surface area enhances their adsorption and target delivery towards different sites (Kashyap et al., 2015). Moreover, the influence of NPs usually depends on various other factors, such as particle size, plant type, species, surface area, etc. (Akbari et al., 2011). The effect of NPs on plants has been broadly studied. To elucidate, TiO_2 NPs minimized UV-mediated oxidative stress markers in spinach by enhancing the photosynthetic attributes, mineral acquisition, and nitrate assimilation, and fixed inorganic nitrogen into an organic form (Lei et al., 2008). TiO_2 NPs also enhanced antioxidant levels in kidney beans (Jacob et al., 2013). Likewise, carbon nanotubes are another important NP with the potential to penetrate tomato seed coats, thereby facilitating the water acquisition needed for seed germination. Along with this, many findings have supported that they also upregulate the stress-responsive genes by enhancing the activities of peroxidases (Smimova et al., 2012).

Nano-remediation is one of the classic strategies that have gained huge attention in the past few years to protect the planet from environmental contamination and clean polluted soils. Stimulation of microbial activities through using nanoparticles is the pivotal step for the elimination of toxic components, also named nano-remediation (Amin et al., 2021). The biggest advantage of this strategy is its effectiveness and ability to work well in assisting with the ongoing processes of bioremediation (Benjamin et al., 2019). The present chapter mainly focuses on the role of green nanotechnology in the remediation of heavy metal stresses along with the sustainable and green synthesis of biogenic nanoparticles to remediate environmental pollution.

8.2 HEAVY METAL IMPACT ON THE ENVIRONMENT AND HUMAN HEALTH

HMs are natural constituents of the environment and are generally available in different concentrations in the earth's crust. Their normal distribution and bioavailability are crucial for the growth and development of plants as well as vital for the survival of ecosystems. However, among these HMs, some undergo continuous accumulation into the environment and further, enter inside living tissues. These HMs include elements like arsenic, lead, nickel, chromium, mercury, cadmium, copper, etc. (Rafique et al., 2022). Their accumulation inside the environment is mainly triggered by urbanization, industrialization, mining, and incorrect agricultural practices (Mitra et al., 2022). Excessive amounts of these HMs and their presence as trace elements may lead to multiple physiological, morphological, and genetic abnormalities (Luo et al., 2012). Food plays a primary role in fulfilling our nutritional requirements and is also found to be the potential source for human exposure to and intake of HMs (90%), whereas 10% comes from skin contact and inhalation (Kim et al., 2009; Khan et al., 2015). Growing food demands are consistently polluting our food chains through the unchecked applications of agrochemicals, and through irrigation methods utilizing wastewater and water containing industrial effluent. Mercury (Hg) is among the hazardous materials that exist in the environment. Their toxic concentrations are consistently rising in the atmosphere due to increasing anthropogenic activities. Mercury undergoes specific interactions with aquatic sediments and leads to the production of highly toxic methyl mercury (Gworek et al., 2020). Methyl mercury enters the human body through the contaminated food chain, from fish, wildlife, other seafood, etc. After it enters the human body, methyl mercury interferes with blood circulation and further, leads to neurological disorders (Rice et al., 2014). Mercury is also known to induce certain skin-related diseases, among them acrodynia (pink disease), a dermatological ailment that leads to pink skin coloration under mercury exposure (Horowitz et al., 2002). Another heavy metal, cadmium (Cd), is available in the environment, and its concentrations are rising due to natural and manmade disturbances. Cd enters aquatic ecosystems through industrial waste, surface runoff, and absorption, and further, is directed into soil sediments. Animals and humans are poisoned by Cd through breathing air, through food, as well as through contaminated water intake (Haider et al., 2021). Cd is found to have severe effects on the CNS (central nervous system) and PNS (peripheral nervous system), with multiple disorders and other clinical manifestations such as learning disabilities, mental retardation, olfactory dysfunctions, peripheral neuropathy, neurological disturbances, behavioral changes, and impaired motor functioning in children and adults (Marchetti, 2014; Miura et al., 2013). Nephrotoxicity is another clinical problem, in which Cd toxicity induces multiple impacts, including glucosuria, phosphaturia, Fanconi-like syndrome, aminoaciduria, etc. (Mitra et al., 2022).

Lead (Pb) is another HM that is non-biodegradable and is available in minute concentrations in the environment. However, anthropogenic activities like the burning of fossil fuels, mining, and other manufacturing activities are increasing the Pb levels in the atmosphere. Children are mainly affected by Pb toxicity when they come in contact with dust containing Pb (Loh et al., 2016). Pb shows its deleterious effects in all body organs, but it has most impact on the kidneys. Chronic Pb symptoms lead to nephropathy, which can be diagnosed as hyperplasia, atrophy of the tubules, renal failure, intestinal fibrosis, etc. (Mitra et al., 2022). Pb also leads to carcinogenic effects, which initiate impaired DNA repair mechanisms, chromosomal abbreviations, as well as upregulation of tumor-inducing genes in multiple cells (Nag et al., 2022). Chronic Pb toxicity in humans may lead to problems such as hypertension, arteriosclerosis, thrombosis, increased inflammation, disrupted vascular smooth muscle Ca^{2+} signaling, etc. (Mitra et al., 2022). Nickel (Ni) is considered to have extensive industrial applications and enters the atmosphere through natural and manmade activities in nature (Li et al., 2019). Ni has multiple harmful impacts on human populations, such as lung cancer, cardiovascular disease, kidney-related issues, and nasal and other potential allergies on inhalation and exposure to contaminated air (Genchi et al., 2020). Ni triggers carcinogenicity by the alteration of multiple regulatory processes like transcription factor management, gene regulation, and ROS production. It is a potential regulator of the expression mechanism of micro RNAs, mRNAs, and long non-coding RNAs. Ni also mediates the promotor methylation process as well as leading to downregulation of the gene 3 (MEG3); however, impairment of these processes can lead to cancer (Zambelli et al., 2016; Zhou et al., 2017). Cobalt (Co) is readily available in the atmosphere and in water, soil, rocks, and vegetation, and is generally used to manufacture alloys. Due to its slow excretion rate and low toxicity, Co has both harmful and beneficial impacts. But at high concentrations, Co is fatal to humans (Doming, 1989). Cobalt toxicity can lead to cardiomyopathy, which might be slow but is fatal (Packer et al., 2016).

Manganese (Mn) is available in different oxidation states in the environment. Being the most plentiful HM, it is also required for multiple physiological functions in different living tissues. The combustion of methylcyclopentadienyl manganese tricarbonyl (MMT) in gasoline releases certain manganese compounds into the environment; thus, excessive amounts will lead to toxicity to humans (Mitra et al., 2022; Loranger and Zayed, 1995; O'Neal and Zheng, 2015). In humans, trace amounts of Mn are potentially required to reduce apoptosis in cells, thus mediating neuroprotective actions. However excessive Mn concentrations will lead to neurological disorders like Parkinson's disease and Alzheimer's disease due to apoptotic cell death and disrupted homeostasis (Goldhaber, 2003). Arsenic (As) is considered to hamper the functioning of the CNS and thus leads to cognitive impairment. This may cause neurological illness and other neurodegenerative diseases (Garza-Lombó et al., 2019). Chronic arsenic concentrations are possible reasons for multiple skin diseases such as hyperpigmentation, some skin cancers, hyperkeratosis, etc. Exposure to high levels of arsenic can trigger symptoms of a specific early skin cancer, namely, Bowen's disease (Huang et al., 2019).

8.3 REACTIVE OXYGEN SPECIES AND OXIDATIVE BURST DUE TO HEAVY METAL TOXICITY

The generation of ROS or oxidative stress markers in plants is an iterative process under natural conditions. The stress markers are usually generated during the critical metabolic processes, including respiration and photosynthesis, occurring in numerous cell organelles, namely, peroxisomes, mitochondria, and chloroplasts. However, their level of production increases enormously when plants grow under the influence of some environmental stress. For instance, the primary toxicological mechanism in plants growing under heavy metal stress is the production of ROS (Khanna et al., 2019; Sharma et al., 2012). Furthermore, oxidative stress brought on by ROS generation offers a potential hazard to plants by causing DNA damage, electrolyte leakage, membrane damage, lipid peroxidation, and eventually, cell apoptosis (Meriga et al., 2004; Sharma et al., 2012).

Moreover, oxidative stress can also induce protein oxidation. It can also be linked to the generation of ROS, which further leads to changes in site-specific amino acids, peptide chain disintegration, cross-linked reaction product aggregation, modified redox, and increased protein sensitivity to proteolysis (Sharma et al., 2012). It has also been documented that the increased level of carbonylated proteins is frequently employed as a marker of protein oxidation during examination of plant tissues affected by oxidative stress (Moller and Kristensen, 2004).

Furthermore, thiol groups containing amino acids are among the most ROS-sensitive types of amino acids. A thiyl radical is generated when activated oxygen removes a protein from the cysteine residue; this thiyl residue can cross-link with another thiyl radical to create a disulfide bridge (Hancock et al., 2006). In plant systems, free radicals generated by various causes act as signaling molecules that activate the antioxidant defense system to minimize the damage inflicted by ROS. The plant antioxidative defense system (enzymatic as well as nonenzymatic antioxidants) makes use of many mechanisms to generate a complex redox equilibrium that restricts ROS overproduction and maintains an oxidative state in plants (Arbona et al., 2003, 2008; Sytar et al., 2013; Akram et al., 2017). The primary enzyme-based antioxidants in plants are superoxide dismutase (SOD), catalase (CAT), ascorbate peroxidase (APX), glutathione reductase (GR), and dehydroascorbate reductase (DHAR), whereas nonenzymatic antioxidants include glutathione, tocopherol, and ascorbate (Foyer and Noctor, 2005; Ahmad et al., 2010; Racchi, 2013). When the environment is disturbed, antioxidant enzymes work together or collaboratively to support the plants in combating oxidative stress. SOD, one of many antioxidant enzymes, not only catalyzes the conversion of superoxide anions into substantially less harmful hydrogen peroxide but also serves as the first line of defense against severely damaging ROS. Following SOD, CAT converts hydrogen peroxide into water and oxygen molecules. Although ascorbate serves as an electron donor, APX also scavenges hydrogen peroxide. The functions of GR and DHAR within the cellular system maintain decreased levels of ascorbate and glutathione, which are crucial for eliminating high concentrations of hydrogen peroxide. Further, in addition to being a critical substrate for the detoxification of harmful peroxidation products of significant biomolecules (proteins and lipids) through glutathione-S-transferase activity, reduced glutathione is yet another vital antioxidant that plays an essential role in the biosynthesis of phytochelatins, which are involved in the removal of ROS (Kalinowska and Pawlik-Skowrónska, 2010).

8.4 NANOPARTICLES AND THEIR BIOSYNTHESIS

In recent times, there has been rapid development of novel approaches for the synthesis of nanomaterials. The process of NP synthesis can be mainly divided into two broad sections: bottom-up and top-down approaches (Figure 8.1). In the top-down methods (physical), NPs are synthesized by systematically breaking large particles into smaller particles, resulting in NP production (Jeevanandam et al., 2018). The techniques mainly used are sputtering, lithographic techniques, etc. The bottom-up

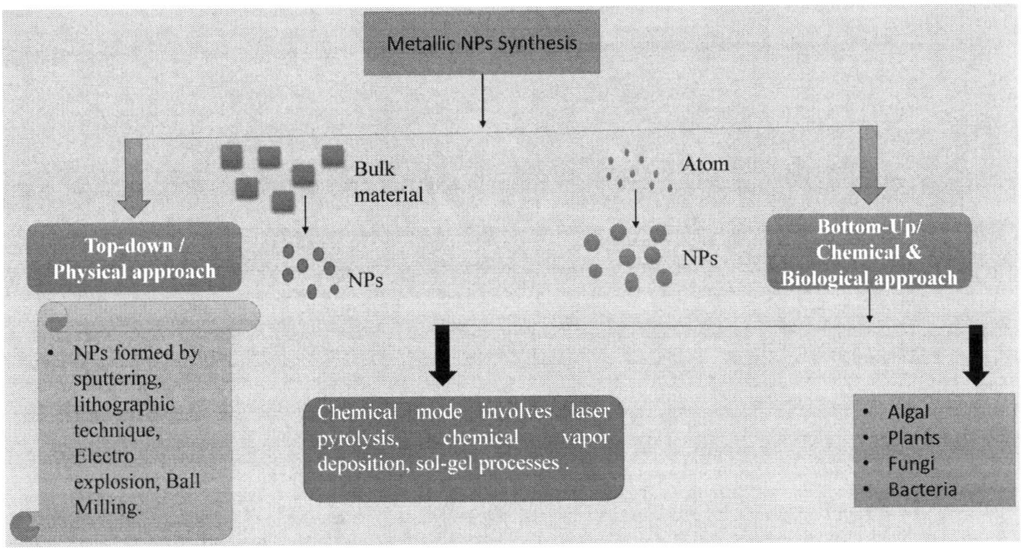

FIGURE 8.1 Synthesis of NPs by various processes.

approach (chemical and biological) includes the synthesis of NPs from the self-assembly of smaller particles by using techniques such as laser pyrolysis, chemical vapor deposition, and sol-gel processes (Kandiah and Chandrasekaran, 2021). Both the physical and chemical methods are expensive, and also, they are accompanied by the formation of toxic by-products (Sherry et al., 2017). The NPs formed as a result of non-biogenic methods involve an additional step of functionalization in which polymers are employed for surface coating. However, during biogenic methods, no additional step of capping is required. So, biogenic NPs have emerged as the best candidates for clean-up of the environment due to their environment-friendly catalytic nature and excellent absorption properties (Kumari et al., 2019). The biogenic synthesis/"green synthesis" approach has overshadowed conventional methods due to its low cost, good polydispersity, dimension, stability, and fewer processing conditions (Ingale et al., 2013). These biogenic NPs find extensive application in the fields of medicine, the food industry, the chemical and electrochemical industries, environmental remediation, and biotechnology (Sharma et al., 2020). The NPs originate biologically from extracts of plants and microbes (Andrade-Zavaleta et al., 2022). In comparison to microbes, the synthesis of NPs via plant-derived extracts is highly favorable and cost-effective. Figure 8.2 summarizes NP synthesis by various biological sources. The mechanism of NP synthesis following the "Green Synthesis" approach includes (Singh et al., 2018):

a) Preparation of extracts (source: plant/microbes) followed by mixing them with a metal salt solution.
b) Biochemical reduction of the metal solution in which monovalent/divalent metal ions are converted into zero-valent states, after which they are nucleated into NPs.
c) Analysis of NPs using UV-VIS spectroscopy and finally characterizing them through transmission electron microscopy (TEM), scanning electron microscopy (SEM), X-ray diffraction (XRD), etc.

Synthesis of NPs by the green approach is environment friendly, as the process of NP synthesis focuses on low waste generation, pollution, and nontoxic solvents and avoids the generation of unwanted by-products (Singh et al., 2018; Kandiah and Chandrasekaran, 2021).

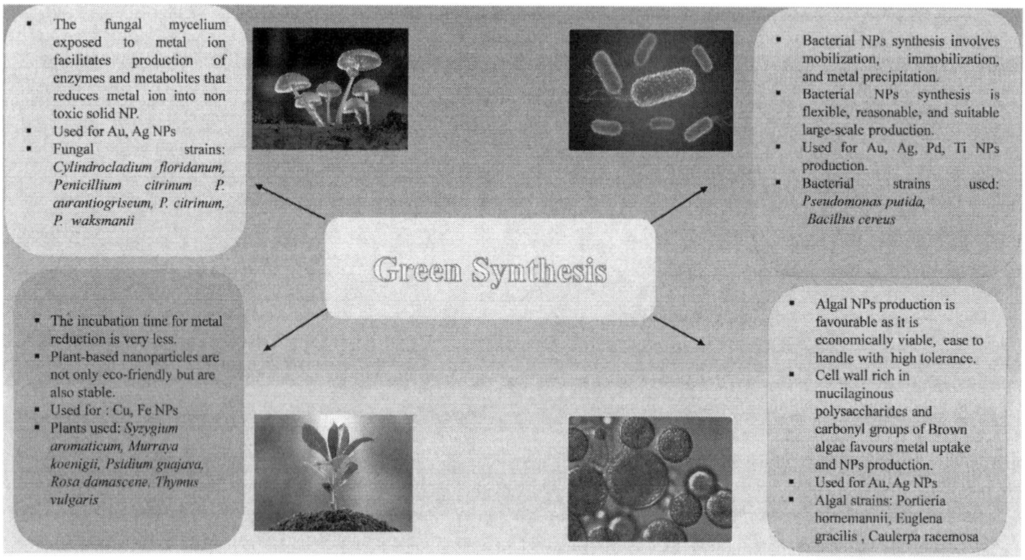

FIGURE 8.2 NP production using biogenic approaches.

8.5 MECHANISTIC PATHWAY OF NP-MEDIATED HEAVY METAL TOXICITY MITIGATION

Numerous investigations have been conducted to determine the nanoparticles' underlying mechanism of action to mitigate heavy metal toxicity. It has been demonstrated in the literature that heavy metal toxicity has an impact on plant physiology, morphology, and biochemistry. According to some reports, NPs controlling plant growth may be detrimental to them at higher concentrations (Vishwakarma et al., 2020). Several strategies have been employed to help plants cope with heavy metal stress, including minimizing the bioavailability of heavy metal from the soil (Moharem et al., 2019), improving the plant defense system, developing protective agents, and regulating the expression of heavy metal transport genes (Cao et al., 2020). NPs get into the cells via permeation or plasma membrane channels. These NPs function as signaling molecules that are involved in inducing the stress-signaling genes and activating the plant defense system.

Moreover, by transforming or absorbing the heavy metals already present in the soil, NPs reduce the accessibility and mobility of heavy metals. For example, as per the report of Sebastian et al. (2019), the application of Fe_3O_4 NPs was observed to reduce the mobility of cadmium. Similarly, another study suggests that under the effect of mercapto Si NPs, cadmium can be transformed into significantly more stable forms (Wang et al., 2020).

Heavy metals bind with NPs in plant cell walls, creating complexes that render the heavy metals inaccessible to biological systems. The complex is then adsorbed to the surface of the plant cell, which prevents the heavy metals from migrating throughout the plant and thus restricts the amount of biological activity they can carry out (Wang et al., 2021). Furthermore, the harm caused by the heavy metal stress can be partially mitigated by chelating the organic acids collected in plants' cell walls. Moreover, there is evidence that NPs stimulate the synthesis of protective agents (Zhou et al., 2020).

Since the production of ROS increases considerably under heavy metal toxicity, the oxidation defense system of plants can also be stimulated to cope with heavy metal toxicity. The key ROS-maintaining enzymes in cells are CAT, SOD, APX, GPX, DHAR, and GR, and as a result, they scavenge ROS when their levels are high. Low molecular weight nonenzymatic metabolites like ascorbic acid, tocopherol, and polyphenols also possibly tend to break down ROS (Ullah et al., 2020; Zhao et al., 2020). Plants can reduce ROS by utilizing antioxidant enzyme–activated NPs

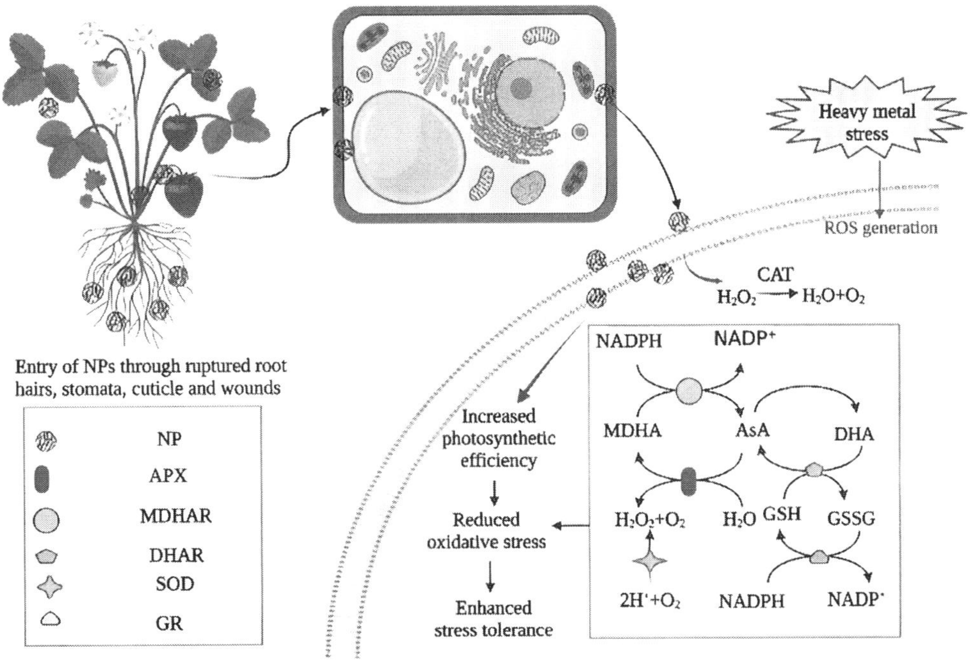

FIGURE 8.3 Diagrammatic illustration of the mechanism involved in mitigating HM stress by NPs.

like Fe_3O_4 NPs, Mn_3O_4 NPs, etc., which would reduce the consequences of crop production losses (Konate et al., 2017; Yao et al., 2018) (Figure 8.3).

8.6 NANOTECHNOLOGY IN REMEDIATION OF METAL-CONTAMINATED SOILS

Growing industrialization is continuously causing soil health to deteriorate by adding toxic heavy metals that pose a threat to the environment and biotic communities. Nanotechnology practices have resulted in improving conventional remediation processes by enhancing the rate of pollutant transformation. Nano-bioremediation is a branch of nanotechnology that removes toxic metals through green nanoparticles (Amin et al., 2021). These nanoparticles are small in size (<100 nm), have a greater surface area, and are adsorptive in nature. The low-cost plant- or microbe-based green nanoparticles break down the toxic substances and recover the soil health (Ashraf et al., 2019). The efficiency of phytoremediation has been upgraded by nanotechnology-based phytoremediation, using nanoparticles to clean the contaminated soil (Awasthi et al., 2022). Specific physicochemical properties of nanoparticles act as catalysts that help to lower the activation energy for the breakdown of toxic compounds and therefore degrade or alter the targeted contaminant. The nano-bioremediation approach is applicable because of three main characteristics: green nanomaterials/nanoparticles, a solution for the elimination of pollutants, and sensors for environmental agents. Various methods of nano-bioremediation such as nano-biosensors, nano-biosurfactants, nanosorbents, and nano-phytoremediation are used for the removal and degradation of heavy metal contaminants from the soil (Mathur et al., 2022).

8.6.1 Nano-biosensors

Nano-bioremediation is the process of eliminating contaminants using biogenic NPs (Amin et al., 2021). It is an economically favorable process that involves breaking down toxic chemicals and refining soil quality (Ashraf et al., 2019). Nano-biosensors are amended biosensors composed of biological elements linked to a transducer. They are highly sensitive for acquiring, processing, and scrutinizing data at atomic levels (Ghaffar et al., 2020). The analytes are sensed at very low concentrations via a transducer, which is very compact and connected to sensitized elements (Scognamiglio et al., 2013). Nano-biosensors offer many captivating features, such as functionalization, immobilization, and miniaturization (Usman et al., 2020). Nano-biosensors offer a great advantage over traditional biosensors. They have improved the specificity of detection and have applications in various fields such as soil nutrient monitoring, environment control, and agriculture. The use of nano-biosensors will add to the efficient management of fertilizers and pesticides in field conditions. Similarly, nano-biosensors can be employed for the detection of toxic heavy metals (Hg, Cd, Pb, As) that result in an imbalance in the prevailing ecosystem. Heavy metals can be detected in soil and water by using nanomaterials (nanoclusters, antibodies, aptamers) and nanodevices (fluorescence resonance energy transfer [FRET], quantum dots, and cantilever biosensors), etc. (Saha and Bhaduri, 2021). Interestingly, heavy metal phytotoxicity can be sensed by measuring the chlorophyll content, plant water status, etc. In this regard, Lew and his coworkers have studied natural arsenic accumulation in *Pteris cretica* by using nano-bionic sensors that are implanted in the plant tissue (Lew et al., 2021). The implanted nano-biosensor detected real-time changes occurring inside the plant as the plant took up Ar. Even though the role of nano-biosensors in remediating heavy metal–contaminated soil is not well illustrated, reports by Rigo et al., (2020a, b) and Ballen et al. (2021) highlight their role in the detection of HM in water and soil. Also, biosensors composed of *Pseudomonas aeruginosa* 9027 and *Bacillus* MS154 have been employed for bioremediation of heavy metals by rhamnolipids and lipopeptide, respectively (Ibrahim et al., 2013; Ravindran et al., 2020).

8.6.2 Nano-biosorbents

Biosorption is the process of removing inorganic and organic contaminants from the environment using material produced by living organisms. When biosorption is combined with nanotechnology for the breakdown of substances, the resulting materials are referred to as "nano-biosorbents." This method is employed to remove heavy metals from water and soil. Utilizing nano-biosorbents is inexpensive and environmentally friendly. The presence of hydroxyl and carboxyl groups has been discovered to be an essential requirement for the adsorption of different contaminants, dyes, radionuclides, and heavy metals. According to research by Mahmoud et al. (2020), waste products such as fly ash, bagasse, and rice husk have a high tendency to behave as biosorbents or nano-biosorbents. In addition, starch, cellulosic material, gums, alginates, chitin, and pectin are used as excellent nano-biosorbent materials. It has been found that ultrafine nano-biosorbents with a large surface area are created by utilizing nanoparticles with a lower size. These nano-biosorbents are distinguished by their high surface area and increased adsorption capacity, chemical activity, decreased internal diffusion resistance, and increased surface binding energy.

The capacity of these nano-biosorbents has been improved by a variety of novel techniques, including immobilization, magnetic alterations, and nanoscale zero-valent metal (NZVM) engineering. These inventions have significant potential for heavy metal removal. In immobilization, a nano-biosorbent is developed in two distinct ways. In the first procedure, a biosorbent is immobilized on a nanoparticle, for instance, the immobilization of *Fusarium verticilloides* on nanosilica and baker's yeast on nano-magnetite, whereas in the second technique, nanoparticles are immobilized on the polymeric material. Immobilization of magnetite nanoparticles in a sodium alginate matrix alone and with saponified orange peel is an example of the second approach. Similarly, the immobilization of chitosan nanoparticles in alginate, iron oxide nanoparticles, and *Agrobacterium fabrum* encased in calcium alginate are examples of this. In another process of magnetic modification, studies indicated the role of chitosan nanoparticles, *Cymopolia barbata*, in stabilization of magnetic nanoparticles. In this process, pectin extracted from apple waste, glutaraldehyde cross-linked pectin, and adipic acid cross-linked pectin are also involved. Other than these, heavy metals are abundantly removed with the help of engineered NZVMs. As a result of their huge surface area for adsorption and excellent reducibility, these NZVMs have the potential for environmental remediation.

8.6.3 Nano-Biosurfactants

Biosurfactants are considered surface-active chemicals and are produced extracellularly by microorganisms. These compounds are amphiphilic due to their hydrophilic and lipophilic properties. These microbial biosurfactants are utilized to remediate copper, zinc, and nickel (Biswas and Raichur, 2008). *Bacillus subtilis*–produced lipopeptide biosurfactant has been reported to remediate hydrocarbons and heavy metals in soil (Reddy et al., 2009). Various microorganisms are responsible for the production of these biosurfactants. *Bacillus* sp. produces the majority of biosurfactants of the lipopolypeptide type, while *Pseudomonas aeruginosa* is responsible for the majority of glycolipid biosurfactants (Ravindran et al., 2020; Varjani et al., 2019). Additional bacteria capable of producing biosurfactants are listed in Table 8.1. Most of these bacteria can grow in soil and water contaminated with oil. They utilize carbon sources for growth and biosurfactant synthesis. These microorganisms obtain much of their carbon from alcohol, hydrocarbons, organic acids, and carbohydrates (Madani et al., 2020).

When nanoparticles are capped with these biosurfactants, they stabilize the complexes formed and help in the remediation of heavy metals. It has been observed that nanoparticles capped with sophorolipids are produced by *Starmerella bombicola* (Sarubbo et al., 2022). Several other studies have been conducted where biosurfactants from various microorganisms have been used for the synthesis of nano-biosurfactants; these are given in Table 8.2. These nano-biosurfactants have a promising future in the field of bioremediation.

TABLE 8.1
Various Microorganisms Producing Biosurfactants

S.No.	Type of biosurfactant	Microorganism	Reference
1.	Lipopeptide	*Bacillus* MS154	Ravindran et al., 2020
2.	Lipopeptide	*Bacillus subtilis* RSL2	Sharma et al., 2020
3.	Lipopeptide	*Achromobacter* sp. A-8	Deng et al., 2020
4.	Lipopeptide	*Paenibacillus* sp. D9	Jimoh et al., 2020
5.	Lipopeptide	*Alcaligenes faecalis*, *Cellulosimicrobium* sp., and *Rhodococcus ruber*	Madani et al., 2020
6.	Trehalolipid	*Rhodococcus ruber*	Krivoruch et al., 2019
7.	Rhamnolipids	*Pseudomonas aeruginosa* NCIM 5514	Varjani et al., 2019
8.	Lipopeptide	*Bacillus cereus*	Durval et al., 2019
9.	Glycolipid	*Pseudoxanthomonas* sp. G3	Astuti et al., 2019
10.	Rhamnolipid (Glycolipid)	*Marinobacter* sp. MCTG107b	Tripathi et al., 2019
11.	Lipopeptide	*Bacillus methylotrophicus* DCS1 strain	Hmidet et al., 2019
12.	Dirhamnolipid (Glycolipid)	*Planococcus maritimus* strain SAMP MCC 3013	Waghmode et al., 2019
13.	Lipopeptide	*Bacillus subtilis* M495086	Datta et al., 2018
14.	Lipopeptide	*Paenibacillus dendritiformis* CN5	Bezza et al., 2017

TABLE 8.2
Nano-biosurfactants synthesised by microbes

S. No.	Biosurfactant	Microorganism	Nano-biosurfactant synthesized	References
1.	Glycolipid	*Brevibacterium casei* MSA19	Silver nanoparticle	Kiran et al., 2016
2.	Surfactin	*Bacillus amyloliquifaciens* strain KSU-109	Cadmium sulfide nanoparticles	Singh et al., 2011
3.	Rhamnolipid	*Pseudomonas aeruginosa*	Zinc sulfide nanoparticle	Narayanan et al., 2010
4.	Lipopeptide Surfactin	*Bacillus subtilis*	Gold nanoparticle	Reddy et al., 2009
5.	Sophorolipid	*Starmerella bombicola*	Cobalt nanoparticle	Kasture et al., 2008

8.6.4 Nano-phytoremediation

Biotechnology has utilized the capability of plants to improvise environmental remediation, called *phytoremediation*. Reports have suggested that plants can remediate HM in sediments, water, and soil. Phytoremediation is a novel approach due to its simplicity and compatibility as well as its sustainable nature (Awasthi et al., 2022). Apart from plants, rhizospheric microorganisms play a critical role in clearing contaminants. Plants along with microbes utilize phytoremediation for degrading various pollutants (Bakshi and Abhilash 2020). Different phytoremediation methods comprise phytoextraction, phytovolatilization, rhizo-degradation, enzymatic transformation, and phytodegradation, while bioremediation provides benefits in the production of biofuels, fibers, stabilization of soil, carbon sequestration, and microbial stabilization (Ali et al., 2013). The main disadvantage of phytoremediation is that it is time-consuming and takes many years or decades to provide results. Moreover, its applicability is also constrained by the quality of soil and phytotoxicity as well as weather conditions (Song et al., 2019).

In the last few years, phytoremediation has gained huge attention due to its efficacy as well as its economically friendly nature after combining with other novel strategies such as nanotechnology as well as microbial remediation (Gerhardt et al., 2017). Quick action along with on-site remediation using nanotechnology is gaining importance as an efficient as well as reasonable way to remove contaminants. Nano-phytoremediation, nanotechnology, and phytotechnology altogether eradicate contaminants. Research has been carried out on carbon-based materials followed by metal-based nanomaterials (Chen et al., 2017). A study conducted by Hussain et al. (2019) depicted that *Triticum vulgare* raised under Cd stress showed Cd remediation using Fe-based NPs. Supplementation of these NPs at a level of 20 ppm reduced the adverse effects of Cd by enhancing plant growth, antioxidant activities, and photosynthetic attributes along with reducing Cd levels in plants. Exogenously applied FeNPs also improved plant morphological attributes like plant biomass, grain yield, and pigment content. Nanomaterials can mitigate heavy metal stress via phytoremediation. A study reported in *Linum usitatissimum* showed that Cu can be removed in huge quantities through phytoremediation (Saleem et al., 2020). In addition, TiO_2 showed protective action towards *Glycine max* in Cd stress alleviation through oxidative stress management and free radical scavenging (Usman et al., 2020). A field study conducted by Liang et al. (2017) depicted that nano-hydroxyapatite and nano-carbon black showed Pb phytoextraction in *Lolium temulentum*. Metal toxicity can be lowered by NPs in order to regulate expression levels of genes encoding redox homeostasis, metal stress, oxidative stress, photosynthetic attributes, etc. (Gopalakrishnan Nair and Chung 2014). NPs induce metal accumulation in plants by regulating cell wall permeability, metal transportation, and gene expression.

8.6.5 Plant- and Microbe-mediated Biogenic Remediation

Phyto-nanotechnology is known to be the most simple, stable, eco-friendly, and economic technique for the biosynthesis of NPs, thereby opening new paths for heavy metal mitigation. Their adaptability and compatibility, as well as pharmaceutical usage of NPs using water as a reducing agent, are aspects of phyto-nanotechnology. The plant-based NPs are made up of plant-based components that are completely toxin-free and meet the rising demands for NPs in various fields for biomedical as well as environmental applications. Using leaf and root extract from *Panax ginseng*, gold and silver NPs have been synthesized (Singh et al., 2016). Many plants have been used to derive different NPs. Moreover, the use of microbes such as viruses, bacteria, fungi, actinomycetes, etc. for NP biosynthesis has also been employed (Huang et al., 2019).

Fungi have emerged as the most substantial source of NP biosynthesis. FeO-NPs coated with polyvinylpyrrolidone applied to Gram-negative bacteria, *Halomonas* enhanced Cd and Pb bioremediation (Alabresm et al., 2018). Besides, considerable Cu, Ni, and Pb removal can be achieved through Fe_3O_4 NPs using *Staphylococcus aureus*. This is mainly possible due to the presence of functional groups present on microbial surfaces that played a critical role in HM mitigation (Mahmoud et al., 2016). Another study conducted on the alleviation of Cd, Pb, Cu, and Cr using *Bacillus cereus* and *Lactobacillus macrolides* depicted that zinc oxide NPs significantly reduced Cr, Cu, and Pb (Baragano et al., 2020). At neutral pH, the negative charge present on ZnO NPs enhances electrostatic attraction to metal cations, although at lower pH, where hydrogen ions vie with adsorbents to encourage binding, the HMs precipitate as hydroxides. *B. cereus* also possesses the ability to reduce Cr ions through an enzymatic mechanism. The reduced form of Cr ions shows higher affinity for the cells in order to coordinate interactions towards functional groups present on the bacterial cell wall (Laslo et al., 2022). The microbial cells use HM pollutants to synthesize biogenic NPs, thus eradicating them from the environment (Ru et al., 2020). The formation of intracellular and extracellular Pd NPs was also achieved using *Enterococcus faecalis*; their size was observed to be 10 nm and further dependent on various environmental factors such as pH, temperature, and biomass. These NPs show higher catalytic efficacy, thereby possessing high potential for the remediation of industrial effluents (Cao et al., 2020). A similar trend was found using riboflavin, where Te NPs were used to treat anaerobic sludge. Tellurite consisting of Te^{4+} oxyanions was used to create NPs with an insoluble elemental form of tellurium, Te^0. It has been revealed that 2-hydroxy-1,4-naphthoquinone enhances Te^{4+} reduction, forming Te^0 NPs, and the process is assisted by *Rhodobacter capsulatus*, where malate is used as an electron donor and riboflavin serves to act as an anaerobic methanogenic granular sludge reducer that reduces Te^{4+} at a rapid rate (Ramos-Ruiz et al., 2017; Borghese et al., 2014).

8.7 NANOTECHNOLOGY FOR MITIGATION OF HEAVY METAL TOXICITY IN PLANTS

Heavy metals not only adversely affect the morphological and physiological functions of plants but also hurt their biochemistry. To curb these pernicious effects, plants have evolved various strategies that restrict the uptake, concentration, and translocation of these metals within them. Various approaches have been attempted to reduce heavy metal–induced toxicity in plants, for instance, decreasing the bioavailability of heavy metals in soil, modulating heavy metal transport genes in plants, improving plants' physiological functions, augmenting their antioxidant defense system, and developing their protective genes (Mengmeng et al., 2019; Cao et al., 2020; Lian et al., 2020). Nanoparticles have been noticed to be immensely effective in ameliorating the detrimental effects of heavy metals and managing their toxicity in plants (Figure 8.4) (Ahmad et al., 2019).

NPs absorb the heavy metals present in the soil and transform them, thereby lowering their mobility as well as bioavailability. For example, the mobility of cadmium and other heavy metals

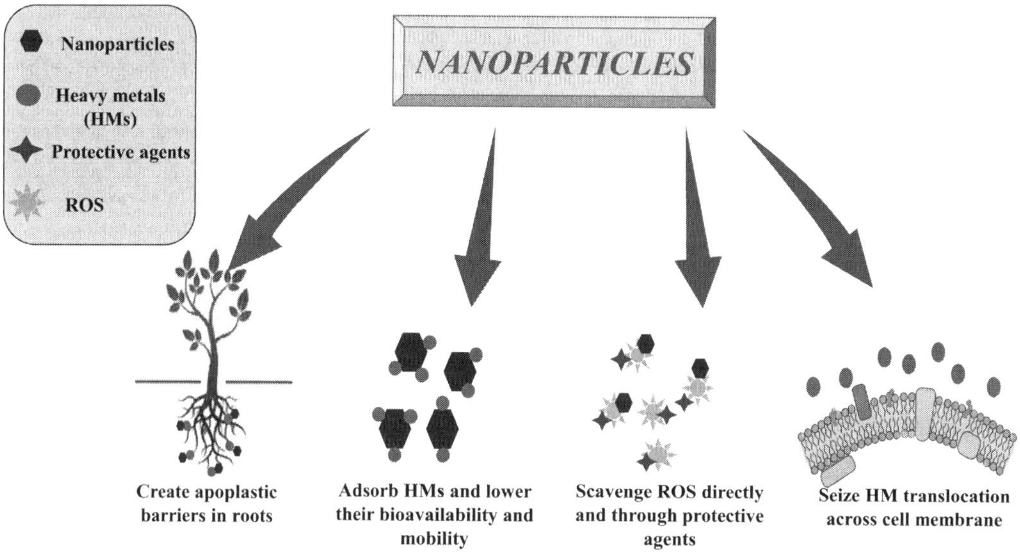

FIGURE 8.4 Various strategies used by NPs for mitigating HM stress in plants.

present in the soil was decreased by the utilization of Fe_3O_4 NPs (Sebastian et al., 2019). It was also reported that under field conditions, cadmium was transformed into a much more stable form by mercapto Si NPs after 3 years (Wang et al., 2020). NPs can also alleviate the ill effects of heavy metals in the soil by improving its properties; for instance, phosphate released by hydroxyapatite NPs can increase the pH of the soil and decrease heavy metal toxicity (Cui et al., 2018). By accumulating in the plant cell walls, NPs can form complexes with heavy metals and become adsorbed on the surface of the plant cells, thus reducing the bioavailability, migration, and activity of heavy metals in the plants (Cui et al., 2017; Wang et al., 2021). Moreover, the impairment triggered by heavy metal toxicity in plants can be diminished by chelating them with organic acids that accumulate in the cell walls of plant leaves and roots (Zhou et al., 2021). NPs increase the formation of protective agents in plants. It has been reported that the exogenous application of Si NPs reduced cadmium stress in plants by enhancing organic acid production (Zhou et al., 2021).

Furthermore, NPs influence the generation of apoplastic barriers in plant roots. These barriers play a protective role by managing water, ion, and oxygen flow in plants and reducing heavy metal concentrations in roots (Yang et al., 2013; Rossi et al., 2017). Additionally, NPs can modulate metal transport genes in plants that interrupt the translocation of heavy metals by strengthening the extracellular barrier in plants. Heavy metal stress can also be ameliorated in plants by activating their antioxidant defense system. ROS produced by plants during respiration, photosynthesis, and other metabolic processes in chloroplasts and other plant parts can be harmful when they accumulate at higher concentrations. They can be scavenged by antioxidative enzymes like CAT, SOD, peroxidase (POD), APX, glutathione peroxidase (GPX), and GR (Mathur et al., 2022). Thus, NPs having antioxidative enzyme activities, such as Mn_3O_4 NPs, CeO_2 NPs, Fe_3O_4 NPs, etc., can be employed to boost plants' ability to minimize the damage caused by ROS (Wu et al., 2017; Konate et al., 2017; Yao et al., 2018). Some studies on the effect of NPs in mitigating heavy metal stress in plants are listed in Table 8.3.

8.8 ENVIRONMENTAL RISKS/FATE OF NPS IN SOIL

Nanotechnology has exhibited various effects against the bioremediation of metal contaminants, but still, this approach has implications regarding the presence of nanoparticles in the

TABLE 8.3
Effect of Some NPs on Plants under Heavy Metal Stress

Nanoparticles	Plant species	Heavy metals	Key effects	References
Cu NPs	Guar	Cu	Lower heavy metal stress	Ahmadi-Nouraldinvand et al., 2022
ZnO NPs	*Oryza sativa*	Cu and Pb	Enhanced plant growth and reduced metal uptake in roots, shoots, and leaves	Akhtar et al., 2021
Fe NPs	*O. sativa*	As	Reduced As toxicity and enhanced antioxidative defense system	Bidi et al., 2021
Iron oxide NPs	*O. sativa*	Cd	Curtailed Cd transporter genes	Ahmed et al., 2021
MgO NPs	*O. sativa*	As	Enhanced plant biomass, decreased ROS, and As translocation	Ahmed et al., 2021
SiO_2 NPs	*Arundinaria pygmaea*	Cu and Mn	Reduced heavy metal stress, increased antioxidant enzyme activities, and improved photosynthesis and plant growth	Emamverdian et al., 2020
TiO_2 NPs	*Vigna radiata*	As	Increased antioxidant enzyme activities and reduced arsenic accumulation	Katiyar et al., 2020
Fe NPs	*Triticum aestivum*	Cd	Improved plant growth and yield, enhanced chlorophyll content and gas exchange parameters	Hussain et al., 2019
Fe_3O_4 NPs	*O. sativa*	Cd	Enhanced crop yield, plant biomass, chlorophyll content, and photosynthetic efficiency	Sebastian et al., 2019
Fe_3O_4 NPs	*O. sativa*	Cd	Reduced oxidative stress caused by cadmium toxicity	Sebastian et al., 2018

environment, its risks, and its toxic effects. Through bioaccumulation or the persistence of biodegradable compounds in soil and water, nanomaterials/nanoparticles pose ecotoxicological risks (Ameen et al., 2021). An increased concentration of nanoparticles changes the properties of soil and interferes with the growth and development of plants. It has been reported that nanoparticles applied to soil showed altered pH, disturbing plant nutrient uptake, microbial community, and therefore, plant health. In another study, it was found that CuO nanoparticles consumed H^+ from the soil, changing the pH and generating Cu ions and $Cu(OH)^+$ complexes. Adverse impacts of nanoparticles in the soil depend on the type and concentration of nanoparticles, the type of soil, and its enzymatic activity. Larger concentrations of nanoparticles can reduce the dehydrogenase activity in the soil, which creates ionic imbalance and changes the nutrient status and soil fertility. Nanoparticles are also considered contaminants that enter into the living system through the food chain and cause bioaccumulation. Three animal species of different trophic levels, i.e., *Danio rero* (zebrafish), *Daphnia pulex* (Daphnia), and microalgae, were examined for nanoparticle toxicity, and it was found that *Daphnia* and microalgae were more vulnerable to nanoparticle toxicity as compared with zebrafish because of their particle filter-feeding habits. Hence, nanoparticles disturb the physiological and biochemical pathways at the molecular level when entering the living system. It has been documented that nanoparticles cause oxidative stress due to ROS generation, DNA damage, protein unfolding, and cell death. Consequently, environmental concerns must be prioritized over the application of nanoparticles in the soil for nano-bioremediation.

8.9 CURRENT TRENDS IN NANOTECHNOLOGY BASED ON BIOTECHNOLOGY

In recent decades, marvelous growth has been witnessed in the field of nanotechnology. The nanosize of NPs and their high surface area to volume ratio change their physical as well as chemical properties, thus making them highly applicable in the fields of agriculture, medicine, environment, biotechnology, and several other sectors (Parisi et al., 2015; Khan et al., 2022). However, much remains to be unraveled in this field. For instance, the functioning of NPs with biomolecules is an area that is worth studying (Vázquez-Núñez et al., 2020). For the recovery or removal of heavy metals, researchers recently developed activated porous carbon and amyloid protein-bearing membranes by changing their tertiary structure to form amyloid fibrils that can capture various heavy metal ions using cysteine moieties (Bolisetty and Mezzenga, 2016). This study showed that a cost-effective source of biomolecules can be found and worked upon by utilizing experimental designs based on natural molecular phenomena.

Biotechnology can actually come in handy for operating NPs in a more sustainable and environment-friendly way. *Komagataeibacter sucrofermentans*, a bacterium used to produce pioneering cellulose-like polymers, was reared with conventional bioreactor techniques on glucose monomers embellished with the required structural modifications (Gao et al., 2019). By using this strategy, one can omit the utilization of complex solvents that can prove hazardous to the environment. Both conventional and next-generation changes like gene editing, protein engineering, or mutagenesis in biotechnological methods can be optimized to rationalize the production of cellulose-based NPs in huge amounts (Sharma et al., 2019). Moreover, nowadays, conventional fungicides are being replaced with RNA-based fungicides. This technology exploits the hybridization of double-stranded RNA with fungal pathogen mRNA, which when sprayed on plants, leads to gene silencing in pathogens (Wang et al., 2016). However, naked RNAs are unstable in the environment and have a short shelf life. To overcome this obstacle, the double-stranded RNA and the probe were protected by using clay nanosheets that enhanced their shelf life and fungicidal potential (Mitter et al., 2017). Currently, apart from being used in aerial plant organ protection, this technology is also being used for protecting plant roots (Vázquez-Núñez et al., 2020).

Molecular biotechnology can also be used in macroscale applications like the desalinization of water. Oxygen-sensitive proteins found in humans, as well as plants, can be exploited to form oxygen biosensors and inducible genetic circuits. Although the original idea was their application under in vivo conditions, they can be developed into nanomaterials that can stoichiometrically react to oxygen levels (Licausi and Giuntoli, 2021). These concepts have also been extended to DNA molecules that can eliminate complex structures like proteins and provide vast opportunities for biosensing and purification (Diederichs et al., 2019). Recently, transmembrane proteins were integrated into membranes and coupled to various transductors in order to be used for pesticide detection, energy harvesting, gas monitoring, and microarrays (Alavarez et al., 2018; Ryu et al., 2019). In the future, to detect highly complex pollutants, enzymes can be included in the aforesaid technological arrangements.

8.10 CONCLUSIONS AND FUTURE PERSPECTIVES

Heavy metals are common elements that are naturally available in the soil environment. Also, anthropogenic activities are contributing to adding substantial amounts of these HMs into the atmosphere. Furthermore, expanding populations are stretching the world's resources, and to maintain these populations, it will be necessary to intensify industrialization and agricultural productivity. Also, incorrect agricultural practices and the use of ample amounts of chemical fertilizers and pesticides are contributing to soil pollution and the entrance of these HMs into the food chains, leading to potential risks to human health as well as carcinogenic effects on humans. Conventional physicochemical methods for the elimination of these HMs are inefficient to treat different ecosystems and have been found to be unable to block the entry of HMs into the food chain. However, nanotechnology has been introduced to counter these HM concentrations in the soil ecosystems and to reduce their toxic impacts on plants and humans. Some other greener approaches have also been implemented through nanoparticles that are quick, cost-effective, practically nontoxic, and highly efficient for the remediation of toxic metals. These plant- and microorganism-based biogenic nanoparticles have variability in their size and shape as well as a large surface area with greater adsorption capacity. The application of such biogenic nanoparticles has changed the approach of applying expensive physiochemical methods for HM remediation. Thus, an integrated approach of implementing bioremediation in combination with nanotechnology has potential advantages over conventional methods in terms of cost as well as effectiveness for the remediation of different ecosystems. However, the commercialization and large-scale effectiveness of these biogenic nanoparticles still require efforts from researchers to achieve sustainable development prospects.

REFERENCES

Ahmad, B., Zaid, A., Jaleel, H., Khan, M. M. A., Ghorbanpour, M. (2019). Nanotechnology for phytoremediation of heavy metals: Mechanisms of nanomaterial-mediated alleviation of toxic metals. In *Advances in Phytonanotechnology* ed. M. Ghorbanpour and S. H. Wani (pp. 315–327). Elsevier: Academic Press.

Ahmad, P., Jaleel, C. A., Salem, M. A., Nabi, G., Sharma, S. (2010). Roles of enzymatic and nonenzymatic antioxidants in plants during abiotic stress. *Critical Reviews in Biotechnology* 30(3): 161–175.

Ahmadi-Nouraldinvand, F., Afrouz, M., Elias, S. G., Eslamian, S. (2022). Green synthesis of copper nanoparticles extracted from guar seedling under Cu heavy-metal stress by *Trichoderma harzianum* and their bio-efficacy evaluation against *Staphylococcus aureus* and *Escherichia coli*. *Environmental Earth Sciences* 81(2): 1–10.

Akbari, B., Tavandashti, M. P., Zandrahimi, M. (2011). Particle size characterization of nanoparticles–a practicalapproach. *Iranian Journal of Materials Science and Engineering* 8(2): 48–56.

Akhtar, N., Khan, S., Rehman, S. U., Rehman, Z. U., Khatoon, A., Rha, E. S., Jamil, M. (2021). Synergistic effects of zinc oxide nanoparticles and bacteria reduce heavy metals toxicity in rice (*Oryza sativa* L.) plant. *Toxics* 9(5): 113.

Akram, N. A., Shafiq, F., Ashraf, M. (2017). Ascorbic acid—A potential oxidant scavenger and its role in plant development and abiotic stress tolerance. *Frontiers in Plant Science* 8: 1–17.

Alabresm, A., Chen, Y. P., Decho, A. W., Lead, J. (2018). A novel method for the synergistic remediation of oil-water mixtures using nanoparticles and oil-degrading bacteria. *Science of the Total Environment* 630: 1292–1297.

Ali, H., Khan, E., Sajad, M. A. (2013). Phytoremediation of heavy metals—Concepts and applications. *Chemosphere* 91(7): 869–881.

Alvarez, P. J., Chan, C. K., Elimelech, M., Halas, N. J., Villagrán, D. (2018). Emerging opportunities for nanotechnology to enhance water security. *Nature Nanotechnology* 13(8): 634–641.

Amin, I., Nazir, R., Rather, M. A. (2021). Nano-bioremediation: An innovative approach for remedying heavy metals using fungi. *Journal of Bioremediation and Biodegradation* 12(487): 2.

Andrade-Zavaleta, K., Chacon-Laiza, Y., Asmat-Campos, D., Raquel-Checca, N. (2022). Green synthesis of superparamagnetic iron oxide nanoparticles with *Eucalyptus globulus* extract and their application in the removal of heavy metals from agricultural soil. *Molecules* 27(4): 1367.

Arbona, V., Flors, V., Jacas, J., García-Agustín, P., Gómez-Cadenas, A. (2003). Enzymatic and nonenzymatic antioxidant responses of *Carrizo citrange*, a salt-sensitive citrus rootstock, to different levels of salinity. *Plant and Cell Physiology* 44(4): 388–394.

Arbona, V., Hossain, Z., López-Climent, M. F., Pérez-Clemente, R. M., Gómez-Cadenas, A. (2008). Antioxidant enzymatic activity is linked to water logging stress tolerance in citrus. *Physiologia Plantarum* 132(4): 452–466.

Ashraf, S., Siddiqa, A., Shahida, S., Qaisar, S. (2019). Titanium-based nanocomposite materials for arsenic removal from water: A review. *Heliyon* 5(5): e01577.

Astuti, D. I., Purwasena, I. A., Putri, R. E., Amaniyah, M., Sugai, Y. (2019). Screening and characterization of biosurfactant produced by *Pseudoxanthomonas* sp. G3 and its applicability for enhanced oil recovery. *Journal of Petroleum Exploration and Production Technology* 9(3): 2279–2289.

Awasthi, G., Nagar, V., Mandzhieva, S.,Minkina, T., Sankhla, M. S., Pandit, P. P., et al. (2022). Sustainable amelioration of heavy metals in soil ecosystem: Existing developments to emerging trends. *Minerals* 12(1): 85. doi: 10.3390/min12010085.

Ballen, S. C., Steffens, J., Steffens, C. (2021). Stability characteristics of cantilever nanobiosensors with simple and complex molecules for determination of cadmium. *Sensors and Actuators. Part A,: Physical* 324: 112686.

Baragaño, D., Forján, R.,Welte, L. and Gallego, J. L. R. (2020). Nanoremediation of as and metals polluted soils by means of graphene oxide nanoparticles. *Scientific Reports* 10(1): 1896. doi: 10.1038/s41598-020-58852-4.

Benjamin, S. R., Lima, F. D., Florean, E. O. P. T., Guedes, M. I. F. (2019). Current trends in nanotechnology for bioremediation. *International Journal of Environment and Pollution* 66(1–3): 19–40.

Bezza, F. A., Chirwa, E. M. N. (2017). Pyrene biodegradation enhancement potential of lipopeptide biosurfactant produced by *Paenibacillus dendritiformis* CN5 strain. *Journal of Hazardous Materials* 321: 218–227.

Bidi, H., Fallah, H., Niknejad, Y., Tari, D. B. (2021). Iron oxide nanoparticles alleviate arsenic phytotoxicity in rice by improving iron uptake, oxidative stress tolerance and diminishing arsenic accumulation. *Plant Physiology and Biochemistry* 163: 348–357.

Biswas, M., Raichur, A. M. (2008). Electrokinetic and rheological properties of Nano zirconia in the presence of rhamnolipid biosurfactant. *Journal of the American Ceramic Society* 91(10): 3197–3201.

Bolisetty, S., Mezzenga, R. (2016). Amyloid–carbon hybrid membranes for universal water purification. *Nature Nanotechnology* 11(4): 365–371.

Borghese, R., Baccolini, C., Francia, F., Sabatino, P., Turner, R. J., Zannoni, D. (2014). Reduction of chalcogen oxyanions and generation of nanoprecipitates by the photosynthetic bacterium *Rhodobacter capsulatus. Journal of Hazardous Materials* 269: 24–30.

Brunner, T. J., Wick, P., Manser, P., Spohn, P., Grass, R. N., Limbach, L. K., Bruinink, A., Stark, W. J. (2006). In vitro cytotoxicity of oxide nanoparticles: Comparison to asbestos, silica, and the effect of particle solubility. *Environmental Science and Technology* 40(14): 4374–4381.

Chaplygin, V. A., Minkina, T. M., Mandzhieva, S. S., Nazarenko, O. G., Zimulina, I. V., Bauer, T. V., Litvinov, Y. A., Rajput, V. (2021). Heavy metals in agricultural crops of Rostov region through the example of soft wheat (*Triticum aestivum*). In *IOP Conference Series. Earth and Environmental Science* (Vol. 624, No. 1, 012204). IOP Publishing.

Chen, M., Zeng, G., Xu, P., Zhang, Y., Jiang, D., Zhou, S. (2017b). Understanding enzymatic degradation of single-walled carbon nanotubes triggered by functionalization using molecular dynamics simulation. Environmental Science: Nano 4(3): 720–727. doi: 10.1039/C7EN00050B.

Cui, H., Shi, Y., Zhou, J., Chu, H., Cang, L., Zhou, D. (2018). Effect of different grain sizes of hydroxyapatite on soil heavy metal bioavailability and microbial community composition. *Agriculture, Ecosystems and Environment* 267: 165–173.

Cui, J., Liu, T., Li, F., Yi, J., Liu, C., Yu, H. (2017). Silica nanoparticles alleviate cadmium toxicity in rice cells: Mechanisms and size effects. *Environmental Pollution* 228: 363–369.

Datta, P., Tiwari, P., Pandey, L. M. (2018). Isolation and characterization of biosurfactant producing and oil degrading *Bacillus subtilis* MG495086 from formation water of Assam oil reservoir and its suitability for enhanced oil recovery. *Bioresource Technology* 270: 439–448.

Deng, Z. H., Ai, H. S., Lu, C. P., Li, J. B. (2020). The Bre1/Rad6 machinery: Writing the central histone ubiquitin mark on H2B and beyond. *Chromosome Research* 28(3): 247–258.

Diederichs, T., Pugh, G., Dorey, A., Xing, Y., Burns, J. R., Hung Nguyen, Q., Tornow, M., Tampé, R., Howorka, S. (2019). Synthetic protein-conductive membrane nanopores built with DNA. *Nature Communications* 10(1): 1–11.

Domingo, J. L. (1989). Cobalt in the environment and its toxicological implications. *Reviews of Environmental Contamination and Toxicology*: 105–132.

Durval, I. J. B., Resende, A. H. M., Figueiredo, M. A., Luna, J. M., Rufino, R. D., Sarubbo, L. A. (2019). Studies on biosurfactants produced using *Bacillus cereus* isolated from seawater with biotechnological potential for marine oil-spill bioremediation. *Journal of Surfactants and Detergents* 22(2): 349–363.

Emamverdian, A., Ding, Y., Mokhberdoran, F., Ahmad, Z., Xie, Y. (2020). Determination of heavy metal tolerance threshold in a bamboo species (*Arundinaria pygmaea*) as treated with silicon dioxide nanoparticles. *Global Ecology and Conservation* 24: e01306.

Foyer, C. H., Noctor, G. (2005). Oxidant and antioxidant signaling in plants: A re-evaluation of the concept of oxidative stress in a physiological context. *Plant, Cell and Environment* 28(8): 1056–1071.

Gao, M., Li, J., Bao, Z., Hu, M., Nian, R., Feng, D., An, D., Li, X., Xian, M., Zhang, H. (2019). A natural in situ fabrication method of functional bacterial cellulose using a microorganism. *Nature Communications* 10(1): 1–10.

Garza-Lombó, C., Pappa, A., Panayiotidis, M. I., Gonsebatt, M. E., Franco, R. (2019). Arsenic-induced neurotoxicity: A mechanistic appraisal. *JBIC, Journal of Biological Inorganic Chemistry* 24(8): 1305–1316.

Genchi, G., Sinicropi, M. S., Lauria, G., Carocci, A., Catalano, A. (2020). The effects of cadmium toxicity. *International Journal of Environmental Research and Public Health* 17(11): 3782.

Gerhardt, K. E., Gerwing, P. D., Greenberg, B. M. (2017). Opinion: taking phytoremediation from proven technology to accepted practice. *Plant Science* 256: 170–185.

Ghaffar, N., Farrukh, M. A., Naz, S. (2020). Applications of nanobiosensors in agriculture. In *Nanoagronomy* (pp. 179–196). Springer, Cham.

Goldhaber, S. B. (2003). Trace element risk assessment: Essentiality vs. toxicity. *Regulatory Toxicology and Pharmacology* 38(2): 232–242.

Gopalakrishnan Nair, P. M., Chung, I. M. (2014). Cell cycle and mismatch repair genes as potential biomarkers in *Arabidopsis thaliana* seedlings exposed to silver nanoparticles. *Bulletin of Environmental Contamination and Toxicology* 92(6): 719–725.

Gworek, B., Dmuchowski, W., Baczewska-Dąbrowska, A. H. (2020). Mercury in the terrestrial environment: A review. *Environmental Sciences Europe* 32(1): 1–19.

Haider, F. U., Liqun, C., Coulter, J. A., Cheema, S. A., Wu, J., Zhang, R., Farooq, M. (2021). Cadmium toxicity in plants: Impacts and remediation strategies. *Ecotoxicology and Environmental Safety* 211: 111887.

Hancock, J., Desikan, R., Harrison, J., Bright, J., Hooley, R., Neill, S. (2006). Doing the unexpected: Proteins involved in hydrogen peroxide perception. *Journal of Experimental Botany* 57(8): 1711–1718.

Hmidet, N., Jemil, N., Nasri, M. (2019). Simultaneous production of alkaline amylase and biosurfactant by *Bacillus methylotrophicus* DCS1: Application as detergent additive. *Biodegradation* 30(4): 247–258.

Horowitz, Y., Greenberg, D., Ling, G., Lifshitz, M. (2002). Acrodynia: A case report of two siblings. *Archives of Disease in Childhood* 86(6): 453–453.

Ibrahim, M. L., Ijah, U. J. J., Manga, S. B., Bilbis, L. S., Umar, S. (2013). Production and partial characterization of biosurfactant produced by crude oil degrading bacteria. *International Biodeterioration and Biodegradation* 81: 28–34.

Ingale, A. G., Chaudhari, A. N. (2013). Biogenic synthesis of nanoparticles and potential applications: An ecofriendly approach. *Journal of Nanomedicine and Nanotechnology* 4(165): 1–7.

Jacob, D. L., Borchardt, J. D., Navaratnam, L., Otte, M. L., Bezbaruah, A. N. (2013). Uptake and translocation of Ti from nanoparticles in crops and wetland plants. *International Journal of Phytoremediation* 15(2): 142–153.

Jeevanandam, J., Barhoum, A., Chan, Y. S., Dufresne, A., Danquah, M. K. (2018). Review on nanoparticles and nanostructured materials: History, sources, toxicity and regulations. *Beilstein Journal of Nanotechnology* 9(1): 1050–1074.

Jimoh, A. A., Lin, J. (2020). Biotechnological applications of *Paenibacillus* sp. D9 lipopeptide biosurfactant produced in low-cost substrates. *Applied Biochemistry and Biotechnology* 191(3): 921–941.

Kalinowska, R., Pawlik-Skowrónska, B. (2010). Response of two terrestrial green microalgae (Chlorophyta, Trebouxiophyceae) isolated from Cu-rich and unpolluted soils to copper stress. *Environmental Pollution* 158(8): 2778–2785.

Kandiah, M., Chandrasekaran, K. N. (2021). Green synthesis of silver nanoparticles using Catharanthusroseus flower extracts and the determination of their antioxidant, antimicrobial, and photocatalytic activity. *Journal of Nanotechnology* 2021: (1–18).

Kashyap, P. L., Xiang, X., Heiden, P. (2015). Chitosan nanoparticle based delivery systems for sustainable agriculture. *International Journal of Biological Macromolecules* 77: 36–51.

Kasture, M. B., Patel, P., Prabhune, A. A., Ramana, C. V., Kulkarni, A. A., Prasad, B. L. V. (2008). Synthesis of silver nanoparticles by sophorolipids: Effect of temperature and sophorolipid structure on the size of particles. *Journal of Chemical Sciences* 120(6): 515–520.

Katiyar, P., Yadu, B., Korram, J., Satnami, M. L., Kumar, M., Keshavkant, S. (2020). Titanium nanoparticles attenuates arsenic toxicity by up-regulating expressions of defensive genes in *Vigna radiata* L. *Journal of Environmental Sciences* 92: 18–27.

Khan, A., Khan, S., Khan, M. A., Qamar, Z., Waqas, M. (2015). The uptake and bioaccumulation of heavy metals by food plants, their effects on plants nutrients, and associated health risk: A review. *Environmental Science and Pollution Research* 22(18): 13772–13799.

Khanna, K., Jamwal, V. L., Sharma, A. (2019). Supplementation with plant growth promoting rhizobacteria (PGPR) alleviates cadmium toxicity in *Solanum lycopersicum* by modulating the expression of secondary metabolites. *Chemosphere* 230: 628–639.

Kim, K. R., Owens, G., Naidu, R. (2009). Heavy metal distribution, bioaccessibility, and phytoavailability in long-term contaminated soils from Lake Macquarie, Australia. *Soil Research* 47(2): 166–176.

Kiran, G. S., Ninawe, A. S., Lipton, A. N., Pandian, V., Selvin, J. (2016). Rhamnolipid biosurfactants: Evolutionary implications, applications and future prospects from untapped marine resource. *Critical Reviews in Biotechnology* 36(3): 399–415.

Konate, A., He, X., Zhang, Z., Ma, Y., Zhang, P., Alugongo, G. M., Rui, Y. (2017). Magnetic (Fe_3O_4) nanoparticles reduce heavy metals uptake and mitigate their toxicity in wheat seedling. *Sustainability* 9(5): 790.

Krivoruchko, A., Kuyukina, M., Ivshina, I. (2019). Advanced *Rhodococcus* biocatalysts for environmental biotechnologies. *Catalysts* 9(3): 236.

Kumari, S., Tyagi, M., Jagadevan, S. (2019). Mechanistic removal of environmental contaminants using biogenic nano-materials. *International Journal of Environmental Science and Technology* 16(11): 7591–7606.

Lal, R. (2015). Restoring soil quality to mitigate soil degradation. *Sustainability* 7(5): 5875–5895.

Laslo, V., Pinzaru, S. C., Zaguła, G., Kluz, M., Vicas, S. I., Cavalu, S. (2022). Synergic effect of selenium nanoparticles and lactic acid bacteria in reduction cadmium toxicity. *Journal of Molecular Structures* 1247: 131325.

Lei, Z., Mingyu, S., Xiao, W., Chao, L., Chunxiang, Q., Liang, C., Hao, H., Xiaoqing, L., Fashui, H. (2008). Antioxidant stress is promoted by nano-anatase in spinach chloroplasts under UV-B radiation. *Biological Trace Element Research* 121(1): 69–79.

Lew, T. T. S., Park, M., Cui, J., Strano, M. S. (2021). Plant nanobionic sensors for arsenic detection. *Advanced Materials* 33(1): 2005683.

Li, C., Zhou, K., Qin, W., Tian, C., Qi, M., Yan, X., Han, W. (2019). A review on heavy metals contamination in soil: Effects, sources, and remediation techniques. *Soil and Sediment Contamination: An International Journal* 28(4): 380–394.

Lian, J., Zhao, L., Wu, J., Xiong, H., Bao, Y., Zeb, A., Tang, J., Liu, W. (2020). Foliar spray of TiO2 nanoparticles prevails over root application in reducing Cd accumulation and mitigating Cd-induced phytotoxicity in maize (*Zea mays* L.). *Chemosphere* 239: 124794.

Liang, S. X., Jin, Y., Liu, W., Li, X., Shen, S. G., Ding, L. (2017). Feasibility of Pb phytoextraction using nano-materials assisted ryegrass: Results of a one-year field-scale experiment. *Journal of Environment Management* 190: 170–175.

Licausi, F., Giuntoli, B. (2021). Synthetic biology of hypoxia. *New Phytologist* 229(1): 50–56.

Loh, M. M., Sugeng, A., Lothrop, N., Klimecki, W., Cox, M., Wilkinson, S. T., Beamer, P. I. (2016). Multimedia exposures to arsenic and lead for children near an inactive mine tailings and smelter site. *Environmental Research* 146: 331–339.

Loranger, S., Zayed, J. (1995). Environmental and occupational exposure to manganese: A multimedia assessment. *International Archives of Occupational and Environmental Health* 67(2): 101–110.

Luo, X. S., Ding, J., Xu, B., Wang, Y. J., Li, H. B., Yu, S. (2012). Incorporating bioaccessibility into human health risk assessments of heavy metals in urban park soils. *Science of the Total Environment* 424: 88–96.

Madani, K. (2020). Biodegradation potential of crude petroleum by hydrocarbonoclastic bacteria isolated from Soummam wadi sediment and chemical-biological proprieties of their biosurfactants. *Journal of Petroleum Science and Engineering* 184: 106554.

Mahmoud, M. E., Abdou, A. E., Mohamed, S. M., Osman, M. M. (2016). Engineered *Staphylococcus aureus* via immobilization on magnetic Fe3O4- phthalate nanoparticles for biosorption of divalent ions from aqueous solutions. *Journal of Environmental and Chemical Engineering* 4(4): 3810–3824.

Mahmoud, M. E., Fekry, N. A., Abdelfattah, A. M. (2020). A novel nanobiosorbent of functionalized graphene quantum dots from rice husk with barium hydroxide for microwave enhanced removal of lead (II) and lanthanum (III). *Bioresource Technology* 298: 122514.

Marchetti, C. (2014). Interaction of metal ions with neurotransmitter receptors and potential role in neurodiseases. *Biometals* 27(6): 1097–1113.

Mathur, S., Singh, D., Ranjan, R. (2022). Remediation of heavy metal (loid) contaminated soil through green nanotechnology. *Frontiers in Sustainable Food Systems* 6: 932424.

Mengmeng, S., Jie, X., Lingling, Z., Dejian, S., Xitong, Z. (2019). Research progress on effects of silicon on plant growth under cadmium stress. *Guangzhou Chemical Industry* 47: 41–43.

Meriga, B., Reddy, B. K., Rao, K. R., Reddy, L. A., Kishor, P. B. K. (2004). Aluminium-induced production of oxygen radicals, lipid peroxidation and DNA damage in seedlings of rice (*Oryza sativa*). *Journal of Plant Physiology* 161(1): 63–68.

Mitra, S., Chakraborty, A. J., Tareq, A. M., Emran, T. B., Nainu, F., Khusro, A., Simal-Gandara, J. (2022). Impact of heavy metals on the environment and human health: Novel therapeutic insights to counter the toxicity. *Journal of King Saud University – Science* 34(3): 101865.

Mitter, N., Worrall, E. A., Robinson, K. E., Li, P., Jain, R. G., Taochy, C., Fletcher, S. J., Carroll, B. J., Lu, G. Q., Xu, Z. P. (2017). Clay nanosheets for topical delivery of RNAi for sustained protection against plant viruses. *Nature Plants* 3(2): 1–10.

Miura, S., Takahashi, K., Imagawa, T., Uchida, K., Saito, S., Tominaga, M., Ohta, T. (2013). Involvement of TRPA1 activation in acute pain induced by cadmium in mice. *Molecular Pain* 9: 1744–8069.

Moharem, M., Elkhatib, E., Mesalem, M. (2019). Remediation of chromium and mercury polluted calcareous soils using nanoparticles: Sorption-desorption kinetics, speciation and fractionation. *Environmental Research* 170: 366–373.

Møller, I. M., Kristensen, B. K. (2004). Protein oxidation in plant mitochondria as a stress indicator. *Photochemistry and Photobiology Sciences* 3(8): 730–735.

Nag, R., Cummins, E. (2022). Human health risk assessment of lead (Pb) through the environmental-food pathway. *Science of the Total Environment* 810: 151168.

Narayanan, J., Ramji, R., Sahu, H., Gautam, P. (2010). Synthesis, stabilisation and characterisation of rhamnolipid-capped ZnS nanoparticles in aqueous medium. *IET Nanobiotechnology* 4(2): 29–34.

O'Neal, S. L., Zheng, W. (2015). Manganese toxicity upon overexposure: A decade in review. *Current Environmental Health Reports* 2(3): 315–328.

Packer, M. (2016). Cobalt cardiomyopathy: A critical reappraisal in light of a recent resurgence. *Circulation: Heart Failure* 9(12): e003604.

Parisi, C., Vigani, M., Rodríguez-Cerezo, E. (2015). Agricultural nanotechnologies: What are the current possibilities? *Nano Today* 10(2): 124–127.

Racchi, M. L. (2013). Antioxidant defenses in plants with attention to *Prunus* and *Citrus* spp. *Antioxidants* 2(4): 340–369.

Rafique, M., Hajra, S., Tahir, M. B., Gillani, S. S. A., Irshad, M. (2022). A review on sources of heavy metals, their toxicity and removal technique using physico-chemical processes from wastewater. *Environmental Science and Pollution Research* 29(11): 16772–16781.

Rai, P. K., Lee, S. S., Zhang, M., Tsang, Y. F., Kim, K. H. (2019). Heavy metals in food crops: Health risks, fate, mechanisms, and management. *Environment International* 125: 365–385.

Rajput, V. D., Minkina, T., Upadhyay, S. K., Kumari, A., Ranjan, A., Mandzhieva, S., Sushkova, S., Singh, R. K., Verma, K. K. (2022). Nanotechnology in the restoration of polluted soil. *Nanomaterials* 12(5): 769.

Ramos-Ruiz, A., Sesma-Martin, J., Sierra-Alvarez, R., Field, J. A. (2017). Continuous reduction of tellurite to recoverable tellurium nanoparticles using an upflow anaerobic sludge bed (UASB) reactor. *Water Research* 108: 189–196.

Reddy, A. S., Chen, C. Y., Chen, C. C., Jean, J. S., Fan, C. W., Chen, H. R., Nimje, V. R. (2009). Synthesis of gold nanoparticles via an environmentally benign route using a biosurfactant. *Journal of Nanoscience and Nanotechnology* 9(11): 6693–6699.

Rice, K. M., Walker Jr, E. M., Wu, M., Gillette, C., Blough, E. R. (2014). Environmental mercury and its toxic effects. *Journal of Preventive Medicine and Public Health* 47(2): 74.

Rigo, A. A., Cezaro, A. M. D., Muenchen, D. K., Martinazzo, J., Manzoli, A., Steffens, J., & Steffens, C. (2020). Heavy metals detection in river water with cantilever nanobiosensor. Journal of Environmental Science and Health, Part B, 55(3): 239–249.

Rossi, L., Zhang, W., Schwab, A. P., Ma, X. (2017). Uptake, accumulation, and in planta distribution of coexisting cerium oxide nanoparticles and cadmium in *Glycine max* (L.) Merr. *Environmental Science and Technology* 51(21): 12815–12824.

Ru, J., Huo, Y., Yang, Y. (2020). Microbial degradation and valorization of plastic wastes. *Frontiers in Microbiology* 11: 442.

Ryu, H., Fuwad, A., Yoon, S., Jang, H., Lee, J. C., Kim, S. M., Jeon, T. J. (2019). Biomimetic membranes with transmembrane proteins: State-of-the-art in transmembrane protein applications. *International Journal of Molecular Sciences* 20(6): 1437.

Saha, A., Bhaduri, D. (2021). Nano-biosensors: Applications in agriculture and allied fields. *Biotica Research Today* 3(12): 1106–1109.

Sarubbo, L. A., Maria da Gloria, C. S., Durval, I. J. B., Bezerra, K. G. O., Ribeiro, B. G., Silva, I. A. (2022). Biosurfactants: Production, properties, applications, trends, and general perspectives. *Biochemical Engineering Journal* 181: 108377.

Scognamiglio, V., Pezzotti, I., Pezzotti, G., Cano, J., Manfredonia, I., Buonasera, K., Giardi, M. T. (2013). A new embedded biosensor platform based on micro-electrodes array (MEA) technology. *Sensors and Actuators B,: Chemical* 176: 275–283.

Sebastian, A., Nangia, A., Prasad, M. N. V. (2018). A green synthetic route to phenolics fabricated magnetite nanoparticles from coconut husk extract: Implications to treat metal contaminated water and heavy metal stress in *Oryza sativa* L. *Journal of Cleaner Production* 174: 355–366.

Sharma, A., Thakur, M., Bhattacharya, M., Mandal, T., Goswami, S. (2019). Commercial application of cellulose nano-composites–A review. *Biotechnology Reports* 21: e00316.

Sharma, P., Bhatt, D., Zaidi, M. G. H., Saradhi, P. P., Khanna, P. K., Arora, S. (2012). Silver nanoparticle-mediated enhancement in growth and antioxidant status of *Brassica juncea*. *Applied Biochemistry and Biotechnology* 167(8): 2225–2233.

Sharma, S., Pandey, L. M. (2020). Production of biosurfactant by *Bacillus subtilis* RSL-2 isolated from sludge and biosurfactant mediated degradation of oil. *Bioresource Technology* 307: 123261.

Sherry Davis, A., Prakash, P., Thamaraiselvi, K. (2017). Nanobioremediation technologies for sustainable environment. In *Bioremediation and Sustainable Technologies for Cleaner Environment* (pp. 13–33). Springer, Cham.

Singh, B. R., Dwivedi, S., Al-Khedhairy, A. A., Musarrat, J. (2011). Synthesis of stable cadmium sulfide nanoparticles using surfactin produced by Bacillus amyloliquifaciens strain KSU-109. *Colloids and Surfaces, B: Biointerfaces* 85(2): 207–213.

Singh, J., Dutta, T., Kim, K. H., Rawat, M., Samddar, P., Kumar, P. (2018). 'Green'synthesis of metals and their oxide nanoparticles: applications for environmental remediation. *Journal of Nanobiotechnology* 16(1): 1–24.

Singh, P., Kim, Y. J., Wang, C., Mathiyalagan, R., Yang, D. C. (2016). The development of a green approach for the biosynthesis of silver and gold nanoparticles by using *panax ginseng* root extract, and their biological applications. *Artificial Cells, Nanomedicine and Biotechnology* 44(4): 1150–1157.

Smirnova, E., Gusev, A., Zaytseva, O., Sheina, O., Tkachev, A., Kuznetsova, E., Lazareva, E., Onishchenko, G., Feofanov, A., Kirpichnikov, M. (2012). Uptake and accumulation of multiwalled carbon nanotubes change the morphometric and biochemical characteristics of *Onobrychis arenaria* seedlings. *Frontiers of Chemical Science and Engineering* 6(2): 132–138.

Song, B., Xu, P., Chen, M., Tang, W., Zeng, G., Gong, J. (2019). Using nanomaterials to facilitate the phytoremediation of contaminated soil. *Critical Reviews in Environmental Science and Technology* 49(9): 791–824.

Suman, J., Uhlik, O., Viktorova, J., Macek, T. (2018). Phytoextraction of heavy metals: A promising tool for clean-up of polluted environment? *Frontiers in Plant Science* 9: 1476.

Sytar, O., Kumar, A., Latowski, D., Kuczynska, P., Strzałka, K., Prasad, M. N. (2013). Heavy metal-induced oxidative damage, defense reactions, and detoxification mechanisms in plants. *Acta Physiologiae Plantarum* 35(4): 985–999.

Tripathi, D. K., Singh, S., Singh, S., Pandey, R., Singh, V. P., Sharma, N. C., Prasad, S. M., Dubey, N. K., Chauhan, D. K. (2017). An overview on manufactured nanoparticles in plants: Uptake, translocation, accumulation and phytotoxicity. *Plant Physiology and Biochemistry* 110: 2–12.

Tripathi, L., Twigg, M. S., Zompra, A., Salek, K., Irorere, V. U., Gutierrez, T., Banat, I. M. (2019). Biosynthesis of rhamnolipid by a *Marinobacter* species expands the paradigm of biosurfactant synthesis to a new genus of the marine microflora. *Microbial Cell Factories* 18(1): 1–12.

Ullah, S., Adeel, M., Zain, M. (2020). Physiological and biochemical response of wheat (*Triticum aestivum*) to TiO_2 nanoparticles in phosphorous amended soil: A full life cycle study. *Journal of Environment Management* 263: 110365.

Usman, M., Farooq, M., Wakeel, A., Nawaz, A., Cheema, S. A., urRehman, H., Sanaullah, M. (2020). Nanotechnology in agriculture: Current status, challenges and future opportunities. *Science of the Total Environment* 721: 137778.

Varjani, S., Upasani, V. N. (2019). Evaluation of rhamnolipid production by a halotolerant novel strain of *Pseudomonas aeruginosa*. *Bioresource Technology* 288: 121577.

Vázquez-Núñez, E., Molina-Guerrero, C. E., Peña-Castro, J. M., Fernández-Luqueño, F., de la Rosa-Álvarez, M. G. (2020). Use of nanotechnology for the bioremediation of contaminants: A review. *Processes* 8(7): 826.

Vishwakarma, K., Singh, V. P., Prasad, S. M., Chauhan, D. K., Tripathi, D. K., Sharma, S. (2020). Silicon and plant growth promoting rhizobacteria differentially regulate AgNP-induced toxicity in *Brassica juncea*: Implication of nitric oxide. *Journal of Hazardous Materials* 390: 121806.

Waghmode, S., Suryavanshi, M., Dama, L., Kansara, S., Ghattargi, V., Das, P., Satpute, S. K. (2019). Genomic insights of halophilic *Planococcus maritimus* SAMP MCC 3013 and detail investigation of its biosurfactant production. *Frontiers in Microbiology* 10: 235.

Wang, K., Wang, Y., Wan, Y., Mi, Z., Wang, Q., Wang, Q., Li, H. (2021). The fate of arsenic in rice plants (*Oryza sativa* L.): Influence of different forms of selenium. *Chemosphere* 264(1): 128417.

Wang, M., Weiberg, A., Lin, F. M., Thomma, B. P., Huang, H. D., Jin, H. (2016). Bidirectional cross-kingdom RNAi and fungal uptake of external RNAs confer plant protection. *Nature Plants* 2(10): 1–10.

Wang, Y., Fang, Z., Kang, Y., Tsang, E. P. (2014). Immobilization and phytotoxicity of chromium in contaminated soil remediated by CMC-stabilized nZVI. *Journal of Hazardous Materials* 275: 230–237.

Wang, Y., Liu, Y., Zhan, W., Zheng, K., Lian, M., Zhang, C., Ruan, X., Li, T. (2020). Long-term stabilization of Cd in agricultural soil using mercapto-functionalized nano-silica (MPTS/nano-silica): A three-year field study. *Ecotoxicology and Environmental Safety* 197: 110600.

Wu, H., Tito, N., Giraldo, J. P. (2017). Anionic cerium oxide nanoparticles protect plant photosynthesis from abiotic stress by scavenging reactive oxygen species. *ACS Nano* 11(11): 11283–11297.

Yang, C., Zhang, X., Liu, G., Zhang, J., Bao, M., Zhou, Z. (2013). Progress on the structure and physiological functions of apoplastic barriers in root. *Bulletin of Botanical Research* 33(1): 114–119.

Yao, J., Cheng, Y., Zhou, M., Zhao, S., Lin, S., Wang, X., Wu, J., Li, S., Wei, H. (2018). ROS scavenging Mn3O4 nanozymes for in vivo anti-inflammation. *Chemical Science* 9(11): 2927–2933.

Zambelli, B., Uversky, V. N., Ciurli, S. (2016). Nickel impact on human health: An intrinsic disorder perspective. *Biochimica et Biophysica Acta (BBA) – Proteins and Proteomics, 1864* 12(12): 1714–1731.

Zhao, L., Lu, L., Wang, A. (2020). Nanobiotechnology in agriculture: Use of nanomaterials to promote plant growth and stress tolerance. *Journal of Agriculture and Food Chemistry* 68(7): 1935–1947.

Zhou, C., Huang, C., Wang, J., Huang, H., Li, J., Xie, Q., Zhu, Q. (2017). LncRNA MEG3 downregulation mediated by DNMT3b contributes to nickel malignant transformation of human bronchial epithelial cells via modulating PHLPP1 transcription and HIF-1α translation. *Oncogene* 36(27): 3878–3889.

Zhou, P., Adeel, M., Shakoor, N. (2020). Application of nanoparticles alleviates heavy metals stress and promotes plant growth: An overview. *Nanomaterials* 11(1): 26.

Zhou, P., Adeel, M., Shakoor, N., Guo, M., Hao, Y., Azeem, I., Li, M., Liu, M., Rui, Y. (2021). Application of nanoparticles alleviates heavy metals stress and promotes plant growth: An overview. *Nanomaterials* 11(1): 26.

9 *In vitro* Production of Plant Bioactive Compounds Enhanced by Nanoparticle Elicitation

Gilberto Velázquez-Juárez, Janet María León-Morales, José Juvencio Castañeda-Nava, Soledad García-Morales

9.1 INTRODUCTION

An elicitor is defined as any substance that is applied to a living biological system in a given concentration and amount to initiate or increase the biosynthesis of a specific secondary metabolite using its metabolism. Thus, the process of inducing or increasing the production of secondary metabolites by the application of a given compound called an elicitor is called elicitation (Ramirez-Estrada et al., 2016; Thakur et al., 2018). The term *elicitor* encompasses chemical compounds and physical factors to which a biological system is exposed (Narayani and Srivastava, 2017). Elicitors can be classified by their origin, i.e., biotic or abiotic: biotic elicitors are those obtained from living organisms such as plant extracts, fungi, bacteria, secondary metabolites, hormones, proteins, or carbohydrates, among others, whereas abiotic elicitors are those that are obtained from the environment, such as mineral particles, light intensity, water availability, etc. (Thakur et al., 2018).

Elicitation is a technique or tool that has been used in different productive sectors because the applications of secondary metabolites have been increasing in recent years. At the same time, the discovery and demonstration of new effects of these bioactive compounds have generated the development of different production and application schemes. Consequently, due to the demand for bioactive compounds, a constant improvement in their procurement is necessary, implementing strategies to increase throughput and production efficiency, giving rise to the use of elicitors to enhance the biosynthesis of plant secondary metabolites (Ramirez-Estrada et al., 2016).

The use of elicitors together with *in vitro* plant cell or tissue culture techniques for the production of secondary metabolites of pharmaceutical interest has had many successful reports. This gives rise to the need to establish different parameters to characterize the performance of elicitors, such as their concentration, selectivity, exposure time of the plant cells, the composition of the growth medium, or the age of the cell culture, among others (Shoja et al., 2022). Most of this characterization has already been reported for biotic or abiotic elicitors; however, recent research is focusing on evaluating these conditions in other types of elicitors, such as those from nanotechnological sources.

Nanotechnology is an emerging area of science whose study and application have been increasing in recent years. Nanomaterials are defined as products with a size between 1 and 100 nm. They can be obtained by chemical or biological synthesis, and their applications are increasingly frequent in different industries such as pharmaceuticals, food, and agriculture due to their unique properties of size, shape, and contact surface (Rivero-Montejo et al., 2021). The use of nanomaterials in the agricultural industry is mainly aimed at developing alternatives to the use of agrochemicals,

 DOI: 10.1201/b23308-9

particularly as an option during the integrated management of plant pests and diseases (Garza-García et al., 2022). In turn, within the biotechnology sector, the study of nanoparticles as elicitors to increase the production of secondary metabolites in *in vitro* plant culture systems is on the rise. Thus, the addition of nanoparticles is a viable option that is gaining importance for the homogeneous, uniform, and continuous large-scale production of phytochemical compounds of economic importance (Lala, 2021).

9.2 SECONDARY METABOLITES

9.2.1 Terpenes and Terpenoids

Terpenes and terpenoids are among secondary metabolites, with about 35,000 different structures characterized to date (Abdallah et al., 2017). These metabolites are considered derivatives of the five-carbon molecule isoprene. This basic unit can be repeated in multiple combinations, giving rise to a considerable variety of compounds. Terpenes are considered exclusively as hydrocarbons, while terpenoids (isoprenoids) may contain oxidized functional groups (Brahmkshatriya and Brahmkshatriya, 2013).

Terpenoids are classified according to the number of repeating isoprene units they contain; hemiterpenes have one isoprene unit, while monoterpenes, sesquiterpenes, and diterpenes would be composed of 2, 3, and 4 isoprene units, respectively. In this classification, they are assumed to chemically bond in a head-to-tail manner; however, non-head-to-tail condensation or head-to-middle fusion is also possible, producing what are termed irregular monoterpenoids (Kobayashi et al., 2019).

The biosynthesis of terpenoids occurs primarily via two independent non-homologous pathways, known as the mevalonate (MVA) pathway and the 2-methyl-D-erythritol-4-phosphate (MEP) pathway (Brock and Dickschat, 2013). The MVA pathway takes place in the cytosol and is responsible for the synthesis of sterols and some sesquiterpenes, occurring mainly in eukaryotes and archaea (Kuzuyama and Seto, 2012). Plants can function with the acetate-mevalonate pathway in the cytoplasm, generating sterols and sesquiterpenes; whereas in plastids, by the MEP pathway, hemi-, mono-, sesqui-, and diterpenes can be synthesized (Hoshino and Gaucher, 2018).

Among the most studied terpenoids with potential therapeutic applications is 3,3-dimethylallyl alcohol (prenol), which in recent studies has shown promising results in neuroprotective processes in *Caenorhabditis elegans* (Phulara et al., 2021). Meanwhile, the group of monoterpenes and monoterpenoids includes myrtenal (Dragomanova et al., 2022) and carvacrol (Zamanian et al., 2021), which have been studied mainly for their potential to inhibit some of the pharmacological targets related to Alzheimer's disease (Wojtunik-Kulesza et al., 2021). On the other hand, antitumor activities of other monoterpenes found in essential oils, such as menthol, borneol, thymol, and myrcene, have been reported (Di Martile et al., 2020).

Within the sesquiterpenes group, anti-inflammatory activity has been evaluated in several compounds, among them farnesol (Moujir et al., 2020), β-caryophyllene (Gushiken et al., 2022), and humulene (Yeo et al., 2021), with excellent prospects. In addition, there is evidence that several sesquiterpene lactones have potential applications in the area of medicine, because in addition to possessing anti-inflammatory activity, some of these molecules also possess anticancer activity.

Moreover, stemodane diterpenes and diterpenoids have shown antilarvicide and antimicrobial activity (Leonelli et al., 2021). In general, terpenes and terpenoids, given their immense diversity, possess multiple biological activities in addition to those mentioned so far, including antiparasitic (Nvau et al., 2020), insecticidal (Dambolena et al., 2016), plant protective (Ninkuu et al., 2021), allelopathic (Li et al., 2020), and antioxidant (Proshkina et al., 2020) activities.

Due to the structural complexity of terpenoids, chemical synthesis requires various complex strategies and catalysts; therefore, several tools are being employed to improve the production of terpenoids that possess some interest in medicine or pharmaceutical research. Among the most commonly used strategies are technologies involving the use of genetic (Wang et al., 2018) and metabolic or biotechnological tools (Helfrich et al., 2019). Among the latter, elicitation strategies with nanoparticles represent an alternative to increase the production of these compounds.

9.2.2 Alkaloids

Alkaloids are a family of natural products derived from secondary metabolism and perform diverse functions in nature. Like terpenes and terpenoids, they possess a wide structural variety, including heterocyclic rings and encompassing more than 20,000 different molecules in organisms. They have been defined as a group of secondary metabolites containing at least one nitrogen atom and can be classified according to the starting molecule of their biosynthetic pathway (Eguchi et al., 2019).

One of the best-known functions of alkaloids is their involvement in the chemical defense of organisms (Ali et al., 2019). In general terms, they are the main pathway for protection against predators or herbivores (Thawabteh et al., 2019). Alkaloids are considered a standard or reference for bitter taste (e.g., caffeine or quinine) when evaluating sensory properties. The bitter taste of alkaloids is also associated with potential toxicity and is an important feature in the interaction of plants with insects (Muñoz et al., 2020).

Additionally, many alkaloids serve analogous functions to amino acid–derived neurotransmitters in central metabolism due to their structural similarity (Hussain et al., 2018). Alkaloids structurally mimic neurotransmitters by binding to neuroreceptors and functioning as agonists or antagonists (Kaserer et al., 2020); also, they have been classified as neurotoxins, and others exhibit cytotoxic properties. Some alkaloids, such as sanguinarine and berberine, can also be intercalating agents, which generate mutations by interaction with DNA (Elissawy et al., 2021).

Despite all these functions, the biosynthesis of alkaloids comes at a high cost for plants because their production is organ-specific, and their content in tissues is regulated and depends on certain phenological stages. For example, plants tend to accumulate these compounds mainly in young leaves, flowers, seeds, or roots in some species. The biosynthetic pathways of alkaloids are extremely complex and varied; however, four fundamental stages are recognized: the accumulation of an amine-type precursor, followed by the accumulation of an aldehyde-type precursor, and then the formation of an imine cation and finally, a Mannich-type reaction. Some of these processes can be carried out intermolecularly or intramolecularly (Lichman, 2021).

Among the most important alkaloids, we can mention isoquinolines, with insecticidal and antifungal properties (Khan and Suresh, 2015), lupins as neurotoxins (Hama and Strobel, 2020), camptothecins and their analog colchicine with antitumor potential (Gracheva et al., 2020), and the analgesics morphine and codeine (Wicks et al., 2021), among others. Other activities that alkaloids may exhibit include antibacterial, antifungal, antiviral, and even herbicidal (Shen et al., 2020).

Similarly to terpenoids, medical or industrially important alkaloids have been studied to improve, optimize, or enhance their synthesis with various strategies, including reports on elicitation by nanoparticles.

9.2.3 Polyphenols

Polyphenols are another group of secondary metabolites present in plants. Like terpenoids and alkaloids, they exhibit a diversity of structures. They are mainly responsible for the organoleptic (smell, taste) and nutritional properties of plant-derived foods and are also used in a variety of applications. Polyphenols are widely distributed in the plant kingdom and are classified into phenolic acids, flavonoids, stilbenes, and lignans (Abbas et al., 2017). One of the most studied groups of polyphenols is flavonoids, which structurally consist of a 15-carbon skeleton and are subdivided into flavonols, flavones, flavanols, flavanones, isoflavones, and proanthocyanidins (Panche et al., 2016).

To date, about 8,000 polyphenolic compounds have been identified in higher plants, and approximately 100 polyphenolic compounds are found in edible plants (Arfaoui, 2021; Iqbal et al., 2022).

Polyphenols are phytochemicals that are generally involved in defense against ultraviolet radiation (Hu et al., 2017) or against pathogens (Stiller et al., 2021).

All phenolic compounds in plants are derived from a common intermediate, phenylalanine, or else from shikimic acid, the latter being a pathway connecting primary and secondary metabolism in higher plants (Rasouli et al., 2017).

Mainly, polyphenols are found in their conjugated form (with one or more residues linked to their hydroxyl groups), associated with carbohydrates (generating the most common form of storage as glycosides) or organic acids. It is also possible to find them associated with amines, lipids, and even other phenols. Polyphenols contribute to improving the organoleptic properties of various foods; for example, in wine, they are responsible for color, astringency, and some other sensory profiles, mainly to avoid browning reactions (Gao et al., 2017).

There are several reports on the biological activity of polyphenolic compounds, highlighting them as antioxidants, anticarcinogens, antimicrobials, and metal chelators (Lakey-Beitia et al., 2021). Other research associates these compounds with protective potential against cardiovascular diseases (Singh et al., 2020), pancreatitis (Gašić et al., 2020), type 2 diabetes (Fernandes et al., 2022), osteoporosis (Chisari et al., 2019), gastrointestinal problems, and even neurodegenerative diseases (Chen et al., 2018).

9.3 NANOPARTICLES AS ELICITORS FOR *IN VITRO* PLANT CULTURE

The biological properties of nanoparticles depend on their physical and chemical characteristics, highlighting their composition, structure, surface area, and size, in addition to other factors such as dissolution, concentration, and manner of exposure (Javed et al., 2017).

In the last decade, metal nanoparticles have been widely used as elicitors for the production of secondary metabolites; it has also been reported that nanoparticles promote plant growth and increase antioxidant activity due to the accumulation of phytochemical compounds. This effect of nanoparticles is related to the concentration applied, generating a hormetic dose–response behavior where generally, low concentrations stimulate growth and the production of bioactive compounds, and high concentrations generate negative effects on biological systems, causing toxicity. High doses of nanoparticles are associated with the formation of reactive oxygen species (ROS) and the release of ions that make up the nanoparticles (Ag^{+}, Cu^{2+}, Fe^{2+}, Zn^{+}, etc.), which pass through the cell wall and membranes to reach the nucleus. During this translocation, ROS causes cell disruption, membrane degradation, lipid peroxidation, DNA damage, and protein denaturation (Ahmad et al., 2020). The accumulation of secondary metabolites occurs as a response, in part, to the oxidative stress caused by nanoparticles; however, when the dose exceeds the hormetic region, both growth and secondary metabolite production decrease significantly.

9.3.1 Silver Nanoparticles (AgNPs)

The combination of AgNPs and plant growth regulators (PGRs) had a positive impact on the proliferation and accumulation of fresh and dry biomass in callus cultures of the medicinal species *Caralluma tuberculata* at an optimum concentration of 60 µg/L. However, the maximum production of phenolic compounds and flavonoids was obtained at the highest concentration of AgNPs (90 µg/L) and without PGRs (Table 9.1). This induction of secondary metabolism was related to the activation of biosynthetic enzymes such as phenylalanine ammonium lyase (PAL) and antioxidant defense enzymes, as well as superoxide dismutase (SOD), peroxidases (POD), catalases (CAT), and ascorbate peroxidases (APX). Thus, the authors propose that a possible cellular mechanism of nanoparticle interaction in plants is through oxidative stress, which would trigger a MAPK kinase cascade, leading to the activation of antioxidant defense and synthesis of secondary metabolites (Ali et al., 2019).

TABLE 9.1
Metal Nanoparticles Are Used in Elicitation Processes for the Production of Secondary Metabolites in *in vitro* Plant Tissue Culture

Nanoparticle type	Nanoparticle properties			Elicitation properties			Reference
	Shape	Average size (nm)	Concentration	Plant species	Bioactive compound	Culture type	
AgNPs	Spherical	40	30, 60, 90 µg/L	*Caralluma tuberculata*	Phenolic compounds and flavonoids	Callus	Ali et al. (2019)
AgNPs	–	10, 40, 100	0.5, 1.0, 5.0 ppm	*Arabidopsis thaliana*	Camalexin	Seedlings	Kruska et al. (2020)
AgNPs-AuNPs	Spherical	25–35	30 µg/L	*Prunella vulgaris*	Phenolic compounds and flavonoids	Suspension cells	Fazal et al. (2019)
AgNP-β-CDs	Spherical	35.3 ± 8.3	10, 20, 30, 40, 50 mg/L	*Salvia miltiorrhiza*	Tanshinone	Hairy root	Ma et al. (2020)
CuONPs	Irregular	40–100	0.1, 1, 10, 100, 1,000 ppm	*Stevia rebaudiana*	Total phenolic and total flavonoid compounds, rebaudioside A and stevioside	Nodal explants	Javed et al. (2017)
CuONPs	Irregular	27	2, 20, 200, 2,000 mg/L	*Stevia rebaudiana*	Rebaudioside A, stevioside, total phenolic, and total flavonoid compounds	Regenerated seedlings	Ahmad et al. (2020)
CuONPs	–	<50	0.05, 0.01, 0.1, 0.2, 0.3 mg/L	*Artemisia annua*	Phenolic compounds and flavonoids	Callus	Fatima et al. (2021)
CuONPs	–	20–50	1, 5, 10, 25, 50, 100 mg/L	*Ocimum basilicum*	Phenolic compounds, flavonoids, rosmarinic acid, cyanidin, and methyl chavicol	Callus	Nazir et al. (2021)
FeONPs	–	–	450, 900, 1,800, 3,600 mg/L	*Hyoscyamus reticulatus*	Hyoscyamine and scopolamine	Hairy root	Moharrami et al. (2017)
Fe_3O_4NPs	Near spherical	10.77, 20.50, 29.33	0.5, 1, 2 ppm	*Ginkgo biloba*	Ginkgolide A, bilobalide, phenolic compounds	Callus	El-Saber et al. (2021)
TiO_2NPs	–	20	20, 40, 60, 80, 100, 200, 400 µL/L	*Stevia rebaudiana*	Rebaudioside A, B, C, and F	Seedlings	Rezaizad et al. (2021)
TiO_2NPs	–	25	10, 15, 30, 60, 80 mg/L	*Paulownia* sp.	Carotenoids, anthocyanins, flavonoids	Seedlings	Dumani et al. (2022)
TiO_2NPs	–	10–15	10, 60, 120 mg/L	*Salvia tebesana*	Phenolic compounds, rosmarinic acid, and flavonoids	Callus	Shoja et al. (2022)
ZnONPs	–	24	0.25, 0.50, 0.75, 1.0 ppm	*Bacopa monnieri*	Bacoside A	Suspension cells	Bhardwaj et al. (2018)
ZnONPs	Hexagonal	35	100 mg/L	*Linum usitatissimum*	Lignans, neolignans, total phenolic, and flavonoid	Suspension cells	Abbasi et al. (2019)
ZnONPs	–	37.8	0.5 mg/L	*Vigna radiata*	Glycosides	Callus	Iqbal et al. (2022)
ZnONPs	Hexagonal	26	2, 20, 200, 2,000 mg/L	*Stevia rebaudiana*	Rebaudioside A and stevioside	Regenerants	Ahmad et al. (2020)

In *Arabidopsis thaliana* seedlings, the size (10, 40, and 100 nm) of AgNPs was found to play an important role in camalexin production. The highest accumulation of this indole alkaloid was observed after 72 h of exposure to 10 nm AgNPs at a concentration of 5 ppm (Table 9.1), although none of the AgNP treatments were comparable to the increase induced by $AgNO_3$ (Kruska et al., 2020).

Most studies focus on evaluating the effect of individual nanoparticles on the production of secondary metabolites. However, Fazal et al. (2019) evaluated the combination of AgNPs and AuNPs (AgAuNPs) in different ratios (1:2, 1:3, 2:1, 3:1) as a synergistic effect on the accumulation of biomass and antioxidant metabolites in suspension cultures of *Prunella vulgaris* compared with AgNPs and AuNPs separately. In the AgAuNPs combination (1:3) in the presence of the growth regulator 1-naphthaleneacetic acid, a 1.69-fold increase in phenolic compound content and a 1.38-fold increase in total flavonoid content were recorded (Table 9.1) compared with the control. But, the maximum biomass accumulation was obtained with the combination of AgAuNPs (3:1) plus the growth regulator, with values of 9.25 g/100 mL fresh weight (FW) and 0.64 g/100 mL dry weight (DW).

Similarly, the 3:1 combination of AgCuNPs stimulated the maximum production of phenolic compounds in callus cultures of *Artemisia absinthium* (Table 9.1), while the 1:3 combination increased the flavonoid content; in both cases, the growth regulator thidiazuron (TDZ) was included. The results obtained with the AgCuNP combinations showed better results than AgNPs and CuNPs, but only in combination with the growth regulator. Another noteworthy feature of this study is that the nanoparticles were obtained by green synthesis using *Moringa oleifera* leaf extract as a reducing and stabilizing agent (Hayat et al., 2021).

In addition to combination with other nanoparticles, AgNPs have been employed in coated form, as reported by Ma et al. (2020), who evaluated the effect of AgNPs-β-cyclodextrin (AgNP-β-CD) on tanshinone production in transformed root cultures of *Salvia miltiorrhiza*. After 7 d of treatment and through the induction of ROS (H_2O_2 and O^{2-}), the concentration of 30 mg/L AgNP-β-CD increased biosynthetic gene expression and tanshinone content 1.8 times more than in the control (Table 9.1), without affecting biomass accumulation.

9.3.2 Copper Oxide Nanoparticles (CuONPs)

CuONPs had a hormetic effect on the production of rebaudioside A, stevioside, total phenolic compounds, and flavonoids in plants regenerated from *Stevia rebaudiana* nodal segment explants (Javed et al., 2017). Concentrations of 0.1, 1, 10, 100, and 1,000 mg/L in Murashige and Skoog (MS) medium were tested, and the highest shoot organogenesis and growth (number of shoots per explant, shoot length, and FW) were found with 10 mg/L CuNPs. The production of steviol glycosides (rebaudioside A and stevioside) and phenolic compounds increased with increasing concentration of CuNPs, reaching a maximum value at the dose of 10 mg/L and then decreasing drastically, with 1,000 mg/L CuNPs causing phytotoxicity (Javed et al., 2017).

In another similar study, Ahmad et al. (2020) evaluated the effect of increasing concentrations of CuONPs (2, 20, 200, and 2,000 mg/L) on the production of steviol glycosides in regenerated seedlings of *S. rebaudiana*. The highest percentage of rebaudioside A, stevioside, total phenol content, total flavonoids, and antioxidant activity were obtained with 20 mg/L. From 200 mg/L, a significant reduction in the content of the secondary metabolites evaluated was obtained, as well as in the growth of regenerated seedlings, roots, number of leaves, and FW.

The chemical composition of the nanoparticles can also determine their effectiveness. In a study by Fatima et al. (2021), they evaluated the effect of different nanoparticles on callus cultures from leaf, root, and shoot explants of *Artemisia annua*. The highest production of flavonoids was obtained by stimulation with 0.1 mg/L ZnONPs in callus cultures obtained from root explants, whereas the highest content of phenolic compounds was recorded in shoot-derived callus cultures exposed to 0.01 mg/L CuONPs (Table 9.1). In the quantification of individual phenolic compounds, rutin recorded the highest content at 0.3 mg/L of ZnONPs (45.81 μg/mg), CuONPs (21.76 μg/mg), and CoONPs (36.26 μg/mg). At the same concentration, the highest content of gallic acid (27.43 μg/mg) was observed with CoONPs and that of caffeic acid (11.72 μg/mg) with ZnONPs.

In contrast, in callus cultures of *Ocimum basilicum*, only CuONPs were effective in increasing the content of phenolic compounds and total flavonoids and biomass accumulation, compared with the control and treatment with MnONPs (Table 9.1). High-performance liquid chromatography (HPLC) analysis identified that rosmarinic acid, chicoric acid, cyanidin, and peonidin recorded higher production. This stimulation in the production of these phenolic compounds coincides with higher *in vitro* and *in vivo* antioxidant activity of the extract. Elicitation of some essential oil components, such as eugenol (0.21 mg/g), methyl eugenol (0.17 μg/g), chavicol (0.30 μg/g), and methyl chavicol (0.17 μg/g), was also observed (Nazir et al., 2021).

9.3.3 Iron Oxide Nanoparticles (FeONPs)

FeONPs proved to be an effective method, as abiotic elicitors, to increase the hyoscyamine and scopolamine content in *Hyoscyamus reticulatus* hairy root cultures. The elicitor effect of FeONPs was a function of concentration (450, 900, 1,800, and 3,600 mg/L) and exposure time (24, 48, and 72 h), resulting in a five-fold increase in hyoscyamine with 900 mg/L at 24 h and for scopolamine with 450 mg/L at 48 h (Moharrami et al., 2017).

Concentration is not the only factor to be considered in elicitation studies with nanoparticles; size also plays an important role. In *Ginkgo biloba* callus cultures, the effect at different concentrations of magnetite nanoparticles (Fe_3O_4NPs) of 10.77, 20.5, and 29.3 nm on the production of ginkgolides and bilobalide, terpene lactones with neuroprotective activity, was evaluated (Feng et al., 2019). The maximum accumulation of ginkgolide A, bilobalide, and phenolic compounds was obtained by treatment with Fe_3O_4NPs of 10.77 nm at a concentration of 2 ppm (Table 9.1). The major phenolics were quercetin, kaempferol, p-coumaric acid, rutin, and caffeic acid, and the accumulation of these phytochemicals was related to the increase in antioxidant activity. In this species, it is confirmed that the smaller the size of Fe_3O_4NPs, the greater the increase in the concentration of bioactive metabolites (El-Saber et al., 2021).

9.3.4 Titanium Dioxide Nanoparticles (TiO_2NPs)

Stevia rebaudiana is a plant of economic importance in the pharmaceutical and nutraceutical industry, and many efforts have been reported to increase the steady production of steviosides and rebaudioside glycosides. One of the strategies is the combination of *in vitro* culture with elicitation. In the latter case, nanoparticles such as TiO_2NPs are being used to increase the production of rebaudioside A, B, C, and F, as well as total phenolic compounds and flavonoids (Rezaizad et al., 2021). In this study, it was found that the content of phenolic compounds, flavonoids, and rebaudioside A and B increased steadily as the dose of TiO_2NPs (20–400 ppm) increased, whereas the production of rebaudioside C and F showed a hormetic effect, whereby the highest concentration of these glycosides was obtained at 80 ppm and then decreased from doses of 100 ppm TiO_2NPs.

The effects of nanoparticles have also been evaluated in tree species propagation methods. Obtaining somatic embryos in three species of *Pawlonia* sp. was successful with the incorporation of TiO_2NPs into the MS medium at a concentration of 60 mg/L. Other advantages of the addition of TiO_2NPs are that they reduce microbial contamination and the time required for explant development to adapted seedlings, as well as vitrification of explants. The *P. shantung* species and leaf explant presented the highest number of embryos and better characteristics (fast growth and more chlorophyll content in leaves) in the regenerated seedlings (Dumani et al., 2022). In the treatment with 60 mg/L TiO_2NPs, the highest levels of antioxidant metabolites such as carotenoids, anthocyanins, and flavonoids were also obtained (Table 9.1).

Elicitation with TiO_2NPs increased the production of O-diphenols, proanthocyanidin, flavonoid, flavone and flavonol, phenolic acid, and total phenolics in *Salvia tebesana* callus culture (Shoja et al., 2022). The accumulation of these secondary metabolites was concentration-dependent, generating a hormetic effect. The highest accumulation of total phenolic compounds, rosmarinic acid, flavone, flavonol, and total O-diphenols was found with 10 mg/L TiO_2NPs and BAP+2,4D (0.5 mg/L of each phytohormone) in callus derived from apical meristems, whereas in leaf-derived callus, treatment with 60 mg/L TiO_2NPs was more effective for the production of flavone and flavonol, rosmarinic acid, proanthocyanidins, phenolic acids, and total O-diphenols.

9.3.5 Zinc Oxide Nanoparticles (ZnONPs)

The highest concentration of ZnONPs (1 ppm) increased the total bacoside A saponin content twofold in *Bacopa monnieri* suspension cells. Furthermore, a linear correlation was found between bacoside A yield and ZnONPs concentration, as well as a regulation in growth, where the lower concentration (0.25 ppm) of ZnONPs enhanced the specific growth rate (Bhardwaj et al., 2018). This work also suggests that ZnONPs might be involved in the regulation of the isoprenoid metabolic pathway at the molecular level by reducing the expression of the *HMG CoA reductase* gene.

The addition of ZnONPs to *Linum usitatissimum* suspension cell cultures at days 0 and 15 stimulated the production of secoisolariciresinol and lariciresinol diglucosides (Table 9.1). On the other hand, elicitation at days 0 and 25 increased the production of phenolic compounds and flavonoids 15.8 and 8.8 times, respectively, compared with the control. The highest contents of dehydrodiconiferyl alcohol glucoside and guaiacylglycerol-β-coniferyl alcohol ether glucoside were observed in this same period (Abbasi et al., 2019). This work highlights elicitation with ZnONPs as a simple and effective method to increase biomass production and accumulation of polyphenolic compounds in a short period, which is easy to manipulate and reproducible. Another work by Zaeem et al. (2020) also found that elicitation with ZnONPs is an effective process for large-scale production of commercially important metabolites (guaiacylglycerol-β-coniferyl alcohol, dehydrodiconiferyl alcohol glucoside, lariciresinol diglucoside, and secoisolariciresinol diglucoside) in both seedlings and *in vitro* culture of *L. usitatissimum.*

In another study, ZnONPs (0.5 mg/L) were used to enhance the production of glycosides and phenolic compounds in the callus culture of three varieties of *Vigna radiata* (Iqbal et al., 2022). ZnONPs were supplied to the MS medium along with 3 mg/L of 2,4-D and increased the phenolic compound content between 22.7% and 26% and the glycoside content between 25% to 50%, depending on the variety. In this work, the application of CuONPs at the same concentration as ZnONPs was also evaluated; however, no significant effect was found in the content of glycosides, but a significant effect was found in the content of phenolic compounds. One of the main differences between the nanoparticles is their average size, with CuNPs being smaller (11.5 nm) than ZnONPs (37.8 nm). These results highlight that the elicitor effect of nanoparticles depends on the plant variety, the type of secondary metabolite to be produced, the plant tissue, and the concentration of growth phytohormones (Iqbal et al., 2022).

A study by Ahmad et al. (2020) evaluated the hormetic effect of ZnONPs at different concentrations (2, 20, 200, and 2,000 mg/L). It was found that the lowest dose (2 mg/L) increased the content of rebaudioside A and stevioside in regenerated *S. rebaudiana* seedlings, whereas increasing the concentration of ZnONPs decreased the amount of these steviol glycosides. It was also found that higher concentrations, from 200 mg/L, of ZnONPs caused adverse effects on plant biomass and antioxidant activity of *S. rebaudiana.*

9.4 APPLICATIONS OF SECONDARY METABOLITES FROM ELICITATION WITH NANOPARTICLES

Most of the studies to increase the production of secondary metabolites applying *in vitro* culture and elicitation techniques have been performed on medicinal plants due to their background in traditional medicine, their multiple uses to improve human health, their low availability, their variability, and the high volume of plant tissue needed when plants are grown in the field or collected from the wild (Nazir et al., 2021). Thus, nanoparticles are becoming an important elicitation strategy to increase the production of bioactive compounds from medicinal plants valuable for their applications or because they are endangered (Ahmad et al., 2020). In turn, metal nanoparticles are commonly used in elicitation processes with a nanotechnological approach due to their physicochemical properties and their concentration-dependent hormetic effect, which allows the activation of plant response mechanisms mainly related to secondary metabolism for the production of bioactive compounds of pharmaceutical, food, or agricultural relevance (Figure 9.1).

9.4.1 Pharmaceutical

Tanshinone is a diterpenoid compound found mainly in the roots of *Salvia miltiorrhiza*. As it has shown a potential effect against cardiocerebral diseases and as a neuroprotectant in *in vivo* studies (Subedi & Gaire, 2021), it is considered a drug candidate. To increase the production of tanshinone, AgNP-β-CDs were found to be an effective elicitor to increase the biosynthesis of this compound compared with other commonly used elicitors (methyl jasmonate or polysaccharide fraction of yeast elicitor) in terms of cost of production and processing (Ma et al., 2020). This work highlighted the biotechnological use of coated nanoparticles to increase the yield of tanshinone production in large-scale root cultures in a bioreactor by optimizing elicitor and plant culture parameters.

Ali et al. (2019) developed a protocol that can be scaled up to industrial production of both biomass and bioactive compounds with pharmacological potential from *Caralluma tuberculata*. This protocol includes the combination of AgNPs (90 μg/L) with 0.5 mg/L 2,4-dichlorophenoxiacetic acid (2,4-D) plus 3.0 mg/L 6-benzyladenine (BA) because this treatment improved the content of antioxidant compounds in the callus culture.

CuONPs increased the production of phenylpropanoid compounds (rosmarinic acid, chicoric acid, caffeic acid, cyanidin, and peonidin) in a concentration-dependent manner applied to *Ocimum basilium* callus culture (Nazir et al., 2021). This species is used commercially and has many uses in the pharmaceutical, cosmetic, and food industries due to its content of essential oils, phenols, and flavonoids. The concentration of these last two groups of secondary metabolites also increased with elicitation with CuONPs, as well as the antioxidant activity *in vitro* and *in vivo*. Due to the presence of these bioactive compounds, *O. basilium* is used to treat diseases such as diabetes, and cancer, as well as cardiovascular and intestinal problems. Compounds such as anthocyanins, rosmarinic acid, chicoric acid, and caffeic acid have enormous commercial value in the pharmaceutical industry.

El-Saber et al. (2021) found that Fe_3O_4NPs increased the content of ginkgolide A (diterpene), bilobalide (sesquiterpene), and phenolic compounds (quercetin, kaempferol, p-coumaric acid, rutin, and caffeic acid) in *Ginkgo biloba* callus in a size- and concentration-dependent manner, with the smallest size (10.77 nm) and highest concentration (2 ppm) being highlighted. Fe_3O_4NPs proved

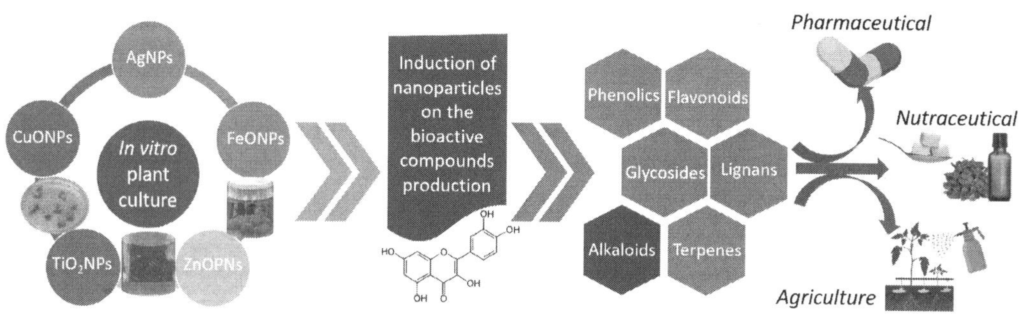

FIGURE 9.1 Metal nanoparticles are used in *in vitro* plant culture as elicitors for the production of secondary metabolites with pharmaceutical, nutraceutical, or agricultural applications.

to be an efficient elicitation technique to increase the production of polyphenols, bilobalide, and ginkgolides. These phytochemical compounds have been used for the treatment of neurological disorders, dementia, Alzheimer's disease, and diabetes (El-Saber et al., 2021).

FeONPs also proved to be an effective strategy to increase the biosynthesis of secondary metabolites of pharmaceutical interest in *Hyoscyamus reticulatus*, namely, the tropane alkaloids hyoscyamine and scopolamine. These bioactive compounds are mainly located in the roots of *H. reticulatus*; they are widely used in traditional medicine, their chemical synthesis is very expensive, and they are commonly obtained from fresh root tissue. Therefore, *in vitro* cultivation of hairy roots and elicitation represents an effective strategy for the production of this type of bioactive compound (Moharrami et al., 2017). Hyoscyamine is an analgesic used for the treatment of the nervous system and digestive and cardiovascular diseases; meanwhile, scopolamine is used as an anesthetic and for the treatment of depression and motion sickness (Shi et al., 2022).

The biosynthesis of secondary metabolites in the callus of *Salvia tebesana* depends on the concentration of TiO_2NPs (10, 60, and 120 mg/L), the origin of the callus (apical meristems or leaf explants), as well as the concentration of phytohormones 6-benzylaminopurine (BAP) and 2,4-D. The content of phenolic compounds and flavonoids increased with the application of TiO_2NPs in the callus culture medium; these phytocompounds are characterized by their antioxidant, anti-inflammatory, and antimicrobial properties, and protection against cardiovascular diseases. Another bioactive compound that increased significantly was rosmarinic acid, one of the most abundant in the genus *Salvia*, which is characterized by its antiviral, antibacterial, and anti-inflammatory health benefits (Shoja et al., 2022).

Flax (*L. usitatissimum*) is a lignan- and neolignan-producing species. These lignans have potential therapeutic applications, mainly against different types of cancer, due to their cytotoxic (Wu and Wu, 2006) and antimitotic activities (Antúnez-Mojica et al., 2019). They have also exhibited antiviral activity against the hepatitis virus (Li et al., 2005) and human immunodeficiency virus (Schroder et al., 1990), as well as antioxidant activity (Fauré et al., 1990). One study evaluated the elicitation of *L. usitatissimum* suspension cells with ZnONPs and found an increased accumulation of lignans and neolignans (Abbasi et al., 2019). The results of this study highlighted the importance of the growth stage of the culture where nanoparticles are added to promote the production of economically important secondary metabolites through a biotechnological approach.

Bacosides (bacoside A and bacoside B) are triterpenoid saponin compounds produced by *Bacopa monnieri* and exhibit neuroprotective activity. For the production of these phytocompounds, the use of ZnONPs in suspension cell culture was reported. With the extracts resulting from the elicitation process, the neuroprotective property was evaluated using N2a neuronal cell lines subjected to oxidative stress. In this study, higher viability of cells treated with methanolic extracts from suspension culture under increasing concentrations of ZnONPs was found. A cytoprotective effect against oxidative stress was also found in *B. monnieri* extracts with ZnONP supply (Bhardwaj et al., 2018).

On the other hand, elicitation with ZnONPs of callus cultures grown in MS medium with 2,4-D increased the content of amygdalin, a cyanogenic glycoside recognized for its anticancer, hepatoprotective, antiviral, anti-inflammatory, and antimicrobial activity. In addition, increased production of phenolic acids was found; these compounds have been associated with antidiabetic and anticancer activities. Both cyanogenic glycosides and phenolic compounds are of great commercial value (Iqbal et al., 2022).

In another investigation with callus cultures, but in *Silybum marianum*, exposure to 50 and 200 mg/L chitosan nanoparticles (ChsNPs) coated with coumarin (40 mg/L) or phenylalanine (40 mg/L) elicited the accumulation of phenolic compounds and some isomers of silymarin (Hassanen et al., 2021). Silymarin is a mixture of flavonolignans that is known for its antioxidant, antifungal, and hepatoprotective activity, it has also been evaluated for use in the treatment of prostate cancer (Janeczko and Kochanowicz, 2019). The increase in flavonolignan production with this treatment was related to high antioxidant capacity and low lipid peroxidation (Hassanen et al., 2021).

9.4.2 Nutraceutical

Steviol glycosides are secondary metabolites found in high levels in *S. rebaudiana* leaves. CuONPs (Javed et al., 2017; Ahmad et al., 2020), ZnONPs (Ahmad et al., 2020), and TiO_2NPs (Rezaizad et al., 2021) have been used as elicitors to increase the production of rebaudioside A and stevioside; these compounds are considered bioactive ingredients in the nutraceutical industry due to their sweetness and taste enhancement; in addition, they are considered calorie-free, representing an alternative for people with conditions such as diabetes, obesity, or cancer (Ahmad et al., 2020). These phytochemicals are also known for their anticancer, antibacterial, and mainly antidiabetic activity (Javed et al., 2017).

CuONPs also proved to be an excellent elicitor to increase the content of essential oil components (eugenol, methyl eugenol, chavicol, and methyl chavicol) in *O. basilicum* callus culture (Nazir et al., 2021). Thus, CuONPs could be used for large-scale production of these high-value metabolites in the food industry.

Phenolic compounds and flavonoids are characterized by their antioxidant power. Elicitation with 10 mg/L TiO_2NPs increased the production of O-diphenols, proanthocyanidin, flavone, and flavonol almost two-fold, whereas with 60 mg/L, the highest quercetin and rutin content was obtained (Shoja et al., 2022). All these compounds act as potent antioxidants for use in the food industry, in addition to their pharmaceutical applications for their health benefits.

9.4.3 Agriculture

It has been shown that AgNPs can act as elicitors for the biosynthesis of phytoalexins such as camalexin and related compounds in *Arabidopsis thaliana* seedlings. This important phytoalexin is produced by several cruciferous species and exhibits antiproliferative and chemopreventive activities (Yamashita et al., 2022), in addition to its role in plant defense. Some derivatives, such as hydroxycamalexin hexoside and malonyl hydroxycamalexin hexoside, were also identified by high-resolution mass spectrometry (Kruska et al., 2020). Because AgNPs induced the production of phytoalexins and other secondary metabolites, the use of these nanoparticles to increase the immune capacity of plants against pathogenic microorganisms could be suggested, optimizing their application. These AgNPs could be used in agriculture as inducers of plant defense against pathogenic microorganisms and could also be used in the field of biotechnology as elicitors for the biosynthesis of bioactive compounds for agricultural purposes.

On the other hand, the production of the compounds hyoscyamine and scopolamine (tropane alkaloids) was increased by FeONP-mediated elicitation in *H. reticulatus* hairy root culture (Moharrami et al., 2017). These secondary metabolites have shown a positive effect on plant growth and have also been reported to have inhibitory activity against insects, bacteria, and weeds, showing potential use in agriculture (Shi et al., 2022). The biological activity of these tropane alkaloids is dose-dependent; this hormetic effect has to be studied in field experiments to identify possible phytotoxic effects.

9.5 CONCLUSIONS

The positive effect of nanoparticles on the growth and production of bioactive metabolites in plant *in vitro* culture depends on their characteristics, such as type of metal ion, concentration, and size; however, in some cases, further experimentation is required to optimize parameters such as concentration and precise duration of nanoparticle treatments. Nanotechnology in combination with plant tissue culture is a promising alternative to modulate biomass accumulation of plant cells and tissues, as well as to regulate metabolic pathways of secondary metabolism and increase the production of high-value bioactive compounds in the pharmaceutical, food, and agricultural industries. Further research is needed to understand the molecular and cellular mechanisms involved in the elicitor effect of nanoparticles during the modulation of development and secondary metabolism for *in vitro* culture of plant tissues, considering the structural composition, cellular transport, and translocation of nanoparticles.

Finally, the main applications of nanotechnology are in the pharmaceutical sector, which is evidenced by the number of practical investigations into *in vitro* plant culture processes for bioactive compounds with therapeutic use. However, this situation represents a challenge and an opportunity to expand the study and potential use in productive sectors that have been less addressed to date.

REFERENCES

Abbas, M., Saeed, F., Anjum, F.M., Afzaal, M., Tufail, T., Bashir, M.S., Ishtiaq, A., Hussain, S., Suleria, H.A.R. (2017). Natural polyphenols: An overview. *International Journal of Food Properties, 20*(8), 1689–1699. https://doi.org/10.1080/10942912.2016.1220393

Abbasi, B.H., Zahir, A., Ahmad, W., Nadeem, M., Giglioli-Guivarc`h, N., Hano, C. (2019). Biogenic zinc oxide nanoparticles-enhanced biosynthesis of lignans and neolignans in cell suspension cultures of *Linum usitatissimum* L. *Artificial Cells, Nanomedicine, and Biotechnology, 47*(1), 1367–1373. https://doi.org/10.1080/21691401.2019.1596942

Abdallah, I.I., Quax, W.J. (2017). A glimpse into the biosynthesis of terpenoids. In *NRLS Conference Proceedings, International Conference on Natural Resources and Life Sciences (2016), KnE Life Sciences*, 81–98. https://doi.org/10.18502/kls.v3i5.981

Ali, A., Mohammad, S., Khan, M.A., Raja, N.I., Arif, M., Kamil, A., Mashwani, Z.-U.-R. (2019). Silver nanoparticles elicited *in vitro* callus cultures for accumulation of biomass and secondary metabolites in *Caralluma tuberculata. Artificial Cells, Nanomedicine, and Biotechnology, 47*(1), 715–724. https://doi.org/10.1080/21691401.2019.1577884

Ahmad, M.A., Javed, R., Adeel, M., Rizwan, M., Ao, Q., Yang, Y. (2020). Engineered ZnO and CuO nanoparticles ameliorate morphological and biochemical response in tissue culture regenerants of Candyleaf (*Stevia rebaudiana*). *Molecules, 25*(6), 1356. https://doi.org/10.3390/molecules25061356

Antúnez-Mojica, M., Rojas-Sepúlveda, A.M., Mendieta-Serrano, M.A., Gonzalez-Maya, L., Marquina, S., Salas-Vidal, E., Álvarez, L. (2019). Lignans from *Bursera fagaroides* affect in vivo cell behavior by disturbing the tubulin cytoskeleton in zebrafish embryos. *Molecules, 24*(1), 8–20. https://doi.org/10.3390/molecules24010008

Arfaoui, L. (2021). Dietary plant polyphenols: Effects of food processing on their content and bioavailability. *Molecules, 26*(10), 2959. https://doi.org/10.3390/molecules26102959

Brahmkshatriya, P.P., Brahmkshatriya, P.S. (2013). Terpenes: Chemistry, biological role, and therapeutic applications. In R. Kishan Gopal M. Jean-Michel (Eds.) *Natural Products: Phytochemistry, Botany and Metabolism of Alkaloids, Phenolics and Terpenes* (pp. 2665–2691). Berlin: Springer Berlin Heidelberg. https://doi.org/10.1007/978-3-642-22144-6_120

Bhardwaj, P., Goswami, N., Narula, P., Jain, C.K., Mathur, A. (2018). Zinc oxide nanoparticles (ZnO NP) mediated regulation of bacosides biosynthesis and transcriptional correlation of HMG-CoA reductase gene in suspension culture of *Bacopa monnieri. Plant Physiology and Biochemistry, 130*, 148–156. https://doi.org/10.1016/j.plaphy.2018.07.001

Brock, N.L., Dickschat, J.S. (2013). Biosynthesis of terpenoids. In: K.G. Ramawat & J.M. Mérillon (Eds.) *Natural Products* (pp. 2693–2732). Berlin: Springer-Verlag Berlin Heidelberg. https://doi.org/10.1007/978-3-642-22144-6_121

Chen, Y.-A., Tsai, J.-C., Cheng, K.-C., Liu, K.-F., Chang, C.-K., Hsieh, C.-W. (2018). Extracts of black garlic exhibits gastrointestinal motility effect. *Food Research International, 107*, 102–109. https://doi.org/10.1016/j.foodres.2018.02.003

Chisari, E., Shivappa, N., Vyas, S. (2019). Polyphenol-rich foods and osteoporosis. *Current Pharmaceutical Design, 25*(22), 2459–2466. https://doi.org/10.2174/1381612825666190722093959

Dambolena, J.S., Zunino, M.P., Herrera, J.M., Pizzolitto, R.P., Areco, V.A., Zygadlo, J.A. (2016). Terpenes: Natural products for controlling insects of importance to human health - A structure-activity relationship study. *Psyche: A Journal of Entomology, 2016*, 4595823. https://doi.org/10.1155/2016/4595823

Di Martile, M., Garzoli, S., Ragno, R., Del Bufalo, D. (2020). Essential oils and their main chemical components: The past 20 years of preclinical studies in melanoma. *Cancers, 12*(9), 2650. https://doi.org/10.3390/cancers12092650

Dragomanova, S. Pavlov, S., Marinova, D., Hodzev, Y., Petralia, M.C., Fagone, P., Nicoletti, F., Lazarova, M., Tzvetanova, E., Alexandrova, A., Kalfin, R., Tancheva, L. (2022). Neuroprotective effects of myrtenal in an experimental model of dementia induced in rats. *Antioxidants*, *11*(2), 374. https://doi.org/10.3390/antiox11020374

Dumani, Y., Mortazavian, S.M.M., Izadi-Darbandi, A., Ramshini, H., Amini, F. (2022). Titanium dioxide nanoparticles affect somatic embryo initiation, development, and biochemical composition in *Paulownia* sp. seedlings. *Industrial Crops and Products*, *176*, 114398. https://doi.org/10.1016/j.indcrop.2021.114398

Eguchi, R., Ono, N., Hirai Morita, A., Katsuragu, T., Nakamura, S., Huang, M., Altaf-Ul-Amin, M., Kanaya, S. (2019). Classification of alkaloids according to the starting substances of their biosynthetic pathways using graph convolutional neural networks. *BMC Bioinformatics*, *20*(1), 380. https://doi.org/10.1186/s12859-019-2963-6

Elissawy, A.M., Soleiman Dehkordi, E., Mehdinezhad, N., Ashour, M.L., Mohammadi Pour, P. (2021). Cytotoxic alkaloids derived from marine sponges: A comprehensive review. *Biomolecules*, *11*(2), 258. https://doi.org/10.3390/biom11020258

El-Saber, M.M., Diab, M.I., Hendawey, M.H., Farroh, K.Y. (2021). Magnetite nanoparticles different sizes effectiveness on growth and secondary metabolites in *Ginkgo biloba* L. callus. *Egyptian Journal of Chemistry*, *64*(9), 5423–5432. https://doi.org/10.21608/EJCHEM.2021.72137.3595

Fatima, K., Abbas, S.R., Zia, M., Sabir, S.M., Khan, R.T., Khan, A.A., Hassan, Z., Zaman, R. (2021). Induction of secondary metabolites on nanoparticles stress in callus culture of *Artemisia annua* L. *Brazilian Journal of Biology*, *81*(2), 474–483. https://doi.org/10.1590/1519-6984.232937

Fauré, M., Lissi, E., Torres, R., Videla, L.A. (1990). Antioxidant activities of lignans and flavonoids. *Phytochemistry*, *29*(12), 3773–3775. https://doi.org/10.1016/0031-9422(90)85329-E

Fazal, H., Abbasi, B.H., Ahmad, N., Ali, M., Ali, S.S., Khan, A., Wei, D.-Q. (2019). Sustainable production of biomass and industrially important secondary metabolites in cell cultures of selfheal (*Prunella vulgaris* L.) elicited by silver and gold nanoparticles. *Artificial Cells, Nanomedicine, and Biotechnology*, *47*(1), 2553–2561. https://doi.org/10.1080/21691401.2019.1625913

Feng, Z., Sun, Q., Chen, W., Bai, Y., Hu, D., Xie, X. (2019). The neuroprotective mechanisms of ginkgolides and bilobalide in cerebral ischemic injury: A literature review. *Molecular Medicine*, *25*(1), 57–64. https://doi.org/10.1186/s10020-019-0125-y

Fernandes, I., Oliveira, J., Pinho, A., Carvalho, E. (2022). The role of nutraceutical containing polyphenols in diabetes prevention. *Metabolites*, *12*(2), 184. https://doi.org/10.3390/metabo12020184

Gao, Y., Ma, S., Wang, M., Feng, X.Y. (2017). Characterization of free, conjugated, and bound phenolic acids in seven commonly consumed vegetables. *Molecules*, *22*(11), 1878. https://doi.org/10.3390/molecules22111878

Garza-García, J.J.O., Hernández-Díaz, J.A., Zamudio-Ojeda, A., León-Morales, J.M., Guerrero-Guzmán, A., Sánchez-Chiprés, D.R., López-Velázquez, J.C., García-Morales, S. (2022). The role of selenium nanoparticles in agriculture and food technology. *Biological Trace Element Research*, *200*(5), 2528–2548. https://doi.org/10.1007/s12011-021-02847-3

Gašić, U., Ćirić, I., Pejčić, T., Radenković, D., Djordjević, V., Radulović, S., Tešić, Ž. (2020). Polyphenols as possible agents for pancreatic diseases. *Antioxidants*, *9*(6), 547. https://doi.org/10.3390/antiox9060547

Gracheva, I.A., Shchegravina, E.S., Schmalz, H.G., Beletskaya, I.P., Fedorov, A.Y. (2020). Colchicine alkaloids and synthetic analogues: Current progress and perspectives. *Journal of Medicinal Chemistry*, *63*(19), 10618–10651. https://doi.org/10.1021/acs.jmedchem.0c00222

Gushiken, L.F.S., Beserra, F.P., Hussni, M.F., Gonzaga, M.T., Ribeiro, V.P., de Souza, P.F., Campos, J.C.L., Massaro, T.N.C., Hussni, C.A., Takahira, R.K., Marcato, P.D., Bastos, J.K., Pellizzon, C.H. (2022). Beta-caryophyllene as an antioxidant, anti-inflammatory and re-epithelialization activities in a rat skin wound excision model. *Oxidative Medicine and Cellular Longevity*, *2022*, 9004014. https://doi.org/10.1155/2022/9004014

Hama, J.R., Strobel, B.W. (2020). Natural alkaloids from narrow-leaf and yellow lupins transfer to soil and soil solution in agricultural fields. *Environmental Sciences Europe*, *32*(1), 126. https://doi.org/10.1186/s12302-020-00405-7

Hassanen, S.A., Hegazi, G.A., Diab, M.I., Mahdi, A.A., Hendawey, M.H., Farroh, K.Y. (2021). Effect of coumarin and phenylalanine conjugated with chitosan nanoparticles on secondary metabolites synthesis in callus of *Silybium marianum*. *Plant Cell Biotechnology and Molecular Biology*, *22*(71&72), 112–130. https://www.ikprress.org/index.php/PCBMB/article/view/7251

Hayat, K., Ali, S., Ullah, S., Fu, Y., Hussain, M. (2021). Green synthesized silver and copper nanoparticles induced changes in biomass parameters, secondary metabolites production, and antioxidant activity in callus cultures of *Artemisia absinthium* L. *Green Processing and Synthesis*, *10*(1), 61–72. https://doi.org/10.1515/gps-2021-0010

Helfrich, E.J.N., Lin, G.M., Voigt, C.A., Clardy, J. (2019). Bacterial terpene biosynthesis: Challenges and opportunities for pathway engineering. *Beilstein Journal of Organic Chemistry, 15*, 2889–2906. https://doi.org/10.3762/bjoc.15.283

Hoshino, Y., Gaucher, E.A. (2018). On the origin of isoprenoid biosynthesis. *Molecular Biology and Evolution, 35*(9), 2185–2197. https://doi.org/10.1093/molbev/msy120

Hu, S., Zhang, X., Chen, F., Wang, M. (2017). Dietary polyphenols as photoprotective agents against UV radiation. *Journal of Functional Foods, 30*, 108–118. https://doi.org/10.1016/j.jff.2017.01.009

Hussain, G., Rasul, A., Anwar, H., Aziz, N., Razzaq, A., Wei, W., Ali, M., Li, J., Li, X. (2018). Role of plant derived alkaloids and their mechanism in neurodegenerative disorders. *International Journal of Biological Sciences, 14*(3), 341–357. https://doi.org/10.7150/ijbs.23247

Iqbal, Z., Javad, S., Naz, S., Shah, A.A., Shah, A.N., Paray, B.A., Gulnaz, A., Abdelsalam, N.R. (2022). Elicitation of the in vitro cultures of selected varieties of *Vigna radiata* L. with zinc oxide and copper oxide nanoparticles for enhanced phytochemicals production. *Frontiers in Plant Science, 13*, 908532. https://doi.org/10.3389/fpls.2022.908532

Janeczko, M., Kochanowicz, E. (2019). Silymarin, a popular dietary supplement shows anti–*Candida* activity. *Antibiotics, 8*(4), 206. https://doi.org/10.3390/antibiotics8040206

Javed, R., Mohamed, A., Yücesan, B., Gürel, E., Kausar, R., Zia, M. (2017). CuO nanoparticles significantly influence in vitro culture, steviol glycosides, and antioxidant activities of *Stevia rebaudiana* Bertoni. *Plant Cell, Tissue and Organ Culture, 131*(3), 611–620. https://doi.org/10.1007/s11240-017-1312-6

Kaserer, T., Steinacher, T., Kainhofer, R., Erli, F., Sturm, S., Waltenberger, B., Schuster, D., Spetea, M. (2020). Identification and characterization of plant-derived alkaloids, corydine and corydaline, as novel Mu Opioid receptor agonists. *Scientific Reports, 10*(1), 13804. https://doi.org/10.1038/s41598-020-70493-1

Khan, A.Y., Suresh Kumar, G. (2015). Natural isoquinoline alkaloids: Binding aspects to functional proteins, serum albumins, hemoglobin, and lysozyme. *Biophysical Reviews, 7*(4), 407–420. https://doi.org/10.1007/s12551-015-0183-5

Kobayashi, M., Kuzuyama, T. (2019). Structural and mechanistic insight into terpene synthases that catalyze the irregular non-head-to-tail coupling of prenyl substrates. *Journal of Chemical BiologyChem, 20*(1), 29–33. https://doi.org/10.1002/cbic.201800510

Kruska, D., Sawikowska, A., Kamalabai Selvakesavan, R., Krajewski, P., Kachlicki, P., Franklin, G. (2020). Silver nanoparticles affect phenolic and phytoalexin composition of *Arabidopsis thaliana*. *Science of the Total Environment, 716*, 135361. https://doi.org/10.1016/j.scitotenv.2019.135361

Kuzuyama, T., Seto, H. (2012). Two distinct pathways for essential metabolic precursors for isoprenoid biosynthesis. *Proceedings of the Japan Academy, Series B, 88*(3), 41–52. https://doi.org/10.2183/pjab.88.41

Lakey-Beitia, J., Burillo, A.M., La Penna, G., Hegde, M.L., Rao, K.S. (2021). Polyphenols as potential metal chelation compounds against Alzheimer's disease. *Journal of Alzheimer's Disease: JAD, 82*(s1), S335–S357. https://doi.org/10.3233/JAD-200185

Lala, S. (2021). Nanoparticles as elicitors and harvesters of economically important secondary metabolites in higher plants: A review. *IET Nanobiotechnology, 15*(1), 28–57. https://doi.org/10.1049/nbt2.12005

Leonelli, F., Migneco, L.M., Valletta, A., Marini Bettolo, R. (2021). Stemodane diterpenes and diterpenoids: Isolation, structure elucidation, biogenesis, biosynthesis, biological activity, biotransformations, metabolites and derivatives biological activity, rearrangements. *Molecules, 26*(9), 2761. https://doi.org/10.3390/ molecules26092761

Li, Y., Fu, L., Yeo, H., Zhu, J.-L., Chou, C.-K., Kou, Y.-H., Yeh, S.-F., Gullen, E., Austin, D., Cheng, Y.-C. (2005). Inhibition of hepatitis B virus gene expression and replication by helioxanthin and its derivative. *Antiviral Chemistry and Chemotherapy, 16*(3), 193–201. https://doi.org/10.1177/095632020501600305

Li, J.Y., Lin, S.X., Zhang, Q., Li, L., Hu, W.W., He, H.B. (2020). Phenolic acids and terpenoids in the soils of different weed-suppressive circles of allelopathic rice. *Archives of Agronomy and Soil Science, 66*(2), 266–278. https://doi.org/10.1080/03650340.2019.1610560

Lichman, B.R. (2021). The scaffold-forming steps of plant alkaloid biosynthesis. *Natural Product Reports, 38*(1), 103–129. https://doi.org/10.1039/D0NP00031K

Ma, Y.J., Xia, J., Wang, Y., Wang, J.W. (2020). Stimulation of tanshinone production in *Salvia miltiorrhiza* hairy roots by β -cyclodextrin-coated silver nanoparticles. *Sustainable Chemistry and Pharmacy, 18*, 100271. https://doi.org/10.1016/j.scp.2020.100271

Moharrami, F., Hosseini, B., Sharafi, A., Farjaminezhad., M. (2017). Enhanced production of hyoscyamine and scopolamine from genetically transformed root culture of *Hyoscyamus reticulatus* L. elicited by iron oxide nanoparticles. *In Vitro Cellular and Developmental Biology – Plant, 53*(2), 104–111. https://doi.org/10.1007/s11627-017-9802-0

Moujir, L., Callies, O., Sousa, P.M.C., Sharopov, F., Seca, A.M.L. (2020). Applications of sesquiterpene lactones: A review of some potential success cases. *Applied Sciences*, *10*(9), 3001. https://doi.org/10.3390/app10093001

Muñoz, I.J., Schilman, P.E., Barrozo, R.B. (2020). Impact of alkaloids in food consumption, metabolism and survival in a blood-sucking insect. *Scientific Reports*, *10*(1), 9443. https://doi.org/10.1038/s41598-020-65932-y

Narayani, M., Srivastava, S. (2017). Elicitation: A stimulation of stress in in vitro plant cell/tissue cultures for enhancement of secondary metabolite production. *Phytochemistry Reviews*, *16*(6), 1227–1252. https://doi.org/10.1007/s11101-017-9534-0

Nazir, S., Jan, H., Zaman, G., Khan, T., Ashraf, H., Meer, B., Zia, M., Drouet, S., Hano, C., Abbasi, B.H. (2021). Copper oxide (CuO) and manganese oxide (MnO) nanoparticles induced biomass accumulation, antioxidants biosynthesis and abiotic elicitation of bioactive compounds in callus cultures of *Ocimum basilicum* (Thai basil). *Artificial Cells, Nanomedicine, and Biotechnology*, *49*(1), 625–633. https://doi.org/10.1080/21691401.2021.1984935

Ninkuu, V., Zhang, L., Yan, J., Fu, Z., Yang, T., Zeng, H. (2021). Biochemistry of terpenes and recent advances in plant protection. *International Journal of Molecular Sciences*, *22*(11), 5710. https://doi.org/10.3390/ijms22115710

Nvau, J.B., Alenezi, S., Ungogo, M.A., Alfayez, I.A.M., Natto, M.J., Gray, A.I., Ferro, V.A., Watson, D.G., de Koning, H.P., Igoli, J.O. (2020). Antiparasitic and cytotoxic activity of Bokkosin, A novel diterpene-substituted Chromanyl benzoquinone from *Calliandra portoricensis*. *Frontiers in Chemistry*, *8*, 574103. https://doi.org/10.3389/fchem.2020.574103

Panche, A.N., Diwan, A.D., Chandra, S.R. (2016). Flavonoids: An overview. *Journal of Nutritional Science*, *5*, e47. https://doi.org/10.1017/jns.2016.41

Phulara, S.C., Pandey, S., Jha, A., Chauhan, P.S., Gupta, P., Shukla, V. (2021). Hemiterpene compound, 3,3-dimethylallyl alcohol promotes longevity and neuroprotection in *Caenorhabditis elegans*. *GeroScience*, *43*(2), 791–807. https://doi.org/10.1007/s11357-020-00241-w

Proshkina, E., Plyusnin, S., Babak, T., Lashmanova, E., Maganova, F., Koval, L., Platonova, E., Shaposhnikov, M., Moskalev, A. (2020). Terpenoids as potential geroprotectors. *Antioxidants*, *9*(6), 529. https://doi.org/10.3390/antiox9060529

Ramirez-Estrada, K., Vidal-Limon, H., Hidalgo, D., Moyano, E., Golenioswki, M., Cusidó, R.M., Palazon, J. (2016). Elicitation, an effective strategy for the biotechnological production of bioactive high-added value compounds in plant cell factories. *Molecules*, *21*(2), 182. https://doi.org/10.3390/molecules21020182

Rasouli, H., Farzaei, M.H., Khodarahmi, R. (2017). Polyphenols and their benefits: A review. *International Journal of Food Properties*, *20*(S2), S1700–S1741. https://doi.org/10.1080/10942912.2017.1354017

Rezaizad, M.S., Abbaspour, H., Hashemi-Moghaddam, H., Gerami, M., Ramezani, M. (2021). Photocatalytic activity of titanium dioxide nanoparticles (TiO_2) on the physiological and phytochemical properties of stevia [*Stevia rebaudiana* (Bertoni) Bertoni]. *Journal of Medicinal Plants and By-Product*, *10*(2), 169–177. https://doi.org/10.22092/jmpb.2020.342676.1205

Rivero-Montejo, S.D.J., Vargas-Hernandez, M., Torres-Pacheco, I. (2021). Nanoparticles as novel elicitors to improve bioactive compounds in plants. *Agriculture*, *11*(2), 134. https://doi.org/10.3390/agriculture11020134

Schroder, H.C., Merz, H., Steffen, R., Müller, W.E.G. (1990). Differential *in vitro* anti-HIV activity of natural lignans. *Zeitschrift fur Naturforschung. C, Journal of biosciences.*, *45*(11–12), 1215–1221. https://doi.org/10.1515/znc-1990-11-1222

Shen, Y., Liang, W.J., Shi, Y.N., Kennelly, E.J., Zhao, D.K. (2020). Structural diversity, bioactivities, and biosynthesis of natural diterpenoid alkaloids. *Natural Product Reports*, *37*(6), 763–796. https://doi.org/10.1039/D0NP00002G

Shi, Z., Zou, W., Zhu, Z., Xion, Z., Li, S., Dong, P., Zhua, Z. (2022). Tropane alkaloids (hyoscyamine, scopolamine and atropine) from genus *Datura*: Extractions, contents, syntheses and effects. *Industrial Crops and Products*, *186*, 115283. https://doi.org/10.1016/j.indcrop.2022.115283

Shoja, A.A., Çirak, C., Ganjeali, A., Cheniany, M. (2022). Stimulation of phenolic compounds accumulation and antioxidant activity in in vitro culture of *Salvia tebesana* Bunge in response to Nano-TiO_2 and methyl jasmonate elicitors. *Plant Cell, Tissue and Organ Culture*, *149*(1–2), 423–440. https://doi.org/10.1007/s11240-022-02251-2

Singh, M., Thrimawithana, T., Shukla, R., Adhikari, B. (2020). Managing obesity through natural polyphenols: A review. *Future Foods*, *1–2*, 100002. https://doi.org/10.1016/j.fufo.2020.100002

Stiller, A., Garrison, K., Gurdyumov, K., Kenner, J., Yasmin, F., Yates, P., Song, B.H. (2021). From fighting critters to saving lives: Polyphenols in plant defense and human health. *International Journal of Molecular Sciences, 22*(16), 8995. https://doi.org/10.3390/ijms22168995

Subedi, L., Gaire, B.P. (2021). Tanshinone IIA: A phytochemical as a promising drug candidate for neurodegenerative diseases. *Pharmacological Research, 169*, 105661. https://doi.org/10.1016/j.phrs.2021.105661

Thakur, M., Bhattacharya, S., Khosla, P.K., Puri, S. (2018). Improving production of plant secondary metabolites through biotic and abiotic elicitation. *Journal of Applied Research on Medicinal and Aromatic Plants, 12*, 1–12. https://doi.org/10.1016/j.jarmap.2018.11.0

Thawabteh, A., Juma, S., Bader, M., Karaman, D., Scrano, L., Bufo, S.A., Karaman, R. (2019). The biological activity of natural alkaloids against herbivores, cancerous cells and pathogens. *Toxins, 11*(11), 656. https://doi.org/10.3390/toxins11110656

Wang, C., Liwei, M., Park, J.B., Jeong, S.H., Wei, G., Wang, Y., Kim, S.W. (2018). Microbial platform for terpenoid production: *Escherichia coli* and yeast. *Frontiers in Microbiology, 9*, 2460. https://doi.org/10.3389/fmicb.2018.02460

Wicks, C., Hudlicky, T., Rinner, U. (2021). Morphine alkaloids: History, biology, and synthesis. *The Alkaloids. Chemistry and Biology, 86*, 145–342. https://doi.org/10.1016/bs.alkal.2021.04.001

Wojtunik-Kulesza, K., Rudkowska, M., Kasprzak-Drozd, K., Oniszczuk, A., Borowicz-Reutt, K. (2021). Activity of selected group of monoterpenes in Alzheimer's disease symptoms in experimental model studies—A non-systematic review. *International Journal of Molecular Sciences, 22*(14), 7366. https://doi.org/10.3390/ijms22147366

Wu, S.-J., Wu, T.-S. (2006). Cytotoxic arylnaphthalene lignans from *Phyllanthus oligospermus*. *Chemical and Pharmaceutical Bulletin, 54*(8), 1223–1225. https://doi.org/10.1248/cpb.54.1223

Yamashita, N., Taga, C., Ozawa, M., Kanno, Y., Sanada, N., Kizu, R. (2022). Camalexin, an indole phytoalexin, inhibits cell proliferation, migration, and mammosphere formation in breast cancer cells via the aryl hydrocarbon receptor. *Journal of Natural Medicines, 76*(1), 110–118. https://doi.org/10.1007/s11418-021-01560-8

Yeo, D., Hwang, S.J., Song, Y.S., Lee, H.J. (2021). Humulene inhibits acute gastric mucosal injury by enhancing mucosal integrity. *Antioxidants, 10*(5), 761. https://doi.org/10.3390/antiox10050761

Zaeem, A., Drouet, S., Anjum, S., Khurshid, R., Younas, M., Blondeau, J.P., Tungmunnithum, D., Giglioli-Guivarc'h, N., Hano, C., Abbasi, B.H. (2020). Effects of biogenic zinc oxide nanoparticles on growth and oxidative stress response in flax seedlings vs. in vitro cultures: A comparative analysis. *Biomolecules, 10*(6), 918. https://doi.org/10.3390/biom10060918

Zamanian, M.Y., Kujawska, M., Nikbakhtzadeh, M., Hassanshahi, A., Ramezanpour, S., Kamiab, Z., Bazmandegan, G. (2021). Carvacrol as a potential neuroprotective agent for neurological diseases: A systematic review article. *CNS and Neurological Disorders - Drug Targets, 20*(10), 942–953. https://doi.org/10.2174/1871527320666210506185042

10 Green Synthesis of Nanoparticles using Endophytes

Mehreen Shah

10.1 INTRODUCTION

Nanotechnology is a field of science that uses interdisciplinary approaches and produces particles of nanoscale for various purposes, such as biomedical, disease diagnosis in plants, bioremediation of the environment, etc. (Cao et al., 2004). Generally, the physical and chemical methods for the synthesis of nanoparticles are usually not safe for the environment and leave us with additional challenges of recovery of these nanoparticles once their use is finished (Biazar et al., 2011). For the synthesis of metal nanoparticles, biological species and nature-derived compounds have been considered due to the effectiveness of biological compounds in plants. Green synthesis using endophytic associations is based on the utilization of natural resources such as plants, fungal extracts, bacteria, actinomycetes, yeast components to reduce metal ions into metallic nanoparticles (Eid et al., 2022). These NPs have the potential to stop the propagation of harmful pathogens, helping us achieve safer environments (Hassan et al., 2018).

10.2 GREEN NANOTECHNOLOGY

The International Union of Pure and Applied Chemistry (IUPAC) in 1997 defined a nanoparticle as a particle that is below 100 nm in dimensions (Iravani et al., 2011). NPs can be generated using different metals, such as iron, copper, gold, etc., or carbon-based such as graphene, etc. (Mihranyan et al., 2012). In the Top-down method, NPs are made through methodologies like sputtering, etching, and ball milling (Cao et al., 2004). The bottom-up approach has many techniques, like atomic/molecular condensation, laser pyrolysis, sol-gel processes, spray pyrolysis, and chemical vapor (Baraton et al., 2003). Eco-friendly routes for the synthesis of nanoparticles can avoid producing pollution of the environment (Thakker et al., 2010). Generation of NPs by conventional means uses large quantities of chemicals and can be carcinogenic, biotoxic, and hazardous to human health at various stages (Biazar et al., 2011). Hence, fusion and reduction with plant extracts are preferred. The green synthesis of nanomaterials can be achieved by biosynthesis from bacterial species via aqueous phase fusion with metallic precursor.. The ease with which bacteria can be multiplied and cultured and propagated easily makes it a great choice for synthesizing nanoparticles (Iravani et al., 2011).

10.3 ENDOPHYTIC MICROORGANISMS

(i) An endophytic association can be defined as that of an organism co-inhabiting with a plant without causing any adversity to the host plant. The relationship can be mutualistic or commensalism, where both the host plant and the endophytic organism receive the benefit at some part of or throughout their life cycle (Stone et al., 2000).

DOI: 10.1201/b23308-10

Endophytes can be introduced to the plant by inoculating the plant seed with the microorganisms and hence allowing the integration of the endophyte with the plant (Johnston-Monje et al., 2011). An endophyte can refer to any organism, which may be bacterial or fungal in its origin, that is found to reside in a host plants. Endophytic fungi are distributed vastly over immense geographic strata in both temporal and spatial aspects, and statistical estimates postulate that over a million endophytic fungi species remain to be discovered (Sun et al., 2012). There are 78 fungal species found in endophytic association with the leaf tissue of *Rhizophora mucronata*. The predominant genera were *Pestalotiopsis*, then *Cladosporium* and *Alternaria* (Hamzah et al., 2018).

10.4 SYNTHESIS OF NANOPARTICLES BY ENDOPHYTES

Green synthesis involves the usage of natural entities for the synthesis of nanoparticles (which refers to synthesis in a way that is sustainable and eco-friendly) (Gaur et al., 2019). They break down metallic ions into metallic nanoparticles and stabilize them subsequently through secondary metabolites and cellular enzymes via intercellular and extracellular processes. This is being studied as a viable option by Li et al. (2011). Endophytic microorganisms have been increasingly viable in expanding the literature on how many avenues are present for nanoparticle synthesis, which has opened newer paths for research (Rana et al., 2020).

Endophytic organisms are present abundantly, found to be residing in almost every plant species, making their usage in the field of nanoparticle synthesis economically beneficial to the researcher; green synthesis is not only viable in monetary aspects but also beneficial to the environment (Ghashghaei et al., 2015). Dong et al. (2017) produced antibacterial Ag-NPs using endophytic bacteria extracted from the root of *Borszczowia aralocaspica* Bunge. Clarence et al. (2019) achieved synthesis of gold nanoparticles in an in-vitro process from *Fusarium solani*, endophytic fungi obtained from host *Chonemorpha fragrans*. Berdy et al. (2005) have identified around 6400 enzymes isolated from fungi. Saqib et al. (2022) achieved the synthesis of zinc oxide nanoparticles using an endophytic bacterial strain of *Enterobacter hormaechei* (Table 10.1).

10.5 NANOPARTICLES SYNTHESIZED BY ENDOPHYTIC BACTERIA

Silver nanoparticles can be synthesized using the endophytic bacteria *Aneurinibacillus migulanus 141*, which was found in the host plant *Mimosa pudica* L. (Syed et al., 2019). Ag-NPs were synthesized using the bacterium *Cytobacillus firmus*, which is an endophytic bacterium on *Terminalia* tree species (Sudarshan et al., 2021). Nirmala et al. (2021) synthesized silver nanoparticles from the endophytic bacterium *Pantoea anthophila*. The synthesized nanoparticles had antibacterial and antimicrobial action against a spectrum of pathogens, and the results were assessed using characterizations. Sunkar et al. (2012) generated Ag-NPs from the endophytic bacterium *Bacillus cereus* extracted from the plant *Garcinia xanthochymus*, which had antibacterial activity. Dong et al. (2017) produced Ag-NPs that were antibacterial in nature, and the endophytic bacteria were extracted from the root of *Borszczowia aralocaspica* Bunge. The strain of bacteria was from the genus *Isoptericola*. Zinc oxide (ZnO) nanoparticles were produced using strains of the endophytic bacterium *Sphingobacterium thalpophilum* (Rajabairavi et al., 2017). Ibrahim et al. (2020) found endophytic bacteria present in garlic and modified them for the stabilization and synthesis of Ag-NPs. Endophytic bacteria were isolated from the ornamental plant *Pennisetum setaceum* and then used in the generation of Ag-NPs (Ahmed et al., 2019).

10.6 NANOPARTICLES SYNTHESIZED BY ENDOPHYTIC FUNGI

From a strain of the fungus *Fusarium oxysporum*, extracellular synthesis of nanoparticles of silver has been achieved. Silver ions in the solution were reduced to a size of around 20–40 nm (Duran et al., 2005). Messaoudi et al. (2020) have proven that fungi have benefits in nanoparticle

TABLE 10.1
Plants and their Endophytes Used for Various Purposes in Nanotechnology Along with Morphological Data

Plants	Endophytes	Size	Types of NPs	Activity of NPs	References
Ocimum sanctum	*Streptomyces coelicolor*	~25 nm	Magnesium	Antimicrobial	El-Molsamy, 2018
Convolvulus arvensis	*Streptomyces capillispiralis* Ca-1	3.5–59 nm	Copper	Insecticidal	Hassan Seld et al., 2018
Erythrophleum fordii	*Alternaria tenuissima*	15–45 nm	ZnO	Antimicrobial, anticancer, and antioxidant	Abdel Hakim et al., 2020
Raphanus sativus	*Alternaria* sp.	4–30 nm	Silver	Antibacterial	Singh et al., 2010
Coffea arabica	*Pseudomonas fluorescens 417*	5–50 nm	Gold	Antibacterial	Syed B et al., 2010
Taxus baccata	*Nemania* sp.	5–70 nm	Silver	Antibacterial	Farsi et al., 2018
Madhuca longifolia	*Pestalotia* sp.	<40 nm	Silver	Antibacterial	Verma et al., 2016
Ocimum tenuiflorum	*Exserohilum rostrata*	10–15 nm	Silver	Anti-inflammatory	Bagur et al., 2019
Pelargonium graveolens	*Colletotrichum* sp.	50 nm	Gold	Antimicrobial	Shankar et al., 2003
Cinnamomum zeylanicum	*Lasiodiplodia theobromae*	~76 nm	Silver	Antibacterial	Ranjani et al., 2020
Pinus densiflora	*Talaromyces purpureogenus*	~25 nm	Silver	Antimicrobial and anticancer	Xiaowen H et al., 2019
Mentha longifolia	*Streptomyces* sp.	2.3–85 nm	Silver	Antimicrobial	El-Gamal et al., 2018
Chonemorpha fragrans	*Fusarium solani ATLOY – 8*	40–45 nm	Gold	Anticancer	Clarence et al., 2019

synthesis by fungal enzymes. Berdy et al. (2005) have identified around 6400 enzymes isolated from fungi. Strains of *Cladosporium cladosporioides* fungus were used for extracellular synthesis of Ag-NPs by Balaji et al.(2009). Mathur et al. (2021) synthesized Fe-NPs from *Penicillium oxalicum*, an endophytic fungus derived from the host plant *Tecomella undulata*. These NPs enabled the degradation of the synthetic dye methylene blue with oxygen radicals formed from peroxide. Sharma et al. (2022) achieved the synthesis of green Ag-NPs using strains of the endophyte *Talaromyces purpureogenus* isolated from the plant *Taxus baccata*. Singh et al. (2014) synthesized Ag-NPs from the endophytic fungus *Penicillium* sp. extracted from the host plant in which it was found, namely, the famous medicinal plant known as *Curcuma longa* (turmeric). Balakumar et al. (2014) prepared in-vitro synthesized silver nanoparticles from the endophytic fungus *Guignardia mangiferae* from several medicinal plants. Clarence et al. (2019) achieved the synthesis of gold nanoparticles in an in-vitro process from *Fusarium solani*, an endophytic fungus obtained from the host *Chonemorpha fragrans*. Devi et al. (2014) synthesized Ag-NPs from endophytic fungi found in the host plant (which was medicinal in nature) called *Gloriosa superba*. Shankar et al. (2003) isolated gold nanoparticles (chloroaurate) from an endophytic fungus found in a symbiotic relationship with the geranium plant. Qian et al. (2013) produced Ag-NPs from the endophytic bacterium *Epicoccum nigrum* and assessed them to be successful against pathogenic activity. Bharathidasan et al. (2012) achieved the synthesis of Ag-NPs using the endophytic fungi *Aspergillus concius*, *Phomosis* sp., and *Penicillium janthinellum*. Ganesan et al. (2020) synthesized zinc oxide nanoparticles using the sol-gel method but incorporating endophytic fungi *Periconium* sp., which possessed excellent antibacterial activity against pathogens such as *Staphylococcus aureus*. Abdel Hafez et al. (2017) achieved biosynthesis of silver nanoparticles from the endophytic fungus *Epicoccum nigrum* using a chemical isolate from the fungi called Curvularin. In another experiment by Abdel hafez et al. (2016), the endophytic fungus *Cladosporium sphaerospermum* was used for the biosynthesis of silver nanoparticles. Bharathidasan et al.(2012) used the endophytic fungi *Aspergillus conius*, *Penicillium janthinellum*, and *Phomopsis* sp. for the synthesis of silver nanoparticles. Vijayanandan et al. (2018) used the endophytic fungus *Aspergillus nidulans* for the biosynthesis of cobalt oxide nanoparticles.

10.7 NANOPARTICLES SYNTHESIZED BY ACTINOMYCETES

Nanoparticles have been successfully synthesized using bacteria from the actinomycetes group, which are very important for human use, as they account for over 70% of antibiotics present in the current industry (Fouda et al., 2019). Endophytic actinomycetes are found to exist in a symbiotic relationship where neither species is harmed (Al-Kahtani et al., 2020).

Hassan et al. (2018) synthesized Cu-NPs from the streptomycete *Streptomyces capillispiralis* Ca-1 derived from the *Convolvulus arvensis* host plant, which is also a medicinal plant. In other research by Hassan SE et al. (2019) using *Oxalis corniculata* L., actinomycetes were isolated and were identified by RNA analysis to be *Streptomyces pseudogriseolus Acv-11* and *Streptomyces zaomyceticus Oc-5*, which were subsequently used for the synthesis of CuO-NPs. Otari et al. (2012) used *Actinomycetes rhodococcus* for the biogenic synthesis of Au-NPs. Sunitha et al. (2013) synthesized Fe and Ag nanoparticles using *Actinomycetes* sp. Ranjhita et al. (2020) synthesized Au-NPs from *Streptomyces griseoruber*, which was isolated as an endophyte from the soil. Ahmad et al. (2003) achieved the synthesis of gold nanoparticles using an extracellular technique with the assistance of the actinomycete species *Rhodococcus*, which is tolerant to a wide range of alkaline pH. Au-NPs were generated extracellularly using *Streptomyces hygroscopicus* sp. by Waghmare et al. (2014). Karthik et al. (2014) achieved the synthesis of silver nanoparticles using *Streptomyces* sp. LK3 by enzymes from the nitrogen cycle, such as α-NADH-dependent nitrate reductase. Prakasham et al. (2014) synthesized silver nanoparticles using the actinomycete *Streptomyces parvulus* SSNP11.

Achillea fragrantissima is a medicinal plant whose root plays host to the actinomycete bacterial strain called *Streptomyces laurentii R-1*, which helped in the biocatalysis of Ag-NPs by Eid et al. (2020) that were beneficial for textile enhancement purposes. These synthesized nanoparticles had antibacterial and antimicrobial uses ranging widely from medicinal use to textile enhancement use. Using green synthesis helps us to create nanoparticles that do not cause harm to the environment, hence allowing us to achieve eco-friendly methods. Produced zinc and copper oxide nanoparticles using *Streptomycetes* sp. to confer antimicrobial properties on textiles. Tsibakhashvili et al. (2011) achieved the synthesis of Ag-NPs by actinomycetes, namely, *Spirulina platensis* and *Streptomyces glaucus*.

10.8 MECHANISMS OF ANTIBACTERIAL ACTION BY NANOPARTICLES

Green nanoparticles show a great affinity for pollutants such as microbes ranging from bacterial cells to protozoan sources. They exhibit greater efficiency against Gram-negative bacteria (Li et al., 2011). Microbes are becoming a problem, particularly the highly contagious ones that are precursors to infections and pandemics (Monowar et al., 2021). In-vitro studies have been performed showing that metallic nanoparticles effectively cause death to microbes by various pathways. Nanoparticles cause cell lysis and apoptosis by making holes in the cell wall (Nirmala et al., 2021). They also cause denaturation of cellular proteins, uncoiling, and damage to cellular DNA. They damage the mRNA. They stop cellular and genetic transcription. They disrupt enzymes by blocking the receptor sites.

There is damage to peptidoglycan and cellular biological cellular membranes (Rana et al,.2020). Many bacterial strains have genes that confer penicillin resistance, sulfonamide resistance, tryptophan resistance, etc. (Singh et al., 2018). When these bacterial strains are treated with antibiotics, they do not die but only become more resistant over time, and pose a great risk to human health and safety. Gold nanoparticles are non-toxic (Syed et al., 2016). Like most nanoparticles, these generate oxygen reactive species, which further enable the antibacterial action by immediate rupture of the peptidoglycan cell wall, disruption of the bacterial replication mechanism, disruption of enzymes, and inhibition of biofilm formation (Shariq et al., 2019).

10.9 MECHANISM FOR ANTIFUNGAL ACTION BY NANOPARTICLES

ROS or reactive oxygen species play a vital role in the fungicidal nature of nanoparticles. .. Due to their strong oxidation potential, excessive ROS generation by nanoparticles can cause damage to biomolecules and organelle structures and lead to DNA/RNA breakage, lipid peroxidation, membrane structure destruction, and protein oxidative carbonylation, which further cause cellular necrosis and apoptosis (Yu et al., 2020). A marker of oxidative stress protein carbonylation was detected by western blot, Fourier transform infrared (FTIR), dinitrophenyl hydrazine (DNPH) binding assay, etc. in fungal samples treated with ZnO.

10.10 ANTIVIRAL MECHANISMS ADAPTED BY NPS IN PLANTS

Gold nanoparticles attach to the coat proteins of potato virus Y (PVY) and tomato mosaic virus (ToMV) and help in attenuating the spread of the virus (El-Dougdoug et al., 2018).

The metallic nanoparticles disarmed the ToMV by damaging the envelope glycoproteins, which led to the disintegration of the virus. The NPs help in disarming the virus mainly by two processes: (i) disruption of the protein coat and (ii) damage to the post-translational modification processes/replication, hence preventing propagation of the virus population in the plant (El-Dougdoug et al., 2018). There is then the subsequent expression of defense genes by the plant, which leads to increased antiviral immunity of the plant against the virus, hence endowing the plant with resistance to the viral siege against it (Cai et al., 2020).

10.11 CONCLUSION

Nanoparticles hold great power for the future of science. To synthesize them, we need to revise conventional methods employing chemicals that pollute the environment. Endophytes are abundant; they are found in every plant, so being distributed widely, they have no restrictions in terms of availability. Endophytic organisms are useful in the field of green synthesis for nanoparticles due to the abundance of natural plant enzymes and extracts. Using this, we can develop newer ways to synthesize nanoparticles using green synthesis methods, i.e., natural methods and also benefit from them disarming harmful pathogens, which may be viral, bacterial, or fungal in origin.

BIBLIOGRPAHY

Abdel-Azeem A, Nada AA, O'Donovan A, Kumar TV, Elkelish A. Mycogenic silver nanoparticles from endophytic Trichoderma atroviride with antimicrobial activity. *Journal of Renewable Materials* 2019; 8(2): 32–37.

Abdel-Hafez SII, Nafady NA, Abdel-Rahim IR, Shaltout AM, Dars JA, Mohamed MA. Biosynthesis of silver nanoparticles using the compound curvularin isolated from the endophytic fungus epicoccum nigrum: Characterization and antifungal activity. *Journal of Pharmaceutical and Applied Chemistry* 2017;3(2):135–146.

Abdel-Hafez Hafez, SII., Nafady, NA, Abdel-Rahim, IR, Shaltout, AM, Mohamed, MA. Biogenesis and optimisation of silver nanoparticles by the endophytic fungus cladosporium sphaerospermum. *Int J Nano Chem* 2016;2(1):11–19.

Abdelhakim HK, El-Sayed ER, Rashidi FB. Biosynthesis of zinc oxide nanoparticles with antimicrobial, anticancer, antioxidant and photocatalytic activities by the endophytic Alternaria tenuissima. *Journal of Applied Microbiology* 2020;128(6):1634–1646.

Ahmad A, Senapati S, Khan MI, Kumar R, Ramani R, Srinivas V, Sastry M. Intracellular synthesis of gold nanoparticles by a novel alkalotolerant actinomycete, Rhodococcus species. *Nanotechnology* 2003;14(7):824–828

Ahmed M, Soundhararajan R, Akther T, Kashif M, Khan J, Waseem M, Srinivasan H. Biogenic AgNPs synthesized via endophytic bacteria and its biological applications. *Environ Sci Pollut Res Int* 2019 Sep;26(26):26939–26946. doi: 10.1007/s11356-019-05869-6.

ALKahtani MDF., Fouda A, Attia KA, Al-Otaibi F, Eid AM, Ewais EE-D, Hijri EE-D., St-Arnaud M., Hassan S.E-D., Khan, N., et al. Isolation and characterization of plant growth promoting endophytic bacteria from desert plants and their application as bioinoculants for sustainable agriculture. *Agronomy* 2020; 10(9):1325. 10.3390/agronomy10091325

Balakumaran, M., Ramachandran, R., Kalaichelvan, P. Exploitation of endophytic fungus, Guignardia mangiferae for extracellular synthesis of silver nanoparticles and their in vitro biological activities. *Microbiol. Res.* 2014; 178:9–17.

Bagur H, Poojari CC, Melappa G, et al. Biogenically synthesized silver nanoparticles using endophyte fungal extract of Ocimum tenuiflorum and evaluation of biomedical properties. *Journal of Cluster Science* 2019;6:30–32.

Balaji DS, Basavaraja S, Deshpande R, Mahesh DB, Prabhakar BK, Venkataraman A. Extracellular biosynthesis of functionalized silver nanoparticles by strains of Cladosporium cladosporioides fungus. *Colloids and Surfaces, Part B: Biointerfaces* 2009;68(1):88–92. 10.1016/j.colsurfb.2008.09.022

Baraton, M-I *Synthesis, Functionalization and Surface Treatment of Nanoparticles.* American Scientific Publishers, 2003.

Beltrán-García MJ, White JF, Prado FM, Prieto KR, Yamaguchi LF, Torres MS et al. Nitrogen acquisition in Agave tequilana from degradation of endophytic bacteria. *Scientific Reports* 2014;4:6938.

Bérdy J. Bioactive microbial metabolites. *Journal of Antibiotics* 2005;58(1):1–26. 10.1038/ja.2005.1

Bharathidasan R, Panneerselvam A. Biosynthesis and characterization of silver nanoparticles using endopyhtic fungus Aspergillus conius, Penicillium janthinellum, and Phomopsis sp. *International Journal of Pharmaceutical Sciences and Research* 2012;3:3163–3169.

Biazar E, Jafarpour M, Montazeri M, Majdi A, Aminifard S, Zafari M, Akbari HR, Rad HG, Rad HG. Nanotoxicology and nanoparticle safety in biomedical designs. *International Journal of Nanomedicine* 2011; 6:2–5.

Cao, G. *Nanastructures and Nanomaterials—Synthesis, Properties and Applications.* World Scientific, 2004; 2:3–7.

Cai L, Liuti C, Jia H, Liu C, Wang D, Sun X. Foliar exposure of Fe3O4 nanoparticles on Nicotiana benthamiana: Evidence for nanoparticles uptake, plant growth promoter and defense response elicitor against plant virus. *Journal of Hazardous Materials* 2020;393:122415.

Clarance, P., Varghese, B., Sales, J., Khusro, A., Agastian, P., Tack, J.-C., Khulaifi, M., Al-Shwaiman, H., Elgorban, A., Syed, A., Kim, H-J. Green synthesis and characterization of Gold Nanoparticles using endophytic fungi Fusarium solani and its in-vitro anticancer and biomedical applications. *Saudi Journal of Biological Sciences* 2019:27. 10.1016/j.sjbs.2019.12.026.

Clarance P, Luvankar B, Sales J, Khusro A, Agastian P, Tack JC, Al Khulaifi MM, Al-Shwaiman HA, Elgorban AM, Syed A, Kim HJ. Green synthesis and characterization of gold nanoparticles using endophytic fungi Fusarium solani and its in-vitro anticancer and biomedical applications. *Saudi Journal of Biological Sciences* 2020; 27:706–712.

Compant S, Clement C, Sessitsch A. Plant growth-promoting bacteria in the rhizo- and endosphere of plants: Their role, colonization, mechanisms involved and prospects for utilization. *Soil Biology and Biochemistry* 2010;42(5):669–678. Curr. Res. Green Sustain. Chem. (2020), Article 100029.

Devi LS, Bareh DA, Joshi SR. Studies on biosynthesis of antimicrobial silver nanoparticles using endophytic fungi isolated from the ethno-medicinal plant Gloriosa superba L. *Proceedings of the National Academy of Sciences India Section B: Biological Sciences* 2014 Dec 1; 84(4):1091–1099.

Dong ZY, Narsing Rao MP, Xiao M, Wang HF, Hozzein WN, Chen W, Li WJ. Antibacterial activity of silver nanoparticles against Staphylococcus warneri synthesized using endophytic bacteria by photo-irradiation. *Frontiers in Microbiology* 2017; 3:3–4.

Dong ZY, Narsing Rao MP, Xiao M, Wang HF, Hozzein WN, Chen W, Li WJ. Antibacterial activity of silver nanoparticles against Staphylococcus warneri synthesized using endophytic bacteria by photo-irradiation. *Frontiers in Microbiology* 2017 Jun 14;8:1090.

Durán, N, Marcato, PD, Alves, OL, De Souza, GIH, Esposito, E. Mechanistic aspects of biosynthesis of silver nanoparticles by several Fusarium oxysporum strains. *Journal of Nanobiotechnology* 2005;3(1):1–7.

Eid AM, Awad MA, Althumayri K, Badr NF, Hamza MF. Endophytic bacterial strain, Brevibacillus brevis-mediated green synthesis of copper oxide nanoparticles, characterization, antifungal, in vitro cytotoxicity, and larvicidal activity. *Green Processing and Synthesis* 2022;11: 931–950.

Eid, AM, Fouda, A, Niedbała, Gniewko, Saad El-Din Hassan, Salem, Salem S, Abdo, Abdullah M, Hetta, Helal F, Shaheen, Tharwat I. Endophytic Streptomyces laurentii mediated green synthesis of ag-NPs with antibacterial and anticancer properties for developing functional textile fabric properties. *Antibiotics* 2020;9(10):641.

El-Dougdoug K.N., Bondok A.M., El-Dougdoug K.A. Evaluation of silver nanoparticles as antiviral agent against ToMV and PVY in tomato plants. *Middle East J Appl Sci* 2018; 8(01):100–111

El-Gamal MS, Salah SS, Salem S, Abdullah A. Biosynthesis, characterization, and antimicrobial activity of silver nanoparticles synthesized by endophytic Streptomyces sp. *Egyptian Journal of Biotechnology* 2018.

El-Moslamy SH. Bioprocessing strategies for cost-effective large-scale biogenic synthesis of Nano-MgO from endophytic Streptomyces coelicolor strain E72 as an anti-multidrug-resistant pathogens agent. *Scientific Reports* 2018.

Farsi M, Farokhi S. Biosynthesis of antibacterial silver nanoparticles by endophytic fungus Nemania sp. isolated from Taxus baccata L. (Iranian yew). *Zahedan Journal of Research in Medical Sciences* 2018; 5: 3–9.

Fouda A, Hassan S, Eid AM, El-Din Ewais, E. *The Interaction between Plants and Bacterial Endophytes under Salinity Stress. Endophytes and Secondary Metabolites.* Springer, Cham; 2019, pp. 1–18.

Fouda A, Hassan SE-D, Abdo AM, El-Gamal MS. Antimicrobial, antioxidant and larvicidal activities of spherical silver nanoparticles synthesized by endophytic Streptomyces spp. *Biological Trace Element Research* 2020;195(2):707–724. 10.1007/s12011-019-01883-4

Ganesan, MH, Vivekanandhan S, Muthuramkumar. Periconium sp. (endophytic fungi) extract mediated sol-gel synthesis of ZnO nanoparticles for antimicrobial and antioxidant applications. *Materials Science in Semiconductor Processing* 2020;105:104739, ISSN 1369-8001

Ghashghaei S, Emtiazi G. The methods of nanoparticle synthesis using bacteria as biological nanofactories, their mechanisms and major applications. *Curr Bionanotechnol* 2015;1(1):3–17. 10.2174/221352940199 9140310104655

Gour A, Jain NK. Advances in green synthesis of nanoparticles. *Artificial Cells, Nanomedicine, and Biotechnology* 2019;47: 844–851.

Guilger-Casagrande M, Lima R. Synthesis of silver nanoparticles mediated by fungi: A review. *Frontiers in Bioengineering and Biotechnology* 2019;7:287. 10.3389/fbioe.2019.00287

Hamzah TN, Lee SY, Hidayat A, Terhem R, Faridah-Hanum I, Mohamed R. Diversity and characterization of endophytic fungi isolated from the tropical mangrove species, Rhizophora mucronata, and identification of potential antagonists against the soil-borne fungus, Fusarium solani. *Frontiers in Microbiology* 2018;9:1707.

Hardoim PR, van Overbeek LS, Berg G, Pirttilä AM, Compant S, Campisano A, et al. The hidden world within plants: Ecological and evolutionary considerations for defining functioning of microbial endophytes. *Microbiology and Molecular Biology Reviews* 2015 Sept 79(3):293–320.

Hassan SE-D, Salem SS, Fouda A, Awad MA, El-Gamal MS, Abdo AM. New approach for antimicrobial activity and bio-control of various pathogens by biosynthesized copper nanoparticles using endophytic actinomycetes. *Journal of Radiation Research and Applied Sciences* 2018;11(3):262–270 . 10.1016/j.jrras.2018.05.003

Hassan SE-D, Fouda A, Radwan AA, Salem SS, Barghoth MG, Awad MA, et al. Endophytic actinomycetes Streptomyces spp mediated biosynthesis of copper oxide nanoparticles as a promising tool for biotechnological applications. *Journal of Biological Inorganic Chemistry* 2019; 4(2): 32–67.

Hassan SELD, Salem SS, Fouda A, Awad MA, El-Gamal MS, Abdo AM. New approach for antimicrobial activity and bio-control of various pathogens by biosynthesized copper nanoparticles using endophytic actinomycetes. *Journal of Radiation Research and Applied Sciences* 2018; 11(3):236–239.

Hassan El-Din, S, Fouda, A, Radwan, AA, Salem, SS, Barghoth, MG, Awad, MA, Abdo, AM, El-Gamal, MS. Endophytic actinomycetes Streptomyces spp mediated biosynthesis of copper oxide nanoparticles as a promising tool for biotechnological applications. *JBIC, Journal of Biological Inorganic Chemistry* 2019;24(3):377–393.

Hulkoti NI, Taranath TC. Biosynthesis of nanoparticles using microbes-A review. *Colloids and Surfaces, Part B: Biointerfaces* 2014;121:474–483. 10.1016/j.colsurfb.2014.05.027

Ibrahim E, Fouad H, Zhang M, Zhang Y, Qiu W, Yan C, ... Chen J. Biosynthesis of silver nanoparticles using endophytic bacteria. 2019; 50:2–3.

Ibrahim E, Zhang M, Zhang Y, Hossain A, Qiu W, Chen Y, Wang Y, Wu W, Sun G, Li B. Green-synthesization of silver nanoparticles using endophytic bacteria isolated from garlic and its antifungal activity against wheat fusarium head blight pathogen Fusarium graminearum. *Nanomaterials* 2020;10(2):219. 10.3390/nano10020219

Iravani S. Green synthesis of metal nanoparticles using plants. *Green Chemistry* 2011;13(10):2638–2650. ISSN 1878-5352. 10.1016/j.arabjc.2016.01.005

Johnston-Monje D, Raizada MN. Conservation and diversity of seed associated endophytes in Zea across boundaries of evolution, ethnography and ecology. *PLoS One* 2011; 6(6):e20396. doi: 10.1371/journal.pone.0020396.

Karthik L, Kumar G, Kirthi AV, Rahuman AA Bhaskara Rao KV. Streptomyces sp. LK3 mediated synthesis of silver nanoparticles and its biomedical application. *Bioprocess and Biosystems Engineering* 2014;37:261–267.

Khan I, Saeed K, Khan I. Nanoparticles: Properties, applications and toxicities. *Arabian Journal of Chemistry* 2019;12(7):908–931.

Li X, Xu H, Chen ZS, Chen G. Biosynthesis of nanoparticles by microorganisms and their applications. *Journal of Nanomaterials* 2011; 2011:270974. 10.1155/2011/270974

Messaoudi, O., Bendahou, M. Biological Synthesis of Nanoparticles Using Endophytic Microorganisms: Current Development. In "BiologicalSynthesis of Nanoparticles Using Endophytic Microorganisms: Current Development" in Nanotechnology and the Environment. Sen, M., Ed. 2020; 1–19.

Mihranyan, A., Ferraz, N., Strømme, M. Current status and future prospects of nanotechnology in cosmetics. *Progress in Materials Science - Prog Mater Sci* 2012:57. 10.1016/j.pmatsci.2011.10.001.

Morisaki H. Endophytic bacteria in the rice plant. *Microbes and Environments* 2008; 23(2): 4–7.

Mathur, P, Saini, S, Paul, E, Sharma, C, Mehtani, P. Endophytic fungi mediated synthesis of iron nanoparticles: Characterization and application in methylene blue decolorization. *Current Research in Green and Sustainable Chemistry*.2021; 4

Monowar, T, Sayedur Rahman Md, Bhore, SJ, Sathasivam, KV. Endophytic bacteria Enterobacter hormaechei fabricated silver nanoparticles and their antimicrobial activity". *Pharmaceutics* 2021;13(4):511. 10.3390/pharmaceutics13040511

Nirmala C, Sridevi M. Characterization, antimicrobial and antioxidant evaluation of biofabricated silver nanoparticles from endophytic Pantoea anthophila. *Journal of Inorganic and Organometallic Polymers and Materials* 31(9):3711–3725 2021. 10.1007/s10904-021-01974-7

Otari SV, Patil RM, Nadaf NH, Ghosh SJ, Pawar SH. Green biosynthesis of silver nanoparticles from an Actinobacteria Rhodococcus sp. *Materials Letters* 2012;72:92–94.

Patil M, Mehta DS, Guvva S. Future impact of nanotechnology on medicine and dentistry. *Journal of Indian Society of Periodontology* 2008 May; 12(2):34–40. 10.4103/0972-124X.44088. PMID: 20142942. PMCID: PMC2813556

Porter AL, Youtie J. How interdisciplinary is nanotechnology? *Journal of Nanoparticle Research* 2009;11(5):1023–1041. 2016,Pages 162–166,ISSN 2213-879X. 10.1016/j.jmau.2016.01.004

Prakasham RS, Kumar BS, Kumar YS, Kumar PK. Production and characterization of protein encapsulated silver nanoparticles by marine isolate Streptomyces parvulus SSNP11. *Indian Journal of Microbiology* 2014;54(3):329–336.

Prathna TC, Lazar Mathew NC, Ashok M, Raichur AM. Biomimetic synthesis of nanoparticles: Science, technology & applicability. In: Amitava M, editor. *Applicability, Biomimetics Learning from Nature*. Rijeka: IntechOpen, 2010.

Qian Y, Yu H, He D, Yang H, Wang W, Wan X, Wang L. Biosynthesis of silver nanoparticles by the endophytic fungus Epicoccum nigrum and their activity against pathogenic fungi. *Bioprocess and Biosystems Engineering*. 2013 Nov 1; 36(11):1613–1619.

Rajabairavi N. et al. (2017). Biosynthesis of novel zinc oxide nanoparticles (ZnO NPs) using endophytic bacteria Sphingobacterium thalpophilum. In: Ebenezar J, editor. *Recent Trends in Materials Science and Applications. Springer Proceedings in Physics*, vol. 189. Cham: Springer. 10.1007/978-3-319-44890-9_23

Rana KL, Kour D, Yadav N, Yadav AN. Endophytic microbes in nanotechnology: Current development, and future biotechnology applications. In: Kumar A, Singh VK, editors. *Microbial Endophytes*. Cambridge, MA: Woodhead Publishing, 2020. pp. 231–262. 10.1016/B978-0-12-818734-0.00010-3

Rana KL, Kour D, Yadav N, Yadav AN *Endophytic Microbes in Nanotechnology: Current Development, and Potential Biotechnology Application* 2020; 4:120–122.

Rani R, Sharma D, Chaturvedi M, Yadav J. Green synthesis, characterization and antibacterial activity of silver nanoparticles of endophytic fungi Aspergillus terreus. *Journal of Nanomedicine and Nanotechnology* 2017; 5:4–8.

Ranjani S, Shariq AM, Mohd A, Senthil KN, Ruckmani K, Hemalatha S. Synthesis, characterization and applications of endophytic fungal nanoparticles. *Inorganic and Nano-Metal Chemistry* 2020.

Ranjitha V, Rai VR. Actinomycetes mediated synthesis of gold nanoparticles from the culture supernatant of Streptomyces griseoruber with special reference to catalytic activity. *3 Biotech.* 2020;7(5):299. 10.1007/s13205-017-0930-3

Saqib S, Nazeer, A, Ali, M, Zaman, W, Younas, M, Shahzad, A, Nisar, M. Catalytic potential of endophytes facilitates synthesis of biometallic zinc oxide nanoparticles for agricultural application. *BioMetals* 2022;35(5):967–985.

Shankar S, Ahmad A, Pasrichaa A, Sastry M. Bioreduction of chloroaurate ions by geranium leaves and its endophytic fungus yields gold nanoparticles of different shapes. *Journal of Materials Chemistry* 2003; 7:11–13.

Shankar SS, Ahmad A, Pasricha R, Sastry M. Bioreduction of chloroaurate ions by geranium leaves and its endophytic fungus yields gold nanoparticles of different shapes. *Journal of Materials Chemistry* 2003;13(7):1822–1826.

Shariq Ahmed M, Soundhararajan R, Akther T. et al. Biogenic AgNPs synthesized via endophytic bacteria and its biological applications. *Environmental Science and Pollution Research International* 2019;26(26):26939–26946. 10.1007/s11356-019-05869-6

Sharma A, Sagar A, Rana J. et al. Green synthesis of silver nanoparticles and its antibacterial activity using fungus Talaromyces purpureogenus isolated from Taxus baccata Linn. *Micro and Nano Systems Letters* 2022;10(1):2. 10.1186/s40486-022-00144-9

Singh A, Singh NÁ, Afzal S, Singh T, Hussain I. Zinc oxide nanoparticles: A review of their biological synthesis, antimicrobial activity, uptake, translocation and biotransformation in plants. *Journal of Materials Science* 2018;53(1):185–201. 10.1007/s10853-017-1544-1

Singh J, Dutta T, Kim KH, Rawat M, Samddar P, Kumar P. 'Green' synthesis of metals and their oxide nanoparticles: Applications for environmental remediation. *Journal of Nanobiotechnology* 2018.

Singh T, Jyoti K, Patnaik A, Singh A, Chauhan R, Chandel SS. Biosynthesis, characterization and antibacterial activity of silver nanoparticles using an endophytic fungal supernatant of Raphanus sativus. *Journal Genetic Engineering & Biotechnology* 2017 4: 4–6.

Singh, D, Rathod, V, Ninganagouda, S, Hiremath, J, Singh, AK, Mathew, J. Optimization and characterization of silver nanoparticle by endophytic fungi Penicillium sp. Isolated from Curcuma longa (turmeric) and application studies against MDR E. coli and S. aureus. *Bioinorganic Chemistry and Applications* 2014:Article ID 408021, 8 pages.

Singh A, Sarkar A, Singh S, S. B. Agrawal. Investigation of supplemental ultraviolet-B-induced changes in antioxidative defense system and leaf proteome in radish (Raphanus sativus L. cv Truthful): An insight to plant response under high oxidative stress. *Protoplasma* 2010;245:75–83.

Soares MA, Li H, Bergen M, White JF. Functional role of an endophytic Bacillus amyloliquefaciens in enhancing growth and disease protection of invasive English ivy (Hedera helix L.). *Plant and Soil* 2015;405:107–123.

Stone JK, Bacon CW, White Jr, JF. *An Overview of Endophytic Microbes: Endophytism Defined* 2000.

Sudarsan S, Kumar Shankar M, Kumar Belagal Motatis A, Shankar S, Krishnappa D, Mohan CD, Rangappa KS, Gupta VK, Siddaiah CN. Green synthesis of silver nanoparticles by Cytobacillus firmus isolated from the stem bark of Terminalia arjuna and their antimicrobial activity. *Biomolecules* 2021;11(2):259.

Sun X, Guo L-D. Endophytic fungal diversity: Review of traditional and molecular techniques. *Mycology* 2012;3:65–76. 10.1080/21501203.2012.656724

Sunitha A, Isaac RSR, Geo S, Sornalekshmi S, Rose A, Praseetha PK. Evaluation of antimicrobial activity of biosynthesized iron and silver nanoparticles using the fungi Fusarium oxysporum and Actinomycetes sp. on human pathogens. *Nano Biomedicine and Engineering* 2013;5:39–45.

Sunkar S, Nachiyar CV. Biogenesis of antibacterial silver nanoparticles using the endophytic bacterium Bacillus cereus isolated from Garcinia Xanthochymus. *Asian Pacific Journal of Tropical Biomedicine* 2012 Dec 1; 2(12):953–959.

Syed B, Prasad NMN, Satish S. Endogenic mediated synthesis of gold nanoparticles bearing bactericidal activity. *Journal of Microscopy and Ultrastructure* 2016; 4(3):162–166.

Syed B, Prasad N, Dhananjaya BL, Yallappa S, Satish S. Synthesis of silver nanoparticles by endosymbiont Pseudomonas fluorescens CA 417 and their bactericidal activity. *Enzyme and Microbial Technology* 2010;95:128–136.

Syed Baker MN Prasad N, Satish S. Synthesis and characterization of silver nanobactericides produced by Aneurinibacillus migulanus 141, a novel endophyte inhabiting Mimosa pudica L. *Arabian Journal of Chemistry* 2019;12(8):3743–3752.

Thakkar KN, Mhatre SS, Parikh RY. Biological synthesis of metallic nanoparticles. *Nanomedicine* 2010; 7:27–29.

Tsibakhashvili NY, Kirkesali EI, Pataraya DT, Gurielidze MA, Kalabegishvili TL, Gvarjaladze DN, Tsertsvadze GI, Frontasyeva MV et al. Microbial synthesis of silver nanoparticles by Streptomyces glaucus and Spirulina platensis. *Advanced Science Letters* 2011;4(11):3408–3417.

Tuan, H, Yih, LS, Asep, H. Diversity and characterization of endophytic fungi isolated from the tropical mangrove species, Rhizophora mucronata, and identification of potential antagonists against the soil-borne fungus, Fusarium solani. *Frontiers in Microbiology* 2018; 9: 7–9.

Usha R, Prabu E, Palaniswamy M, Venil CK, Rajendran R. Synthesis of metal oxide Nano particles by Streptomyces sp. for development of antimicrobial textiles. *Global Journal of Biotechnology and Biochemistry* 2010;5:153–160.

Verma SK, Gond SK, Mishra A, Sharma VK, Kumar J et al. Biofabrication of antibacterial and antioxidant silver nanoparticles (Agnps) by an endophytic fungus Pestalotia sp. isolated from Madhuca longifolia. *Journal of Nanomaterials and Molecular Nanotechnology* 2016; 5:6–9.

Verma SK, Kingsley K, Bergen M, English C, Elmore M, Kharwar RN et al. Bacterial endophytes from rice cut grass (Leersia oryzoides L.) increase growth, promote root gravitropic response, stimulate root hair formation, and protect rice seedlings from disease. *Plant and Soil* 2017;422(1–2):223–238.

Vert M, Doi Y, Hellwich KH, Hess M, Hodge P, Kubisa P, Rinaudo M, Schué FO. Terminology for biorelated polymers and applications (IUPAC Recommendations 2012). *Pure and Applied Chemistry* 2012; 84:4–6.

Vijayanandan, RMB. Biosynthesis of cobalt oxide nanoparticles using endophytic fungus Aspergillus nidulans. *Journal of Environmental Management* 2018; 20–25

Waghmare SS, Deshmukh AM, Sadowski Z. Biosynthesis, optimization, purification and characterization of gold nanoparticles. *African Journal of Microbiology Research* 2014;8(2):138–146.

White JF, Crawford H, Torres MS, Mattera R, Irizarry I, Bergen M. A proposed mechanism for nitrogen acquisition by grass seedlings through oxidation of symbiotic bacteria. *Symbiosis* 2012;57(3):161–171.

Xiaowen H, Saravankumar K, Jin T, Myeong-Hyeon W. Mycosynthesis, characterization, anticancer and antibacterial activity of silver nanoparticles from endophytic fungus Talaromyces purpureogenus. *International Journal of Nanomedicine* 2019; 7:3–5.

Yadav KK, Singh JK, Gupta N, Kumar V. A review of nanobioremediation and its technologies for environmental cleanup: A novel biological approach. *Journal of Materials and Environmental Science* 2017; 8:740.

Yu, Z., Qi, L., Jing, W., Yali, Y., Yin, W., Qihui, Z., Peifeng, L.. "Reactive oxygen species-related nanoparticle toxicity in the biomedical field." *Nanoscale research letters* 2020; 15(1)::115.

Zhou-Yan, D, Prabhu, NRM, Min, X, Hong-Fei, W, Hozzein Wael N, Wei, C, Wen-Jun, L. Antibacterial activity of silver nanoparticles against Staphylococcus warneri synthesized using endophytic bacteria. *Frontiers in Microbiology* 2017; 8:20–23.

11 Phyto-carrier Systems Based on Nanoparticles

Adina-Elena Segneanu

11.1 INTRODUCTION

Pollution, radiation, smoking, and consumption of food with a high content of synthetic dyes and chemical additives are among the main risk factors for health, leading to the generation of oxidative stress, responsible for triggering cancer, neurodegenerative diseases, etc.[1,2] In addition, the resistance to antibiotics developed in recent years has determined the increase in mortality by considerably reducing the effectiveness of conventional antibiotics.[3] From these considerations emerges the urge for new performance therapy approaches, and the new phyto-carrier systems based on different types of nanoparticles may represent a viable solution.

Healing plants are part of the history of humanity. Ethnomedicine is used as the first treatment option in many areas globally.[4] Most of the time, modern studies of chemical screening and investigation of the pharmacological activity of different medicinal plants have confirmed therapeutic recommendations from traditional medicine. Moreover, approximately half of the current pharmaceutical products are based on plant-bioactive compounds.[4] Although natural products do not have the many side effects of synthetic drugs, expanding the range of new pharmaceutical products based on natural products is challenging, mainly due to the highly complex chemical structure and the presence of numerous stereocenters.[5–7] In addition, the composition and proportion of phytoconstituents depend on various biotic and abiotic factors.[5,6] Moreover, despite their very high biological activity, many of the secondary metabolites, due to their size, present an impediment related to reduced bioavailability.[5–7] Therefore, innovative assemblies and carriers capable of ensuring drug-tagged delivery, controlled release, and maximizing chemical stability and biocompatibility of biomolecules are required.[5–9]

Special attention was given to the development of nanocarriers as transport systems with dimensions between 10 and 100 nm and very high adsorption capacity, made by encapsulating phytoconstituents from different types of bio-nanoparticles.[5–9] The latest research has shown that the distinctive physico-chemical characteristics of nanoparticles through functionalization with different biomolecules lead to newly engineered phyto-carrier systems outperforming pharmacological activity and ensuring controlled dosage, prolonged release, and specific target delivery.[11–14]

11.2 NANOCARRIERS

Nanocarriers represent new nano-therapeutic strategies for the selective and rapid transport of molecules with a high pharmacological activity that overcomes the limitations of conventional drug delivery systems.[6,15] The advantages of nanocarriers are numerous: improved efficiency, increased bioavailability, high chemical stability, reduced immunogenicity, membrane transport, and controlled release.[6,15] Also, the design of the morpho-structural and chemical properties (such as specific surfaces, shape, and chemical composition) could enhance the nanocarrier's efficiency.[6,15] Numerous nanocarriers have been developed, but the best yields and implicitly, a high specificity

 DOI: 10.1201/b23308-11

were obtained if there was a high ratio between the specific surface area and volume. The main nanocarriers that fulfill this condition are inorganic, organic, and hybrid (Figure 11.1).[6,15]

11.3 ORGANIC NANOCARRIERS

Research has shown that organic nanocarriers display high biocompatibility, and low toxicity and can deliver targeted and simultaneous transport of some hydrophobic and hydrophilic drugs.[15,16] In Figure 11.2 is shown the classification of organic nanocarriers. The organic types are divided into two main categories: Polymer-based and lipid-based nanocarriers.[8–9]

The *polymer-based nanocarriers* (polymer nanoparticles) present a series of significant advantages (very high photo- and chemical stability, controlled release, selectivity, minimization of side effects, and improved therapeutic action), which allows their use in many types of diseases.[9] They are spherical and up to 200 nm in size, and the polymeric membrane can be natural (chitosan, alginate, albumin, gelatin, modified starch, etc.) or synthetic (polyethylene glycol [PEG], polylactide-polyglycolide, polylactide, polycaprolactones, polyacrylates, etc.). Depending on the arrangement of the polymer membrane, polymer nanoparticles are classified into a) nanocapsules with a core-shell structure and b) nanospheres in which biomolecules are encapsulated or adsorbed in a spherical polymer matrix.[8,9]

The selection of the optimal method for the polymeric membrane preparation depends on several factors: The solubility of the biomolecule, the size of the nanocarriers, porosity, chemical stability, bioavailability, etc. The most common methods are non-precipitation, simple emulsion evaporation, double emulsion evaporation, ionic gelation, and supercritical fluid.[8,9]

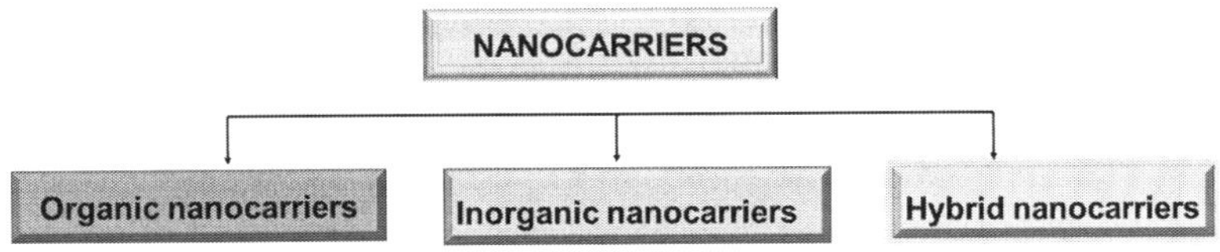

FIGURE 11.1 The main types of nanocarriers

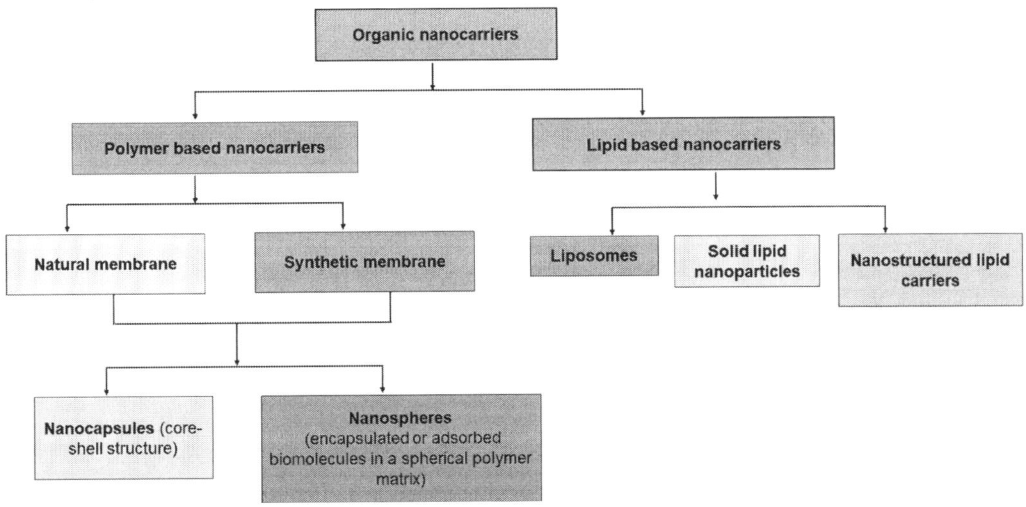

FIGURE 11.2 Schematic representation of organic nanocarriers

Liposomes and solid lipid nanoparticles are considered the most relevant types of lipid-based nanocarriers.[8,9] The importance of lipid-based nanocarriers lies in the fact that lipids allow the solubilization of biomolecules in the gastric fluid and then, their diffusion into the lymphatic system.[8,9]

The liposome is a spherical lipid nanoparticle that encapsulates an aqueous core in one or more lipid layers.[8,9] These lipid membranes can be natural (cholesterol, phosphatidylcholine) or synthetic (polyethylene glycol, poly(lactic-co-glycolic acid)).[8,9] There are several classifications of liposomes.[8,9] Depending on the size and composition of the lipid membrane, liposomes are classified into a) small, unilaminar liposomes, b) large unilaminar liposomes, and c) multilaminar liposomes. The type of charge on the surface of the liposome determines their classification into cationic, anionic, or neutral (zwitterionic).[8,9]

The main advantage of these lipid carriers is the ability to include hydrophilic molecules in the aqueous phase and lipophilic molecules in the lipid membrane.[8,9] These particular structural properties give them physical stability, biocompatibility, and high biodegradability.[8,9]

The most frequently used liposome preparation techniques are:

a) Hydration of a thin lipid film, which consists of several stages: i) dissolution of the lipid compound in the solvent, ii) evaporation of the solvent with the formation of the lipid film, iii) obtaining dispersion liposomes by introducing the lipid film into an aqueous medium or buffer solution, iv) ultrasonication, extrusion to unify and reduce particle sizes.[9]
b) Detergent dialysis, which involves i) the solubilization of phospholipids in an aqueous medium in the presence of detergent micelles, ii) the formation of phospholipid-rich micelles, coalescence, and the removal of the detergent by dialysis or chromatographic methods. The method allows obtaining liposomes of uniform shape and good reproducibility.[9]
c) Solvent injection technique (ensures obtaining liposomes with sizes below 100 nm).[9]
d) Reverse-phase evaporation technique (preparation of liposomes with a high aqueous phase: lipid ratio).[9]
e) Membrane contactor technology, which allows good control of liposome sizes.[9]

Solid lipid nanoparticles are spherical particles, with dimensions between 50 and 1000 nm, consisting of surfactants or co-surfactants and a lipid membrane. [8,9]

The lipid core is a solid lipid nucleus composed of fatty acids, waxes, triglycerides, fatty alcohols, etc.[9] The essential condition imposed on the lipid matrix is high biocompatibility and stability at physiological temperature.[8,9] Having very low toxicity and high biocompatibility, this type of lipid carrier represents a superior alternative to polymer nanocarriers.[8,9]

The solid lipid core structure depends on the chosen preparation technique.[9]

Several methods of obtaining solid lipid nanoparticles are known:

a) Microemulsion ensures the encapsulation of different types of biomolecules in mild experimental conditions and is solvent-free.[8–9]
b) High-pressure homogenization leads to very homogeneous, small-sized lipid carriers.[9]
c) Solvent emulsification evaporation takes place without a temperature rise, allowing obtain a highly stable final product to be obtained.[9]
d) Solvent emulsification diffusion occurs in hydrophobic solvents, which makes it dangerous to administer the resulting lipid carrier to humans.[9]

Nanostructured lipid carriers are prepared from a combination of solid lipids (stearic acid, cetyl alcohol, hydrogenated palm oil, glyceryl monostearate) and liquid lipids (cod liver oil, olive oil, castor oil, mustard oil).[8] This presents superior advantages in relation to solid lipid carriers: increased capacity for encapsulation of biomolecules, better stability, improved drug release control, and higher adsorption capacity.[8]

11.4 INORGANIC CARRIERS

Inorganic carriers have been considered for a long time in the development of drug delivery systems because they present multiple assets with a predefined size and shape and stability in biological environments. Recent studies have reported that these carriers, under certain conditions, exhibit toxicity *in vivo*.[15–17]

The main categories are metallic (iron, silver, gold, platinum, palladium, cobalt, copper, zinc, etc.) and metal oxides (iron oxide, silicon dioxide, titanium oxide, zinc oxide, aluminum oxide, cerium oxide, etc.) nanoparticles.[17] In Figure 11.3 are presented the main types of inorganic nanocarriers.

11.5 METALLIC NANOPARTICLES

The principal metallic nanoparticles used in various biomedical applications are iron/magnetite, gold, silver, and platinum.[18–22]

According to the recent classification of the preparation methods of these nanoparticles, they are divided into top-down and bottom-up techniques. The first category includes mechanical milling, laser ablation, sputtering, inert gas condensation, spark discharge, etc. The second involves hydrothermal, sol-gel, polyol, sonochemical, electrochemical, coprecipitation, microemulsions, thermal decomposition, electrodeposition chemical vapor deposition, microwave irradiation, laser pyrolysis, gamma radiation, spinning, chemical reduction, and green synthesis.[18–22] Physical methods ensure the rapid synthesis of metallic nanoparticles with predefined shapes and sizes, of high purity, in the absence of toxic reagents.[23] The main disadvantages are low efficiency, high cost, high energy consumption, extreme temperature and pressure, radiation exposure, modification of the NPs' surface and physical-chemical characteristics, and a high volume of unwanted secondary products.[16,23] Other drawbacks are: NPs are not thermally stable, and it is very challenging to adjust the particle size and shape.[23]

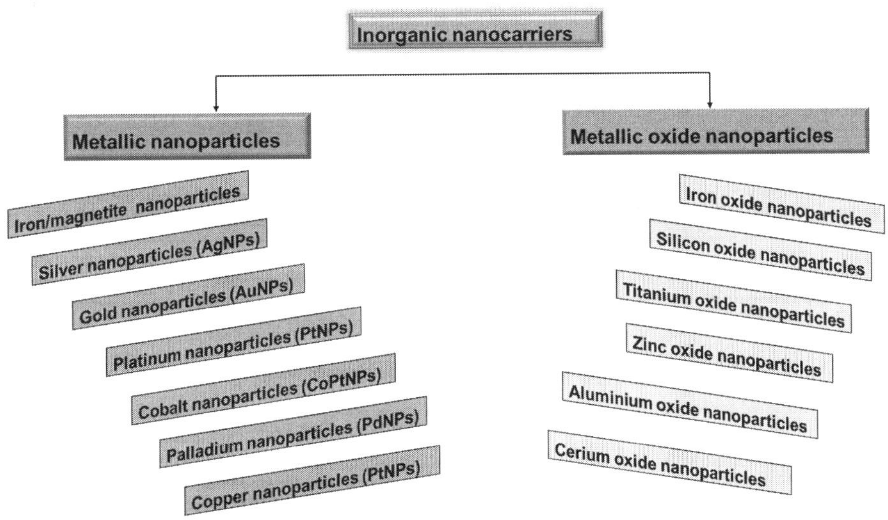

FIGURE 11.3 Schematic representation of inorganic nanocarriers

Generally, metal nanoparticle synthesis through chemical methods requires the presence of three reagent categories: a) Metal precursor, b) reducing agent, and c) capping/stabilizing agent.[17]

The latest studies have addressed green synthesis methods due to their superior advantages (ecological, predefined properties, reproducibility, simple rescale) compared with the conventional approach of obtaining nanoparticles.[19]

Biological methods, or the synthesis of nanoparticles assisted by biomolecules, ensure the synthesis of biocompatible particles with predetermined shapes and dimensions by controlling the following reaction parameters: pH, temperature, time, and amount of reducing agent.[22] It takes place under the action of various biomolecules: Secondary metabolites from plant extracts, different microorganisms (bacteria, fungi), enzymes, etc.[23] In the case of the synthesis of nanoparticles in plant extracts, certain types of phytoconstituents fulfill the role of reducing and capping agents.[23] It is considered that the synthesis mechanism includes the following stages: i) Bioreduction of metal ions, ii) formation and aggregation of particles in thermally stable nanoparticles, iii) production of NPs with predefined shape and size, and iv) purification of NPs.[23] The biosynthesis of metal nanoparticles assisted by microorganisms occurs through an intra- or extracellular reduction mechanism. This approach is expensive and time-consuming.[23]

11.6 MAGNETIC NANOPARTICLES (MNPs)

The new generations of magnetic nanoparticles, such as magnetite (Fe_3O_4) and maghemite (γ-Fe_2O_3), present many advantages: non-toxic, economic efficiency, high chemical stability, biocompatibility, and magnetic properties, which allow their application in the biomedical field (magnetic resonance imaging, hyperthermia, etc.).[18–22,24] Another benefit of magnetic nanoparticles consists of selectivity and vectorization, which makes them useful in the design of targeted drug delivery, thus affecting only tumor tissue, without affecting healthy cells. This approach is more effective and eliminates the secondary effects of conventional, invasive therapeutic techniques such as chemotherapy and radiotherapy.[24,25] Nevertheless, the principal drawback of magnetic nanoparticles is the consequence of the formation of aggregates, even sizeable, which can limit their use in biomedical applications. That's why different approaches are used to increase the stability of the dispersion in an aqueous medium, preventing the formation of large aggregates.[25,26] The most well-known strategies that have been developed are magnetic nanoparticle coating through entrapment or encapsulation: Organic coating (chitosan, carrageenan, lipids, peptides, etc.), inorganic coating (silica, carbon, gold), polymers (polyethylene glycol, polyvinyl alcohol, polyvinylpyrrolidone, polylactic-co-glycolic acid, poly(ethyleneglycol)-poly(acrylic acid), poly(ethyleneoxide)-b-polystyrene, poly(acrylic)-b-polystyrene, poly(4-vinylpyridine)-b-polystyrene, poly(ε-caprolactone)-b-poly(ethylene glycol), etc.) or the ligand exchange method (dendrimers, dextrans, polyethylene glycol, polyethylene oxides).[25,26]

Unlike MNPs, drug delivery systems based on magnetic nanoparticles with a high magnetic moment improved the performance of early detection methods, diagnosis, prognosis, and monitoring of the evolution of cancer treatment. In addition, through the complementary action of biomolecules and magnetic nanoparticles, therapeutic efficiency increases.[26] Additionally, a nanocarrier based on magnetic nanoparticles ensures high permeability and drug retention in the target tissue, thus enhancing their efficiency.[12]

There are several techniques to obtain these drug delivery systems, such as self-organization, systems with straight conformation or template-assisted conformation in the external field, and the use of different fragments (biomolecules, low molecular weight compounds, inorganic materials, and organic polymers).[26] However, assemblies with polymer moieties are more advantageous from the point of view of medical applications due to their predefined physico-chemical properties (molecular weight, sequence of monomers, electrical charge, etc.) and the possibility to incorporate several types and varieties of biomolecules.[26] Furthermore, the structure of these systems can be easily adjusted to ensure simple handling and correspond to a wide range of biomedical applications.[26] A precise spatial arrangement of magnetic nanoparticles occurs within systems with a guided, spherical, "worm" type structure, etc.[26]

Current research focuses on the main drawback of magnetic nanoparticles, represented by a possible immune response due to phagocyte attack. A feasible solution is encapsulation into a polymeric capsule, obtaining a better therapeutic activity performance.[12,21,22,24,25]

The magnetic properties of a nanocarrier based on magnetic nanoparticles depend on the size of the nanoparticles, dipole–dipole interactions, the structure and thickness of the polymer matrix used, etc.[22,24,25] Various studies reported that by coating magnetic nanoparticles with polyethylene glycol, chitosan, carrageenan, and alginate, there is an improvement in chemical stability and adsorption capacity.[10,25–31]Although the encapsulation of nanoparticles with polyethylene glycol considerably increases chemical stability and minimizes agglomeration and the body's immune response, applications are limited due to the oxidative degradation of the polymer.[29]

On the other hand, research has demonstrated that coating magnetic nanoparticles with chitosan, carrageenan, or alginate enhances therapeutic efficiency significantly as an effect of the complementary biological action of these organic compounds.[28–31] Magnetic nanoparticles coated with inorganic materials have applications mainly in tumor treatment through hyperthermia.[27]

Magnetic nanoparticles coated with silica show increased stability in the biological environment, and can be used in various applications: Detection, diagnosis, drug delivery, and tumor treatment through hyperthermia.[27] Nonetheless, at room temperature, a drop in the magnetic properties of silica-coated MNPs compared with bare magnetic nanoparticles has been reported.[27]

MNPs-graphene-oxide sheet systems are highly efficient in tumor diseases and display superior chemical and thermal stability.[27]

Gold could represent the preferred choice for covering MNPs, considering its very high biocompatibility, but it induces a decrease in the magnetic properties of the system.[27]

11.7 GOLD NANOPARTICLES (AuNPs)

Gold has been used since ancient times, all over the world, as a remedy for various ailments: Inflammation of the joints, lung ailments, skin lesions, cancer, diabetes, and as an antidote for mercury poisoning, syphilis, etc.[13]

Numerous studies have demonstrated that gold has antitumor, anti-proliferative, anti-inflammatory, anti-rheumatic, antibacterial, and anti-parasitic activity.[13]

AuNPs present a series of engaging properties from the point of view of biomedical research: Flexible, highly biocompatible, deliquescent, inactive, and providing lower-toxicity biological environments.[13,32] Recent research has reported various biomedical applications of AuNPs: Antitumor agents, drug delivery systems, ophthalmology, neurodegenerative diseases, medical imaging, photothermic therapy, tissue engineering, and regenerative medicine.[13]

AuNPs have unique optical and geometric properties: Surface plasmon resonance, the high ratio between volume and surface area of the particle, different shapes (spherical, tetrahedral, cubic, triangles, prisms, nanorods, irregular shapes, etc.) and sizes between 1 nm and 1 μm, depending on the preparation method used.[32–34] Preparation of AuNPs of a predetermined size is possible through chemical, physical, and ecological strategies.[32–34]

The *chemical processes* proceed through the reduction of $HAuCl_4$. The most frequently used ones are as follows: [32–34]

i) Through the *Turkevitch method*, particles with dimensions of about 20 nm are obtained. The reaction uses a mild surfactant (trisodium citrate, ascorbic acid, tannic acid), which is necessary to control the primary reaction parameters (pH, temperature, concentration).[32–34]

ii) The Brust and Schiffrin method leads to particles with much smaller dimensions (1.5 to 5 nm). The reaction uses sodium borohydride ($NaBH_4$) as the reducing agent in the presence of tetraoctylammonium bromide (TOAB) and dodecanethiol.[32–34]
iii) Particles with sizes between 5 and 40 nm are obtained through the seeding growth method, which is simple, fast, and cheap. The various block co-polymers used as templates allow the synthesis of stable particles with a size of 30 nm. By the intra-molecular reduction of sodium gold sulfite $Na_3Au(SO_3)_2$ in the presence of cetyltrimethylammonium bromides (CTAB), gold particles of about 10 nm were synthesized.[32–34]

Ligand place-exchange strategies are used for the synthesis of AuNPs functionalized with different biomolecules.[32,33] Some examples of such methods are:

i) Reduction of HauCl4 with sodium borohydride ($NaBH_4$) in the presence of PEG-attached dendrimers.[32,33]
ii) Modified burst method in which $HAuCl_4$ is reduced in methanol in the presence of the thiol-ligands PEG 500 and PEG 2000, with the synthesis of particles with sizes between 4 and 15 nm.[32,33]
iii) Photochemical reduction of $HAuCl_4$ with citrate, malate, or tartrate ligands.[32–35]

The most common physical processes for the synthesis of AuNPs functionalized with various biomolecules are:

i) γ-Irradiation – particles with sizes between 2 and 40 nm are prepared in the presence of alginate.[32–34] The particle sizes were up to 10 nm at different ratios of $HAuCl_4$:hyaluro nan.[32–34]
ii) UV-irradiation – used for the synthesis of AuNPs functionalized with amino acids.[32–34]
iii) Microwave irradiation – synthesis of AuNPs in the presence of a reducing agent (citric acid) and ammonium bromide (CTAB).[32–34] Other methods are sonochemical, thermolysis, radiolysis, NIR-irradiation, etc.[32–34]

Biological methods proceed in the presence of microorganisms, enzymes, etc.[32–34]

Green synthesis procedures are simple, cheap, and ecological methods that use a wide range of plant extracts, different types of biomaterials, or sunlight.[30] These synthesis strategies have received particular attention in recent years.[34]

11.8 SILVER NANOPARTICLES (AgNPs)

Silver has been used since ancient times for its antimicrobial, antiviral, and antifungal properties.[14,35–37]

Silver nanoparticles were synthesized for medical applications, considering their outstanding therapeutic properties (antibacterial, antiviral, antifungal, antitumor, antioxidant) confirmed by various studies.[34–37] In addition, the physico-chemical properties (high specific surface area, optical, electrical conductivity, thermal, catalytic) endow them with remarkable biological characteristics: Flexibility, biocompatibility, high stability in living environments, and the ability to cross the blood–brain barrier (BBB).[34–37] Much research has demonstrated the enormous biologic potential of AgNPs, from detection, diagnosis, and drug delivery to therapeutic and antioxidant agents, etc.[34–37]

A series of studies reported that AgNPs are toxic to different tissues or organs.[34–37] The genotoxicity and cytotoxicity mechanism of AgNPs has not yet been elucidated.[34–37] The biological activity of AgNPs is altered by diverse physical and chemical parameters (size, shape, charge, concentration, surface functionalization, etc.).[34–37] Coating/encapsulation or functionalization of AgNPs can minimize toxicity.[34–37]The AgNP synthesis can be achieved through chemical, physical, or ecological methods.[34–37]

Chemical methods proceed by reducing silver ions to metallic silver in the presence of various organic or inorganic compounds (reducing agents).[36,37]The most frequently used are sodium citrate, sodium ascorbate, sodium borohydride, N,N-dimethylformamide, etc.[36,37]

The particle dispersion and size control can be achieved using capping agents and stabilizers (surfactants) such as dodecanethiol, PEG, chitosan, cellulose, polyvinylpyrrolidone, polymethacrylic acid, etc.[36,37]

The best-known methods of stabilizing AgNPs in suspensions are i) Electrostatic repulsion of the coated AgNP surface with anions (halides, citrates, carboxylates, etc.) or cations (polyethyleneimine); ii) steric repulsion (organic polymers, alkylammonium cations).[36,37]

The most frequently used *physical methods* are laser ablation and evaporation-condensation.[36,37]

The *biological strategy* involves the reduction of silver ions under the action of microorganisms (bacteria, fungi, etc.), enzymes, proteins, and vegetable extracts.[36,37]

11.9 PLATINUM NANOPARTICLES (PtNPs)

Platinum exhibits unique biologically active properties: Antitumor, antioxidant, and antimicrobial.[23,38,40] Therefore, recently, PtNPs have been developed, and their potential in various biomedical applications has been explored (detection, diagnosis, medical imaging, drug delivery, and treatment).[23,38,40]

Several studies have reported that PtNPs can present toxicity in biological environments, depending on different physico-chemical parameters: Particle size, concentration, and exposure time. [23,38,40] PtNP synthesis can be achieved through chemical, physical, or biological techniques.[17,40]

The *chemical methods* for PtNP synthesis are classified into several categories: Wet chemical reduction, coprecipitation, hydrolysis, photochemical reduction, microemulsion, thermal decomposition, chemical vapor deposition, electrochemical technique, confined reaction, and sono-decomposition.[17,40]

A specific approach is preferred, following the criteria: i) The dispersion of the particles, ii) the main predetermined physical characteristics of the particles (size and shape), iii) the reaction parameters (temperature, reaction medium, concentration), as well as iv) the type of reducing/capping agent used.[17]

A wet chemical reduction technique is used for fixed dimensions of metal nanoparticle synthesis.[17] Colloidal PtNPs are obtained from metal ions under the action of a reducing agent (sodium borohydride, ascorbate, citrate, tartrate, methoxy PEG, etc.).[17]

The most used *physical methods* are ball milling, flame pyrolysis, arc discharge, vapor deposition, melt mixing, sputter deposition, and laser ablation.[17]

Biological methods for PtNP synthesis use vegetable or microorganism extracts (sulfate-reducing bacteria, fungi).[17,23,40]

11.10 METHODS OF CHARACTERIZATION OF NANOPARTICLES

Several analytical techniques are used to investigate the physico-chemical characteristics of nanoparticles: Specific surface area, pore volume, particle size distribution, morphology, size, shape, chemical composition, surface charge, presence of surface plasmon resonance in the UV-Vis spectrum (AuNPs, AgNPs, PtNPs, etc.), magnetic properties (magnetic nanoparticles), coating/encapsulation efficiency, and drug loading capacity.[36,41–43] Thus, scanning electron microscopy (SEM) or transmission electron microscopy (TEM) provides information on morpho-structural features: Shape, size, orientation, etc.[17,36] Brunauer–Emmett–Teller (BET) analysis

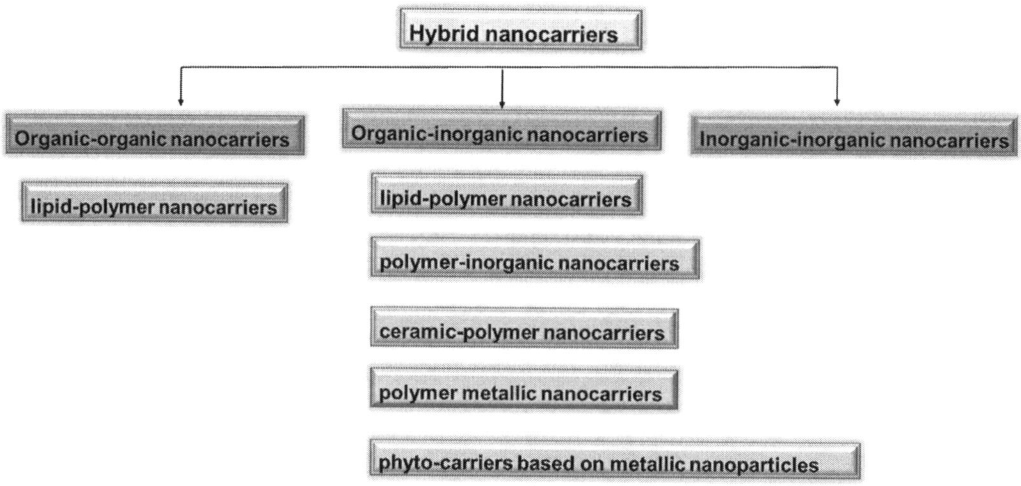

FIGURE 11.4 Schematic representation of the hybrid nanocarriers

allows the determination of the specific surface area and the pore size distribution. X-ray diffraction spectroscopy (XRD) gives details on the crystalline structure.[17,36] The data related to the chemical structure, the type of chemical bonds, and the arrangement of atoms are determined by Fourier transform infrared (FT-IR) spectroscopy.[17] Through the UV-Vis spectroscopy technique, it is possible to obtain information about the optical properties and even approximate data about the shape and size of the nanoparticles.[17] Zeta potential establishes the electric charge of the nanoparticles, and data about the dispersion of the nanoparticles can be obtained by dynamic light spectroscopy (DLS).[17]

11.11 HYBRID CARRIERS

Hybrid nanocarriers are considered to be an improved alternative nano-therapeutic strategy, as they combine properties of different components (organic-organic, inorganic-organic, inorganic-inorganic, etc.) in a single system with superior performance and multiple applications (detection and treatment, etc.) (Figure 11.4).[6,15] The most used hybrid nanocarrier types are lipid-polymer, ceramic-polymer, polymer-inorganic, polymer-metal particles (Au, Ag, Fe), silica-polymer, and others.[6,15,45] The main drawback of organic hybrid nanocarriers is their reduced chemical stability *in vivo*.[6,15,45] The main methods for preparing *organic-organic hybrid nanocarriers* are:

a) *The procedures in the core and the outer shell are performed separately*, followed by the assembly with the formation of hybrid nanocarriers. In this way, the polymer core and the lipid shell are obtained (liposomes, lipo particles, etc.).[45]
b) *The self-assembly procedure* involves obtaining a single or multilayered shell by extrusion, sonication, vortexing, or high-pressure homogenization, and a polymeric core by sonication, solvent diffusion, dissolution, emulsification-solvent evaporation, nanoprecipitation, or high-pressure homogenization.[45] This method is used if the core components of

the nanocarrier (polymers, silica, etc.) are soluble in the solvent and compatible with the therapeutic agent that will be loaded.[45] If the components are not soluble in the solvent, the double emulsification technique is used to obtain hybrid nanocarriers of large dimensions, which can be reduced by ultrasonication or extrusion by high-pressure homogenization.[45]

c) *Ecological synthesis* methods are cheap and ensure products with a smooth coating, better stability, drug release, and biocompatibility.[45]

Preparation methods of organic/inorganic hybrid nanocarriers involve different strategies:

a) polymerization (monomers, metal oxides, organosilane compounds, others).[45]
b) Sself-assembly of organic and inorganic components in nano blocks.
c) functionalization.[45]
d) preparing the organic core, followed by covering with silica and metal nanoparticles.[45]

Obtaining hybrid nanocarriers is influenced by several factors:

a) *The ratio between the lipid component and the polymer.*[45] When this ratio has a low value, the lipid layer does not ensure complete coverage, determining the formation of carriers of high dimensions. When this ratio is too high, it leads to the formation of liposomes, and decreases the homogeneity and yield of obtaining the final compound.[45]
b) *PEG-lipid (PEGylation) incorporation* is used to modify the surface of hybrid carriers and increase colloidal stabilization.[45] Various studies have reported that the molar ratio between PEG and lipid and the chain length of PEG-lipid influences the stability of the hybrid carriers.[45] Thus, a longer chain and a high molar ratio improved product stability.[45]
c) *Polymer properties* (density, surface charge) affect the stability and size of hybrid nanocarriers. High density of the polymer decreases the stability of the product.[45]

11.12 PHYTO-CARRIER SYSTEM BASED ON METALLIC NANOPARTICLES

Results related to the toxic effects of metallic nanoparticles under certain conditions (concentration, contact time, dimensions, etc.) require limitation of biomedical applications.[13,16,25,26,32,35–37] Recent studies have focused on various methods of co-coating/co-encapsulation with different biomolecules as a new carrier development for biomedical applications.[16,45] Another limitation of current drug delivery systems is the limited retention and dissemination capacity of nanoparticles and associated drugs in the tumor tissue. Moreover, these associated molecules have a short retention time in the body and are "memorized" by the immune system, determining the drug resistance of cancer cells.[16]

On the other hand, many secondary metabolites from medicinal plants show remarkable biological effects (antitumor, antibacterial, antioxidant, anti-inflammatory, etc.) and can be considered ideal candidates in therapy for severe disease.[14,42–44,46–50]

Nonetheless, various researches have reported that some phytoconstituents have reduced bioavailability *in vivo* or cannot cross different biological barriers due to their size.[14,51–52] Therefore, a new approach is required for the development of innovative drug delivery systems with improved performance adsorption capacity, target delivery, and the controlled release of secondary metabolites.

Specific phyto-carriers based on nanoparticle development that combines the biological activity of secondary metabolites and metallic nanoparticles represent a new strategy for obtaining performant multifunctional therapeutic platforms. In this way, synergistic therapeutic effects are achieved and could enable overcoming the resistance of bacteria to antibiotics and the drug resistance of cancer cells.

Table 11.1 shows examples of biomolecule–metallic nanoparticle carrier systems and their biomedical applications.

TABLE 11.1
Biomolecules–Metallic Nanoparticle Carrier Systems and Their Biomedical Applications

Nr crt	Biomolecule–metallic nanoparticle carrier system	Biomedical application	Ref.
1	Alginate-coated caseinate NPs	Anticancer drug delivery	50
2	Chitosan/sodium alginate functionalized graphene oxide–based nanocomposites	Anticancer drug delivery	50
3	Sodium alginate and hydroxyapatite bi-coated iron oxide NPs	Anticancer drug delivery	50
4	Fe_3O_4/calcium phosphate/alginate core-shell-corona NPs	Targeted chemotherapy	50
5	1 – Lipid-based nanoparticles2 – Metal and metal oxide NPs/3-resveratrol-zinc NPs	COVID-19; SARS-Cov-2 viral diseaseCOVID-19	51
6	Chitosan-based nanocarriers of AgNPs	Antitumoral	35
7	*Artemisia annua*-magnetite nanoparticles	Antitumoral	
8	*Curcuma longa*-AgNPs	Neurodegenerative disorders	14
9	*Viscum album*-AuNPs	Antitumoral	13
10	AuNPs-quercetin conjugate	Antibiotic	52
11	AuNPs-kaempferol	Antibiotic	52
12	AuNPs-chrysin	Antibiotic	52
13	Curcumin-AuNPs-liposome	Antitumoral (melanoma)	53
14	Epigallocatechin-3-gallate-AuNPs	Antitumoral (melanoma)	53
15	Genistein-AuNPs	Antitumoral	53
16	Armoracia rusticana–kaolinite carrier system	Antioxidant, antitumoral	5

11.13 CONCLUSIONS

Phyto-carrier systems based on metallic nanoparticles open up new opportunities for obtaining novel therapeutic platforms with superior biological properties (improved retention time and bioavailability, transmembrane transport capacity, non-specific interactions with serum proteins, minimized toxicity) and increased therapeutic efficiency in contrast to the individual components. These new types of drug delivery systems could offer enormous biomedical potential.

REFERENCES

1. Sinha K.B., 2013, Roles of free radicals in the toxicity of environmental pollutants and toxicants. *Journal of Clinical Toxicology*, 01: s12.
2. Pham-Huy L.A., He H., Pham-Huy C., 2008, Free radicals, antioxidants in disease and health. *International Journal of Biomedical Sciences*, 4(2): 89–96.
3. Urban-Chmiel R., Marek A., Stepien-Pysniak D., Wieczorek K., Dec M., Nowaczek A., Osek J., 2022, Antibiotic resistance in bacteria—A review. *Antibiotics*, 11(8): 1079.
4. Mehta P., Shah R., Lohidasan S., Mahadik K.R., 2015, Pharmacokinetic profile of phytoconstituent(s) isolated from medicinal plants—A comprehensive review. *Journal of Traditional and Complementary Medicine*, 5(4): 207–227.
5. Nath R., Roy R., Barai G., Bairagi S., Manna S., Chakraborty R., 2021, Modern developments of Nano based grug delivery system by combined with phytochemicals-presenting new aspects. *International Journal of Scientific & Technology Research*, 8: 107–129.
6. Patra J.K., Das G., Fraceto L.F., Campos E.V.R., del Pilar Rodriguez-Torres M., Acosta-Torres L.S., Diaz-Torres L.A., Grillo R., Swamy M.K., Sharma S., et al. 2018, Nano based drug delivery systems: Recent developments and future prospects. *Journal of Nanobiotechnology*, 16(1):71.
7. Atanasov A.G., Zotchev S.B., Dirsch V.M., et al. 2021, Natural products in drug discovery: Advances and opportunities. *Nature Reviews. Drug Discovery*, 20(3): 200–216.

8. Rahman H.S., Othman H.H., Hammadi N.I., Yeap S.K., Amin K.M., Abdul Samad N., Alitheen N.B., 2020, Novel drug delivery systems for loading of natural plant extracts and their biomedical applications. *International Journal of Nanomedicine*, 15: 2439–2483.
9. Lammari N., Tarhini M., Miladi K., Louaer O., Meniai A.H., Sfar S., Fessi H., Elaïssari A., 2021, *Encapsulation Methods of Active Molecules for Drug Delivery. Drug Delivery Devices and Therapeutic Systems*, Chappel E. (Ed), Elsevier, pp. 289–306. 978–970-12-819838-4.
10. Fouad D., Bachra Y., Ayoub G., Ouaket A., Bennamara A., Knouzi N., Berrada M., 2020, A novel drug delivery system based on nanoparticles of magnetite Fe3O4 embedded in an auto cross-linked chitosan. *Chitin and Chitosan - Physicochemical Properties and Industrial Applications*, Berrada M. (Ed), IntechOpen. ISBN 978-1-78984-425-2.
11. Sanzari I., Leone A., Ambrosone A., 2019, Nanotechnology in plant science: To make a long story short. *Frontiers in Bioengineering and Biotechnology*, 7(12): 120
12. Segneanu A.-E., Marin C.N., Ghirlea I.O.F., Feier C.V.I., Muntean C., Grozescu I., 2021, *Artemisia annua* growing wild in Romania—A metabolite profile approach to target a drug delivery system based on magnetite nanoparticles. *Plants*, 10(11): 2245.
13. Segneanu A.-E., Marin C.N., Herea D.D., Stanusoiu I., Muntean C., Grozescu I., 2022, Romanian *Viscum album* L.—Untargeted low-molecular metabolomic approach to engineered viscum–AuNPs carrier assembly. *Plants*, 11(14): 1820.
14. Segneanu A.-E., Vlase G., Lukinich-Gruia A.T., Herea D.D., Grozescu I., 2022, Untargeted metabolomic approach of *Curcuma longa* to neurodegenerative phytocarrier system based on silver nanoparticles. *Antioxidants*, 11(11): 2261.
15. Chamundeeswari M., Jeslin J., Verma M.L., 2018, Nanocarriers for drug delivery applications. *Environmental Chemistry Letters*, 17(2): 849–865.
16. Xie P., Wang Y., Wei D., Zhang L., Zhang B., Xiao H., … Mao X., 2021, Nanoparticle-based drug delivery systems with platinum drugs for overcoming cancer drug resistance. *Journal of Materials Chemistry B*, 9(26): 5173–5194.
17. Jeyaraj M., Gurunathan S., Qasim M., Kang M.H., Kim J.H., 2019, A comprehensive review on the synthesis, characterization, and biomedical application of platinum nanoparticles. *Nanomaterials (Basel)*, 2(12): 1719.
18. Cele T., 2020, Preparation of nanoparticles. *Engineered Nanomaterials-Health and Safety*. IntechOpen. Crossref, doi:10.5772/intechopen.90771.
19. Baig N., Kammakakam I., Falath W., 2021, Nanomaterials: A review of synthesis methods, properties, recent progress, and challenges. Materials advances, 2(6): 1821–1871.
20. Rane A.V., Kanny K., Abitha V.K., Thomas S., 2018, Methods for Synthesis of Nanoparticles and Fabrication of Nanocomposites In Micro and Nano Technologies. Mohan Bhagyaraj S., Oluwafemi O.S., Kalarikkal N., Thomas S. (Eds.) *Synthesis of Inorganic Nanomaterials* (pp. 121–139).
21. Chopra H., Bibi S., Singh I., Hasan M.M., Khan M.S., Yousafi Q., Baig A.A., Rahman M.M., Islam F., Emran T.B., Cavalu S., 2022, Green metallic nanoparticles: Biosynthesis to applications. *Frontiers in Bioengineering and Biotechnology*, 10: 874742.
22. Socoliuc V., Peddis D., Petrenko V.I., Avdeev M.V., Susan-Resiga D., Szabó T., Turcu R., Tombácz E., Vékás L., 2020, Magnetic nanoparticle systems for nanomedicine-A materials science perspective. *Magnetochemistry*, 6(1): 2.
23. Abed A., Derakhshan M., Karimi M., Shirazinia M., Homayonfal M., Hamblin M.R., Mirzaei S.A., Soleimanpour H., Dehghani S., Dehkordi F.F., Mirzaei H., 2022, Platinum nanoparticles in biomedicine: Preparation, anti-cancer activity, and drug delivery vehicles. *Frontiers in Pharmacology*, 13.
24. Mohammadi M., Pourseyed Aghaei F., 2020, Magnetite Fe_3O_4 surface as an effective drug delivery system for cancer treatment drugs: Density functional theory study. *Journal of Biomolecular Structure and Dynamics*: 1–12: 797804.
25. Gómez-Sotomayor R., Ahualli S., Viota J.L., Rudzka K., Delgado Á.V., 2015, Iron/magnetite nanoparticles as magnetic delivery systems for antitumor drugs. *Journal of Nanoscience and Nanotechnology*, 15(5): 3507–3514.
26. Li Y., Wang N., Huang X., Li F., Davis T.P., Qiao R., Ling D., 2019, Polymer-assisted magnetic nanoparticle assemblies for biomedical applications. *ACS Applied Bio Materials*, 3(1): 121–142.
27. Wu K., Su D., Liu J., Saha R., Wang J.P., 2019, Magnetic nanoparticles in nanomedicine: A review of recent advances. *Nanotechnology*, 30(50): 502003.
28. Pacheco-Quito E.M., Ruiz-Caro R., Veiga M.D., 2020, Carrageenan: Drug delivery systems and other biomedical applications. *Marine Drugs*, 18(11): 583.

29. Shanmuga S.I., Singhal M., Sen S., 2015, Synthesis and characterization of carrageenan coated magnetic nanoparticles for drug delivery applications. *Translational Biomedicine*, 6: 3.
30. Nguyen D.T., Nguyen N.M., Vu D.M., Tran M.D., Ta V.T., 2021, On-demand release of drug from magnetic nanoparticle-loaded alginate beads. *Journal of Analytical Methods in Chemistry*, 2021: Article ID 5576283, 7 pages.
31. Hariyadi D.M., Islam N., 2020, Current status of alginate in drug delivery. *Advances in Pharmacological and Pharmaceutical Sciences*, 2020: Article ID 8886095, 16 pages.
32. Yafout M., Ousaid A., Khayati Y., El Otmani I.S., 2021, Gold nanoparticles as a drug delivery system for standard chemotherapeutics: A new lead for targeted pharmacological cancer treatments. *Scientific African*, 11: e00685.
33. Khan A., Rashid R., Murtaza G., Zahra A., 2014, Gold nanoparticles: Synthesis and applications in drug delivery. *Tropical Journal of Pharmaceutical Research*, 13(7): 1169.
34. Hussain S., Amjad M., 2021, A review on gold nanoparticles (GNPs) and their advancement in cancer therapy. *International Journal of Nanotechnology and Nanomedicine*, 7(1): 019–025.
35. Hussein H.A., Abdullah M.A., 2022, Novel drug delivery systems based on silver nanoparticles, hyaluronic acid, lipid nanoparticles and liposomes for cancer treatment. Applied nanoscience. *Applied Nanoscience*, 12(11): 3071–3096.
36. Gomes H.I.O., Martins C.S.M., Prior J.A.V., 2021, Silver nanoparticles as carriers of anticancer drugs for efficient target treatment of cancer cells. *Nanomaterials*, 11(4): 964.
37. Galatage S.T., Hebalkar S.A., Dhobale S.V., Mali O.R., Kumbhar P.S., Nikade S.V., Killedar S.G., 2021, Silver nanoparticles: Properties, synthesis, characterization, applications and future trends in Silver micro-nanoparticles - Properties, synthesis, characterization, and applications Kumar S., Kumar P., Shakher Pathak C. (Eds.), IntechOpen. ISBN 978-1-83968-660-3.
38. Chandrakala V., Aruna V., Angajala G., 2022, Review on metal nanoparticles as nanocarriers: Current challenges and perspectives in drug delivery systems. *Emergent Mater*, 5: 1593–1615.
39. Naoi M., Shamoto-Nagai M., Maruyama W., 2019, Neuroprotection of multifunctional phytochemicals as novel therapeutic strategy for neurodegenerative disorders: Antiapoptotic and antiamyloidogenic activities by modulation of cellular signal pathways. *Future Neurology*, 17: 1149639.
40. Zhang Q., Kuang G., Zhang L., Zhu Y., 2023, Nanocarriers for platinum drug delivery. *Biomedical Technology*, 2: 77–89.
41. Nguyen K.C., Seligy V.L., Massarsky A., Moon T.W., Rippstein P., Tan J., Tayabali A.F., 2013, Comparison of toxicity of uncoated and coated silver nanoparticles. *Journal of Physics: Conference Series*, 429: 012025.
42. Jahangeer M., Fatima R., Ashiq M., Basharat A., Qamar S.A., Bilal M., Iqbal H.M., 2021, Therapeutic and biomedical potentialities of terpenoids—A review. *Journal of Pure and Applied Microbiology*, 15(2): 471–483.
43. Mazalovska M., Calvin Kouokam J., 2020, Plant-Derived Lectins as potential cancer therapeutics and diagnostic tools. *BioMed Research International*. Article ID 1631394, 13 pages.
44. Chiangjong W., Chutipongtanate S., Hongeng S., 2020, Anticancer peptide: Physicochemical property, functional aspect and trend in clinical application (Review). *International Journal of Oncology*, 57(3): 678–696.
45. Madni A., Tahir N., Rehman M., Raza A., Mahmood M.A., Khan M.I., Kashif P.M., 2017, Hybrid nanocarriers for potential drug delivery. In *Advanced Technology for Delivering Therapeutics*, Maiti S., Sen K.K. (Eds.), Tech ISBN 978-953-51-3122-9. 53–86.
46. Segneanu A.-E., Grozescu I., Cziple F., Berki D., Damian D., Niculite C.M., Florea A., Leabu M., 2015, *Helleborus purpurascens*—Amino acid and peptide analysis linked to the chemical and antiproliferative properties of the extracted compounds. *Molecules*, 20(12): 22170–22187.
47. Küpeli Akkol E., Genç Y., Karpuz B., Sobarzo-Sánchez E., Capasso R., 2020, Coumarins and coumarin-related compounds in pharmacotherapy of cancer. *Cancers (Basel)*, 12(7): 1959.
48. Coyoy-Salgado A., Segura-Uribe J.J., Guerra-Araiza C., Orozco-Suárez S., Salgado-Ceballos H., Feria-Romero I.A., Gallardo J.M., Orozco-Barrios C.E., 2019, The importance of natural antioxidants in the treatment of spinal cord injury in animal models: An overview. *Oxidative Medicine and Cellular Longevity*, 2019: 3642491.
49. Wang J.Y., Wen L.L., Huang Y.N., Chen Y.T., Ku M.C., 2006, Dual effects of antioxidants in neurodegeneration: Direct neuroprotection against oxidative stress and indirect protection via suppression of gliamediated inflammation. *Current Pharmaceutical Design*, 12(27): 3521–3533.
50. Iravani S., Varma R.S., 2022, Alginate-based micro- and nanosystems for targeted cancer therapy. *Marine Drugs*, 20(10): 598.

51. Afzal O., Altamimi A.S.A., Nadeem M.S., Alzarea S.I., Almalki W.H., Tariq A., Mubeen B., Murtaza B.N., Iftikhar S., Riaz N., et al. 2022, Nanoparticles in drug delivery: from history to therapeutic applications. *Nanomaterials*, 12(24): 4494.
52. Alhadrami H.A., Orfali R., Hamed A.A., Ghoneim M.M., Hassan H.M., Hassane A.S.I., Rateb M.E., Sayed A.M., Gamaleldin N.M., 2021, Flavonoid-coated gold nanoparticles as efficient antibiotics against gram-negative bacteria-evidence from in silico-supported in vitro studies. *Antibiotics (Basel)*, 10(8): 968.
53. Cunha C., Daniel-da-Silva A.L., Oliveira H., 2022, Drug delivery systems and flavonoids: Current knowledge in melanoma treatment and future perspectives. *Micromachines*, 13(11): 1838.

12 Silver Nanoparticles in Agriculture
Synthesis and Applications

Iqra Munir, Sunbal Khalil Chaudhari, Madiha Manzoor, Sundas Asghar, and Ghazala Mustafa

12.1 INTRODUCTION TO NANOPARTICLES

Nanotechnology is an emerging field in which we study particles at the nanometer scale, that is, approximately 1 to 100 nm (Mulfinger et al., 2007). Silver (Ag) is the 67th most abundant transition element, having metallic white luster and durability (Yu et al., 2013). Silver nanoparticles are very desirable due to their stability, relative abundance, low cost, and physical properties (Desireddy et al., 2013). There are 20 to 15000 atoms present in silver nanoparticles, and they have a large surface-to-volume ratio, which remarkably enhances their various physical and chemical properties even at low concentrations (Yin et al., 2020).

Silver nanoparticles have many antifungal and antimicrobial properties. They hinder the growth of different Gram-positive and Gram-negative bacteria, like *Pseudomonas aeruginosa*, *Staphylococcus aureus*, and *Escherichia coli* (Birla et al., 2009), and fungi, such as *Aspergillus fumigatus*, *Saccharomyces cerevisiae*, *Candida tropicalis*, and *Staphylococcus aureus*. Silver nanoparticles have a large surface-to-volume ratio, so they can easily enter the cell membrane of microorganisms, altering their membrane structure, which eventually leads to cell death (Yin et al., 2020).

The role of silver nanoparticles in the biomedical field is vital; they are used for the diagnosis and treatment of cancer (Zhang et al., 2016). They are also used in several other medical applications, like wound dressings, surgical instruments, and personal hygiene products. Silver nanoparticles can also be used as a good alternative to antibiotics, particularly against those bacterial strains that produce biofilms and are resistant to conventional antibiotics (Sarahan and King, 2010).

Silver nanoparticles can be synthesized through various methods. Pyrolysis and spark discharging are conventional physical methods for the preparation of silver nanoparticles (Pluym et al., 1993; Tien et al., 2008). These conventional methods are less hazardous but produce low yields (Al-Warthan et al., 2010; Shameli et al., 2010). On the other hand, chemical methods are more expensive but produce a high yield (Tsuji et al., 2005). Chemical methods give better yields, but they are time-consuming, expensive, and hazardous, because the chemicals used in this process have strong reducing properties and are harmful to human life (Malik et al., 2002).

Silver nanoparticles are getting attention from agricultural scientists, who have already attained high yields using nanomaterials (DeRosa et al., 2010). The antimicrobial role of silver nanoparticles also protects crops against various plant pathogens. They are also used in pest control, as nano-fertilizers, and for the recognition of plant diseases (Kale et al., 2021).

12.2 SYNTHESIS OF AGNPS

There are two types of approaches used for the synthesis of metallic nanoparticles: one is called "bottom-up" and the other is "top-down" (Deepak et al., 2011). In top-down approaches, silver

 DOI: 10.1201/b23308-12

atoms are broken down into nanosized solid or aerosolized particles, which are highly stable (Pleus, 2012). Physical methods are included in this category, such as laser ablation, sputtering, and ball milling (Gurunathan et al., 2009; Zodrow et al., 2009), while in the bottom-up approach, materials are built up from basic units to large structures (Abid et al., 2021) using biological and chemical techniques. These processes are very expensive, and the materials used, such as thio-glycerol, 2-mercaptoethanol, borohydride, and citrate, are hazardous and poisonous (Mallick et al., 2004). To minimize the issues of a bottom-up approach, the focus should be on the green synthesis of Ag nanoparticles using biological methods, which are non-toxic, cost-effective, and environmentally friendly. Different biological systems have been used for the high-yield manufacturing of silver nanoparticles, such as vitamins, amino acids, bacteria, fungi, and plant extracts (Kalimuthu et al., 2008; Kalishwaralal et al., 2008). These biological systems are also used for the production of many other nanoparticles, like graphene and gold (Gurunathan et al., 2014a; Gurunathan et al., 2013c). Several microorganisms have been reported that are involved in the biological synthesis of silver nanoparticles, which include *Lactobacillus* strains (Nair and Pradeep, 2002), *Pseudomonas stutzeri* AG259 (Klaus et al., 1999), *Escherichia coli*, *Bacillus licheniformis* (Kalimuthu et al., 2008), and *Brevibacterium casei* (Kalishwaralal et al., 2010). Some fungal species are also used for this purpose, like *Ganoderma neo-japonicum* Imazeki (Gurunathan et al., 2013a) and *Fusarium oxysporum* (Shankar et al., 2003). In addition to microbes, some biopolymers (Leung et al., 2010), amino acids (Shankar and Rhim, 2015), starch (Kumar et al., 2014), and fibrinolytic enzymes (Deepak et al., 2011) are also utilized.

There are three factors upon which the synthesis of nanoparticles depends: the reducing agent, solvent, and non-toxic materials. Nanoparticles prepared from biological methods have controlled size and shape, which make them feasible for different biomedical applications (Gurunathan et al., 2014b). To achieve uniform size and the desired shape during nanoparticle synthesis, strong reducing agents such as sodium borohydride ($NaBH_4$) have been used (Pyatenko et al., 2007).

12.3 CHARACTERIZATION OF NANOPARTICLES

Characterization of silver nanoparticles is essential for analyzing their proficiency, safety, and functional evaluation. Several analytical techniques are used for the characterization of silver nanoparticles, such as X-ray diffractometry (XRD), dynamic light scattering (DLS), UV-Vis spectroscopy, X-ray photoelectron spectroscopy (XPS), atomic force microscopy (AFM), and transmission electron microscopy (TEM) (Zhang et al., 2016).

For the characterization of Ag nanoparticles, UV-visible spectroscopy is considered a reliable technique that monitors the stability and synthesis of nanoparticles (Sastry et al., 1998). Unique optical properties of Ag nanoparticles interact strongly with light of specific wavelengths (Philip, 2008). This technique is fast, sensitive, and needs no calibration for the characterization of colloidal suspension particles (Huang et al., 2007; Tomaszewska et al., 2013). During Ag nanoparticle analysis, it is observed that electrons move freely between the valence band and the conduction band, as the distance between these two bands is very small. These free-moving electrons produce resonance with a wavelength of light, and as a result, a surface plasmon resonance absorption band (SPR) is produced (Das et al., 2009; Link and El-Sayed, 2003; Taleb et al., 1998). Dielectric medium, chemical surroundings, and particle size are some factors upon which the absorption of silver nanoparticles depends (Nath et al., 2007; Taleb et al., 1998).

To examine the molecular and crystalline structures of nanoparticles, the X-ray diffraction technique is used (Das et al., 2009; Waseda et al., 2011). This popular analytical technique can be used for isomorphic substitution (Ananias et al., 2013), measuring particle size (Singh et al., 2013), calculating the degree of crystallinity (Dey et al., 2009), qualitative identification of different compounds (Ivanisevic, 2010), quantitative determination of species resonance (Cabral et al., 2013), etc. Many diffraction patterns are formed when X-ray light falls and then reflects on any crystal. These diffraction patterns can reflect various physical and chemical characteristics of the crystal structures.

A wide range of materials can be analyzed through the XRD technique, including superconductors, inorganic catalysts, glasses, biomolecules, polymers, etc. (Macaluso, 2009). Every material has a unique diffraction beam which can easily be compared with the reference database of the JCPDS (Joint Committee on Powder Diffraction Standards) library (Khan et al., 2013; Vaia and Liu, 2002).

FTIR spectroscopy is another useful analytical technique used to detect small adsorption changes and provide a satisfactory signal-to-noise ratio. Through this technique, one can easily differentiate between minor absorption bands of functionally active functional groups from the large absorption background of the whole protein (Gerwert, 1999; Jung, 2000; Kim and Barry, 2001). With the help of this technique, we can confirm whether or not biomolecules are used in the synthesis process of nanoparticles (Perevedentseva et al., 2010; Shang et al., 2007). Many recent advancements have been made in the conventional FTIR technique, and its new version is called attenuated total reflection (ATR)-FTIR spectroscopy (Harrick and Beckmann, 1974; Hind et al., 2001). In short, FTIR is a non-invasive, valuable, simple, and cost-effective technique used to detect the presence and role of biological molecules in silver nitrate reduction into silver (Kim and Barry, 2001; Mäntele et al., 1988; Zscherp and Barth, 2001).

The latest advanced technique used for Ag nanoparticle characterization is scanning electron microscopy, in which a beam of high-energy electrons is used to examine objects on a very small scale (Romero et al., 2007; Wang, 2000). This surface imaging technique can determine the particle size, shape, distribution, and surface morphology of synthesized Ag nanoparticles (Hall et al., 2007; Lemons et al., 1996). We can derive histograms from the images obtained through SEM with the help of different software. The morphology of nanoparticles below the level of 10 nm can be identified with the help of high-resolution scanning electron microscopy (Fissan et al., 2014). Transmission electron microscopy (TEM) is also frequently used for the characterization of Ag nanoparticles. This technique provides high spatial resolution as compared with SEM but requires high vacuum and thin sample preparation, which is time-consuming (Hall et al., 2007; Williams and Carter, 1996).

12.4 ANTIMICROBIAL ACTIVITY OF AGNPS

AgNPs show high antimicrobial activity and can reduce the resistance of bacterial strains against antibiotics. The antibacterial activity of silver nanoparticles is due to their crystallographic surface structure and large surface-to-volume ratio. Silver nanoparticles can penetrate the cell wall of *E. coli* and make gaps and pits, which lead to cell death (Sondi and Salopek-Sondi, 2004). In the same *E. coli* strain, particles having a small size and large surface-to-volume ratio are more efficient antibacterial agents than large particles (Baker et al., 2005). Silver nanoparticles synthesized using four different saccharides having an average size of 25 nm showed great antibacterial activity against *Staphylococcus aureus*, which is extremely resistant to methicillin. This antibacterial activity was due to the silver nanoparticles' size and shape. Strong antibacterial activity was also observed against Gram-positive, Gram-negative, and *Staphylococcus aureus* bacteria (Pal et al., 2007). The inhibitory concentration of $AgNO_3$, the ligand (CHX), the gold standard, and silver sulfadiazine is higher than that of nanocrystalline silver chlorhexidine (AgCHX) (Pal et al., 2009).

The biological synthesis of silver nanoparticles and their utilization as antimicrobial agents is common practice nowadays. Many studies have been conducted in this regard. A comprehensive study was carried out to find out the effectiveness of the antimicrobial properties of silver nanoparticles against *E. coli*, *Staphylococcus aureus*, and yeast. The results of this study showed that the presence of a small number of silver nanoparticles can completely inhibit the growth of *E. coli* and yeast, while a minor effect was seen in *Staphylococcus aureus* (Kim et al., 2007). High microbial activity was observed in the chitosan-Ag nanoparticle composite as compared with the individual components at their respective concentrations. This is because one-pot synthesis favors the creation of small silver nanoparticles that are attached to the polymer and can be dispersed in media of pH ≤ 6.3 (Sanpui et al., 2008). In the same way, Ag nanoparticles have been synthesized by using guava leaf extracts, which showed great stability and antimicrobial activity against *E. coli* as compared with chemically synthesized silver nanoparticles. This is because biomolecules can be adsorbed on the surface of these nanoparticles (Parashar et al., 2011). The activity of different human pathogenic bacteria like *Salmonella typhi*, *Candida albicans*, *S. aureus*, and *E. coli* can be stopped with the help of Ag nanoparticles synthesized biologically from *Cryphonectria* sp. (Besinis et al., 2014). Antibacterial and anti-biofilm properties can be enhanced by using a combination of antibiotics and silver nanoparticles against *Shigella flexneri*, *Streptococcus pneumoniae*, *Pseudomonas aeruginosa*, and *Staphylococcus aureus* instead of using them individually (Gurunathan et al., 2014b).

A limited number of antifungal drugs are available, so there is an urgent need to grow antifungal mediators that are environmentally friendly, non-toxic, and biocompatible. For this purpose, Ag nanoparticles can be used, which are effective antifungal agents against various fungal strains (Kim et al., 2008). Strong antifungal activity of nano-Ag particles was observed against ATCC strains of *Candida species* and *Trichophyton mentagrophytes* having a concentration of 1–7 μg/mL (Esteban-Tejeda et al., 2009). The effect of fluconazole (an antifungal drug) against *Phoma herbarum*, *Fusarium semitectum*, *Trichoderma* sp., *Candida albicans*, and *Phoma glomerata* can be enhanced if combined with biologically synthesized Ag nanoparticles (Gajbhiye et al., 2009). Antifungal activity of biologically synthesized Ag nanoparticles with a concentration of 15 mg proved effective against various phytopathogenic fungi, such as *Sclerotinia sclerotiorum*, *Alternaria alternata*, *Macrophomina phaseolina*, *Rhizoctonia solani*, *Curvularia lunata*, and *Botrytis cinerea* (Gajbhiye et al., 2009). It is interesting to know that the growth of indoor fungal species like *Aspergillus fumigatus*, *Penicillium brevicompactum*, *Cladosporium cladosporoides*, *Chaetomium globosum*, *Mortierella alpine*, and *Stachybotrys chartarum* can also be inhibited with the help of silver nanoparticles (Ogar et al., 2015). Xiang et al., (2011) reported that Ag nanoparticles have inhibitory properties against the hepatitis virus and immunodeficiency virus. They also stated that the percentage of infection, virus concentration, and severity of disease caused by Bean yellow mosaic virus can be reduced if treated with AgNPs for about 24 hours.

12.5 ANTICANCER ACTIVITY OF AGNPS

There are many efforts in progress to develop nanomaterials that can specifically kill tumor cells without causing systemic side effects (Thorley and Tetley, 2013). During the investigation of the molecular mechanism of silver nanoparticles, it was found that a specific concentration of Ag nanoparticles can cause apoptosis. Collective effects of non-uracil phosphoribosyltransferase–expressing cells and uracil phosphoribosyltransferase–expressing cells in the presence of FU (fluorouracil) can cause programmed cell death (Gopinath et al., 2008). AshaRani et al. (2009) found anticancer properties in starch-coated AgNPs when treated with normal human glioblastoma cells and human lung fibroblast cells. This is because silver nanoparticles can change the cell structure, decrease cell metabolism, and enhance oxidative stress, which causes damage to mitochondria and produces more reactive oxygen species that lead to DNA damage.

Cells treated with silver nanoparticles showed different abnormalities, like up-regulation and down-regulation of metallothionein and filamin, respectively, and mitotic arrest (AshaRani et al., 2009). Silver nanotriangles coated with chitosan exhibit a greater rate of cell mortality (Boca et al., 2011). Silver nanoparticles of various sizes showed a significant cytotoxic effect in acute myeloid leukemia (Guo et al., 2013). Ag nanoparticles produced through bacterial and fungal extracts were observed to have anticancer properties in human breast cancer MDA-MB-231 cells (Gurunathan et al., 2013b). Various therapeutic approaches have been used for the treatment of cancer with the help of silver nanoparticles. Plasmonic magnetic nanoparticles were made to increase magnetic resonance imaging (MRI) contrast by using various components of multiple nanoparticles at a single stage, producing monolayer silver-gold coated magnetic nanoparticles, which showed highly efficient killing of SKBr3 (a human breast cancer cell line) at a low exposure of 12.7 W/cm at 808 nm

(Lim et al., 2007). Along with this, photo-based nanomedicine has also gained significant importance for cancer treatment (Rai et al., 2010).

The heterogeneity of tumor cells is a great challenge for nanotechnologists, and they are trying to develop strategies using Ag nanoparticles that can target cancer cells more precisely. Nanotechnologists are also trying to achieve biocompatibility, safety, reduction in toxicity, and better efficacy of cancer therapy (Wicki et al., 2015).

12.6 ROLE OF SILVER NANOPARTICLES IN AGRICULTURE

Bioavailability, toxicity, and transport of silver nanoparticles in soil are greatly affected by the soil's physiochemical characteristics (Benoit et al., 2013). Soil pH, soil texture, soil organic matter, and soil cation exchange ability direct the mobility, transport, and absorption of silver nanoparticles in soil (Geranmayeh Oromieh, 2011). Silver nanoparticles have been used to improve crop productivity. The nitrogen content of crops directly affects crop yield (Iqra et al., 2018). Various researchers have reported that the appropriate concentration of AgNPs plays a significant role in the germination of seeds (Shelar and Chavan, 2015), the growth of plants (Kaveh et al., 2013), improving fertilizer use efficiency (Lu et al., 2002), and improving chlorophyll content and photosynthetic quantum efficiency (Hatami and Ghorbanpour, 2013). Silver nanoparticles have been used as a pesticide to remove unwanted and harmful pests from crops. Foliar applications of Ag nanoparticles showed great preventive effects on microorganisms that cause downy and powdery mildew disease (Kale et al., 2021). Furthermore, they also promote disease resistance, stress resistance, and physiological activity (Sharon et al., 2010). Silver nanoparticles can be used as an effective seed-protecting agent, as they protect stored rice for up to 2 months after treatment with AgNPs. In addition to this, silver nanoparticles can be used as nano packages to improve the shelf life of fresh agricultural products, like different vegetables and fruits (Sameen et al., 2014). Now, scientists should focus on the actual and long-term effects of silver nanoparticles on human health and environment before using them in the mass production of agricultural products.

12.7 CONCLUSION

Silver nanoparticles are attracting increasing attention for a wide range of new applications in various fields of industry. Different types of methods and processes are available for the synthesis of silver nanoparticles, including biological, physical, and chemical methods. However, the green synthesis of silver nanoparticles is less hazardous and environmentally friendly. The applications of AgNPs in the biomedical field have increased over the years, as they are efficient and effective against pathogens. The application of nanotechnology, specifically AgNPs, could be one of the best solutions in disease control, infection control, anticancer therapy, dental applications, coating systems, wound healing, and many more biomedical applications with minimal or no side effects.

REFERENCES

Abid, N., Khan, A.M., Shujait, S., Chaudhary, K., Ikram, M., Imran, M., Haider, J., Khan, M., Khan, Q., and Maqbool, M. (2021). Synthesis of nanomaterials using various top-down and bottom-up approaches, influencing factors, advantages, and disadvantages: A review. *Advances in Colloid and Interface Science*, *300*, 102597.

Al-Warthan, A., Kholoud, M., El-Nour, A., Eftaiha, A., and Ammar, R. (2010). Synthesis and applications of silver nanoparticles. *Arabian Journal of Chemistry* 3(3), 135–140.

Ananias, D., Paz, F.A.A., Carlos, L.D., and Rocha, J. (2013). Chiral microporous rare-earth silico-germanates: Synthesis, structure and photoluminescence properties. *Microporous and Mesoporous Materials* 166, 50–58.

AshaRani, P., Low Kah Mun, G., Hande, M.P., and Valiyaveettil, S. (2009). Cytotoxicity and genotoxicity of silver nanoparticles in human cells. *ACS Nano* 3(2), 279–290.

Baker, C., Pradhan, A., Pakstis, L., Pochan, D.J., and Shah, S.I. (2005). Synthesis and antibacterial properties of silver nanoparticles. *Journal of Nanoscience and Nanotechnology* 5(2), 244–249.

Benoit, R., Wilkinson, K.J., and Sauvé, S. (2013). Partitioning of silver and chemical speciation of free Ag in soils amended with nanoparticles. *Chemistry Central Journal* 7(1), 1–7.

Besinis, A., De Peralta, T., and Handy, R.D. (2014). The antibacterial effects of silver, titanium dioxide and silica dioxide nanoparticles compared to the dental disinfectant chlorhexidine on Streptococcus mutans using a suite of bioassays. *Nanotoxicology* 8(1), 1–16.

Birla, S., Tiwari, V., Gade, A., Ingle, A., Yadav, A., and Rai, M. (2009). Fabrication of silver nanoparticles by Phoma glomerata and its combined effect against Escherichia coli, Pseudomonas aeruginosa and Staphylococcus aureus. *Letters in Applied Microbiology* 48(2), 173–179.

Boca, S.C., Potara, M., Gabudean, A.-M., Juhem, A., Baldeck, P.L., and Astilean, S. (2011). Chitosan-coated triangular silver nanoparticles as a novel class of biocompatible, highly effective photothermal transducers for in vitro cancer cell therapy. *Cancer Letters* 311(2), 131–140.

Cabral, M., Pedrosa, F., Margarido, F., and Nogueira, C. (2013). End-of-life Zn–MnO2 batteries: Electrode materials characterization. *Environmental Technology* 34(9–12), 1283–1295.

Das, R., Nath, S., Chakdar, D., Gope, G., and Bhattacharjee, R. (2009). Preparation of silver nanoparticles and their characterization. *Journal of Nanotechnology* 5, 1–6.

Deepak, V., Umamaheshwaran, P.S., Guhan, K., Nanthini, R.A., Krithiga, B., Jaithoon, N.M.H., and Gurunathan, S. (2011). Synthesis of gold and silver nanoparticles using purified URAK. *Colloids and Surfaces, Part B: Biointerfaces* 86(2), 353–358.

DeRosa, M.C., Monreal, C., Schnitzer, M., Walsh, R., and Sultan, Y. (2010). Nanotechnology in fertilizers. *Nature Nanotechnology* 5(2), 91–91.

Desireddy, A., Conn, B.E., Guo, J., Yoon, B., Barnett, R.N., Monahan, B.M., Kirschbaum, K., Griffith, W.P., Whetten, R.L., and Landman, U. (2013). Ultrastable silver nanoparticles. *Nature* 501(7467), 399–402.

Dey, A., Mukhopadhyay, A.K., Gangadharan, S., Sinha, M.K., and Basu, D. (2009). Characterization of microplasma sprayed hydroxyapatite coating. *Journal of Thermal Spray Technology* 18(4), 578–592.

Esteban-Tejeda, L., Malpartida, F., Esteban-Cubillo, A., Pecharromán, C., and Moya, J. (2009). The antibacterial and antifungal activity of a soda-lime glass containing silver nanoparticles. *Nanotechnology* 20(8), 085103.

Fissan, H., Ristig, S., Kaminski, H., Asbach, C., and Epple, M. (2014). Comparison of different characterization methods for nanoparticle dispersions before and after aerosolization. *Analytical Methods* 6(18), 7324–7334.

Gajbhiye, M., Kesharwani, J., Ingle, A., Gade, A., and Rai, M. (2009). Fungus-mediated synthesis of silver nanoparticles and their activity against pathogenic fungi in combination with fluconazole. *Nanomedicine: Nanotechnology, Biology and Medicine* 5(4), 382–386.

Geranmayeh Oromieh, A. (2011). Evaluating solubility, aggregation and sorption of nanosilver particles and silver ions in soils. https://stud.jpgilon.slu.se/

Gerwert, K. (1999). Molecular reaction mechanisms of proteins monitored by time-resolved FTIR-spectroscopy. *Biol. Chem. 380*, 931–935.

Gopinath, P., Gogoi, S.K., Chattopadhyay, A., and Ghosh, S.S. (2008). Implications of silver nanoparticle induced cell apoptosis for in vitro gene therapy. *Nanotechnology* 19(7), 075104.

Guo, D., Zhu, L., Huang, Z., Zhou, H., Ge, Y., Ma, W., Wu, J., Zhang, X., Zhou, X., and Zhang, Y. (2013). Anti-leukemia activity of PVP-coated silver nanoparticles via generation of reactive oxygen species and release of silver ions. *Biomaterials* 34(32), 7884–7894.

Gurunathan, S., Han, J., Park, J.H., and Kim, J.-H. (2014a). A green chemistry approach for synthesizing biocompatible gold nanoparticles. *Nanoscale Research Letters* 9, 1–11.

Gurunathan, S., Han, J.W., Kwon, D.-N., and Kim, J.-H. (2014b). Enhanced antibacterial and anti-biofilm activities of silver nanoparticles against Gram-negative and Gram-positive bacteria. *Nanoscale Research Letters* 9(1), 1–17.

Gurunathan, S., Han, J.W., Dayem, A.A., Eppakayala, V., Park, J.H., Cho, S.-G., Lee, K.J., and Kim, J.-H. (2013a). Green synthesis of anisotropic silver nanoparticles and its potential cytotoxicity in human breast cancer cells (MCF-7). *Journal of Industrial and Engineering Chemistry* 19(5), 1600–1605.

Gurunathan, S., Han, J.W., Eppakayala, V., Jeyaraj, M., and Kim, J.-H. (2013b). Cytotoxicity of biologically synthesized silver nanoparticles in MDA-MB-231 human breast cancer cells. *BioMed Research International 8*, 2719–2732.

Gurunathan, S., Han, J.W., and Kim, J.-H. (2013c). Green chemistry approach for the synthesis of biocompatible graphene. *International Journal of Nanomedicine* 8, 2719.

Gurunathan, S., Lee, K.-J., Kalishwaralal, K., Sheikpranbabu, S., Vaidyanathan, R., and Eom, S.H. (2009). Antiangiogenic properties of silver nanoparticles. *Biomaterials* 30(31), 6341–6350.

Hall, J.B., Dobrovolskaia, M.A., Patri, A.K., and McNeil, S.E. (2007). Characterization of nanoparticles for therapeutics *Nanomed. Nanotechnol. Biol. Med. 2*, 789–803.

Harrick, N., and Beckmann, K. (1974). *Internal Reflection Spectroscopy. Characterization of Solid Surfaces*. Boston: Springer.

Hatami, M., and Ghorbanpour, M. (2013). Effect of nanosilver on physiological performance of pelargonium plants exposed to dark storage. *Journal of Horticultural Research* 21, 15–20.

Hind, A.R., Bhargava, S.K., and McKinnon, A. (2001). At the solid/liquid interface: FTIR/ATR—The tool of choice. *Advances in Colloid and Interface Science* 93(1–3), 91–114.

Huang, X., Jain, P.K., El-Sayed, I.H., and El-Sayed, M.A. (2007). Gold nanoparticles: Interesting optical properties and recent applications in cancer diagnostics and therapy. *Nanomed. Lond. 2*, 681–693.

Iqra, M., Chaudhari, S.K., Roha, R., Hira, F., Noshin, S., and Akbar, K.F. (2018). Evaluation of various strategies for measuring nitrogen content in wheat. *Fresenius Environmental Bulletin* 27, 8658–8664.

Ivanisevic, I. (2010). Physical stability studies of miscible amorphous solid dispersions. *Journal of Pharmaceutical Sciences* 99(9), 4005–4012.

Jung, C. (2000). Insight into protein structure and protein–ligand recognition by Fourier transform infrared spectroscopy. *Journal of Molecular Recognition* 13(6), 325–351.

Kale, S.K., Parishwad, G.V., and Patil, A.S.H.A.S. (2021). Emerging agriculture applications of silver nanoparticles. *ES Food & Agroforestry* 3, 17–22.

Kalimuthu, K., Babu, R.S., Venkataraman, D., Bilal, M., and Gurunathan, S. (2008). Biosynthesis of silver nanocrystals by Bacillus licheniformis. *Colloids and Surfaces, B: Biointerfaces* 65(1), 150–153.

Kalishwaralal, K., Deepak, V., Pandian, S.R.K., Kottaisamy, M., BarathManiKanth, S., Kartikeyan, B., and Gurunathan, S. (2010). Biosynthesis of silver and gold nanoparticles using Brevibacterium casei. *Colloids and Surfaces, B: Biointerfaces* 77(2), 257–262.

Kalishwaralal, K., Deepak, V., Ramkumarpandian, S., Nellaiah, H., and Sangiliyandi, G. (2008). Extracellular biosynthesis of silver nanoparticles by the culture supernatant of Bacillus licheniformis. *Materials Letters* 62(29), 4411–4413.

Kaveh, R., Li, Y.-S., Ranjbar, S., Tehrani, R., Brueck, C.L., and Van Aken, B. (2013). Changes in Arabidopsis thaliana gene expression in response to silver nanoparticles and silver ions. *Environmental Science and Technology* 47(18), 10637–10644.

Khan, A., Asiri, A.M., Rub, M.A., Azum, N., Khan, A.A.P., Khan, S.B., Rahman, M.M., and Khan, I. (2013). Synthesis, characterization of silver nanoparticle embedded polyaniline tungstophosphate-nanocomposite cation exchanger and its application for heavy metal selective membrane. *Composites Part B: Engineering* 45(1), 1486–1492.

Kim, J.S., Kuk, E., Yu, K.N., Kim, J.-H., Park, S.J., Lee, H.J., Kim, S.H., Park, Y.K., Park, Y.H., and Hwang, C.-Y. (2007). Antimicrobial effects of silver nanoparticles. *Nanomedicine: Nanotechnology, Biology and Medicine* 3(1), 95–101.

Kim, K.-J., Sung, W.-S., Moon, S.-K., Choi, J.-S., Kim, J.-G., and Lee, D.-G. (2008). Antifungal effect of silver nanoparticles on dermatophytes. *Journal of Microbiology and Biotechnology* 18(8), 1482–1484.

Kim, S., and Barry, B.A. (2001). *Reaction-Induced FT-IR Spectroscopic Studies of Biological Energy Conversion in Oxygenic Photosynthesis and Transport (ACS Publications)*, pp. 4072–4083.

Klaus, T., Joerger, R., Olsson, E., and Granqvist, C.-G. (1999). Silver-based crystalline nanoparticles, microbially fabricated. *Proceedings of the National Academy of Sciences* 96(24), 13611–13614.

Kumar, B., Smita, K., Cumbal, L., Debut, A., and Pathak, R.N. (2014). Sonochemical synthesis of silver nanoparticles using starch: A comparison. *Bioinorganic Chemistry and Applications 2014*, 784268.

Lemons, J., Ratner, B., Hoffman, A., and Schoen, F. (1996). *Biomaterials Science: An Introduction to Materials in Medicine*. Academic P, Toronto, ON, Canada.

Leung, T.C.-Y., Wong, C.K., and Xie, Y. (2010). Green synthesis of silver nanoparticles using biopolymers, carboxymethylated-curdlan and fucoidan. *Materials Chemistry and Physics* 121(3), 402–405.

Lim, J., Tilton, R.D., Eggeman, A., and Majetich, S.A. (2007). Design and synthesis of plasmonic magnetic nanoparticles. *Journal of Magnetism and Magnetic Materials* 311(1), 78–83.

Link, S., and El-Sayed, M.A. (2003). Optical properties and ultrafast dynamics of metallic nanocrystals. *Annual Review of Physical Chemistry* 54, 331–366.

Lu, C., Zhang, C., Wen, J., Wu, G., and Tao, M. (2002). *Research of the Effect of Nanometer Materials on Germination and Growth Enhancement of. Glycine max*, pp. 168–172.

Macaluso, R.T. (2009). Introduction to Powder Diffraction and Its Application to Nanoscale and Heterogeneous Materials. *Nanotechnol. Undergrad. Educ. 1010*, 75–86.

Malik, M.A., O'Brien, P., and Revaprasadu, N. (2002). A simple route to the synthesis of core/shell nanoparticles of chalcogenides. *Chemistry of Materials* 14(5), 2004–2010.
Mallick, K., Witcomb, M., and Scurrell, M. (2004). Polymer stabilized silver nanoparticles: A photochemical synthesis route. *Journal of Materials Science* 39(14), 4459–4463.
Mäntele, W., Wollenweber, A., Nabedryk, E., and Breton, J. (1988). Infrared spectroelectrochemistry of bacteriochlorophylls and bacteriopheophytins: Implications for the binding of the pigments in the reaction center from photosynthetic bacteria. *Proceedings of the National Academy of Sciences* 85(22), 8468–8472.
Mulfinger, L., Solomon, S.D., Bahadory, M., Jeyarajasingam, A.V., Rutkowsky, S.A., and Boritz, C. (2007). Synthesis and study of silver nanoparticles. *Journal of Chemical Education* 84(2), 322.
Nair, B., and Pradeep, T. (2002). Coalescence of nanoclusters and formation of submicron crystallites assisted by Lactobacillus strains. *Crystal Growth and Design* 2(4), 293–298.
Nath, S., Chakdar, D., and Gope, G. (2007). Synthesis of CdS and ZnS quantum dots and their applications in electronics. *Nanotrends–J, Nanotechnol*, *2*, 20–28.
Ogar, A., Tylko, G., and Turnau, K. (2015). Antifungal properties of silver nanoparticles against indoor mould growth. *Science of the Total Environment* 521, 305–314.
Pal, S., Tak, Y.K., and Song, J.M. (2007). Does the antibacterial activity of silver nanoparticles depend on the shape of the nanoparticle? A study of the gram-negative bacterium Escherichia coli. *Applied and Environmental Microbiology* 73(6), 1712–1720.
Pal, S., Yoon, E.J., Tak, Y.K., Choi, E.C., and Song, J.M. (2009). Synthesis of highly antibacterial nanocrystalline trivalent silver polydiguanide. *Journal of the American Chemical Society* 131(44), 16147–16155.
Parashar, U.K., Kumar, V., Bera, T., Saxena, P.S., Nath, G., Srivastava, S.K., Giri, R., and Srivastava, A. (2011). Study of mechanism of enhanced antibacterial activity by green synthesis of silver nanoparticles. *Nanotechnology* 22(41), 415104.
Perevedentseva, E.V., Su, F., Su, T., Lin, Y., Cheng, C.L., Karmenyan, A., Priezzhev, A.V.E., and Lugovtsov, A.E. (2010). Laser-optical investigation of the effect of diamond nanoparticles on the structure and functional properties of proteins. *Quantum Electronics* 40, 1089.
Philip, D. (2008). Synthesis and spectroscopic characterization of gold nanoparticles. *Spectrochimica Acta. Part A: Molecular and Biomolecular Spectroscopy* 71(1), 80–85.
Pleus, R. (2012). *Nanotechnologies-Guidance on Physicochemical Characterization of Engineered Nanoscale Materials for Toxicologic Assessment*. ISO: Geneva.
Pluym, T., Powell, Q., Gurav, A., Ward, T., Kodas, T., Wang, L., and Glicksman, H. (1993). Solid silver particle production by spray pyrolysis. *Journal of Aerosol Science* 24(3), 383–392.
Pyatenko, A., Yamaguchi, M., and Suzuki, M. (2007). Synthesis of spherical silver nanoparticles with controllable sizes in aqueous solutions. *The Journal of Physical Chemistry C* 111(22), 7910–7917.
Rai, P., Mallidi, S., Zheng, X., Rahmanzadeh, R., Mir, Y., Elrington, S., Khurshid, A., and Hasan, T. (2010). Development and applications of photo-triggered theranostic agents. *Advanced Drug Delivery Reviews* 62(11), 1094–1124.
Romero, E., Gómez, F., and Iregui, M. (2007). *Modern Research and Educational Topics in Microscopy Virtual Microscopy in Medical Images: A Survey*, 996–1006.
Sameen, A., Fathima, S.J., Ramlal, S., Kumar, S., and Khanum, F. (2014). Nanopackaging of silver using spice extract and their characterization. *Science, Technology and Arts Research Journal* 3(3), 52–56.
Sanpui, P., Murugadoss, A., Prasad, P.D., Ghosh, S.S., and Chattopadhyay, A. (2008). The antibacterial properties of a novel chitosan–Ag-nanoparticle composite. *International Journal of Food Microbiology* 124(2), 142–146.
Sarahan, P.C., and King, G. (2010). University Nano labs: Assessing and minimizing environmental, health, and safety risks. *JC & UL* 37, 589.
Sastry, M., Patil, V., and Sainkar, S. (1998). Electrostatically controlled diffusion of carboxylic acid derivatized silver colloidal particles in thermally evaporated fatty amine films. *The Journal of Physical Chemistry B* 102(8), 1404–1410.
Shameli, K., Ahmad, M.B., Yunus, W.M.Z.W., Rustaiyan, A., Ibrahim, N.A., Zargar, M., and Abdollahi, Y. (2010). Green synthesis of silver/montmorillonite/chitosan bionanocomposites using the UV irradiation method and evaluation of antibacterial activity. *International Journal of Nanomedicine, 2010*, 875–887.
Shang, L., Wang, Y., Jiang, J., and Dong, S. (2007). PH-dependent protein conformational changes in albumin: Gold nanoparticle bioconjugates: A spectroscopic study. *Langmuir* 23(5), 2714–2721.
Shankar, S., and Rhim, J.-W. (2015). Amino acid mediated synthesis of silver nanoparticles and preparation of antimicrobial agar/silver nanoparticles composite films. *Carbohydrate Polymers* 130, 353–363.

Shankar, S.S., Ahmad, A., and Sastry, M. (2003). Geranium leaf assisted biosynthesis of silver nanoparticles. *Biotechnology Progress* 19(6), 1627–1631.

Sharon, M., Choudhary, A.K., and Kumar, R. (2010). Nanotechnology in agricultural diseases and food safety. *Journal of Phytology 2*, 875–887.

Shelar, G.B., and Chavan, A.M. (2015). Myco-synthesis of silver nanoparticles from Trichoderma harzianum and its impact on germination status of oil seed. *Biolife* 3, 109–113.

Singh, D., Pandey, D., Yadav, R., and Singh, D. (2013). A study of ZnO nanoparticles and ZnO-EG nanofluid. *Journal of Experimental Nanoscience* 8(5), 731–741.

Sondi, I., and Salopek-Sondi, B. (2004). Silver nanoparticles as antimicrobial agent: A case study on E. coli as a model for Gram-negative bacteria. *Journal of Colloid and Interface Science* 275(1), 177–182.

Taleb, A., Petit, C., and Pileni, M. (1998). Optical properties of self-assembled 2D and 3D superlattices of silver nanoparticles. *The Journal of Physical Chemistry B* 102(12), 2214–2220.

Thorley, A.J., and Tetley, T.D. (2013). New perspectives in nanomedicine. *Pharmacology and Therapeutics* 140(2), 176–185.

Tien, D., Liao, C., Huang, J., Tseng, K., Lung, J., Tsung, T., Kao, W., Tsai, T., Cheng, T., and Yu, B. (2008). Novel technique for preparing a nano-silver water suspension by the arc-discharge method. *Reviews on Advanced Materials Science* 18, 752–758.

Tomaszewska, E., Soliwoda, K., Kadziola, K., Tkacz-Szczesna, B., Celichowski, G., Cichomski, M., Szmaja, W., and Grobelny, J. (2013). Detection limits of DLS and UV-vis spectroscopy in characterization of polydisperse nanoparticles colloids. *Journal of Nanomaterials, 2013*, 60–60.

Tsuji, M., Hashimoto, M., Nishizawa, Y., Kubokawa, M., and Tsuji, T. (2005). Microwave-assisted synthesis of metallic nanostructures in solution. *Chemistry–A European Journal* 11(2), 440–452.

Vaia, R.A., and Liu, W. (2002). X-ray powder diffraction of polymer/layered silicate nanocomposites: Model and practice. *Journal of Polymer Science Part B: Polymer Physics* 40(15), 1590–1600.

Wang, Z. (2000). Transmission Electron Microscopy of Shape-Controlled Nanocrystals and Their Assemblies. *The Journal of Physical Chemistry B, 104*(6), 1153–1175.

Waseda, Y., Matsubara, E., and Shinoda, K. (2011). *X-ray Diffraction Crystallography: Introduction, Examples and Solved Problems*. Springer Science & Business Media.

Wicki, A., Witzigmann, D., Balasubramanian, V., and Huwyler, J. (2015). Nanomedicine in cancer therapy: Challenges, opportunities, and clinical applications. *Journal of Controlled Release* 200, 138–157.

Williams, D.B., and Carter, C.B. (1996). The transmission electron microscope. In *Transmission Electron Microscopy*. Springer, pp. 3–17.

Xiang, D.-X., Chen, Q., Pang, L., and Zheng, C.-L. (2011). Inhibitory effects of silver nanoparticles on H1N1 influenza A virus in vitro. *Journal of Virological Methods* 178(1–2), 137–142.

Yin, I.X., Zhang, J., Zhao, I.S., Mei, M.L., Li, Q., and Chu, C.H. (2020). The antibacterial mechanism of silver nanoparticles and its application in dentistry. *International Journal of Nanomedicine* 15, 2555.

Yu, S.-j., Yin, Y.-g., and Liu, J.-f. (2013). Silver nanoparticles in the environment. *Environmental Science: Processes and Impacts* 15(1), 78–92.

Zhang, X.-F., Liu, Z.-G., Shen, W., and Gurunathan, S. (2016). Silver nanoparticles: Synthesis, characterization, properties, applications, and therapeutic approaches. *International Journal of Molecular Sciences* 17(9), 1534.

Zodrow, K., Brunet, L., Mahendra, S., Li, D., Zhang, A., Li, Q., and Alvarez, P.J. (2009). Polysulfone ultrafiltration membranes impregnated with silver nanoparticles show improved biofouling resistance and virus removal. *Water Research* 43(3), 715–723.

Zscherp, C., and Barth, A. (2001). Reaction-induced infrared difference spectroscopy for the study of protein reaction mechanisms. *Biochemistry* 40(7), 1875–1883.

13 Green Synthesis of Silver Nanoparticles Using Plant Extracts for Photocatalytic and Antibacterial Activity

Muhammad Nazim Lakhan, Abdul Hanan, Irfan Ali Soomro, Anisa Ratnasari, Altaf Hussain, Kishore Chand, and Mukhtiar Ahmed

13.1 INTRODUCTION

Nanotechnology is a swiftly expanding and diverse field of contemporary study that focuses on the synthesis, characterization, strategies, and manipulation of 1–100-nm particle structures (Wang, Lombi, Zhao, & Kopittke, 2016). Environmental pollution is a global threat due to the extensive usage of hazardous chemicals in industries and daily lives and also contributes to environmental contamination (Chand, Cao, Eldin Fouad et al., 2020). The United States Environmental Protection Agency (US-EPA) listed about 129 dyes, including Rose Bengal (RB), Methyl Blue (MB), Methyl Red (MR), Congo Red (CR), Methyl Orange (MO), Direct Blue-24, 4-nitrophenol, and numerous others as carcinogenic and hazardous to humans and the ecosystem. As a result, getting rid of these pollutants is crucial. Previous research has used a wide variety of methods, including ozonation, adsorption, coagulation, and chemical and biological treatments. These techniques are time-consuming and resource-intensive because they use solvents, expensive chemicals, and substantial amounts of energy (Chand et al., 2021). Nanoparticles (NPs) have been created and utilized to remove different dyes in ecologically and economically favorable ways (Hassan, Carlson, Mohanty, Sirajuddin, & Canlier, 2018). NPs contain distinct characteristics, such as a greater surface-to-volume ratio, that distinguish them from bulk qualities and allow them to be employed in a wide range of real-world applications, including biological and industrial applications (Arokiyaraj et al., 2017). In recent years, numerous scientists have reported metal nanoparticles (MNPs) due to their fascinating morphologies and sizes, which make them different from others. Numerous MNPs, such as zinc, gold, cobalt, titanium oxide, palladium, and platinum, have been produced and employed for a variety of purposes, including antibacterial and catalytic activities (Nair, Sajini, & Mathew, 2022; Ying et al., 2022). Silver nanoparticles (AgNPs) are especially good at breaking down dyes and killing germs because their surface area-to-volume ratio is higher than that of other metal NPs (Lakhan et al., 2020). Generally, AgNPs can be easily synthesized through physical and chemical techniques, including photochemical, radiolysis, electrochemical, radiation, etc. However, these techniques are unable to stop the synthesis protocol's production of harmful by-products (Hashemi, Mizwari, Mohammadi-Aghdam, Mortazavi-Derazkola, & Ali Ebrahimzadeh, 2022). The utilization of biological assets present in nature, especially microorganisms and plants, has attracted significant attention for the efficient and quick production of AgNPs. The biological approach has emerged as a green alternative that is affordable, eco-friendly, and energy-efficient (Ahmed et al., 2023). The synthesis of AgNPs

DOI: 10.1201/b23308-13

using plants is undeniably a straightforward method. In most cases, the production of metallic salts from plant extracts requires just a short time. Plant-mediated manufacturing eliminates the requirement for microbial cultures and the associated danger of leakage, reducing potential negative effects on both human and environmental health (Hashemi et al., 2022). For the production of AgNPs, extracts of different plant components have been used in previous studies, including flowers, latex, fruits, peels, roots, leaves, and seeds. Various additional biomolecules present in plants, such as carbohydrates, sugars, alkaloids, proteins, lipids, terpenoids, flavonoids, enzymes, phenols, and tannins, are also known to be effective in the efficient synthesis of AgNPs due to their promotion of the interaction of NPs with the substrates employed (Alabdallah & Hasan, 2021). For example; Kishore Chand and others (Chand, Cao, Eldin Fouad, et al., 2020) used extracts of onion (O), acacia catechu (C), tomato (T), and a combination of these to synthesize AgNPs. The transmission electron microscopy (TEM) and X-ray diffraction (XRD) analysis showed that all of the extracts contained crystalline AgNPs with particle sizes ranging from 5 to 100 nm. Three distinct kinds of liquid dyes (MO, MR, and CR) were degraded using the produced AgNPs as catalysts. When comparing the efficacy of several products for dye degradation, COT-synthesized products were shown to be superior. Mehwish Younas and others (Younas et al., 2023) focused on the production of AgNPs utilizing fresh *Moringa oleifera* plant leaf extract. AgNPs were investigated for antibacterial activity against *Acinetobacter baumannii*, *Klebsiella pneumoniae*, *Pseudomonas aeruginosa*, and *Escherichia coli*, among other bacteria. Inhibitory zones measured in an agar-well test ranged from 7 to 15 mm at a minimum inhibitory concentration (MIC) of 250 g/ml, indicating that the strains are susceptible to growth inhibitory circumstances. This chapter provides a concise overview of the process for synthesizing AgNPs from plant extracts and their subsequent photocatalytic and antibacterial applications.

13.2 SIGNIFICANCE OF AGNPS

Silver (Ag) is a fundamental element that is scarce yet naturally occurring. It is considerably ductile and malleable and is somewhat firmer than gold. Pristine Ag is the metal with the highest electrical and thermal conductivity and the smallest contact resistance (Arokiyaraj et al., 2017). It can be discharged into the air and water by natural events, including the weathering of minerals, or by human activities, such as ore processing, cement production, and fossil fuel combustion. AgNPs are typically between 1 and 100 nm in size with special electrical, optical, and magnetic properties that have a broad range of applications. Metallic Ag ions are inert, but when they come into contact with a reducing agent, they undergo ionization and become active. Ionic silver is a type of silver that is active and binds to bacterial cell walls, causing significant structural alterations in cell shape (Rai, Yadav, & Gade, 2009). Due to its bactericidal activity at low concentrations, Ag is also known as an oligodynamic. Because of this, it has found widespread use in pharmaceuticals.

13.3 DIFFERENT SYNTHESIS APPROACHES OF AGNPS

The development of AgNPs with antibacterial properties that can be used in a variety of applications has garnered considerable attention. Generally, synthesis approaches of AgNPs are characterized into two main classes: top-down and bottom-up (Ramanathan, Gopinath, Arshad, Poopalan, & Perumal, 2021). With the top-down synthesis method, a bigger molecule (bulk material) is broken down into smaller molecules. This method is referred to as a "physical approach." On the other hand, the bottom-up method starts with a simple molecule and builds up to bigger ones. It is the reverse of the top-down production approach. This bottom-up synthesis includes techniques such as physical, chemical, biological, and an integrated approach. Physical approaches include laser ablation, ultraviolet irradiation, and microwave-assisted synthesis. Femtosecond laser ablation has been employed to synthesize AgNPs using extremely high energy, with the direct band gap energy value equal to 3.82–3.97 eV (Menazea, 2020). As shown in Figure 13.1(a), exceptionally high energy is focused at

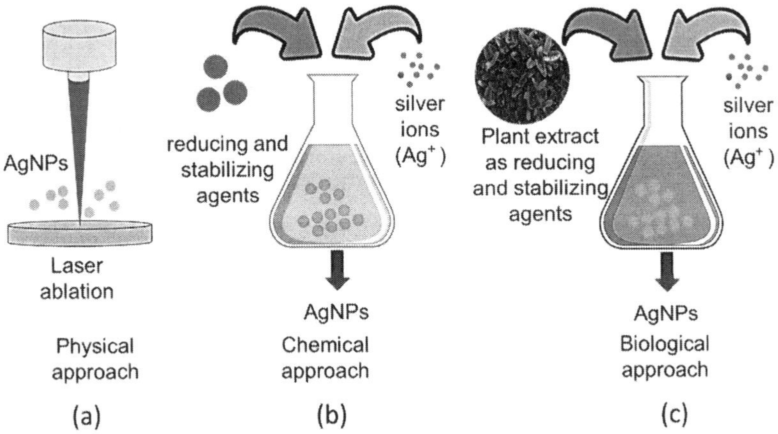

FIGURE 13.1 Different synthesis techniques of AgNPs: (a) physical, (b) chemical, and (c) biological technique.

a particular place on a solid target as part of the laser ablation process to manufacture AgNPs from the silver material surface. When a laser wave strikes the surface of a bulk material, electrons on the surface propagate electromagnetic radiation. This electromagnetic radiation is then delivered to the vibrating lattice of the material. As the pulse energy increases, ablation efficiency tends to increase as well. Several researchers (Khan, Kumari, Shameli, Selamat, & Sazili, 2019; Calamak & Ermis, 2021; Phan et al., 2019; Khan et al., 2019) utilized ultraviolet irradiation for the synthesis of AgNPs, producing a face-centered cubic (FCC) crystalline structure with a particle size of 6.02 ± 2.37 nm. Similarly, extremely monodisperse AgNPs with a particle size of 20 ± 2.2, 30 ± 3.6, and 50 ± 4.8 nm were produced via in-situ ultraviolet irradiation (Calamak & Ermis, 2021). Another study used AgNPs from $AgNO_3$ solution attached to cellulose nanofibrils to evaluate antibacterial properties against *E. coli* and *Bacillus subtilis* (Phan et al., 2019). As part of the chemical approach, a chemical compound must be present that promotes AgNP formation, for instance, sodium borohydride ($NaBH_4$), trisodium citrate (TSC), tyrosine, and β-1,4-poly-D-glucosamine. In AgNP formation, $NaBH_4$ (Thiruvengadam & Bansod, 2020) and β-1,4-poly-D-glucosamine (Laghrib et al., 2020), ethanol (Thiruvengadam & Bansod, 2020), polyethylene glycol (PEG) (Muhammad et al., 2016; Silva et al., 2022), TSC (Halder, Ahmed, Gafur, Seong, & Hossain, 2021; Quintero-Quiroz et al., 2019), and tyrosine (Saragih, Arifin, Rusdiarso, Suyanta, & Santosa, 2020) can serve as stabilizing and reducing agents. Stabilizing agents prevent AgNPs from aggregating, whereas reducing agents convert Ag^+ to Ag^0 (as seen in Figure 13.1(b)). The formation process of AgNPs using a chemical approach is unaffordable, and the accumulation of chemical waste after the formation process is not environmentally friendly.

Plant extracts are used biologically as environmentally friendly reducing and stabilizing agents, as shown in Figure 13.1(c). There are several reducing and stabilizing agents found in plant extracts, such as terpenoids, alkaloids, phenol, tannins, and vitamins (Jini, Sharmila, Anitha, Pandian, & Rajapaksha, 2022). These compounds are easily found in *Artemisia ciniformis* extract (Aslany, Tafvizi, & Naseh, 2020), *Beauveria bassiana* (Tyagi, Tyagi, Gola, Chauhan, & Bharti, 2019), aloe vera leaf extract (Anju et al., 2021), *Ageratum conyzoidez*, and *Mikania micrantha* leaf extract (Fadlilah et al., 2020; Ratnasari, Endarko, & Syafiuddin, 2020). It has been determined that the presence of plant extracts such as *Populus ciliata* leaf, Copperpod plant, *Lysiloma acapulcensis*, and *Chlorella vulgaris* resulted in the formation of AgNPs whose particle size ranged between 1.20 and 55.06 nm (Garibo et al., 2020; Hafeez et al., 2021; Pannerselvam et al., 2021; Rajkumar, Ezhumalai, & Gnanadesigan, 2021). Sheep blood serum was used to synthesize AgNPs, resulting in the particle size range of 20–50 nm (Kakakhel et al., 2021).

13.4 GREEN SYNTHESIS OF AGNPS USING VARIOUS PLANTS

Plant extracts may be utilized to manufacture AgNPs in a cost-effective and ecologically friendly way, since they need less time and resources to culture than other techniques such as bacteria, algae, and fungi-mediated synthesis. Therefore, a plant-based approach is good for the ecosystem due to the various biological compounds present in plants. In previous studies, various plant parts have been used as stabilizing and reducing agents for the successful synthesis of AgNPs, as depicted in Table 13.1. For example; Bappi Paul and others (Paul, Bhuyan, Purkayastha, & Dhar, 2016) studied a green method for producing AuNPs and AgNPs from dried *Parkia roxburghii* leaf biomass. The AuNPs were monodispersed, spherical in form, and varied in size from 5 to 25 nm, according to the TEM image. The degradation of the dyes Rhodamine B (RhB) and MB is significantly influenced by the photocatalytic activity of Au and Ag nanoparticles when exposed to sunlight. *Staphylococcus aureus* and *E. coli* were utilized as test subjects for the antibacterial activity of the synthesized NPs. The activity of Au and AgNPs on *S. aureus* was marginally higher than that on *E. coli*. Muthupandi Kasithevar and others (Kasithevar et al., 2017) offered a one-pot synthesis of AgNPs by employing plant extract of *Alysicarpus monilifer* leaves and demonstrated its antibacterial efficacy against methicillin-resistant *S. aureus* (MRSA) and coagulase-negative *Staphylococci* (CoNS) isolates from HIV patients that are resistant to several treatments. Within an hour, stable, precisely defined AgNPs with a mean size of 15 $\pm$ 2 nm were produced. Additionally, green-produced AgNPs showed strong dose-dependent antibacterial activity against MRSA and CoNS isolates. Vasantharaj Seerangaraj and co-authors (Kumar, Palanichamy, & Roopan, 2014) made the AgNPs using the leaf extract of *Ruellia tuberosa*. Morphological analysis showed that the NPs had a circular shape and were 55.65 nm in size on average.

Furthermore, the antibacterial activity of green-fabricated AgNPs toward Gram-negative and Gram-positive pathogenic bacteria was investigated, revealing promising bactericidal action. AgNPs were subjected to tests for photocatalytic degradation using Coomassie brilliant blue (CBB) and crystal violet (CV). The AgNPs showed higher degradation of the photocatalytic and biological applicability. V.N. Anjana and others (Anjana, Koshy, & Mathew, 2020) used a novel and simple method to produce AgNPs in an environmentally acceptable manner using *Azolla caroliniana* extract. The AgNPs had a crystalline structure and a spherical shape, and their diameter was on average 23 nm. Significant antibacterial activity was also shown by the AgNPs produced against the various pathogens that were examined. Ramesh Vinayagam and others (Sadeghi & Gholamhoseinpoor, 2015) created AgNPs using *Bridelia retusa* fruit extract. The crystalline nature of AgNPs was corroborated by XRD measurements, which indicated a strong peak at 38.52° matching the (111) plane and an average crystallite size of 22.48 nm. The rapid reduction of CR dye in the presence of SNPs, with a degradation rate constant of 0.056 min^{-1}, demonstrated the nanoparticles' catalytic capacity. Manal A. Awad and others (Ulug, Haluk Turkdemir, Cicek, & Mete, 2015) employed *Trigonella foenum-graecum* (fenugreek) in a green procedure for the synthesis of AgNPs with low scale size and ultra-stable properties. The bulk of the NPs generated were spherical, non-aggregated, and had an average diameter of 82.53 nm, as validated by XRD and TEM. AgNPs' photocatalytic activity was evaluated by deterioration of Rhodamine B dye when exposed to light. Rhodamine B dye went through a time-dependent photodegradation process, which ended in complete photodegradation (almost 93%) and decolorization after 216 hours had passed. The zones of inhibition for *E. coli*, *S. aureus*, and *B. cereus* were determined to be (14, 2.0 mm), (5.0, 2.0 mm), and (10, 0.9 mm) respectively, after using three different bacterial strains for the experiment.

13.5 STRUCTURAL AND MORPHOLOGICAL PROPERTIES OF AGNPS

AgNPs exhibit remarkable shape- and size-dependent properties and are used in antimicrobial activities, electrometers, computer transistors, wireless electronic logic, chemical sensors, memory systems, biosensors, medical imaging, catalysts, and optics (Huang, Jiang, Guo, & Gu, 2011; Kumar

TABLE 13.1
Different Plants Used for the Synthesis of AgNPs

Sr. No	Plant	Metal	Plant part	Size and shape	Applications	References
1	*Parkia roxburghii*	Ag	Leaf	Spherical, 5–25 nm	Photocatalytic and antibacterial activities	Paul et al., 2016
2	*Ougeinia oojeinensis*	Ag	Leaf	Spherical, 5–100 nm	Antimicrobial activity	D. Kumar, Arora, Abdullah, & Danish, 2019
3	*Calliandra haemacephala*	Ag	Leaf	Spherical, 70 nm	Antibacterial activity	Raja, Ramesh, & Thivaharan, 2017
4	*Alysicarpus monilifer*	Ag	Leaf	Spherical hexagonal, 5–45 nm	MRSA/CoNS antibacterial activity	Kasithevar et al., 2017
5	*Cleome viscosa*: Cleomaceae	Ag	Fruit	Spherical, some irregular, 5–30 nm	Acts against germs and cancer	Lakshmanan et al, 2018
6	*Carica papaya*	Ag	Fruit peels	Spherical, 16–20 nm	Good antibiotic properties against human pathogens like *E. coli* and *S. aureus*	Balavijayalakshmi & Ramalakshmi, 2017
7	Zingiberaceae: *Amomum villosum/Elettaria cardamomum*	Ag	Fruit	Spherical, 5–10 nm (AuNPs) and 5–15 nm (AgNPs)	Antimicrobial against *E. coli* and *S. aureus*/AuNPs outperformed AgNPs against breast cancer cells (MCF-7)	Soshnikova et al., 2018
8	*Boletus edulis*	Ag	Fruit	–	Antibacterial, antifungal, and cancer-fighting properties	Kaplan et al., 2021
	Diospyros lotus: Ebenaceae	Ag	Leaf	Spherical, 20 nm	Excellent antibacterial action against *E. coli* at varying doses	Hamedi & Shojaosadati, 2019
9	*Azolla caroliniana*	Ag	Leaf	–	Cytotoxic, catalytic, and antimicrobial actions	Anjana et al., 2020
10	*Moringa oleifera*: Moringaceae	Ag	Leaves	Spherical, 15.22–29.45 nm	Antimicrobial and antioxidant properties	Mohammed et al., 2022
11	*Decalepis hamiltonii*	Ag	Root	Spherical, 34 nm	Antibacterial activity	Rashmi & Sanjay, 2017
12	*Piper retrofractum*: Piperaceae	Ag	Fruits	Spherical, 1–5 nm	Promising antibacterial activity	Suci Amaliyah & 2022
13	*Myrtus communis*	Ag	Leaf	Spherical, 5–30 nm	Activity against microorganisms (Disc diffusion method)	Alyousef, Arshad, AlAkeel, & Alqasim, 2019
14	*Capsicum chinense*	Ag/Au	Leaves	Spherical, 16.76 ± 0.32 nm	Antibacterial activity	Lomelí-Rosales et al., 2022

(Continued)

TABLE 13.1 (CONTINUED)
Different Plants Used for the Synthesis of AgNPs

Sr. No	Plant	Metal	Plant part	Size and shape	Applications	References
15	*Sida cordifolia*	Ag	Whole plant	Spherical, 3–8 nm	Antibacterial activity	Pallela, Ummey, Ruddaraju, Pammi, & Yoon, 2018
16	Purple sweet potato	Ag	Roots	Polygonal, 30–120 nm	Tests for antibacterial efficacy (well diffusion method)	L. Wang et al., 2016
17	*Citrus sinensis*	Ag	Fruit	–	Antimicrobial, antimycotic, and catalytic activities	Anwar & Alghamdi, 2020
18	*Fumaria officinalis* L.: Papaveraceae	Ag	Aerial part	Spherical, 25 nm	Antimicrobial properties	Milorad et al., 2018
19	Purple sweet potato	Ag	Roots	Polygonal, 30–120 nm	Antibacterial tests (well diffusion method)	Castro, Blázquez, Muñoz, González, & Ballester, 2013
20	Green tea	Ag	Leaf	Aggregates, 15–33 nm	Antimicrobial activity	Widatalla et al., 2022
21	*Piper retrofractum*	Ag	Fruits	Spherical, 1–5 nm	Potentially antibacterial activity	Suci Amaliyah & 2022
22	*Alternanthera dentate*	Ag	Leaves	Spherical, 50–100 mm	Antimicrobial activity	D. A. Kumar et al., 2014
23	*Ruellia tuberosa*	Ag	Leaf	–	Catalytic, antibacterial, and anticancer activities	Nakkala, Mata, Gupta, & Sadras, 2014
24	*Ziziphora tenuior*	Ag	Leaves	Spherical, 8–40 mm	–	Sadeghi & Gholamhoseinpoor, 2015
25	*Rehmannia glutinosa* high crystallinity	AgNPs	–	30 mm	Stronger *E. coli* antibacterial activity than chemically produced AgNPs	Sadeghi & Gholamhoseinpoor, 2015
26	*Bridelia retusa*	Ag	Fruit	–	Anticoagulant and catalytic activities	Ulug et al., 2015
27	*Trigonella foenum-graecum*	Ag	Seed	–	Catalytic and antibacterial activities	Yong et al., 2021
28	*Ocimum sanctum*: Lamiaceae	Ag	Leaf	Spherical 10–20 nm	Antibacterial activity (disc diffusion method)	Vinayagam, Varadavenkatesan, & Selvaraj, 2017
29	*Vitex negundo*	Ag	Leaves	Spherical, 5 and 10–30 mm	Antibacterial activity	Jain & Mehata, 2017
30	*Sida cordifolia*: Malvaceae	Ag	Whole plant	Spherical, 3–8 nm	Antibacterial activity	Zargar et al., 2011
31	*Nelumbo nucifera*	Ag	Leaves	Spherical/triangular, 25–80 nm	Biological	Santhoshkumar et al., 2011

& Yadav, 2009). Materials with superior properties may be designed and manufactured by changing their shape and size on a nano scale. AgNPs' morphological features are also influenced by their size. AgNPs' surface area-to-volume ratio rises as their size decreases, resulting in enhanced reactivity as well as changes in overall optical characteristics (Duran et al., 2016; Hamad, Khashan, & Hadi, 2020; Pulit-Prociak, Stoklosa, & Banach, 2015).

The surface characteristics of AgNPs are also an important factor in their morphological properties. The surface of AgNPs may be changed with different functional groups, such as thiol, amino, or carboxylic acids, which affect their surface chemistry and reactivity. A substitution reaction was used to create Ag dendrites. This produced simple dendrites with fractal structures including, after many iterations of micrometer-long generations of branches, dendritic rods with a cactus-shaped framework formed of numerous generations of rods, including dendritic flowers with a dendritic lattice festooned with numerous unique flower petals (Fu, Sokiranskya, Wang, Lai, & Yu, 2016; Ivanovic, Nikolic, & Radmilovic, 2015; Li, Wang, Guo, & Wang, 2007). In the acidic aqueous solution of the polynaphthalene system, silver nitrate reacted with ascorbic acid to form flower-like AgNP superstructures. Each petal of the three-dimensional flower-like structure consists of strands of silver particles encased in a polymer sheet (Hasan et al., 2022; S. Kumar et al., 2021; Suber & Campi, 2012). As a novel element, Ag nanocubes have proven unmatched. The nanocube does not have a spherical shape. By rendering the six sides of a nanocube hydrophilic or hydrophobic, the layout of the resultant structures may be altered. Nanocubes of Ag were produced when $AgNO_3$ was reduced with ethylene glycol (EG) in the presence of polyvinylpyrrolidone (PVP) and hydrochloric acid (Pal, Lee, & Cho, 2020; Zhang, Xiao, & Fang, 2018). Core-shell Ag@Au nanoprisms were generated with the use of a seed-mediated approach that did not need the use of any surfactants by exploiting the asymmetric structure of the Ag nanoprisms as the seeds. A coating of gold was applied to the surface of the silver nanoprisms by using hydroxylamine as a mild reducing agent (Vokhidova & Rashidova, 2022; Zhang, 2015). High-quality Ag nanoplates have a (111) surface, making them strongly orientated single crystals. For the crystallization of homogeneous Ag nanoplates and ring-shaped gold nanoparticles in a solution of N,N-dimethylformamide (DMF), a straightforward sonochemical approach was established. Plate-like structures were produced by preferential growth on separate crystal planes in the vicinity of PVP, in conjunction with processes based on Ostwald ripening processes. These conditions led to the formation of plate-like structures. In the presence of PVP, the reduction of $AgNO_3$ with DMF resulted in the production of Ag nanoplates. A high-intensity ultrasonic instrument was inserted effectively into the reaction to provide ultrasound irradiation (Jin et al., 2001). In the process of fabricating nanoscale electronics, it is anticipated that one-dimensional nanostructures would play an important part. Ag nanorods and nanowalls (NWs) with very good size are particularly interesting because such structures display high electrical and thermal conductivity and exceptional optical properties (Shenashen, El-Safty, & Elshehy, 2014). Several studies utilizing ascorbic acid as the reducing agent to regulate the size of Ag nanospheres have been published. Utilizing ascorbic acid as the reductant and citrate as the stabilizer, Ag nanospheres were formed on immersion in water at 30 °C (Qin et al., 2010; Sarkar, Kapoor, & Mukherjee, 2010; Zhang, Hu, Li, & Gajaraj, 2016). In summary, the morphological properties of AgNPs are critical in determining their physicochemical properties, including their optical, electronic, and catalytic properties. The shape, size, and surface characteristics of AgNPs can be tailored to achieve specific functional properties, making them promising candidates for various applications in fields such as medicine, catalysis, and electronics. Desired features including crystallinity, as well as structures, are depicted in Figure 13.2 (a) and (b). Crystallinity is revealed by XRD patterns within Figure 13.2 (a), and the overall schematic representation of crystallinity and structures of AgNPs are shown in Figure 13.2 (b). Various morphologies are shown through TEM images, such as flower, dendritic, cubes, and so on (Figure 13.3).

The crystalline structure of AgNPs influences their structural characteristics. The structure of AgNPs is important for their physicochemical characteristics, which include optical, magnetic, and electronic properties. Different crystal forms, such as body-centered cubic (BCC), hexagonal

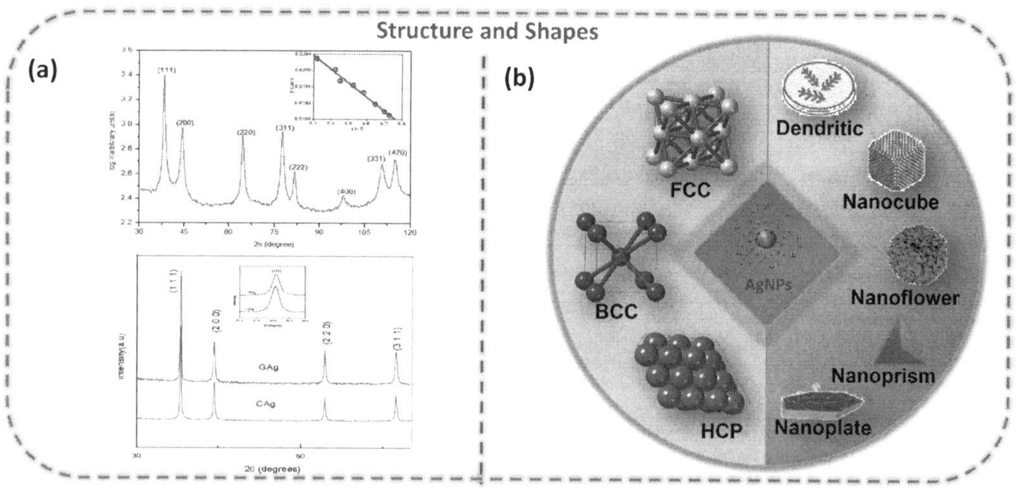

FIGURE 13.2 (a) XRD pattern, and (b) different structural shapes of AgNPs.

close-packed (HCP), and FCC, may be found in AgNPs (Terenteva, Apyari, Kochuk, Dmitrienko, & Zolotov, 2017; Zafar, Rizvi, & Mahmood, 2019). The FCC structure, which has a high degree of symmetry, is the most often seen in AgNPs. While the BCC structure is uncommon in AgNPs, it is seen in extended or anisotropic nanoparticles. AgNPs' optical characteristics, such as absorbance, scattering, as well as refractive index, are influenced by their crystal structure. AgNPs' FCC structure exhibits a significant surface plasmon resonance (SPR) region in the visible spectrum as an outcome of the conduction band electrons' collective oscillation (Majeed Khan, Kumar, Ahamed, Alrokayan, & Alsalhi, 2011). Depending on the particle's aspect ratio, the HCP structure may show a shift in the SPR band toward the blue or red area. Due to its poor symmetry, the BCC structure features a less noticeable SPR band. The electrical characteristics of AgNPs, such as their conductivity and surface energy are also influenced by their crystal structure. Due to its great symmetry, the FCC structure of AgNPs has excellent conductivity, but the HCP structure may have anisotropic conductivity (Umadevi & Jegatha Christy, 2013). The poor symmetry of the BCC structure results in low conductivity. Additionally, the magnetic properties and anisotropy of AgNPs are impacted by their crystal structure. While AgNPs with an HCP structure show a significant level of magnetic anisotropy, those with an FCC structure have little magnetic anisotropy. AgNPs show moderate magnetic anisotropy in their BCC structure (Wickramasinghe et al., 2015; Zhang, Cohen, & Chou, 1987). To summarize, the structural features of AgNPs are crucial to their physicochemical properties because the crystal structure of AgNPs can be altered to produce specific functional properties, and they are intriguing candidates for use in nanoelectronics, catalysis, biomedicine, antibacterial uses, composite fibers, biosensor materials, electronic components, cryogenic superconductors, and cosmetic items.

13.6 PHOTOCATALYTIC AND ANTIBACTERIAL ACTIVITIES OF AGNPS

Since photocatalysis is a highly efficient, energy-saving technology that does not generate secondary pollutants, it can be used to effectively control the degradation of dyes contained in wastewater (Dong et al., 2022; Liu et al., 2022; Zhou et al., 2022). Green synthesized AgNPs have shown excellent photocatalytic competence in the removal of noxious dyes (Dayanidhi, Vadivel, Jothi, & Eusuff, 2020; Díaz-Cruz, Alonso Nuñez, Espinoza-Gómez, & Flores-López, 2016). In particular, they have been used to effectively remove MO (Naseem et al., 2022), MB (Ringwal, Bartwal, & Sati, 2021), malachite green (Jaast & Grewal, 2021), and aniline blue (Eswaran, Narayan, & Vasimalai, 2021). It has been discovered that the reaction time and percentage of elimination achieved by

FIGURE 13.3 Morphological results of AgNPs (a)–(f) at different magnifications with various shapes.

the photocatalytic process are both affected by the size of the AgNPs utilized (Ajitha et al., 2016; Bordbar & Mortazavimanesh, 2018). Recent studies have shown that the photocatalytic efficiency of bulk photocatalysts may be significantly improved by combining plasmonic Ag@AgCl photocatalysts with a variety of semiconductors such as $H_2WO_4.H_2O$. This effect is shown when the photocatalysts are exposed to visible light (Wang, Li, Ma, Yu, & Yu, 2011), ZnO (Xu et al., 2011), TiO_2 (Wen & Ding, 2011), or $BiVO_4$ (Z. Zhou, Long, Cai, & Cai, 2012). In order to degrade RhB under visible light irradiation, Cui and others (Cui, Wang, Liu, Liang, & McEvoy, 2013) created Ag@AgCl intercalated $K_4Nb_6O_{17}$ photocatalysts, which showed greater photocatalytic activity than that of $K_4Nb_6O_{14}$/AgCl and an equal quantity of Ag@AgCl. For photocatalytic degradation of MB, an environmentally friendly approach for synthesizing $Zn_{0.95}Ag_{0.05}O$ NPs, including an extract from rosemary leaves, was experimentally performed (Hojjati-Najafabadi, Davar, Enteshari, & Hosseini-Koupaei, 2021). Figure 13.4 (a) depicts the photocatalytic degradation of MB with and without ZnAgO NPs. It is noticeable that photocatalytic degradation of MB increases to about 98.5% in the presence of Ag-doped ZnO NPs after 100 minutes of exposure, whereas it decreases to 70% in the

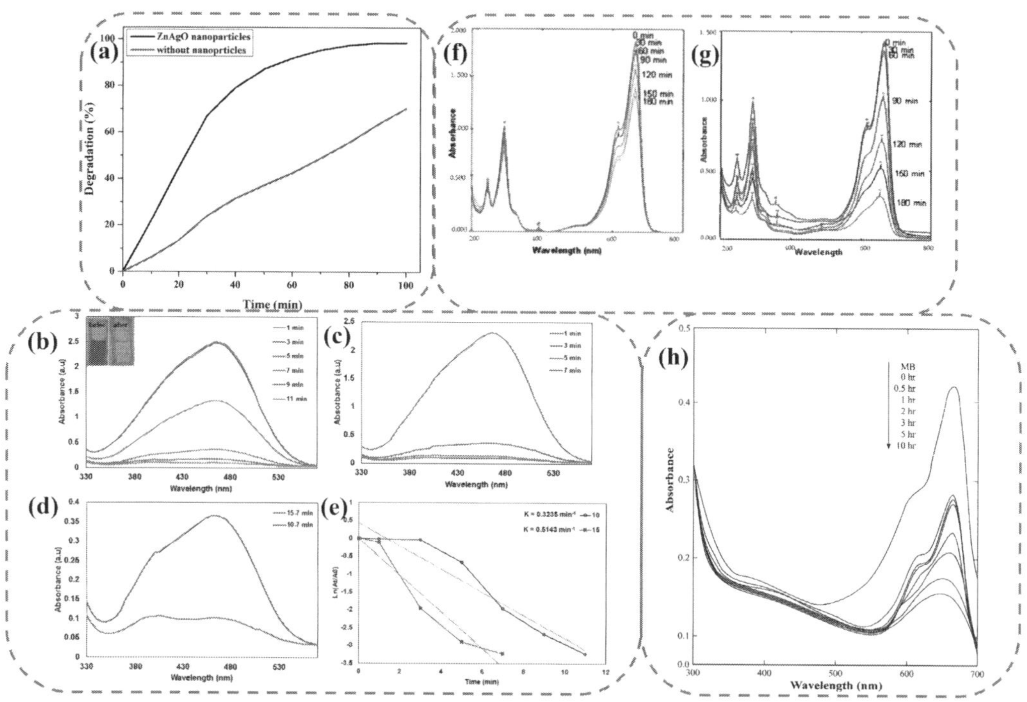

FIGURE 13.4 (a) Photocatalytic degradation of MB versus time with doped ZnO NPs and without ZnO NPs; UV–Vis absorption spectra of reduction of MO in the presence of $NaBH_4$ and AgNPs as a catalyst at (b) 10 µl AgNPs and (c) 15 µl AgNPs; (d) comparison of 10 and 15 µl AgNPs and (e) degradation kinetics of MO; (f) degradation of MB under solar irradiation in the absence of PAgNPs; (g) degradation of MB under solar irradiation in the presence of PAgNPs; (h) UV–Vis spectra for photocatalytic degradation of MB (ethanol extract–mediated AgNPs). (Panel (a) from Hojjati-Najafabadi, A. et al., *Ceram. Int.*, 47(22), 31617–31624, 2021; panels (b)–(e) from Hashemi, Z. et al., *Arab. J. Chem.*, 15(1), 103525, 2022; panels (f) and (g) from Ravichandran, V. et al., *Results Phys.*, 15, 102565, 2019; panel (h) from Sharma, R., *Materials Today: Proceedings*, 44, 1995–1998, 2021.)

absence of NPs under the same conditions. Doping zinc oxide with silver changed its band structure, and flaws inserted into the ZnO lattice resulted in electron and hole separation when exposed to UV light. For creating metallic AgNPs with enhanced antibacterial, anticancer, and photocatalytic properties, Zehra and others (Hashemi et al., 2022) used the extract of the *Sambucus ebulus* (AgNPs@SEE) plant in a biosynthetic process. MO was used as a dye containment degradation to examine the photocatalytic efficiency of green synthesized AgNPs in the presence of $NaBH_4$. In Figure 13.4 (b)–(e), it is shown that the orange hue of MO as an azo dye was converted into colorless hydrazine derivatives. With time, the 470 nm peak intensity produced by the degradation process faded. This is because the surface area of the active sites of the Ag nano catalyst grows in proportion to the amount of AgNPs present. In light of this, the response rate gradually increased. This demonstrated that the extract-mediated AgNPs had strong photocatalytic activity in the breakdown of MO as a dye impurity. In another study on aqueous extract of *Parkia speciosa* leaves, Ravichandran and others (Ravichandran et al., 2019) bio-reduced silver nitrate to create PAgNPs. MB degradation in the presence of PAgNPs under solar irradiation was used to quantify the photocatalytic activity of PAgNPs (Sahu & Patel, 2016). The photocatalytic degradation efficiency of PAgNPs as a function of time in the sun is shown in Figure 13.4(f–g). After 180 minutes of irradiation, a reduction of 84% in the typical absorption peak at 662 nm was detected (Figure 13.4 (g). The results recommended that PAgNPs are an auspicious MB-degrading agent, with degradation occurring as a result of photogeneration of electron-hole pairs between valence bands and conduction, or because it accepts electrons and transforms into colorless leucomethylene blue (LMB), both of which are consistent with MB's electroactivity and ability to accept electrons. Previously, MB's degradation time was found to be 72 h (Vanaja et al., 2014), but in a recent study carried out by Rama and others (Sharma, 2021) on AgNPs synthesized by *Terminalia bellirica* fruit ethanolic extract, the results of the photocatalytic degradation reaction were detected as a drop in absorption intensity over time under sunlight, with the decline originating from the surface resonance of AgNPs. When the degradation process was tracked for 10 hours, the peak intensity of MB gradually declined, as depicted in Figure 13.4(h). Visual examination reveals that the AgNPs produced by *T. bellirica* fruit extract are potent photocatalytic agents for the degradation of MB dye in the presence of sunshine.

Scientific investigation into innovative therapeutic compounds with broad antibacterial activity is urgently needed to combat the worldwide rise of antibiotic-resistant microorganisms. There are already a number of antibacterial compounds on the market, which are used to reduce the prevalence of germs in a variety of consumer goods. Although effective against germs, these drugs often come with unwanted side effects, including hazardous levels of toxicity, exorbitant prices, and poor solubility (Ahmad et al., 2017). This highlights the need for research on safe and efficient antibacterial compounds (Ameen et al., 2019; Kouhbanani et al., 2019). To combat drug-resistant microorganisms, AgNPs offer great promise as a new platform on which to base antimicrobial agent development. It has been previously noted that AgNPs and their derivatives have the best antibacterial action and the lowest cytotoxicity on human cells (Qais et al., 2019). For the most part, the antibacterial action of Ag is induced by using the nitrate form of the element, but when AgNPs are used, the surface area accessible to the microorganisms is greatly increased. AgNPs show significant differences in antibacterial bioactivity between Gram-positive *S. aureus* and Gram-negative *E. coli* (Rajeshkumar & Bharath, 2017). A large number of reactive oxygen species (ROS), including hydroxyl radical, singlet oxygen, hypochlorous acid, hydrogen peroxide, and superoxide anion, are credited with AgNPs' antibacterial action. According to studies, smaller AgNPs had more antibacterial bioactivity and may generate more apoptotic agents, necrotic factors, and ROS. AgNPs' antimicrobial properties include blocking the activity of respiratory chain dehydrogenases and ultimately, producing supplemental ROS, both of which are necessary for bacterial respiration and growth (Rizzello & Pompa, 2014). Several different methods for the environmentally friendly production of AgNPs have been recently discovered for their potential antibacterial action (Maleki et al., 2021; Miri, Sarani, Rezazade Bazaz, & Darroudi, 2015). Studies have shown that cow urine may be used as a reducing and capping agent to produce AgNPs that are manufactured in a green manner (Santhosh, Sandeep, Manukumar, Mahesh, & Kumara Swamy, 2021). The antimicrobial activity of these NPs is shown by their ability to significantly reduce the growth of bacteria such as *E. coli* and methicillin-resistant *S. aureus*. At different temperatures, gelatine was used as a polymerization agent in the sol-gel method to create green Ag/Ag_2O-NPs (Khandan Nasab, Sabouri, Ghazal, & Darroudi, 2020). Both Gram-positive and Gram-negative bacteria were repressed by these biosynthesized NPs, demonstrating their efficacy as antimicrobial agents. Antibacterial studies on AgNPs combining various plant extracts were synthesized by Rafique and others (Rafique, Sadaf, Rafique, & Tahir, 2017) using extracts from *Albizia procera* leaves, and the findings demonstrated that the AgNPs revealed potent antibacterial action against *S. aureus* and *E. coli*. The effect was greater against *S. aureus* than against *E. coli*. To produce spherical AgNPs smaller than 20 nm in size, Ganaie and others (Ganaie, Zahoor, & Singh, 2022) used leaf extract of *Proteus vulgaris* as a reducing and capping agent to synthesize AgNPs in an ecologically friendly manner. Improved antibacterial and antifungal action against target microorganisms was seen in the produced nanoparticles, which had a very stable FCC structure. The leaves of the herbal plant known as *Bryophyllum pinnatum* are known to have antibacterial characteristics (Akinpelu, 2000; Liao, Li, & Tjong, 2019). The antimicrobial and antibacterial properties of *Bryophyllum pinnatum* may be attributed to the existence of phytochemicals such as flavonoids, polyphenols, terpenoids, alkaloids, and organic acids in the plant extract. When they interact with bacterial cell membranes, they produce highly reactive free radical species that disrupt cellular processes (A. Hassan & Ullah, 2019; Pant et al.,

2018). The antibacterial effects of AgNPs generated from *Bryophyllum pinnatum* leaf extract are enhanced, in part because of the reactivity and diminutive size of the resulting particles. AgNPs were tested for antibacterial effectiveness against Gram-negative bacteria using the disk diffusion technique (Klančnik, Piskernik, Jeršek, & Možina, 2010). For the first time, $Zn_{0.95}Ag_{0.05}O$ NPs were made in a sustainable manner by Akbar and others (Hojjati-Najafabadi et al., 2021) utilizing extract from rosemary leaves.

As shown in Figure 13.5(a), ZnAgO nanoparticles' antibacterial efficiency was relatively increased in comparison to that of pure ZnO nanoparticles thanks to Ag doping into the ZnO lattice. Hence, $Zn_{0.95}Ag_{0.05}O$ nanoparticles are a potential contender due to their antibacterial and photocatalytic degradation properties. Aurlina and others (Aurlina et al., 2022) successfully manufactured AgNPs through bio-reduction of argon trioxide utilizing an extract of Strychnos potatorum seed, which has a high concentration of alkaloids, as shown by the presence and strength of COOH, hence stabilizing the nanoparticles. The antibacterial activity of AgNPs has been tested against Gram-positive and Gram-negative bacteria, including *S. aureus*, *Klebsiella pneumoniae*, and *E. coli*. Because Gram-positive and Gram-negative bacteria have fundamentally different cell wall structures, the effects of AgNPs on these bacteria are quite different. As illustrated in Figure 13.5(b), the robust peptidoglycan cell wall of Gram-positive strains limits the entry of foreign materials, whereas the thinner cell wall of Gram-negative bacteria exposes the AgNPs to severe cell wall destruction (Aurlina et al., 2022). For antibacterial application of AgNPs, the antibacterial activity of AgNPs@SEE was tested against *E. coli*, *K. pneumoniae*, *Proteus mirabilis*, *Pseudomonas aeruginosa*, and *Acinetobacter baumannii*, as well as two Gram-positive bacteria (*Enterococcus faecalis* and *S. aureus*). Antibacterial activity was highest against *A. baumannii* (3.25 g/ml) and *Ps.*

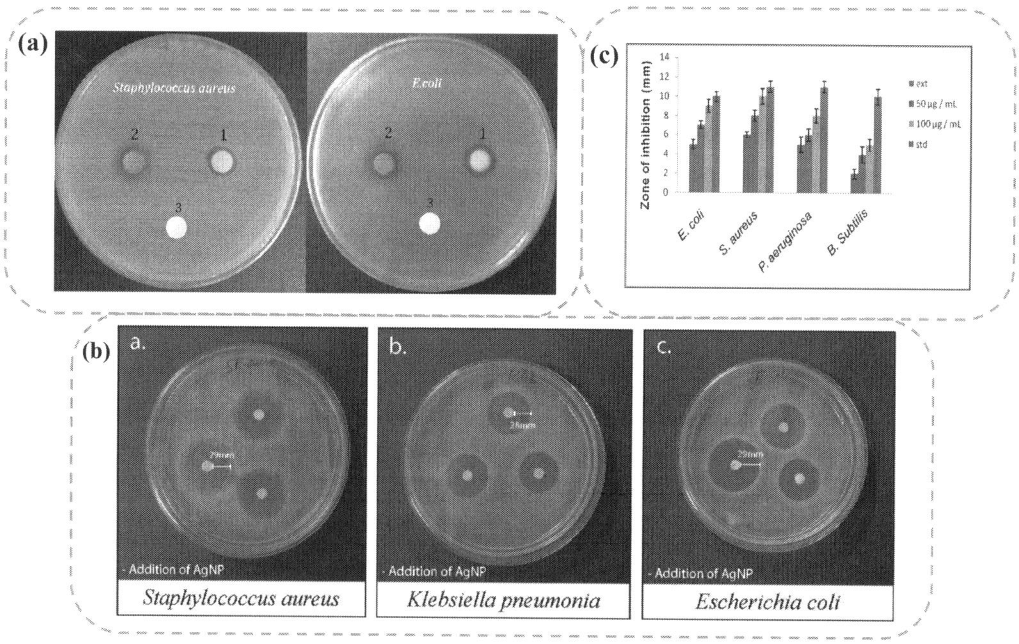

FIGURE 13.5 (a) The antibacterial activity of ZnAgO and ZnO NPs against *S. aureus* and *E. coli* standard bacteria. Sample disc 1 is ZnAgO NPs, sample disc 2 is ZnO NPs, and sample disc 3 is the negative control. (b) Antibacterial activity of AgNPs on S. aureus, K. pneumonia, and *E. coli*. The inhibition zones are marked and mentioned in the figure. (c) Antimicrobial activity (zone of inhibition) of PAgNPs against human pathogens (values ± SD, n = 3). (Panel (a) from Hojjati-Najafabadi, A. et al., *Ceram. Int.*, 47(22), 31617–31624, 2021; panel (b) from Aurlina, R. et al., *Materials Today: Proceedings*, 68, 448–453, 2022; panel (c) from Ravichandran, V. et al., *Results Phys.*, 15, 102565, 2019; panel (h) from Sharma, R., *Materials Today: Proceedings*, 44, 1995–1998, 2021.)

aeruginosa (1.5 g/ml), isolated from wounds and mucus, respectively (Hashemi et al., 2022). In a study conducted by Ravichandran and others (Ravichandran et al., 2019), it was shown that AgNPs derived from an aqueous extract of petai leaves had moderate to strong antibacterial activity against a panel of pathogenic bacteria. Figure 13.5(c) shows that the most effective antibacterial activity was shown against *S. aureus*, followed by *B. subtilis*, *E. coli*, and *Ps. aeruginosa*. As compared with pure $AgNO_3$ and petai leaf aqueous extract, the produced AgNPs showed enhanced efficacy against *E. coli*. AgNPs' germicidal effect in terms of toxic mechanism is yet unclear. Although there are various theories regarding how AgNPs kill bacteria, they have the propensity to cling to the bacterial cell wall, damage its integrity, and cause DNA to be destroyed and cellular components to flow out. Bacterial cells become more absorbent, and DNA damage finally stops bacterial growth and results in bacterial death, as shown in Figure 13.6(a) (Younas et al., 2023). The proposed mechanism of photocatalytic activity is depicted in Figure 13.6(b), which is in good agreement with earlier reported studies (Chand, Cao, Fouad, et al., 2020; Mechouche et al., 2022).

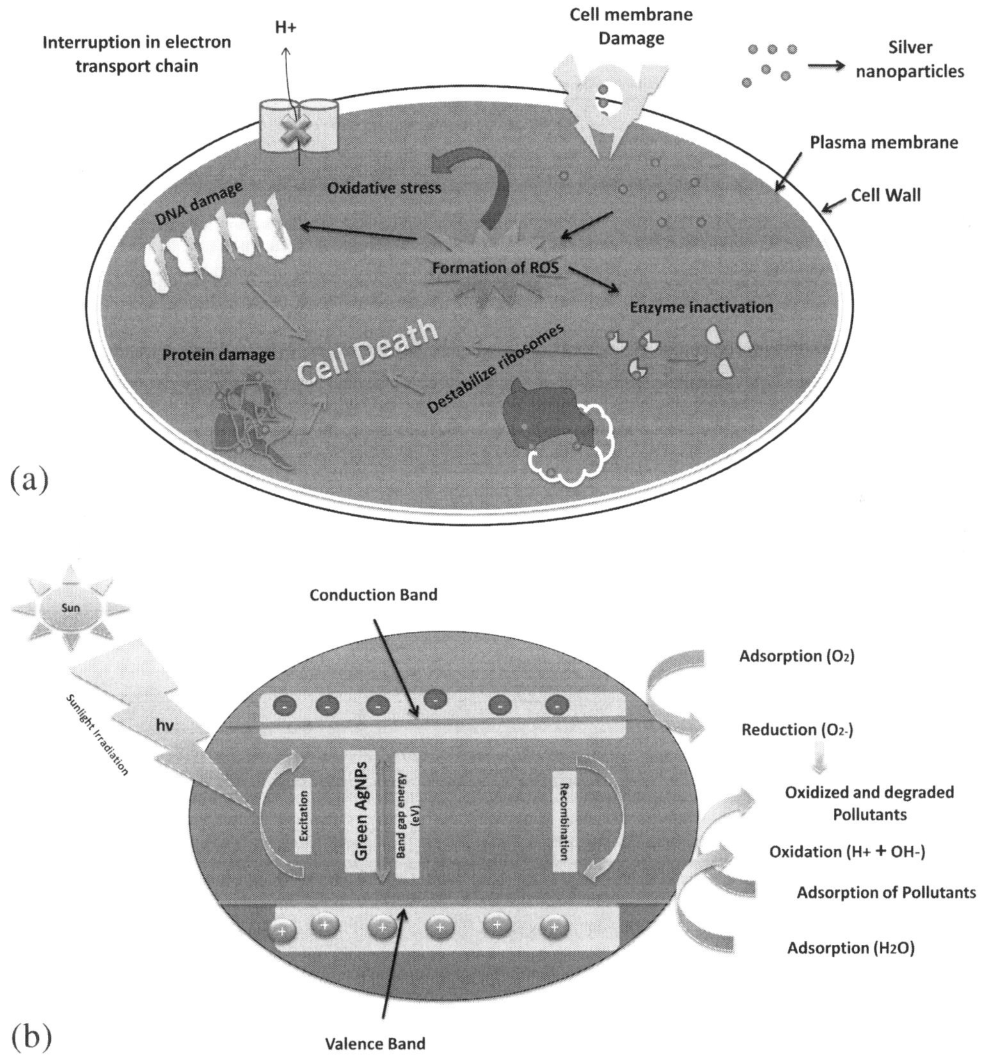

FIGURE 13.6 (a) Proposed mechanism of plant-mediated AgNPs' antibacterial action (Younas et al., 2023), and (b) Proposed mechanism of plant-mediated AgNPs' photocatalytic activity (Chand, Cao, Fouad, et al., 2020; Mechouche et al., 2022).)

13.7 CONCLUSION AND FUTURE PERSPECTIVES

Over the decades, the synthesis of AgNPs has been done mainly through chemical and physical methods. However, the cost of such synthetic method is high due to the use of different harmful chemicals. This chapter demonstrates that significant efforts have been made over the past ten years to provide a straightforward, affordable, and ecofriendly green synthesis of nanoparticles. In addition, it is observed that green synthesis is favored due to its unique physical, biological, and chemical characteristics as opposed to the physical and chemical approaches. Henceforward, plants play a significant role in the synthesis of metallic nanoparticles, particularly for photocatalytic and antibacterial activities. These plant-mediated nanoparticles can successfully address the problems of mutation and medication resistance while also offering a potential option to fight infections and pollutants. Additionally, the focus must be given to expanding the usage of AgNPs derived from plants in related sectors.

REFERENCES

Ahmad, A., Wei, Y., Syed, F., Tahir, K., Rehman, A. U., Khan, A., … Yuan, Q. (2017). The effects of bacteria-nanoparticles interface on the antibacterial activity of green synthesized silver nanoparticles. *Microbial Pathogenesis*, *102*, 133–142.

Ahmed, A., Usman, M., Ji, Z., Rafiq, M., Yu, B., Shen, Y., & Cong, H. (2023). Nature-inspired biogenic synthesis of silver nanoparticles for antibacterial applications. *Materials Today Chemistry*, *27*, 101339.

Ajitha, B., Reddy, Y. A. K., Reddy, P. S., Suneetha, Y., Jeon, H.-J., & Ahn, C. W. (2016). Instant biosynthesis of silver nanoparticles using Lawsonia inermis leaf extract: Innate catalytic, antimicrobial and antioxidant activities. *Journal of Molecular Liquids*, *219*, 474–481.

Akinpelu, D. A. (2000). Antimicrobial activity of Bryophyllum pinnatum leaves. *Fitoterapia*, *71*(2), 193–194.

Alabdallah, N. M., & Hasan, M. M. (2021). Plant-based green synthesis of silver nanoparticles and its effective role in abiotic stress tolerance in crop plants. *Saudi Journal of Biological Sciences*, *28*(10), 5631–5639.

Alyousef, A. A., Arshad, M., AlAkeel, R., & Alqasim, A. (2019). Biogenic silver nanoparticles by Myrtus communis plant extract: Biosynthesis, characterization and antibacterial activity. *Biotechnology and Biotechnological Equipment*, *33*(1), 931–936.

Amaliyah, S., Sabarudin, A., Masruri, M., & Sumitro, S. B. (2022). Characterization and Antibacterial Application of Biosynthesized Silver Nanoparticles Using Piper retrofractum Vahl Fruit Extract as Bioreductor. *Journal of Applied Pharmaceutical Science 12*(3), 103–114.

Ameen, F., Srinivasan, P., Selvankumar, T., Kamala-Kannan, S., Al Nadhari, S., Almansob, A., … Govarthanan, M. (2019). Phytosynthesis of silver nanoparticles using Mangifera indica flower extract as bioreductant and their broad-spectrum antibacterial activity. *Bioorganic Chemistry*, *88*, 102970.

Anjana, V. N., Koshy, E. P., & Mathew, B. (2020). Facile synthesis of silver nanoparticles using Azolla caroliniana, their cytotoxicity, catalytic, optical and antibacterial activity. *Materials Today: Proceedings*, *25*, 163–168.

Anju, T. R., Parvathy, S., Valiya Veettil, M., Rosemary, J., Ansalna, T. H., Shahzabanu, M. M., & Devika, S. (2021). Green synthesis of silver nanoparticles from Aloe vera leaf extract and its antimicrobial activity. *Materials Today: Proceedings*, *43*, 3956–3960.

Anwar, Y., & Alghamdi, K. M. (2020). Imparting antibacterial, antifungal and catalytic properties to cotton cloth surface via green route. *Polymer Testing*, *81*, 106258.

Arokiyaraj, S., Vincent, S., Saravanan, M., Lee, Y., Oh, Y. K., & Kim, K. H. (2017). Green synthesis of silver nanoparticles using Rheum palmatum root extract and their antibacterial activity against Staphylococcus aureus and Pseudomonas aeruginosa. *Artificial Cells, Nanomedicine and Biotechnology*, *45*(2), 372–379.

Aslany, S., Tafvizi, F., & Naseh, V. (2020). Characterization and evaluation of cytotoxic and apoptotic effects of green synthesis of silver nanoparticles using Artemisia Ciniformis on human gastric adenocarcinoma. *Materials Today Communications*, *24*, 101011.

Aurlina, R., Gopi, R. R., Ebenezer, T., Joy Prabu, H., & Johnson, I. (2022). Antibacterial study of silver nanoparticles synthesized using Strychnos potatorum(linn) – Green synthesis method. *Materials Today: Proceedings*, *68*, 448–453.

Balavijayalakshmi, J., & Ramalakshmi, V. (2017). Carica papaya peel mediated synthesis of silver nanoparticles and its antibacterial activity against human pathogens. *Journal of Applied Research and Technology*, *15*(5), 413–422.

Bordbar, M., & Mortazavimanesh, N. (2018). Biosynthesis of waste pistachio shell supported silver nanoparticles for the catalytic reduction processes. *IET Nanobiotechnology, 12*(7), 939–945.

Calamak, S., & Ermis, M. (2021). In situ silver nanoparticle synthesis on 3D-printed polylactic acid scaffolds for biomedical applications. *Journal of Materials Research, 36*(1), 166–175.

Castro, L., Blázquez, M. L., Muñoz, J. A., González, F., & Ballester, A. (2013). Biological synthesis of metallic nanoparticles using algae. *IET Nanobiotechnology, 7*(3), 109–116.

Chand, K., Cao, D., Eldin Fouad, D., Hussain Shah, A., Qadeer Dayo, A., Zhu, K., … Dong, S. (2020). Green synthesis, characterization and photocatalytic application of silver nanoparticles synthesized by various plant extracts. *Arabian Journal of Chemistry, 13*(11), 8248–8261.

Chand, K., Jiao, C., Lakhan, M. N., Shah, A. H., Kumar, V., Fouad, D. E., … Cao, D. (2021). Green synthesis, characterization and photocatalytic activity of silver nanoparticles synthesized with Nigella sativa seed extract. *Chemical Physics Letters, 763*, 138218.

Cui, W., Wang, H., Liu, L., Liang, Y., & McEvoy, J. G. (2013). Plasmonic Ag@AgCl-intercalated K4Nb6O17 composite for enhanced photocatalytic degradation of rhodamine b under visible light. *Applied Surface Science, 283*, 820–827.

Dayanidhi, K., Vadivel, P., Jothi, S., & Eusuff, N. S. (2020). Facile synthesis of Silver@Eggshell nanocomposite: A heterogeneous catalyst for the removal of heavy metal ions, toxic dyes and microbial contaminants from water. *Journal of Environmental Management, 271*, 110962.

Díaz-Cruz, C., Alonso Nuñez, G., Espinoza-Gómez, H., & Flores-López, L. Z. (2016). Effect of molecular weight of PEG or PVA as reducing-stabilizing agent in the green synthesis of silver-nanoparticles. *European Polymer Journal, 83*, 265–277.

Dong, X., Li, Y., Li, D., Liao, D., Qin, T., Prakash, O., … Liu, J. (2022). A new 3D 8-connected Cd(ii) MOF as a potent photocatalyst for oxytetracycline antibiotic degradation. *CrystEngComm, 24*(39), 6933–6943.

Duran, N., Duran, M., de Jesus, M. B., Seabra, A. B., Favaro, W. J., & Nakazato, G. (2016). Silver nanoparticles: A new view on mechanistic aspects on antimicrobial activity. *Nanomedicine: Nanotechnology, Biology and Medicine, 12*(3), 789–799.

Eswaran, S. G., Narayan, H., & Vasimalai, N. (2021). Reductive photocatalytic degradation of toxic aniline blue dye using green synthesized banyan aerial root extract derived silver nanoparticles. *Biocatalysis and Agricultural Biotechnology, 36*, 102140.

Fadlilah, D., Endarko, E., Ratnasari, A., Hozairi, H., Yusop, Z., & Syafiuddin, A. J. B. R. A. C. (2020). Enhancement of antibacterial properties of various polymers functionalized with silver nanoparticles, *10*, 5592–5598.

Fu, L., Sokiranskya, M. M., Wang, J., Lai, G. S., & Yu, A. M. (2016). Development of Ag dendrites-reduced graphene oxide composite catalysts via galvanic replacement reaction. *Physica E: Low-Dimensional Systems and Nanostructures, 83*, 146–150.

Lakshmanan, G., Sathiyaseelan, A., Kalaichelvan, P. T., & Murugesan, K. (2018). Plant-mediated synthesis of silver nanoparticles using fruit extract of Cleome viscosa L.: Assessment of their antibacterial and anticancer activity. *Karbala International Journal of Modern Science, 4*(1), 61–68.

Ganaie, S. A., Zahoor, I., & Singh, R. (2022). Prunella vulgaris leaf extract assisted green synthesis of silver nanoparticles: Antimicrobial activity. *Materials Today: Proceedings, 79*(1), 107–112.

Garibo, D., Borbón-Nuñez, H. A., de León, J. N. D., García Mendoza, E., Estrada, I., Toledano-Magaña, Y., … Susarrey-Arce, A. (2020). Green synthesis of silver nanoparticles using Lysiloma acapulcensis exhibit high-antimicrobial activity. *Scientific Reports, 10*(1), 12805.

Hafeez, M., Zeb, M., Khan, A., Akram, B., Abdin, Z.-u., Haq, S., … Ali, S. (2021). Populus ciliata mediated synthesis of silver nanoparticles and their antibacterial activity, *84*(3), 480–488.

Halder, S., Ahmed, A. N., Gafur, M. A., Seong, G., & Hossain, M. Z. (2021). Size-controlled facile synthesis of silver nanoparticles by chemical reduction method and analysis of their antibacterial performance, *6*(36), 9714–9720.

Hamad, A., Khashan, K. S., & Hadi, A. (2020). Silver nanoparticles and silver ions as potential antibacterial agents. *Journal of Inorganic and Organometallic Polymers and Materials, 30*(12), 4811–4828.

Hamedi, S., & Shojaosadati, S. A. (2019). Rapid and green synthesis of silver nanoparticles using Diospyros lotus extract: Evaluation of their biological and catalytic activities. *Polyhedron, 171*, 172–180.

Hasan, K. M. F., Liu, X. Y., Zhou, S. Q., Horvath, P. G., Bak, M., Bejo, L., … Alpar, T. (2022). Functional silver nanoparticles synthesis from sustainable point of view: 2000 to 2023-A review on game changing materials. *Heliyon, 8*(12), e12322.

Hashemi, Z., Mizwari, Z. M., Mohammadi-Aghdam, S., Mortazavi-Derazkola, S., & Ali Ebrahimzadeh, M. (2022). Sustainable green synthesis of silver nanoparticles using Sambucus ebulus phenolic extract (AgNPs@SEE): Optimization and assessment of photocatalytic degradation of methyl orange and their in vitro antibacterial and anticancer activity. *Arabian Journal of Chemistry, 15*(1), 103525.

Hassan, A., & Ullah, H. (2019). Antibacterial and antifungal Activities of the Medicinal Plant Veronica biloba. *Journal of Chemistry, 2019*, 5264943.

Hassan, S. S., Carlson, K., Mohanty, S. K., Sirajuddin, & Canlier, A. (2018). Ultra-rapid catalytic degradation of 4-nitrophenol with ionic liquid recoverable and reusable ibuprofen derived silver nanoparticles. *Environmental Pollution, 237*, 731–739.

Hojjati-Najafabadi, A., Davar, F., Enteshari, Z., & Hosseini-Koupaei, M. (2021). Antibacterial and photocatalytic behaviour of green synthesis of Zn0.95Ag0.05O nanoparticles using herbal medicine extract. *Ceramics International, 47*(22), 31617–31624.

Huang, Z. H., Jiang, X. L., Guo, D. W., & Gu, N. (2011). Controllable synthesis and biomedical applications of silver nanomaterials. *Journal of Nanoscience and Nanotechnology, 11*(11), 9395–9408.

Ivanovic, E. R., Nikolic, N. D., & Radmilovic, V. R. (2015). Randomly oriented twin domains in electrodeposited silver dendrites. *Journal of the Serbian Chemical Society, 80*(1), 107–113.

Jaast, S., & Grewal, A. (2021). Green synthesis of silver nanoparticles, characterization and evaluation of their photocatalytic dye degradation activity. *Current Research in Green and Sustainable Chemistry, 4*, 100195.

Jain, S., & Mehata, M. S. (2017). Medicinal plant leaf extract and pure flavonoid mediated green synthesis of silver nanoparticles and their enhanced antibacterial property. *Scientific Reports, 7*(1), 15867.

Jin, R., Cao, Y., Mirkin, C. A., Kelly, K. L., Schatz, G. C., & Zheng, J. G. (2001). Photoinduced conversion of silver nanospheres to nanoprisms. *Science, 294*(5548), 1901–1903.

Jini, D., Sharmila, S., Anitha, A., Pandian, M., & Rajapaksha, R. M. H. (2022). In vitro and in silico studies of silver nanoparticles (AgNPs) from Allium sativum against diabetes. *Scientific Reports, 12*(1), 22109.

Kakakhel, M. A., Wu, F., Feng, H., Hassan, Z., Ali, I., Saif, I., ... Wang, W. (2021). Biological synthesis of silver nanoparticles using animal blood, their preventive efficiency of bacterial species, and ecotoxicity in common carp fish, *84*(8), 1765–1774.

Kaplan, Ö., Gökşen Tosun, N., Özgür, A., Erden Tayhan, S., Bilgin, S., Türkekul, İ., & Gökce, İ. (2021). Microwave-assisted green synthesis of silver nanoparticles using crude extracts of Boletus edulis and Coriolus versicolor: Characterization, anticancer, antimicrobial and wound healing activities. *Journal of Drug Delivery Science and Technology, 64*, 102641.

Kasithevar, M., Saravanan, M., Prakash, P., Kumar, H., Ovais, M., Barabadi, H., & Shinwari, Z. K. (2017). Green synthesis of silver nanoparticles using Alysicarpus monilifer leaf extract and its antibacterial activity against MRSA and CoNS isolates in HIV patients, *2*(2), 131–141.

Khan, M. J., Kumari, S., Shameli, K., Selamat, J., & Sazili, A. Q. (2019). Green synthesis and characterization of pullulan mediated silver nanoparticles through ultraviolet irradiation, *12*(15), 2382.

Khandan Nasab, N., Sabouri, Z., Ghazal, S., & Darroudi, M. (2020). Green-based synthesis of mixed-phase silver nanoparticles as an effective photocatalyst and investigation of their antibacterial properties. *Journal of Molecular Structure, 1203*, 127411.

Klančnik, A., Piskernik, S., Jeršek, B., & Možina, S. S. (2010). Evaluation of diffusion and dilution methods to determine the antibacterial activity of plant extracts. *Journal of Microbiological Methods, 81*(2), 121–126.

Kouhbanani, M. A. J., Beheshtkhoo, N., Yazdanpanah, S., Zomorodian, K., Taghizadeh, S., & Amani, A. M. (2019). *Green Synthesis of Spherical Silver Nanoparticles Using Ducrosia Anethifolia Aqueous Extract and Its Antibacterial Activity, 7(3),* 461–466.

Kumar, D., Arora, S., Abdullah, & Danish, M. (2019). Plant Based Synthesis of Silver Nanoparticles from Ougeinia oojeinensis Leaves Extract and their Membrane Stabilizing, Antioxidant and antimicrobial activities. *Materials Today: Proceedings, 17*, 313–320.

Kumar, D. A., Palanichamy, V., & Roopan, S. M. (2014). Green synthesis of silver nanoparticles using Alternanthera dentata leaf extract at room temperature and their antimicrobial activity. *Spectrochimica Acta. Part A: Molecular and Biomolecular Spectroscopy, 127*, 168–171.

Kumar, S., Basumatary, I. B., Sudhani, H. P. K., Bajpai, V. K., Chen, L., Shukla, S., & Mukherjee, A. (2021). Plant extract mediated silver nanoparticles and their applications as antimicrobials and in sustainable food packaging: A state-of-the-art review. *Trends in Food Science and Technology, 112*, 651–666.

Kumar, V., & Yadav, S. K. (2009). Plant-mediated synthesis of silver and gold nanoparticles and their applications. *Journal of Chemical Technology and Biotechnology, 84*(2), 151–157.

Laghrib, F., Houcini, H., Khalil, F., Liba, A., Bakasse, M., Lahrich, S., & El Mhammedi, M. A. (2020). Synthesis of silver nanoparticles using chitosan as stabilizer agent: Application towards electrocatalytical reduction of p-nitrophenol, *5*(3), 1220–1227.

Lakhan, M. N., Chen, R., Shar, A. H., Chand, K., Shah, A. H., Ahmed, M., … Wang, J. (2020). Eco-friendly green synthesis of clove buds extract functionalized silver nanoparticles and evaluation of antibacterial and antidiatom activity. *Journal of Microbiological Methods*, *173*, 105934.

Li, S., Wang, X. K., Guo, W. L., & Wang, J. G. (2007). Sonochemical synthesis of silver dendrites nanostructures from AgNO3 Aqueous solution. *Chinese Journal of Inorganic Chemistry*, *23*(10), 1824–1828.

Liao, C., Li, Y., & Tjong, S. C. (2019). Bactericidal and cytotoxic properties of silver nanoparticles. *International Journal of Molecular Sciences*, *20*(2), 449.

Liu, X., Yang, H., Diao, Y., He, Q., Lu, C., Singh, A., … Lan, Q. (2022). Recent advances in the electrochemical applications of Ni-based metal organic frameworks (Ni-MOFs) and their derivatives. *Chemosphere*, *307*(2), 135729.

Lomelí-Rosales, D. A., Zamudio-Ojeda, A., Reyes-Maldonado, O. K., López-Reyes, M. E., Basulto-Padilla, G. C., Lopez-Naranjo, E. J., … Velázquez-Juárez, G. (2022). Green synthesis of gold and silver nanoparticles using leaf extract of Capsicum chinense Plant. *Molecules*, *27*(5), 1692.

Majeed Khan, M. A., Kumar, S., Ahamed, M., Alrokayan, S. A., & Alsalhi, M. S. (2011). Structural and thermal studies of silver nanoparticles and electrical transport study of their thin films. *Nanoscale Research Letters*, *6*(1), 434.

Maleki, P., Nemati, F., Gholoobi, A., Hashemzadeh, A., Sabouri, Z., & Darroudi, M. (2021). Green facile synthesis of silver-doped cerium oxide nanoparticles and investigation of their cytotoxicity and antibacterial activity. *Inorganic Chemistry Communications*, *131*, 108762.

Mechouche, M. S., Merouane, F., Messaad, C. E. H., Golzadeh, N., Vasseghian, Y., & Berkani, M. (2022). Biosynthesis, characterization, and evaluation of antibacterial and photocatalytic methylene blue dye degradation activities of silver nanoparticles from Streptomyces tuirus strain. *Environmental Research*, *204*(D), 112360.

Menazea, A. A. (2020). Femtosecond laser ablation-assisted synthesis of silver nanoparticles in organic and inorganic liquids medium and their antibacterial efficiency. *Radiation Physics and Chemistry*, *168*, 108616.

Milorad, C., Glišić, S., Cvetković, D., Cvetinov, M., Stanojević, L., Danilović, B., & Cakić, K. (2018). Green synthesis, characterization and antimicrobial activity of silver nanoparticles produced fromFumaria officinalis L. Plant extract. *Colloid Journal*, *80*(6), 803–813.

Miri, A., Sarani, M., Rezazade Bazaz, M., & Darroudi, M. (2015). Plant-mediated biosynthesis of silver nanoparticles using Prosopis farcta extract and its antibacterial properties. *Spectrochimica Acta. Part A: Molecular and Biomolecular Spectroscopy*, *141*, 287–291.

Mohammed, A. B. A., Mohamed, A., El-Naggar, N. E.-A., Mahrous, H., Nasr, G. M., Abdella, A., … Ali, A. S. (2022). Antioxidant and antibacterial activities of Silver Nanoparticles biosynthesized by *Moringa oleifera* through Response Surface Methodology. *Journal of Nanomaterials*, *2022*, 9984308.

Muhammad, Z., Raza, A., Ghafoor, S., Naeem, A., Naz, S. S., Riaz, S., … Rana, N. F. (2016). PEG capped methotrexate silver nanoparticles for efficient anticancer activity and biocompatibility. *European Journal of Pharmaceutical Sciences*, *91*, 251–255.

Nair, G. M., Sajini, T., & Mathew, B. (2022). Advanced green approaches for metal and metal oxide nanoparticles synthesis and their environmental applications. *Talanta Open*, *5*, 100080.

Nakkala, J. R., Mata, R., Gupta, A. K., & Sadras, S. R. (2014). Biological activities of green silver nanoparticles synthesized with Acorous calamus rhizome extract. *European Journal of Medicinal Chemistry*, *85*, 784–794.

Naseem, K., Ali, F., Tahir, M. H., Afaq, M., Yasir, H. M., Ahmed, K., … Habila, M. A. (2022). Investigation of catalytic potential of sodium dodecyl sulfate stabilized silver nanoparticles for the degradation of methyl orange dye. *Journal of Molecular Structure*, *1262*, 132996.

Pal, N., Lee, J. H., & Cho, E. B. (2020). Recent trends in morphology-controlled synthesis and application of mesoporous silica nanoparticles. *Nanomaterials*, *10*(11), 2122.

Pallela, P., Ummey, S., Ruddaraju, L. K., Pammi, S. V. N., & Yoon, S. G. (2018). Ultra Small, mono dispersed green synthesized silver nanoparticles using aqueous extract of Sida cordifolia plant and investigation of antibacterial activity. *Microbial Pathogenesis*, *124*, 63–69.

Pannerselvam, B., Thiyagarajan, D., Pazhani, A., Thangavelu, K. P., Kim, H. J., & Rangarajulu, S. K. (2021). Copperpod plant synthesized AgNPs enhance cytotoxic and apoptotic effect in cancer cell lines, 9(5), 888.

Pant, B., Park, M., Ojha, G. P., Kim, D.-U., Kim, H.-Y., & Park, S.-J. (2018). Electrospun salicylic acid/ polyurethane composite nanofibers for biomedical applications. *International Journal of Polymeric Materials and Polymeric Biomaterials*, *67*(12), 739–744.

Paul, B., Bhuyan, B., Purkayastha, D. D., & Dhar, S. S. (2016). Photocatalytic and antibacterial activities of gold and silver nanoparticles synthesized using biomass of Parkia roxburghii leaf. *Journal of Photochemistry and Photobiology, Part B: Biology, 154*, 1–7.

Phan, D.-N., Dorjjugder, N., Khan, M. Q., Saito, Y., Taguchi, G., Lee, H., … Kim, I.-S. (2019). Synthesis and attachment of silver and copper nanoparticles on cellulose nanofibers and comparative antibacterial study. *Cellulose, 26*(11), 6629–6640.

Pulit-Prociak, J., Stoklosa, K., & Banach, M. (2015). Nanosilver products and toxicity. *Environmental Chemistry Letters, 13*(1), 59–68.

Qais, F. A., Shafiq, A., Khan, H. M., Husain, F. M., Khan, R. A., Alenazi, B., … Ahmad, I. (2019). Antibacterial Effect of Silver Nanoparticles Synthesized Using *Murraya koenigii* (L.) against multidrug-Resistant Pathogens. *Bioinorganic Chemistry and Applications, 2019*, 4649506.

Qin, Y., Ji, X., Jing, J., Liu, H., Wu, H., & Yang, W. (2010). Size control over spherical silver nanoparticles by ascorbic acid reduction. *Colloids and Surfaces. Part A: Physicochemical and Engineering Aspects, 372*(1), 172–176.

Quintero-Quiroz, C., Acevedo, N., Zapata-Giraldo, J., Botero, L. E., Quintero, J., Zárate-Triviño, D., … Pérez, V. Z. (2019). Optimization of silver nanoparticle synthesis by chemical reduction and evaluation of its antimicrobial and toxic activity. *Biomaterials Research, 23*(1), 27.

Rafique, M., Sadaf, I., Rafique, M. S., & Tahir, M. B. (2017). A review on green synthesis of silver nanoparticles and their applications. *Artificial Cells, Nanomedicine and Biotechnology, 45*(7), 1272–1291.

Rai, M., Yadav, A., & Gade, A. (2009). Silver nanoparticles as a new generation of antimicrobials. *Biotechnology Advances, 27*(1), 76–83.

Raja, S., Ramesh, V., & Thivaharan, V. (2017). Green biosynthesis of silver nanoparticles using Calliandra haematocephala leaf extract, their antibacterial activity and hydrogen peroxide sensing capability. *Arabian Journal of Chemistry, 10*(2), 253–261.

Rajeshkumar, S., & Bharath, L. V. (2017). Mechanism of plant-mediated synthesis of silver nanoparticles – A review on biomolecules involved, characterisation and antibacterial activity. *Chemico-Biological Interactions, 273*, 219–227.

Rajkumar, R., Ezhumalai, G., & Gnanadesigan, M. (2021). A green approach for the synthesis of silver nanoparticles by Chlorella vulgaris and its application in photocatalytic dye degradation activity. *Environmental Technology and Innovation, 21*, 101282.

Ramanathan, S., Gopinath, S. C. B., Arshad, M. K. M., Poopalan, P., & Perumal, V. (2021). 2 - Nanoparticle synthetic methods: Strength and limitations. In S. C. B. Gopinath & F. Gang (Eds.), *Nanoparticles in Analytical and Medical Devices* (pp. 31–43). Elsevier.

Rashmi, V., & Sanjay, K. R. (2017). Green synthesis, characterisation and bioactivity of plant-mediated silver nanoparticles using Decalepis hamiltonii root extract. *IET Nanobiotechnology, 11*(3), 247–254.

Ratnasari, A., Endarko, E., & Syafiuddin, A. (2020). A green method for the enhancement of antifungal properties of various Textiles functionalized with silver nanoparticles, *Biointerface Research in Applied Chemistry, 10*(6), 7284–7294.

Ravichandran, V., Vasanthi, S., Shalini, S., Shah, S. A. A., Tripathy, M., & Paliwal, N. (2019). Green synthesis, characterization, antibacterial, antioxidant and photocatalytic activity of Parkia speciosa leaves extract mediated silver nanoparticles. *Results in Physics, 15*, 102565.

Ringwal, S., Bartwal, A. S., & Sati, S. C. (2021). Photo-catalytic degradation of different toxic dyes using silver nanoparticles as photo-catalyst, mediated via Citrus aurantium peels extract. *Journal of the Indian Chemical Society, 98*(12), 100221.

Rizzello, L., & Pompa, P. P. (2014). Nanosilver-based antibacterial drugs and devices: Mechanisms, methodological drawbacks, and guidelines. *Chemical Society Reviews, 43*(5), 1501–1518.

Sadeghi, B., & Gholamhoseinpoor, F. (2015). A study on the stability and green synthesis of silver nanoparticles using Ziziphora tenuior (Zt) extract at room temperature. *Spectrochimica Acta. Part A: Molecular and Biomolecular Spectroscopy, 134*, 310–315.

Sahu, M. K., & Patel, R. K. (2016). Novel visible-light-driven cobalt loaded neutralized red mud (Co/NRM) composite with photocatalytic activity toward methylene blue dye degradation. *Journal of Industrial and Engineering Chemistry, 40*, 72–82.

Santhosh, A. S., Sandeep, S., Manukumar, H. M., Mahesh, B., & Kumara Swamy, N. (2021). Green synthesis of silver nanoparticles using cow urine: Antimicrobial and blood biocompatibility studies. *JCIS Open, 3*, 100023.

Santhoshkumar, T., Rahuman, A. A., Rajakumar, G., Marimuthu, S., Bagavan, A., Jayaseelan, C., … Kamaraj, C. (2011). Synthesis of silver nanoparticles using Nelumbo nucifera leaf extract and its larvicidal activity against malaria and filariasis vectors. *Parasitology Research, 108*(3), 693–702.

Saragih, D. I., Arifin, D. C. V., Rusdiarso, B., Suyanta, S., & Santosa, S. J. (2020). *Synthesis of silver nanoparticles using tyrosine as reductor and capping agent.* Paper presented at the Key Engineering Materials.

Sarkar, A., Kapoor, S., & Mukherjee, T. (2010). Synthesis and characterisation of silver nanoparticles in viscous solvents and its transfer into non-polar solvents. *Research on Chemical Intermediates, 36*(4), 411–421.

Sharma, R. (2021). Synthesis of Terminalia bellirica fruit extract mediated silver nanoparticles and application in photocatalytic degradation of wastewater from textile industries. *Materials Today: Proceedings, 44*, 1995–1998.

Shenashen, M. A., El-Safty, S. A., & Elshehy, E. A. (2014). Synthesis, morphological control, and properties of silver nanoparticles in potential applications, *31*(3), 293–316.

Silva, M. R. P. Matos, d.R. S., Monteiro, M. D. S., Santos, S. B., Filho, H. D. F., Andrade, G. R. S., … Almeida, L. E. (2022). Exploiting the physicochemical and antimicrobial properties of PHB/PEG and PHB/PEG/ALG-e blends loaded with Ag nanoparticles, *15*(21), 7544.

Soshnikova, V., Kim, Y. J., Singh, P., Huo, Y., Markus, J., Ahn, S., … Yang, D. C. (2018). Cardamom fruits as a green resource for facile synthesis of gold and silver nanoparticles and their biological applications. *Artificial Cells, Nanomedicine and Biotechnology, 46*(1), 108–117.

Suber, L., & Campi, G. (2012). Hierarchic self-assembling of silver nanoparticles in solution. *Nanotechnology Reviews, 1*(1), 57–78.

Terenteva, E. A., Apyari, V. V., Kochuk, E. V., Dmitrienko, S. G., & Zolotov, Y. A. (2017). Use of silver nanoparticles in spectrophotometry. *Journal of Analytical Chemistry, 72*(11), 1138–1154.

Thiruvengadam, V., Bansod, A. V. J. B. R. A. C. (2020). Characterization of silver nanoparticles synthesized using chemical method and its antibacterial property, *10*, 7257–7264.

Tyagi, S., Tyagi, P. K., Gola, D., Chauhan, N., & Bharti, R. K. (2019). Extracellular synthesis of silver nanoparticles using entomopathogenic fungus: Characterization and antibacterial potential. *SN Applied Sciences, 1*(12), 1545.

Ulug, B., Haluk Turkdemir, M., Cicek, A., & Mete, A. (2015). Role of irradiation in the green synthesis of silver nanoparticles mediated by fig (Ficus carica) leaf extract. *Spectrochimica Acta. Part A: Molecular and Biomolecular Spectroscopy, 135*, 153–161.

Umadevi, M., & Jegatha Christy, A. (2013). Optical, structural and morphological properties of silver nanoparticles and its influence on the photocatalytic activity of TiO2. *Spectrochimica Acta. Part A: Molecular and Biomolecular Spectroscopy, 111*, 80–85.

Vanaja, M., Paulkumar, K., Gnanajobitha, G., Rajeshkumar, S., Malarkodi, C., & Annadurai, G. (2014). Herbal Plant Synthesis of antibacterial Silver Nanoparticles by *Solanum trilobatum* and Its Characterization. *International Journal of Metals, 2014*, 692461.

Vinayagam, R., Varadavenkatesan, T., & Selvaraj, R. (2017). Evaluation of the anticoagulant and catalytic activities of the Bridelia retusa Fruit extract-functionalized silver nanoparticles. *Journal of Cluster Science, 28*(5), 2919–2932.

Vokhidova, N. R., & Rashidova, S. S. (2022). The influence of synthesis conditions on the film morphology of chitosan-stabilized silver nanoparticles. *Polymer Bulletin, 79*(6), 3419–3436.

Wang, L., Xu, H., Gu, L., Han, T. T., Wang, S., & Meng, F. B. (2016). Bioinspired synthesis, characterization and antibacterial activity of plant-mediated silver nanoparticles using purple sweet potato root extract. *Materials Technology, 31*(8), 437–442.

Wang, P., Lombi, E., Zhao, F.-J., & Kopittke, P. M. (2016). Nanotechnology: A new opportunity in plant sciences. *Trends in Plant Science, 21*(8), 699–712.

Wang, X., Li, S., Ma, Y., Yu, H., & Yu, J. (2011). H2WO4·H2O/Ag/AgCl composite nanoplates: A plasmonic Z-scheme visible-light photocatalyst. *The Journal of Physical Chemistry C, 115*(30), 14648–14655.

Wen, Y., & Ding, H. (2011). Preparation and Photocatalytic Activity of Ag@AgCl Modified natase TiO2 nanotubes. *Chinese Journal of Catalysis, 32*(1), 36–45.

Wickramasinghe, S., Atnagulov, A., Yoon, B., Barnett, R. N., Griffith, W. P., Landman, U., Bigioni, T. P. (2015). M3Ag17(SPh)12 nanoparticles and their structure prediction. *Journal of the American Chemical Society, 137*(36), 11550–11553.

Widatalla, H. A., Yassin, L. F., Alrasheid, A. A., Rahman Ahmed, S. A., Widdatallah, M. O., Eltilib, S. H., & Mohamed, A. A. (2022). Green synthesis of silver nanoparticles using green tea leaf extract, characterization and evaluation of antimicrobial activity. *Nanoscale Advances, 4*(3), 911–915.

Xu, Y., Xu, H., Li, H., Xia, J., Liu, C., & Liu, L. (2011). Enhanced photocatalytic activity of new photocatalyst Ag/AgCl/ZnO. *Journal of Alloys and Compounds, 509*(7), 3286–3292.

Ying, S., Guan, Z., Ofoegbu, P. C., Clubb, P., Rico, C., He, F., & Hong, J. (2022). Green synthesis of nanoparticles: Current developments and limitations. *Environmental Technology and Innovation, 26*, 102336.

Yong, D. W. Y., Lieu, Z. Z., Cao, X., Yong, X. E., Wong, J. Z. L., Cheong, Y. S., … Chin, W. S. (2021). Biogenic synthesis of silver nanoparticles with high antimicrobial and catalytic activities using Sheng Di Huang (Rehmannia glutinosa). 16, *3*, 237–246.

Younas, M., Rasool, M. H., Khurshid, M., Khan, A., Nawaz, M. Z., Ahmad, I., & Lakhan, M. N. (2023). Moringa oleifera leaf extract mediated green synthesis of silver nanoparticles and their antibacterial effect against selected gram-negative strains. *Biochemical Systematics and Ecology*, *107*, 104605.

Zafar, A., Rizvi, R., & Mahmood, I. (2019). Biofabrication of silver nanoparticles from various plant extracts: Blessing to nanotechnology. *International Journal of Environmental Analytical Chemistry*, *99*(14), 1434–1445.

Zargar, M., Hamid, A. A., Bakar, F. A., Shamsudin, M. N., Shameli, K., Jahanshiri, F., & Farahani, F. (2011). Green synthesis and antibacterial effect of silver nanoparticles using Vitex negundo L. *Molecules*, *16*(8), 6667–6676.

Zhang, H. (2015). Ultrathin two-dimensional nanomaterials. *ACS Nano*, *9*(10), 9451–9469.

Zhang, S. B., Cohen, M. L., & Chou, M. Y. (1987). Electronic shell structure of simple metal heteroclusters. *Physical Review. Part B, Condensed Matter*, *36*(6), 3455–3458.

Zhang, W. C., Xiao, B. D., & Fang, T. (2018). Chemical transformation of silver nanoparticles in aquatic environments: Mechanism, morphology and toxicity. *Chemosphere*, *191*, 324–334.

Zhang, C. Q., Hu, Z. Q., Li, P., & Gajaraj, S. (2016). Governing factors affecting the impacts of silver nanoparticles on wastewater treatment. *Science of the Total Environment*, *572*, 852–873.

Zhou, J., Fan, G., Ruan, F., Li, Y., Fan, D., & Chen, Q. (2022). Construction of Bi2Sn2O7/Ag/Ag3PO4 heterojunction and its photocatalytic degradation properties. *Journal of the Taiwan Institute of Chemical Engineers*, *138*, 104443.

Zhou, Z., Long, M., Cai, W., & Cai, J. (2012). Synthesis and photocatalytic performance of the efficient visible light photocatalyst Ag–AgCl/BiVO4. *Journal of Molecular Catalysis A: Chemical*, *353–354*, 22–28.

14 Physiological Responses of Plants to Nanoparticles and Chelating Agents

Soleyman Dayani, Maryam Mazaheri-Tirani, and Ramin Hosseini

14.1 INTRODUCTION

Nanoparticles (NPs) are ultra-small forms of elements with one dimension <100 nm in size (Thakkar et al. 2010). Technically, there are three types of nanoparticles, natural, incidental, and engineered NPs that differ based on their source properties (Lin et al. 2010; Navarro et al. 2008; Roco 2011). The main types of engineered NPs are carbon-based materials (i.e., single-walled and multi-walled carbon nanotubes) (Cha et al. 2013), metal-based materials (i.e., Au NPs, CuO NPs, ZnO NPs, Al NPs, etc.) (Dreizin 2009), dendrimers (nano-sized polymers), and composite materials that are composed of different components including nanoparticles and larger micrometer-scale particles (Kolahalam et al. 2019; Monica and Cremonini 2009). NPs have gained increasing attention because of their physiochemical properties, such as a large specific surface area, which leads to a higher reaction activity (Jayarambabu et al. 2014). Nanotechnology, regardless of the material type produced, has revolutionized the application and properties of conventional materials in many aspects. The interesting properties of NPs, such as their ultra-small size and very high chemical activity have attracted a lot of attention in biology and environmental communities due to their mainly un-defined effects on living organisms and environment. Nanoparticles can easily cross biological membranes, such as those of plant and animal cells. The higher surface-to-volume ratio of nanoparticles is due to their ultra-small size, which leads to different chemical and physical properties with higher biological activities compared with Micro particles (MPs) (Kolahalam et al. 2019).

Among the many contributions of nanotechnology to different industries, it offered feasible solutions for various aspects of modern agricultural and biological practices in terms of systems and materials that empowered the concept of sustainable agriculture (Gangwar et al. 2023; Hazarika et al. 2022). Nanotechnology-derived advances in crop plant nutrition include nano-fertilizers (Kumar et al. 2023; ul Ain et al. 2023), nano-carriers/nanocapsules (Seabra et al. 2014), smart delivery systems, controlled/slow-release fertilizers (Firmanda et al. 2022), and clay nanotubes (Wang et al. 2020). Regarding plant protection and pathology, various solutions using nano-bactericides, nano-insecticides, nano-pesticides, nano-fungicides, and nano-herbicides have been presented (Elmer and White 2018; Sharma et al. 2022). Livestock and poultry are major sections in agriculture that have benefited from nano-products for animal feed (Swain et al. 2015), health and disease diagnoses and treatment (Ahangari et al. 2023), animal fertility and re-production (Olaniyan et al. 2023), and farm product quality and safety (Aggarwal and Idrishi 2023). Precision agriculture is an emerging and fast-growing field that encompasses several state-of-the-art technologies to provide various near and remote sensing methods by sensors to monitor and report crop plant conditions at multiple growth and development levels (Shafi et al. 2019). Nano-bio-ano-bio-sensors (Singh 2021), nano-barcodes (Ahuekwe et al. 2023), nano-chips, and nanoparticle-mediated material delivery to plants has significantly contributed to precision agriculture (Bhagat et al. 2023; Duhan et al. 2017). Soil

DOI: 10.1201/b23308-14

and water quality maintenance has benefited from the progress of nanotechnology (Huang et al. 2021; Kulanthaivel et al. 2021). At the environmental level, nano-delivered solutions offer valuable and innovative advantages for wastewater treatment, agro-waste management (Rani et al. 2023; Roy 2023), bioremediation of salt-affected or contaminated soils (Singh et al. 2020), reduction of mass chemical load into the environment, etc. (Shahi et al. 2021). The plant breeding and development programs have also utilized nanotechnology solutions, notably for removing some bottlenecks in genetic engineering (Squire et al. 2023; Yan et al. 2022), gene delivery systems (Garcia-Guerra et al. 2018), inducing predictable changes in plant growth and development, diagnostic tools (Dar et al. 2020), and tissue culture (Mazaheri-Tirani and Dayani 2020).

The use of agro-fertilizers in NP form has been an advantageous solution to provide the required nutrients to plants. This method can be enhanced when techniques for gradual and controlled release are applied and may alleviate the concern over NPs' uncontrolled release into the environment (Laware and Raskar 2014). Slow-release NP fertilizers are great alternatives to traditional fertilizers that greatly contribute to the efficient use of micronutrients (Jatav and De 2013; Manjunatha et al. 2016). The most common metal trace nutrients applied in nano form include FeO, CuO, and ZnO (Mazaheri-Tirani and Dayani 2022); Panwar et al. 2012; Sabir et al. 2014; Singh et al. 2018).

14.2 PLANTS AND MICRONUTRIENTS

Our understanding of the uptake, accumulation, and molecular mechanisms of the various trace elements, such as Zn, Mg, Fe, Se, and Ca in plants is profound. This resulted in, for example, developing practical molecular (DNA/RNA/protein) markers to speed up the pace of conventional breeding programs, and also as genetic engineering projects for the development of bio-fortified crops containing significantly higher mineral concentrations. Nonetheless, the application of widely used cultivation techniques, such as hydroponic cultures, can be a simple way to manage trace/micronutrients and to guide the plants to produce valuable secondary metabolites (Dayani and Sabzalian 2017). Connolly (2008) reported a method to increase the calcium content in plants through a modified calcium/proton antiporter (sCAX1), which increases Ca transport into carrot vacuoles. It was shown that the consumption of Ca-fortified crops resulted in lower human Ca deficiency. Therefore, the expression of a high-capacity transporter has a high potential for elevating a plant's nutritional value. There are other transgenic methods recommended to increase the bioavailability of minerals in edible portions of crop plants. For example, the concentrations of metal-binding proteins, such as lactoferritin and ferritin, can be increased, and anti-nutrient agents, such as phytate and tannins, can be reduced (Dayani and Sabzalian 2018; Lonnerdal 2003).

Plant breeding programs have mainly focused on a few main goals, including the improvement of yield and its components, enhancing crop tolerance against biotic and abiotic factors, and improving the nutritional efficiency and food functionality of the crops. In fact, breeding more versatile plant varieties that can withstand poor-nutrient soils may reduce production costs, reduce environmental pollution, and induce plant resistance to other stress conditions (Medina-Lozano and Díaz 2022; Singh et al. 2016). This potential generally comes from species and genotypic differences in crop plants for the uptake and use of nutrients.

Zinc and iron are the most commonly deficient micronutrients in calcareous and alkaline soils due to their low mobility. The societies of about 1–2 billion people that rely on such agro-produce may suffer from malnutrition (Natasha et al. 2022; Zimmermann and Hurrell 2007). Anemia, which is due to low hemoglobin in the blood is a common symptom of iron deficiency among other health issues due to iron deficiency among human societies. Therefore, those plants that can mobilize Fe and make it more available could be considered as "Fe-efficient" compared with plants that do not have this ability (Riaz and Guerinot 2021; Zuo and Zhang 2011).

Zinc is a critical micronutrient and its plant tissue content is correlated with Fe content. In a study of Zn deficiencies, Wissuwa et al. (2008) suggested that soil zinc content was the major factor that affected rice grain zinc concentrations. He reported that the rice plants' genotype was the second critical factor in grain Zn content. However, the bio-fortification of crop grain and any edible plant parts to accumulate higher Zn content through breeding new crop varieties is among the most promising and cost-effective strategies (Alloway 2009). In rice, wheat, and maize, high variation in zinc and iron content has been found among different accessions of cultivated and among wild diploid varieties (White and Broadley 2005). This reveals the huge genetic potential on which to base breeding programs for developing high accumulators of valuable nutritional minerals in crop plants. A similar high genetic variation, and high heritability values, for seed Ca concentrations have been found among wild and cultivated wheat, beans, and peas (White and Broadley 2005).

14.3 PLANT SPECIES AND NUTRIENT DEFICIENCY

It is well established that in crop plant species and cultivars, variation within the species and their interaction with the surrounding biotic and abiotic factors have a critical impact on fertilizer efficiency (Baligar et al. 2001). Under poor nutrient conditions, tolerant plants are those that can assimilate, translocate, or use higher concentrations of nutrients compared with other plant species under the same soil nutrition deficiency or relatively low nutrient availability. Moreover, tolerant plant species can produce higher dry matter or overall harvestable mass per unit of cultivation time and area and applied nutrients. These species show fewer nutrient deficiency symptoms and sometimes accumulate higher concentrations of mineral nutrients compared with other plant species under similar soil and fertilizer conditions (Fageria et al. 2008).

Fageria et al. (2008) believed that the differences in nutrient uptake and utilization among crop plants are influenced by root geometry and the plant's ability to take up sufficient amounts of nutrients from poor or subsoil mineral concentrations. They also indicated that the plant rhizosphere's potential to solubilize nutrients, mineral transport, distribution, and assimilation in plant parts to create a balanced source–sink relationship are critical factors. Plant breeders have paid considerable attention to roots to increase yields without causing environmental damage, and hence, roots are considered the path to the second green revolution. This is because extensive and vigorous plant root systems can expand over or penetrate deeper into the soil and explore higher soil volumes, which leads to absorbing more water and nutrients, especially under low nutrient conditions. This capacity results in increasing overall crop yield. There is increasing evidence that genotype diversity for root characteristics exists among many crop species and within the species that are responsive to environmental and soil nutrition conditions. This may encourage interest in considering this significant potential diversity for improving nutrient-use efficiency and overall agricultural production.

The chemical modifications in the rhizosphere are the second mechanism that significantly influences the solubility of nutrients and their uptake by plants. For example, plant roots may excrete hydrogen (H^+) at their surface to enhance the uptake of some minerals, such as zinc. The reducing activity at the plant root surface can be considered a response to iron deficiency in some plants, including a few monocotyledonous and dicotyledonous species (Tripathi et al. 2015). However, iron-efficient plants also have the ability to respond to iron deficiency by activating biochemical reactions that release compounds (phytosiderophores) to enhance Fe uptake (Tripathi et al. 2015). The pigeon pea and *Lupinus albus* (white lupin) can assimilate fixed soil phosphorus by exudation of organic acids in their rhizosphere. When facing P-deficiency conditions, white lupin and pigeon peas exude large quantities of citrate and malonic/piscidic acids into their rhizosphere zone, respectively. This can significantly increase phosphorus bioavailability by decreasing soil acidity. The chelation of other minerals, such as Al and Fe that naturally bound to P in soil, and limiting the re-adsorption or precipitation of organic phosphorus into the soil, can be another mechanism that increases phosphorus bio-availability for these plants (Keerthisinghe et al. 2001). Generally, those plants that have higher nitrogen and phosphorus accumulation in their grain are tolerant to mineral deficiencies and improve their yield, which leads to higher nutrient-use efficiency. Alongside partitioning minerals, higher distribution of dry matter in grains and shoots, and balanced source and sink relationships, are also generally associated with higher yields and consequently, higher nutrient-use efficiency, leading to fully exploiting crop yield potentials.

14.4 AGRICULTURAL FERTILIZERS

The application of fertilizers resulted in a 50% increase in agricultural plants' yields worldwide after the green revolution and during the 20th century (Fageria et al. 2008). In grain crops, such as rice and wheat, nitrogen and phosphorus have about 85% and 90% share of the variation in the yields, respectively. In the 21st century, human health and crop yield are much more dependent on plant nutrition. This is partly due to the processes that have decreased the productivity and fertility of the soil in many parts of the world. It is estimated that more than 950 million hectares worldwide are now affected by salt. The crop yield under such conditions may become lower than half. The low productivity of crop plants in salt-affected soils may be mainly due to the combinations of deficiency or unavailability of essential macro- and micronutrients, and elemental toxicities (Grattan and Grieve 1999).

It is estimated that the world's food demand will double by 2030, while we are already occupying valuable land available for agricultural expansion to support this demand. Moreover, abiotic stresses are causing dramatic yield losses in the agricultural sector worldwide. The negative effect of abiotic stresses has increased due to global climate change and unprecedented anthropogenic destructive activities in nature. This will put huge pressure on arable lands, and farmers will need to use higher rates of agricultural fertilizers and pesticides to gain more from their land. This is while the process of incorporating new land for cultivation is very slow due to various geographical, urban, and industrial factors, and more importantly, access to water sources. Therefore, the addition of fertilizers and some other amendments is critical to maximizing yields in these soils. The advent of modern agricultural fertilizers led to optimizing plant nutrition and resulted in a marked increase in agricultural production. Agricultural fertilizers are now well-established products that are regulated by national and global standards at the formulation, production, application, and food quality stages. The issue is that the efficiency of commonly used chemical fertilizers in terms of acquisition, transport, and utilization of nutrients is sometimes very low. The new trend of specialty agricultural fertilizers, including nano-fertilizers, may present interesting advantages beyond simply meeting the plants' nutritional demands. These compounds can significantly contribute to production efficiency in terms of environmental safety by reducing the chemical load into the farmlands, the economy of production by lowering the overall costs, and the nutrient assimilation efficiency through higher absorption rates.

14.5 NANOPARTICLES' EFFECT ON PLANTS

Oxide nanoparticles are the most commonly used forms of nanoparticles in different sectors of industry. Oxide NPs have received considerable attention among environmental, medical, and biological communities due to their potential negative ecological and biological impacts on living organisms and ecosystems (Lin et al. 2010; Navarro et al. 2008; Warheit et al. 2008). Recently, the application of NPs in agriculture is attracting a lot of interest (Laware and Raskar 2014). One of the main applications of nanoscale particles in agriculture is fertilizers in NP forms. The mineral in nano forms is absorbed and assimilated at significantly higher rates when compared with conventional ionic or micro forms. Increasing the production of NPs and consequently, the release, entry, and accumulation of engineered NPs in the environment and agricultural soils might induce irreversible effects (Ghormade et al. 2011; Nair et al. 2014). However, our understanding of their negative side effects over their advertised benefits is still a matter of disagreement in scientific communities. Therefore, to understand the mechanisms by which nanomaterials might affect biological systems, the first step is to investigate the penetration and transport mechanisms of NPs in plants (Monica and Cremonini 2009).

Priester et al. (2012) used the transmission electron microscopy (TEM) technique to study the bio-accumulation behavior of ZnO NPs in soybean tissues. They observed that the particles could be found in almost all plant tissues, especially in the leaves as "hot spots" of ZnO NP treatment. The zinc concentrations in the plant root, root nodule, stem, pod, and leaves treated with 0.5 g kg^{-1} ZnO NPs were approximately four, two, six, three, and four times greater than in the control group, respectively. The amount of Zn in the leaves was three times higher than that in the plant stems. However, the plant growth was negatively affected due to ZnO NP treatments (Priester et al. 2012). The CuO NP bio-accumulation in *Lemna gibba* tissues resulted in CuO NP toxicity due to Cu^{2+} released from NPs (Perreault et al. 2014). The size of NPs is a critical factor that determines the NPs' penetration depth into plant tissue, such that NPs smaller than 5 nm may easily pass through cell wall pores with a diameter of 5–8 nm. NPs up to 20 nm can also penetrate the cells through endocytosis and move via the plasmodesmata between cells. In addition to the size of the NPs, the status of chemical groups on the surface determines the ability of a molecule to enter the cells (Dietz and Herth 2011). The stability, bio-accumulation, and effects of nanoparticles in plant physiological processes, such as development and metabolism mainly depend on the nano forms' particle size, concentration, chemical composition, and surface energy. Moreover, the plant's physiological state and the target plant tissue's chemical status are influencing factors when applying NPs (Boonyanitipong et al. 2011; Dietz and Herth 2011).

NPs actively interfere in a wide range of plant survival and growth processes, including the photosynthesis process, photosynthetic pigmentation, protein biosynthesis, and antioxidant capacity (Kumar et al. 2015; Nair et al. 2014; Sanjay et al. 2015). Studies have shown that different forms of zinc (MPs or NPs) are similar in terms of quality and kinetic effects of diffusion. However, their solubility and the rate of absorption and transfer of NPs in the plant were different from MPs. On the other hand, the larger nanomaterial might more rapidly aggregate and be fixed in the environment, so that it cannot be absorbed by plants (Dimkpa et al. 2012). The aggregation of NPs is influenced by the organic matter and pH strength in the environment. The agglomeration or decrease of the NP suspension in water can decrease the NP's availability to plants (Han et al. 2016; Montes-Burgos et al. 2010; Vaseem et al. 2010).

There are different reports about the role of NPs in germination and plant growth. On the one hand, some reports showed positive effects of NPs on plants. Ghormade et al. (2011) showed that NP forms of micronutrients could increase crop yield. The exogenous application of iron and magnesium NPs improved the yield and efficient photosynthesis compared with control in soybean (Delfani et al. 2014). The application of iron nano-fertilizer (1 g L^{-1}) in *Dracocephalum moldavica* improved plant resistance against drought stress. Iron nanoparticles also increased *D. moldavica*'s oil yield and dry mass under drought stress (Yousefzadeh and Sabaghnia 2016). Sheykhbaglou et al. (2010) observed that 0.75 g L^{-1} iron oxide NPs increased pod dry and leaf weight and the total yield. The highest grain yield, with a 48% increase compared with the control, belonged to soybean seed treated with 0.5 g L^{-1} iron oxide NPs. Peanut seeds treated with ZnO NPs at 1000 ppm also showed increased growth of stems and roots of seedlings (Prasad et al. 2012).

On the other hand, negative effects of NPs on plant growth and yield have been reported by several studies. The phytotoxicity of five types of nanoparticles were studied by Lin and Xing (2007); these included multi-walled carbon nanotubes, zinc and zinc oxide (Zn, ZnO), aluminum, and alumina (Al, Al_2O_3) and their effects on seed germination and seedling root growth in *Cucumis sativus*, *Brassica napus*, *Zea mays*, *Lactuca sativa*, *Lolium multiflorum*, and *Raphanus sativus*. They showed that the application of NPs reduced seed germination and root growth indices. Many studies reported the toxicity of various oxide nanoparticles in plants, such as rare earth oxide (nano-CeO_2, nano-La_2O_3, nano-Gd_2O_3, and nano-Yb_2O_3) nanoparticles (Ma et al. 2010), TiO_2 (Ruffini Castiglione et al. 2010), FeO (Libralato et al. 2016), CuO (Dimkpa et al. 2012), and ZnO (Dimkpa et al. 2012; Sanjay et al. 2015). Lin and Xing (2007) investigated the effect of five types of NPs, including nano-multi-walled carbon nanotubes, nano-Al, nano-Al_2O_3, nano-Zn, and nano-ZnO, under different treatments (i.e., seed soaking or incubation after the soaking) on seed germination and root growth. They showed that the inhibitory effect of NPs occurred during the incubation after the soaking in the NP process instead of at the seed soaking stage. Yang and Watts (2005) showed that aluminum oxide NPs (nano Al_2O_3) at 2000 mg L^{-1} inhibited the root length of seedlings in *Daucus carota*, *Glycine max*, *Brassica oleracea*, *Cucumis sativus*, and *Zea mays*. Crop plants' response to NP types and concentrations varies widely even under similar treatments. The NPs increased root and shoot growth of green gram; however, at the same concentrations, they inhibited the indices in cucumber (Kumar et al. 2015). Treatment of peanut seeds with ZnO NPs at 2000 ppm concentration resulted in a significant decrease in the average number of leaves, fresh and dry weights of roots, and leaves of seedlings compared with the control group (Priester et al. 2012). Kumar et al. (2015) investigated the effect of ZnO NPs on root and shoot growth in four plant species. They reported that NPs reduced the root and shoot length of cucumber at concentrations above 50 mg L^{-1}, while in green gram, the reduction occurred at NP concentrations above 100 mg L^{-1}. However, NP treatment inhibited root and shoot growth of *Cucumis sativus* and *Vigna radiata* at all the studied concentrations. Rahmani et al. (2016) studied the effect of ZnO NPs at 10 mg ml^{-1} level on *Brassica napus* growth and observed significantly improved growth and early flowering ten days earlier than the control group. However, NP concentrations of 100 and 1000 mg ml^{-1} resulted in toxic responses in the plant. It was reported that CuO NP utilization might lead to pollution of aquatic ecosystems and suppression of photosynthetic activity via inactivation of photosystem II reaction centers, electron transport reduction, and thermal energy dissipation enhancement in the macrophyte plant *Lemna gibba* L. (Perreault et al. 2014). Sanjay et al. (2015) found that nano-ZnO shows a positive effect on the *Solanum melongena* plant only at 50 ppm, and at higher levels (50 and 100 ppm) of MPs and NPs, ZnO shows its phytotoxic activity. At the highest concentrations of Zn, plants exhibited a different adaptation strategy, closing stomata, further decreasing growth, and signs of enhancing photosynthetic energy dissipation through the violaxanthin cycle pigments (Sagardoy et al. 2009). In another study, the inhibitory effect of ZnO NPs treatments on *Brassica napus* was lower than for ZnO MPs (MousaviKouhi et al. 2014). The phytotoxicity of nanoparticles and their polluting effects on the environment are still a problem and need more investigation (Bandyopadhyay et al. 2015; Lin and Xing 2008).

The toxicity level of NPs in crop plants compared with similar concentrations of non-nano forms of minerals should be investigated. Unfortunately, this issue has not been considered in most of the studies on NP's effect on crops. Few reports have compared and reported the effect of non-nano and nanoparticle forms of fertilizers (Adhikari et al. 2016; MousaviKouhi et al. 2014; Sanjay et al. 2015; Tirani et al. 2018). Libralato et al. (2016) reported that both MP and NP forms of iron oxide could cause a significant reduction in root and leaf fresh weight, but the shoot FW was not significantly affected in *Lepidium sativum*, *Sinapis alba*, and *Sorghum saccharatum*. This suggests that NPs even have different effects on different tissues of the same plant. The effect of ZnO NPs and CuO NPs and MPs in wheat was compared by Dimkpa et al. (2012). They reported that both particle forms were accumulated in similar concentrations in plant tissues. However, the NPs showed more negative effects on the studied plant's growth. This result could be due to NPs' higher biochemical activity. Therefore, it seems that seed germination and plant growth can vary depending on the plant species, plant tissue, and NP size, type, and concentration.

14.6 FERTILIZER CHELATES

At high pH and unfavorable soil conditions, metal ions, such as Fe, Mg, Cu, and Zn, become insoluble and are not absorbed by plants. In fertilizer technology, to solve this problem, the metal mineral is covered in synthetic or natural compounds called chelators that improve the mineral availability to plants (Meers et al. 2005; Sekhon 2003). The term *chelate* is derived from the Greek word *chelé*, which means "lobster's claw". A chelate indicates that a metal nutrient ion is encircled by the larger organic molecule (the claw), known as a ligand or chelator. Metal–ethylene diamine-tetra-acetic acid (EDTA) complexes are very stable (Meers et al. 2009). Chelates strongly modulate the availability of metals, especially under unfavorable soil conditions (Meers et al. 2005; Sekhon 2003). There are different types of chelator compounds used in agricultural fertilizers, such as citric acid, EDTA, N-(hydroxyethyl) ethylene-diamine-triacetic acid (HEEDTA), ethylene diamine disuccinate (EDDS), nitrilo-tri-acetic acid (NTA), and ethylene-diamine-di-(o-hydroxyphenylacet ic acid) (EDDHA) (Labanowski et al. 2008; Meers et al. 2005; Sekhon 2003).

Wang et al. (2013) reported that the highest concentrations of Zn in plant tissues grown in soil or solution media were found when Zn was bound to citrate chelates. The potential of some chelating agents, such as citrate, was shown to be higher, even when compared with NP forms. Under the same experimental conditions, lower Zn concentrations in plant tissues were recorded when Zn was bound to phytate, histidine, phosphate, or ZnO NPs. Citrate is a carboxylic acid that is abundant in plant tissues, and its role in Zn hyper-accumulation in plant cells is well documented (Callahan et al. 2006; Küpper et al. 2004; Liu et al. 2010; Wójcik et al. 2006). Citrate is known to increase water transport in the xylem of cut flowers and act as a pH regulator (van Doorn 2012). The citrate complex facilitates iron (Fe) ion mobility inside plant tissues (Hell and Stephan 2003). Moreover, citrate was able to recover the green color of leaves affected by chlorosis (Álvarez-Fernández et al. 2004). However, it has been indicated that the effectiveness of Fe^{3+}-EDTA in arsenic-treated rice was much higher than that of Fe^{3+}-citrate in chlorosis remediation (Shaibur and Kawai 2017). Some trials have shown that chelates may interfere with the inhibitory effects of released Zn^{2+} from ZnO NPs and decrease its toxic effects (Landa et al. 2012).

Despite the beneficial effects of chelators on nutrient absorption, a few reports have revealed the negative impact of chelators on plant seedling germination (İlbaş et al. 2006; Székely et al. 2011). Some reports have shown that the presence of free EDTA could be toxic to plants because it may result in imbalance of minerals, such as Cu, Fe, Zn, and Ca, which consequently leads to cell metabolism disturbances and destabilization of biological membranes (Ruley et al. 2006). For instance, it has been observed that the addition of EDTA to soil caused a significant increase in the amount of copper in the above- and below-ground parts of plants, but caused phosphorus deficiency as a nutritional imbalance (Pastor et al. 2007). EDTA was shown to reduce mitotic division via chromosome aberrations during cell division in *Phaseolus vulgaris* and *Hordeum vulgare* (Fischbein et al. 1970). Székely et al. (2011) reported that the photosynthetic pigments (chlorophyll and carotenoids) were negatively affected under EDTA treatment and that EDTA may cause chlorosis in treated *Sorghum bicolor* and sudangrass seedlings. Low pigment content could be due to oxidative stress by EDTA. Moreover, chlorophyll formation might be affected by Mg-EDTA, leading to lower central metal ion availability (Székely et al. 2011). High doses of the chelating agents EDTA and DTPA did not cause serious damage to plants, although they reduced the biomass production (Pastor et al. 2007). It also has been reported that EDTA, applied separately or combined with Cu, increased the biomass accumulation of sudangrass seedlings, while the pigment content declined in the leaf tissues (Székely et al. 2011).

14.7 NANOPARTICLES AND CHELATES

There are various reports on the effects of different concentrations of metal MPs and NPs on plants, while the combined effects of the two forms of metal with and without chelating agents are still in question. The effect of NPs depends on their size and aggregation state. NPs have a high surface-to-volume ratio (Jayarambabu et al. 2014), and the aggregation of small particles could be accelerated in aquatic solutions. The processes of aggregation and releasing of nanoparticles such as zinc oxide NPs and iron oxide NPs could be related to ionic strength, natural colloids, and natural organic matter that could affect nanoparticle transport in aqueous environments and consequently, their uptake into plants. The size and size distribution of nanoparticles are particularly important. Several factors, such as pH, ionic strength, natural colloids, and natural organic matter, could influence the aggregation and solubility of nanoparticles, and their transport in an aqueous environment and uptake by plants (Baalousha 2009; Bedi and Arshdeep 2015; Han et al. 2016; Lin et al. 2010).

Chelates may influence the nanoparticles' environmental stabilization and bioavailability to plants (Pastor et al. 2007). The application of iron nano-chelate in appropriate concentrations increased quantitative and qualitative plant growth and was more effective and efficient compared with traditional fertilizers (Hossain et al. 2016). In the presence of citric acid, the formation of a stable suspension of ZnO NPs has been observed (Mudunkotuwa et al. 2012). Citrate and EDTA chelates improved the ZnO NPs' dissolution rate and made Zn(II)-EDTA or Zn(II)-citrate complexes soluble (Feng et al. 2016). The constant dissolution rate of ZnO with EDTA (2.24 min^{-1}) was significantly higher than that of citrate (0.37 min^{-1}) or the control with no chelate substance (0.09 min^{-1}). Tirani et al. (2019) reported a decrease in the color of tobacco cotyledon in the presence of EDTA due to oxidative stress when applying ZnO NPs. The appearance of main leaves, hypocotyl status, radicle shape, and cotyledon color were more sensitive to ZnO NPs compared with ZnO MPs. They were also more sensitive to EDTA than citrate treatment. Chlorophyll formation may also be affected by EDTA by chelating Mg^{2+} and lead to a decrease in the central metal ion availability (Székely et al. 2011).

NP metal toxicity could be reduced by chelators. Sabella et al. (2014) speculate that metal-containing NPs, such as semiconductor NPs, metal oxide, and metallic forms, may utilize different toxicity mechanisms that are mainly related to their corresponding toxic ions released in the environment or accumulated in plant tissues. However, nanoparticles combined with different types of chelating agents do not release metal ions to toxic levels because of NPs' stable coating and strong bonds with the chelating agent, which does not cause damage to intracellular environments. EDTA materials, which are synthetic polyaminocarboxylic acid compounds, could affect the toxic effects of ZnO. Although many previous studies have shown that NPs are commonly more toxic than their MPs counterparts, high concentrations (50 ppm) of ZnO MPs-EDTA caused higher cell death in 21-day *Nicotiana tobaccum* seedlings. This might indicate that the form of ZnO, the type of chelator, as well as the amount of ZnO are very important and definitive (Tirani et al. 2018). The nano-citrate form in ZnO NP treatments reduced the activity of the superoxide dismutase (SOD) enzyme and DPPH radical scavenging potential, and reduced cell death. Thus, it could mitigate the negative effects of ZnO NPs. EDTA with ZnO NP treatments increased the amount of reactive oxygen species (ROS), malondialdehyde (MDA), and cell death, and reduced antioxidant activity, including FRAP, DPPH, and SOD activities, which led to increasing ZnO stress effects (Tirani et al. 2018).

14.8 CONCLUSION

The application of nanoparticles as a novel combination of micronutrients seems to be a new way of supplying the necessary micro and macro elements for crops. Unfortunately, few studies have been conducted on the application of NPs in agriculture. On the other hand, the location of bioaccumulation of NPs in humans, animals, and plants and the number of useful NPs is still unclear and requires further investigation.

REFERENCES

Adhikari T, Kundu S, Rao AS (2016) Zinc delivery to plants through seed coating with Nano zinc oxide particles. *Journal of Plant Nutrition, 39*, 136–146.

Aggarwal D, Idrishi R (2023) Nanotechnology applications for food traceability. In Arun Sharma, P.S. Vijayakumar, Er. Pramod K Prabhakar, Ritesh Kumar (eds),*Nanotechnology Applications for Food Safety and Quality Monitoring*. Academic Press, pp. 457–472.

Ahangari A, Mahmoodi P, Mohammadzadeh A (2023) Advanced Nano biosensors for rapid detection of zoonotic bacteria. *Biotechnology and Bioengineering* 120(1):41–56.

Ahuekwe EF Isibor, Patrick O., Akinduti, Paul A., Salami, Abimbola O., Onuselogu, Chinedu C., Oshamika, Oyewumi O., Oyesola, Olusola L., Ichor, Tersagh, Ezekiel, Olawale M., Oziegbe, Olubukola, Akinyosoye, Abimbola D., Akinhanmi, Fadekemi, Oyewale, John O., Taiwo, Olugbenga S., Akinwunmi, Ruth A., Ajiboye, Ibukun, Adekeye, Bosede T., Akinpelu, Sharon O., Kuye, Alice

D., Bello, Adetutu O., George, Dango Z., Ojo-Omoniyi, Olusola A., Popoola, Taiwo S., Akinyemi, Oluwatobi D., Adebayo, Glory P., Oniha, Margaret I., Iheagwam, Franklyn N., Adelodun, Comfort A., Orukotan, Kesioluwa E., Bilewu, Olayemi F., Onibokun, Elizabeth A., Fasuyi, Nifemi O., Nwinyi, Obinna C., Ayanda, Isaac O., Omonhinmin, Conrad A., Akinnola, Olayemi O., Oranusi, Solomon U., Eni, Angela O., Popoola, Jacob O., Olasehinde, Grace I., Obembe,Olawole O. (2023) Application of nanochitosan in tagging and nano-barcoding of aquatic and animal meats. In Charles Adetunji, Daniel Hefft, Jaison Jeevanandam, Michael Danquah (eds), *Next Generation Nanochitosan*, Academic Press,pp. 483–498.

Alloway B (2009) Soil factors associated with zinc deficiency in crops and humans. *Environmental Geochemistry and Health* 31(5):537–548.

Álvarez-Fernández A, García-Laviña P, Fidalgo C, Abadía J, Abadía A (2004) Foliar fertilization to control iron chlorosis in pear (Pyrus communis L.) trees. *Plant and Soil* 263(1):5–15.

Baalousha M (2009) Aggregation and disaggregation of iron oxide nanoparticles: Influence of particle concentration, pH and natural organic matter. *Science of the Total Environment* 407(6):2093–2101.

Baligar V, Fageria N, He Z (2001) Nutrient use efficiency in plants. *Communications in Soil Science and Plant Analysis* 32(7–8):921–950.

Bandyopadhyay S, Plascencia-Villa G, Mukherjee A, Rico CM, José-Yacamán M, Peralta-Videa JR, Gardea-Torresdey JL (2015) Comparative phytotoxicity of ZnO NPs, bulk ZnO, and ionic zinc onto the alfalfa plants symbiotically associated with Sinorhizobium meliloti in soil. *Science of the Total Environment* 515:60–69.

Bedi PS, Arshdeep K (2015) An overview on uses of zinc oxide nanoparticles. *World Journal of Pharmacy and Pharmaceutical Sciences* 4:1177–1196.

Bhagat R, Ingle AP, Chen H (2023) Nanosensors and nanobiosensors for sustainable agriculture. In *Nanotechnology in Agriculture and Agroecosystems*. Elsevier, pp. 93–112.

Boonyanitipong P, Kositsup B, Kumar P, Baruah S, Dutta J (2011) Toxicity of ZnO and TiO2 nanoparticles on germinating rice seed. *International Journal of Bioscience, Biochemistry and Bioinformatics* 1:282–285.

Callahan DL, Baker AJ, Kolev SD, Wedd AG (2006) Metal ion ligands in hyperaccumulating plants. *JBIC, Journal of Biological Inorganic Chemistry* 11(1):2–12.

Cha C, Shin SR, Annabi N, Dokmeci MR, Khademhosseini A (2013) Carbon-based nanomaterials: Multifunctional materials for biomedical engineering. *ACS Nano* 7(4):2891–2897.

Connolly EL (2008) Raising the bar for biofortification: Enhanced levels of bioavailable calcium in carrots. *Trends in Biotechnology* 26(8):401–403.

Dar FA, Qazi G, Pirzadah TB (2020) Nano-biosensors: NextGen diagnostic tools in agriculture In *Nanobiotechnology in Agriculture: An Approach Towards Sustainability*, pp. 129–144.

Dayani S, Sabzalian MR (2017) Production of secondary metabolites in medicinal plants through hydroponic system. In *Controlled Environmental Agriculture-Production of Specialty Crops Providing Human Health Benefits through Hydroponics*. Nova Science, Inc, New York, pp. 33–53.

Dayani S, Sabzalian MR (2018) Genetically modified plants as sustainable and economic sources for RUTFs. In Alina Maria Holban, Alexandru Mihai Grumezescu (eds) *Handbook of Food Bioengineering, Genetically Engineered Foods*. Academic Press, pp. 49–84.

Delfani M, Baradarn Firouzabadi M, Farrokhi N, Makarian H (2014) Some physiological responses of black-eyed pea to iron and magnesium nanofertilizers. *Communications in Soil Science and Plant Analysis* 45(4):530–540.

Dietz KJ, Herth S (2011) Plant nanotoxicology. *Trends in Plant Science* 16(11):582–589 doi:10.1016/j.tplants.2011.08.003.

Dimkpa CO et al. (2012) CuO and ZnO nanoparticles: Phytotoxicity, metal speciation, and induction of oxidative stress in sand-grown wheat. *Journal of Nanoparticle Research* 14(9):1–15.

Dreizin EL (2009) Metal-based reactive nanomaterials. *Progress in Energy and Combustion Science* 35(2):141–167.

Duhan JS, Kumar R, Kumar N, Kaur P, Nehra K, Duhan S (2017) Nanotechnology: The new perspective in precision agriculture. *Biotechnology Reports* 15:11–23 doi:10.1016/j.btre.2017.03.002.

Elmer W, White JC (2018) The future of nanotechnology in plant pathology. *Annual Review of Phytopathology* 56:111–133.

Fageria N, Baligar V, Li Y (2008) The role of nutrient efficient plants in improving crop yields in the twenty first century. *Journal of Plant Nutrition* 31(6):1121–1157.

Feng X et al. (2016) Enhanced dissolution and transformation of ZnO nanoparticles: The role of inositol hexakisphosphate. *Environmental Science and Technology* 50(11):5651–5660.

Firmanda A, Fahma F, Syamsu K, Suryanegara L, Wood K (2022) Controlled/slow-release fertilizer based on cellulose composite and its impact on sustainable agriculture: Review. *Biofuels, Bioproducts and Biorefining* 16(6):1909–1930.

Fischbein L, Flamm W, Falk H (1970) *Chemical Mutagens: Environmental Effects on Biological Systems*. Academic Press, New York .

Gangwar J, Kadanthottu Sebastian J, Puthukulangara Jaison J, Kurian JT (2023) Nano-technological interventions in crop production—A review. *Physiology and Molecular Biology of Plants*, *29*(1): 1–15.

Garcia-Guerra A, Dunwell TL, Trigueros S (2018) Nano-scale gene delivery systems: Current technology, obstacles, and future directions. *Current Medicinal Chemistry* 25(21):2448–2464.

Ghormade V, Deshpande MV, Paknikar KM (2011) Perspectives for nano-biotechnology enabled protection and nutrition of plants. *Biotechnology Advances* 29(6):792–803.

Grattan S, Grieve C (1999) Mineral nutrient acquisition and response by plants grown in saline environments. Handbook of plant and crop stress 2:203–229.

Han Y et al. (2016) Transport, retention, and long-term release behavior of ZnO nanoparticle aggregates in saturated quartz sand: Role of solution pH and biofilm coating. *Water Research* 90:247–257.

Hazarika A, Yadav M, Yadav DK, Yadav HS (2022) An overview of the role of nanoparticles in sustainable agriculture. *Biocatalysis and Agricultural Biotechnology* 43:102399 doi:10.1016/j.bcab.2022.102399.

Hell R, Stephan UW (2003) Iron uptake, trafficking and homeostasis in plants. *Planta* 216(4):541–551.

Hossain Z, Mustafa G, Sakata K, Komatsu S (2016) Insights into the proteomic response of soybean towards Al 2 O 3, ZnO, and Ag nanoparticles stress. *Journal of Hazardous Materials* 304:291–305.

Huang W, Zhou C, Liu Z, Sun H, Du J, Zhang L (2021) Improving soil-water characteristics and pore structure of silty soil using nano-aqueous polymer stabilisers. *KSCE Journal of Civil Engineering* 25(9):3298–3305.

İlbaş Aİ, Eroğlu Y, Eroğlu HE (2006) Effects of dosages and application periods of EDTA on morphological and cytogenetic characters of barley (Hordeum vulgare L.) seedlings. *Turkish Journal of Biology* 30:59–63.

Jatav G, De N (2013) Application of nano-technology in soil-plant system. *An Asian Journal of Soil Science* 8:176–184.

Jayarambabu N, Kumari BS, Rao KV, Prabhu Y (2014) Germination and Growth Characteristics of Mungbean Seeds (Vigna radiata L.) affected by Synthesized zinc oxide Nanoparticles. *International Journal of Current Engineering and Technology* 4:3411–3416.

Keerthisinghe G, Zapata F, Chalk P, Hocking P (2001) Integrated approach for improved P nutrition of plants in tropical acid soils. In *Plant Nutrition*. Springer, pp. 974–975.

Kolahalam LA, Viswanath IK, Diwakar BS, Govindh B, Reddy V, Murthy Y (2019) Review on nanomaterials: Synthesis and applications. *Materials Today: Proceedings* 18:2182–2190.

Kulanthaivel P, Soundara B, Velmurugan S, Naveenraj V (2021) Experimental investigation on stabilization of clay soil using nano-materials and white cement. *Materials Today: Proceedings* 45:507–511.

Kumar N, Samota SR, Venkatesh K, Tripathi S (2023) Global trends in use of nano-fertilizers for crop production: Advantages and constraints–A review. *Soil and Tillage Research* 228:105645.

Kumar S, Patra AK, Datta SC, Rosin KG, Purakayastha TJ (2015) Phytotoxicity of nanoparticles to seed germination of plants. *International Journal of Advanced Research* 3:854–865.

Küpper H, Mijovilovich A, Meyer-Klaucke W, Kroneck PM (2004) Tissue-and age-dependent differences in the complexation of cadmium and zinc in the cadmium/zinc hyperaccumulator Thlaspi caerulescens (Ganges ecotype) revealed by X-ray absorption spectroscopy. *Plant Physiology* 134(2):748–757.

Labanowski J, Monna F, Bermond A, Cambier P, Fernandez C, Lamy I, Van Oort F (2008) Kinetic extractions to assess mobilization of Zn, Pb, Cu, and Cd in a metal-contaminated soil: EDTA vs. citrate. *Environmental Pollution* 152(3):693–701.

Landa P et al. (2012) Nanoparticle-specific changes in Arabidopsis thaliana gene expression after exposure to ZnO, TiO2, and fullerene soot. *Journal of Hazardous Materials* 241:55–62.

Laware S, Raskar S (2014) Influence of zinc oxide nanoparticles on growth, flowering and seed productivity in onion. *International Journal of Current Microbiology and Applied Sciences* 3:874–881.

Libralato G et al. (2016) Phytotoxicity of ionic, micro-and nano-sized iron in three plant species. *Ecotoxicology and Environmental Safety* 123:81–88.

Lin D, Tian X, Wu F, Xing B (2010) Fate and transport of engineered nanomaterials in the environment. *Journal of Environmental Quality* 39(6):1896–1908.

Lin D, Xing B (2007) Phytotoxicity of nanoparticles: Inhibition of seed germination and root growth. *Environmental Pollution* 150(2):243–250 doi:10.1016/j.envpol.2007.01.016.

Lin D, Xing B (2008) Root uptake and phytotoxicity of ZnO nanoparticles. *Environmental Science and Technology* 42(15):5580–5585.

Liu Z, Ya J, Lei E (2010) Effects of substrates and seed layers on solution growing ZnO nanorods. *Journal of Solid State Electrochemistry* 14(6):957–963.

Lonnerdal B (2003) Genetically modified plants for improved trace element nutrition. *The Journal of Nutrition* 133:1490S–1493S.

Ma X, Geiser-Lee J, Deng Y, Kolmakov A (2010) Interactions between engineered nanoparticles (ENPs) and plants: Phytotoxicity, uptake and accumulation. *Science of the Total Environment* 408(16):3053–3061.

Manjunatha S, Biradar D, Aladakatti YR (2016) Nanotechnology and its applications in agriculture: A review. *Journal of Farm Science* 29:1–13.

Mazaheri-Tirani M, Dayani S (2020) In vitro effect of zinc oxide nanoparticles on Nicotiana tabacum callus compared to ZnO micro particles and zinc sulfate (ZnSO4). *Plant Cell, Tissue and Organ Culture (PCTOC)* 140:279–289 doi:10.1007/s11240-019-01725-0

Medina-Lozano I, Díaz A (2022) Applications of genomic tools in plant breeding: Crop biofortification. *International Journal of Molecular Sciences* 23(6):3086.

Meers E, Qadir M, De Caritat P, Tack F, Du Laing G, Zia M (2009) EDTA-assisted Pb phytoextraction. *Chemosphere* 74(10):1279–1291.

Meers E, Ruttens A, Hopgood M, Samson D, Tack F (2005) Comparison of EDTA and EDDS as potential soil amendments for enhanced phytoextraction of heavy metals. *Chemosphere* 58(8):1011–1022.

Mazaheri-Tirani M, Dayani S (2022) Growth, Flowering and Physiological Response of Trachyspermum ammi L. to Zinc Oxide Micro- and Nanoparticles. *Russian Journal of Plant Physiology* 69:14 doi:10.1134/S1021443722010125

Monica RC, Cremonini R (2009) Nanoparticles and higher plants. *Caryologia* 62(2):161–165.

Montes-Burgos I, Walczyk D, Hole P, Smith J, Lynch I, Dawson K (2010) Characterisation of nanoparticle size and state prior to nanotoxicological studies. *Journal of Nanoparticle Research* 12(1):47–53.

MousaviKouhi SM, Lahoutia M, Ganjealia A, Entezarib MH (2014) Comparative phytotoxicity of ZnO nanoparticles, ZnO microparticles,and Zn2+ on rapeseed Brassica napus L.: Investigating a wide range of concentrations. *Toxicological and Environmental Chemistry, 96*: 861–868.

Mudunkotuwa IA, Rupasinghe T, Wu CM, Grassian VH (2012) Dissolution of ZnO nanoparticles at circumneutral pH: a study of size effects in the presence and absence of citric acid. *Langmuir 28*(1):396–403.

Nair PMG, Kim S-H, Chung IM (2014) Copper oxide nanoparticle toxicity in mung bean (Vigna radiata L.) seedlings: Physiological and molecular level responses of in vitro grown plants. *Acta Physiologiae Plantarum* 36(11):2947–2958.

Natasha N et al. (2022) Zinc in soil-plant-human system: A data-analysis review. *Science of the Total Environment* 808:152024.

Navarro E et al. (2008) Environmental behavior and ecotoxicity of engineered nanoparticles to algae, plants, and fungi. *Ecotoxicology* 17(5):372–386.

Olaniyan OT, Nagpure G, Singh RP, Adetunji CO (2023) Bionanomaterials' utility for enhancing livestock fertility. In *Nanobiotechnology for the Livestock Industry*. Elsevier, pp. 243–252.

Panwar J, Jain N, Bhargaya A, Akhtar M, Yun Y (2012) Positive effect of zinc oxide nanoparticels on tomato plants: A step towards developing "Nano-fertilizers" 01/2012. In Proceeding of 3rd International Conference on Environmental Research and Technology (ICERT).

Pastor J, Aparicio AM, Gutierrez-Maroto A, Hernández AJ (2007) Effects of two chelating agents (EDTA and DTPA) on the autochthonous vegetation of a soil polluted with Cu, Zn and Cd. *Science of the Total Environment* 378(1–2):114–118.

Perreault F, Samadani M, Dewez D (2014) Effect of soluble copper released from copper oxide nanoparticles solubilisation on growth and photosynthetic processes of Lemna gibba L. *Nanotoxicology* 8(4):374–382.

Prasad T et al. (2012) Effect of nanoscale zinc oxide particles on the germination, growth and yield of peanut. *Journal of Plant Nutrition* 35(6):905–927.

Priester JH et al. (2012) Soybean susceptibility to manufactured nanomaterials with evidence for food quality and soil fertility interruption. *Proceedings of the National Academy of Sciences* 109(37):E2451–E2456.

Rahmani F, Peymani A, Daneshvand E, Biparva P (2016) Impact of zinc oxide and copper oxide nano-particles on physiological and molecular processes in Brassica napus L. *Indian Journal of Plant Physiology 21*:1–7.

Rani GM et al. (2023) Agro-waste to sustainable energy: A green strategy of converting agricultural waste to nano-enabled energy applications. *Science of the Total Environment 875*:162667.

Riaz N, Guerinot ML (2021) All together now: Regulation of the iron deficiency response. *Journal of Experimental Botany* 72(6):2045–2055.

Roco MC (2011) The long view of nanotechnology development: The National Nanotechnology Initiative at 10 years. *Journal of Nanoparticle Research* 13(2):427–445.

Roy P (2023) Nothing is 'Waste'in agriculture: From nanotechnology and bioprocesses perspectives. In *Agriculture Waste Management and Bioresource: The Circular Economy Perspective*, pp. 131–148.

Ruffini Castiglione M, Giorgetti L, Geri C, Cremonini R (2010) The effects of Nano-TiO2 on seed germination, development and mitosis of root tip cells of Vicia narbonensis L. and Zea mays L. *Journal of Nanoparticle Research* 13(6):2443–2449 doi:10.1007/s11051-010-0135-8.

Ruley AT, Sharma NC, Sahi SV, Singh SR, Sajwan KS (2006) Effects of lead and chelators on growth, photosynthetic activity and Pb uptake in Sesbania drummondii grown in soil. *Environmental Pollution* 144(1):11–18.

Sabella S et al. (2014) A general mechanism for intracellular toxicity of metal-containing nanoparticles. *Nanoscale* 6(12):7052–7061.

Sabir S, Arshad M, Chaudhari SK (2014) Zinc oxide nanoparticles for revolutionizing agriculture: Synthesis and applications. *The Scientific World Journal 2014*:925494.

Sagardoy R, Morales F, López-Millán AF, Abadía A, Abadía J (2009) Effects of zinc toxicity on sugar beet (Beta vulgaris L.) plants grown in hydroponics. *Plant Biology* 11(3):339–350.

Sanjay SS, Pandey AC, Singh M, Prasad MS (2015) Effects of functionalized ZnO nanoparticles on the phytohormones: Growth and development of Solanum melongena L. (brinjal) plant. *World Journal of Pharmaceutical Research* 5:1990–2009.

Seabra AB, Rai M, Durán N (2014) Nano carriers for nitric oxide delivery and its potential applications in plant physiological process: A mini review. *Journal of Plant Biochemistry and Biotechnology* 23(1):1–10.

Sekhon B (2003) Chelates for micronutrient nutrition among crops. *Resonance* 8(7):46–53.

Shafi U, Mumtaz R, García-Nieto J, Hassan SA, Zaidi SAR, Iqbal N (2019) Precision agriculture techniques and practices: From considerations to applications. *Sensors* 19(17):3796.

Shahi MP, Kumari P, Mahobiya D, Shahi SK (2021) Nano-bioremediation of environmental contaminants: Applications, challenges, and future prospects. in Vineet Kumar, Gaurav Saxena, Maulin P. Shah (eds) *Bioremediation for Environmental Sustainability*, Elsevier, 83–98.

Shaibur M, Kawai S (2017) Effectiveness of Fe3+-citrate and Fe3+-EDTA in ameliorating arsenic-toxicity (chlorosis) in hydroponic rice. *Journal of Plant Nutrition 40*(11):1515–1525.

Sharma B, Lakra U, Sharma R, Sharma SR (2022) Chapter 16 - A comprehensive review on nanopesticides and nanofertilizers—A boon for agriculture. In Ghorbanpour M, MA Shahid (eds) *Nano-enabled Agrochemicals in Agriculture*. Academic Press, pp. 273–290. doi:10.1016/B978-0-323-91009-5.00026-4.

Sheykhbaglou R, Sedghi M, Shishevan MT, Sharifi RS (2010) Effects of nano-iron oxide particles on agronomic traits of soybean. *Notulae Scientia Biologicae* 2(2):112–113.

Singh A, Singh N, Afzal S, Singh T, Hussain I (2018) Zinc oxide nanoparticles: A review of their biological synthesis, antimicrobial activity, uptake, translocation and biotransformation in plants. *Journal of Materials Science* 53(1):185–201.

Singh R, Behera M, Kumar S (2020) Nano-bioremediation: An innovative remediation technology for treatment and management of contaminated sites. In *Bioremediation of Industrial Waste for Environmental Safety Volume II: Biological Agents and Methods for Industrial Waste Management*, pp. 165–182.

Singh RP (2021) Recent trends, prospects, and challenges of nanobiosensors in agriculture In *Biosensors in Agriculture: Recent Trends and Future Perspectives*, pp. 3–13.

Singh U, Praharaj C, Chaturvedi S, Bohra A (2016) Biofortification: Introduction, approaches, limitations, and challenges. In *Biofortification of Food Crops*, pp. 3–18.

Squire HJ, Tomatz S, Voke E, González-Grandío E, Landry M (2023) The emerging role of nanotechnology in plant genetic engineering. *Nature Reviews Bioengineering 1*:314–328.

Swain PS, Rajendran D, Rao S, Dominic G (2015) Preparation and effects of Nano mineral particle feeding in livestock: A review. *Veterinary World* 8(7):888.

Székely Á, Poór P, Bagi I, Csiszár J, Gémes K, Horváth F, Tari I (2011) Effect of EDTA on the growth and copper accumulation of sweet sorghum and Sudangrass seedlings. *Acta Biologica Szeged* 55:159–164.

Thakkar KN, Mhatre SS, Parikh RY (2010) Biological synthesis of metallic nanoparticles. *Nanomedicine: Nanotechnology, Biology and Medicine* 6(2):257–262.

Tirani MM, Haghjou MM, Sulieman S, Colville L, Ismaili A (2019) The influence of organic and inorganic chelators on the toxicity of bulk and nanoparticles of zinc oxide during germination and seedling growth of Nicotiana tabacum L. *Plant Biosystems – An International Journal Dealing with all Aspects of Plant Biology 153*(3):436–449. doi:10.1080/11263504.2018.1498402

Tripathi DK, Singh S, Singh S, Mishra S, Chauhan D, Dubey N (2015) Micronutrients and their diverse role in agricultural crops: Advances and future prospective. *Acta Physiologiae Plantarum* 37(7):139.

ul Ain Q et al. (2023) Use of nano-fertilizers to improve the nutrient use efficiencies in plants. In Tariq Aftab, Khalid Rehman Hakeem (eds), *Sustainable Plant Nutrition*. Academic Press, pp. 299–321.

van Doorn WG (2012) Water relations of cut flowers: An update. *Horticultural Reviews* 40:55–106.

Vaseem M, Umar A, Hahn Y-B (2010) *ZnO Nanoparticles: Growth, Properties, and Applications* (Vol 5). American Scientific Publishers, New York.

Wang C et al. (2020) Chitosan-modified halloysite nanotubes as a controlled-release nanocarrier for nitrogen delivery. *Applied Clay Science* 198:105802.

Wang P et al. (2013) Fate of ZnO nanoparticles in soils and cowpea (Vigna unguiculata). *Environmental Science and Technology* 47(23):13822–13830.

Warheit DB, Sayes CM, Reed KL, Swain KA (2008) Health effects related to nanoparticle exposures: Environmental, health and safety considerations for assessing hazards and risks. *Pharmacology and Therapeutics* 120(1):35–42.

White PJ, Broadley MR (2005) Biofortifying crops with essential mineral elements. *Trends in Plant Science* 10(12):586–593.

Wissuwa M, Ismail AM, Graham RD (2008) Rice grain zinc concentrations as affected by genotype, native soil-zinc availability, and zinc fertilization. *Plant and Soil* 306(1–2):37–48.

Wójcik M, Skórzyńska-Polit E, Tukiendorf A (2006) Organic acids accumulation and antioxidant enzyme activities in Thlaspi caerulescens under Zn and Cd stress. *Plant Growth Regulation* 48(2):145–155.

Yan Y, Zhu X, Yu Y, Li C, Zhang Z, Wang F (2022) Nanotechnology strategies for plant genetic engineering. *Advanced Materials* 34(7):2106945.

Yang L, Watts DJ (2005) Particle surface characteristics may play an important role in phytotoxicity of alumina nanoparticles. *Toxicology Letters* 158(2):122–132.

Yousefzadeh S, Sabaghnia N (2016) Nano-iron fertilizer effects on some plant traits of dragonhead (Dracocephalum moldavica L.) under different sowing densities. *Acta Agriculturae Slovenica* 107(2):429–437.

Zimmermann MB, Hurrell RF (2007) Nutritional iron deficiency. *The Lancet* 370(9586):511–520.

Zuo Y, Zhang F (2011) Soil and crop management strategies to prevent iron deficiency in crops. *Plant and Soil* 339(1–2):83–95.

15 The Exploitation of Marine Seaweeds for Nanoparticle Biosynthesis and Applications

Pragya Tiwari, Muhammad Adil, Subir Kumar Bose, Jen-Tsung Chen, and Laurent Dufosse

15.1 SEAWEEDS AS THE BIOFACTORIES FOR NANOPARTICLE BIOSYNTHESIS

The recent era has witnessed a convergence in the integration of nanotechnologies, green chemistry, and bio-based technologies, with the objective of developing clean manufacturing technologies. New innovative routes are undertaken to synthesize novel nanomaterials. The bio-based synthesis of nanoparticles (NPs) presents distinct advantages, in terms of cost-effective, non-toxic, and high yields, compared with the conventional chemical synthesis procedures (Bansal et al. 2011, Makarov et al. 2014, Tiwari et al. 2022). In this direction, biological organisms are studied for their potential to synthesize NPs, with key literature studies discussing the cost-effective plant-based and microbial biosynthesis of NPs. The well-studied potential of prokaryotes for metal ion reduction is attributed to the presence of anionic functional groups, sugar hydroxyl groups, and amide groups in peptides found on the cell wall (Shaw 1990, Jiang et al. 2004), while the eukaryotes (fungi, algae, plants, etc.) are efficient reducers of inorganic metal ions to metal NPs, attributed to the functional moieties in the extracts (Prasad et al. 2011, Gracia-Sevillano et al. 2015). Nanometer-scale materials showcase unique physical, optical, thermal, biological, and chemical properties due to their size, shape, and small size distribution (Fawcett et al. 2017). The biological synthesis of NPs offers a more sustainable and feasible method and employs the reducing properties of compounds present in biological organisms. During synthesis, the biological compounds cover the NPs' surface, and this leads to the improved properties of green NPs (Javed et al. 2020) and prevents contamination of green synthesized NPs with toxic by-products. However, substantial effort and time need to be put into the optimization of a biological method. Studies have documented the green synthesis of NPs; some studies discuss parameters involved in the green synthesis of NPs (Rana et al. 2020), while others emphasize specifically a single type of NPs (Xu et al. 2021).

Marine organisms constitute approximately 70% of the earth's surface and remain largely unexplored for their socio-economic applications. While plants of marine origin demonstrate remarkable properties as producers of NPs, marine seaweeds are less investigated and studied among major producers of NPs. In a quest for natural product–mediated drug discovery, several marine algae are prolific producers of different types of NPs. Seaweeds, a particular group of interesting marine algae, have been consumed as nutritional supplements and in healthcare by coastal communities; however, seaweeds as a source of potential NPs, are an area that remains unexplored. Seaweeds are defined as multicellular, macroscopic, benthic plants, regarded as the only source of commercially important substances like iodine, mannitol, sodium alginate, agar, and carrageenan, including others (Holdt and Kraan 2011). In addition, seaweeds also demonstrate pharmacological activities, including antiviral, antifungal, anticancer, antimicrobial, etc., highlighting the emerging

 DOI: 10.1201/b23308-15

importance of seaweeds in the socio-economic context (Pereira et al. 2012). The classification of seaweeds is based on the occurrence and distribution of pigments. They are classified into Phaeophyta (brown algae), Chlorophyta (green algae), and Rhodophyta (red algae). A marine-based habitat has been reported for red and brown algae, while some red algal species are found in low light intensities (Fawcett et al. 2017). In recent times, studies have highlighted the bioactive prospects of crude extracts and compounds from seaweeds as potent pharmacological agents; however, very few have been clinically validated and commercialized. Alternative therapies and medical treatments have highlighted the bioactive potential of seaweed-derived compounds (Chanda et al. 2010, Wada et al. 2011, Mohamed et al. 2012). Studies have documented the inhibitory and anti-inflammatory properties of extracts from seaweeds in treating cancer and other health conditions (Khan et al. 2008, Namwar et al. 2012). In addition, antifouling properties of secondary metabolites from seaweeds are also reported (Da Gama et al. 2002). Seaweed as a potential source of gold (Au) and silver (Ag) NPs defines new paradigms in the biological origin and production of NPs. The NPs biosynthesized from marine seaweeds display significant prospects in diverse biological applications. Hence, the chapter explores the potential and recent prospects in marine-derived biosynthesis of NPs and their potential applications in socio-economic applications.

15.2 NANOPARTICLE BIOSYNTHESIS USING RED, GREEN, AND BROWN SEAWEEDS

Seaweed-mediated biosynthesis has been explored in the recent era and is regarded as an eco-friendly and non-toxic approach, resulting in NPs with different sizes, physicochemical properties, and compositions. The cell wall in seaweeds comprises polysaccharides that consist of hydrophilic groups, namely, sulfate, carboxyl, and hydroxyl groups (Venkatesan et al. 2016). The presence of biomolecules, like proteins, enzymes, and polysaccharides, namely, alginate, fucoidan, agar, carrageenan, and laminarin (Barros et al. 2013), has attracted attention from researchers globally, and there are applications in the food sector, biomedicine, and as a source of bioenergy (Panchanathan and Se-Kwon 2015, Venkatesan et al. 2015).

15.2.1 Biosynthesis of NPs from Rhodophyta (Red Seaweeds)

The eukaryotic algae Rhodophyta are a source of pigments, namely, carotene, phycoerythrin, lutein, phycobilins, etc., with more than 6000 species identified. The biosynthesis of AgNPs employing extract from the seaweed *Acanthophora spicifera* was reported (Kumar et al. 2012), and they inhibited exopolysaccharide production by the pathogens *Salmonella typhii*, *Staphylococcus aureus*, *Escherichia coli*, and *Shigella flexneri*. Kumar and coworkers (2013) discussed AgNP synthesis by *Gracilaria corticata*, and it inhibited the growth of *Candida glabrata* and *C. albicans* (Kumar et al. 2013). El-Kassas and Attis (2014) reported the biosynthesis of AgNPs from *Pterocladiella capillacea* and their cytotoxicity in the HepG2 cell line. The cytotoxic effect of AuNPs from *Corallina officinalis* was documented (El-Kassas and El-Sheekh 2014). Moreover, Vivek and coworkers (2011) studied an extract of the red seaweed *Gelidiella acerosa* for the biosynthesis of AgNPs. The AgNPs showed potent antifungal activity against *Trichoderma reesei*, *Mucor indicus*, *Humicola insolens*, and *Fusarium dimerum* (Vivek et al. 2011). AuNPs synthesized from an extract of *Galaxaura elongata* showed promising antibacterial functions against methicillin-resistant *S. aureus* (MRSA), *Pseudomonas aeruginosa*, *Klebsiella pneumoniae*, and *E. coli* (Abdel-Raouf et al. 2013).

15.2.2 Biosynthesis of NPs from Chlorophyta (Green Seaweeds)

Chlorophyta or green seaweeds are classified into the Plantae kingdom and found mainly in freshwater and the sea. Studies have documented the biosynthesis of NPs from green seaweeds and their potential applications. The biosynthesis of AgNPs by *Ulva lactuca* was achieved and used for the photo-catalytic degradation of methyl orange dye (Kumar et al. 2013). The AgNPs were used for the degradation of dyes and organic compounds in an eco-friendly way. Raja et al. (2012) studied the AgNPs synthesized from *U. lactuca* for antibacterial effect against *E. coli*, *S. aureus*, and *Ps. aeruginosa* (Raja et al. 2012). Studies have also documented the agricultural and biomedical applications of AgNPs synthesized from extracts (fresh and dry) of *Codium capitatum* (Kannan et al. 2013). In addition, AgNPs from *U. lactuca* and *Caulerpa scalpelliformis* were synthesized and evaluated for mosquitocidal activity (Murugan et al. 2015a, Murugan et al. 2015b).

15.2.3 Biosynthesis of NPs from Phaeophyta (Brown Seaweeds)

Brown algae are classified in the Phaeophyta phylum and include approximately 2000 species. Singaravelu et al. (2007) discussed the AuNP biosynthesis from *Sargassum wightii* (Singaravelu et al. 2007). AgNPs were synthesized from *Sargassum longifolium* and tested (at different concentrations) against Hep-2 cancer cell lines (Devi et al. 2013). Another study discussed the synthesis and antibacterial activity of AgNPs from *S. wightii*, showing potent efficacy against *K. pneumoniae*, *S. aureus*, and *Salmonella typhi* (Shanmugam et al. 2013). Kumar and coworkers reported AgNP biosynthesis from an aqueous extract of *Sargassum ilicifolium*, and it demonstrated inhibition of pathogenic strains of *S. typhi*, *Vibrio cholerae*, *K. pneumoniae*, and *E. coli*. Furthermore, biofunctionalized AuNPs were synthesized via 1-cyclopentyl-4(3-cyclopentyl propyl) dodecane and *Sargassum myriocystum* (Stalin Dhas et al. 2012). Thangaraju and coworkers (2012) employed a crude extract of *Sargassum polycystum* for AgNP synthesis. In another study, a single-step procedure was used for *Turbinaria conoides*–mediated AuNP synthesis (Rajeshkumar et al. 2013). A similar study showed that *Turbinaria conoides* (aqueous extract) was used for the synthesis of AgNPs and AuNP, that showed potent antibiofilm activity against bacteria (Vijayan et al. 2014). The AgNPs derived from *Sargassum cinereum* showed potent antimicrobial effects against *S. typhi*, *Enterobacter aerogenes*, and *Proteus vulgaris* (Namvar et al. 2015). Moreover, silver chloride ($AgCl_2$) NPs (from *Sargassum plagiophyllum*) showed an antibacterial effect (Dhas et al. 2014), and AuNPs (from *Sargassum swartzii*) displayed a cytotoxic effect (Mohandass et al. 2013).

At present, the exact mechanisms of NP formation by seaweeds are practically unknown; however, studies have documented that multiple biomolecules in seaweed cell walls catalyze the reduction of metal salts into metal oxide NPs (Mahdavi et al. 2013), and large biomolecules act as surfactants for growth of NPs (Stalin Dhas et al. 2012). In addition, biomolecules found in seaweeds affect the particle morphology, size, and physicochemical properties, including others, and these determine the NP quality. At present, parameters like pH of the solution, reaction time, seaweed concentration, temperature, etc. are being studied for NP formation due to their effect on the properties and quality of NPs (Ai et al. 2011) (Figure 15.1).

15.3 TYPES OF NANOPARTICLES DERIVED FROM MARINE SEAWEEDS

Basically, three types of NPs are synthesized from marine seaweeds: carbon-based (graphene and nanotubes), organic (polymers and liposomes), and inorganic (metal oxides, quantum dots, ceramics) (Mauricio et al. 2018, Matteis and Rinaldi 2018).

15.3.1 Carbon-based Nanoparticles

These NPs possess biological compatibility and low toxicity; therefore, carbon-based NPs like graphene oxide, graphite, fullerenes, etc. are widely used in biomedical applications, therapies, and diagnostics (Yadav et al. 2022). The key properties of carbon-based NPs include- conduction, semi-conduction, dielectric, and superconduction, among others (Safiyev and Abdullayeva 2021) (Figure 15.2).

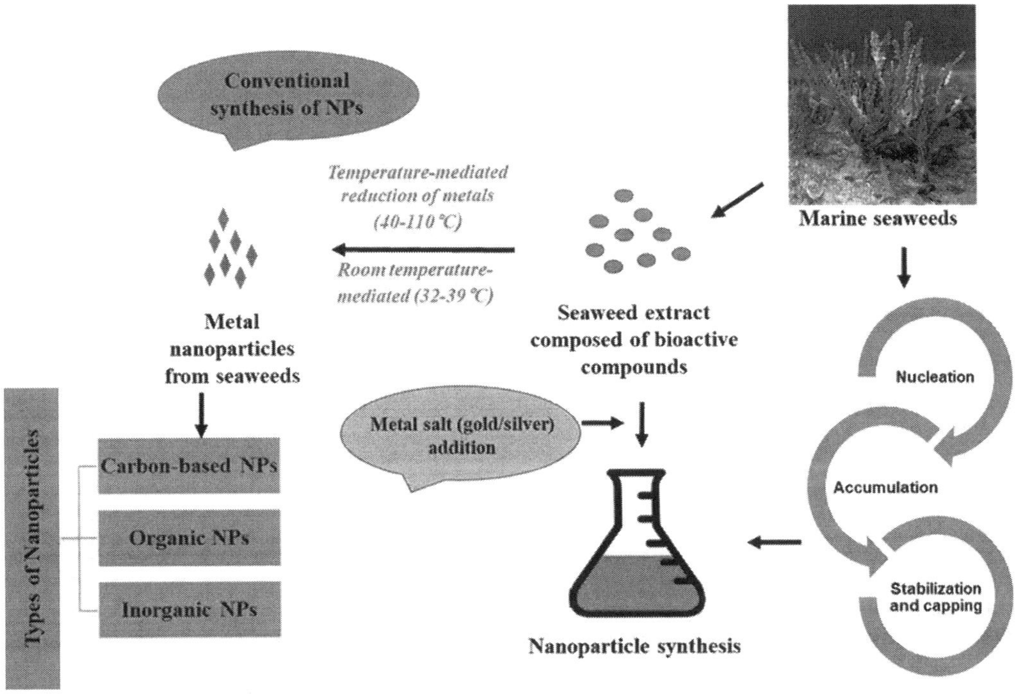

FIGURE 15.1 Conventional and bio-based methods in diverse nanoparticle synthesis.

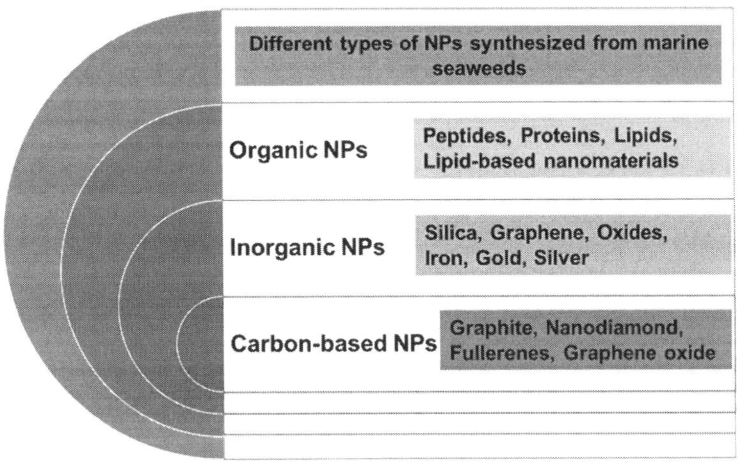

FIGURE 15.2 Different types of nanoparticles synthesized from seaweeds.

15.3.2 Organic Nanoparticles

These NPs are commonly used in biomedical applications, are effective, and may be injected into a specific location in the body, known as targeted drug delivery (Ealia et al. 2017). Organic NPs include lipids, peptides, and proteins, and nanomaterials including liposomes, proteins, or lipid-based nanomaterials are classified in this category (Yadav et al. 2022).

15.3.3 Inorganic Nanoparticles

Inorganic NPs are less cytotoxic, biocompatible, and nonbiodegradable, and therefore safer than organic NPs. These can be easily molded into different shapes due to their distinct electrical and optical properties. Key examples of inorganic NPs include magnesium (Mg), silica (SiO_2), silver (Ag), oxides, graphene, etc., among others (Yadav et al. 2022).

15.4 POTENTIAL APPLICATIONS OF SEAWEED-DERIVED NANOPARTICLES

Marine seaweeds are an attractive source of "high-value" substances necessary for human health and include carbohydrates, vitamins, fibers, proteins, and micro- and macronutrients. Seaweed is considered a single-cell protein (SCP) and is easily available in the sea; these seaweeds have lots of medicinal properties and are a good resource for marine bionanotechnology (Ramkumar et al. 2016). Scientists are exploring various properties of marine seaweeds, including NP-based medicinal and other properties such as antimicrobial, antitumor, antidiabetic, antifouling, bio-fabrication, etc. (Singaravelu et al. 2007, Dhas et al. 2016, Manjunath et al. 2017).

15.4.1 ANTIMICROBIAL

An antimicrobial is frequently used as a therapeutic agent that restricts microbial growth and prevents infection. For example, antibiotics and antifungals are used against bacteria and fungi. Modern new-generation antibiotics have many side effects and damage the gut microbial flora. The increasing prevalence of multidrug-resistant microbes is raising key health concerns, necessitating urgent action (Tiwari et al. 2021, Tiwari et al. 2023). Considering the multidrug resistance and harmful effects of microbes, scientists are exploring nanoparticle synthesis and its beneficial effects from diverse biological sources. Green synthesis of spherical AgNPs from the marine seagrass *Syringodium isoetifolium* has been documented and analyzed by UV–Vis, dynamic light scattering (DLS), X-ray diffraction (XRD), and atomic force microscopy (AFM) instruments for the identification and characterization of polydispersed (2–50 nm) spherical AgNPs (Ahila et al. 2016). AgNPs synthesized through marine mangrove, *Rhizophora mucronata*, in combination with itraconazole and fluconazole (antifungals) displayed potent activity against human fungal pathogens, including *Candida albicans*, *Aspergillus fumigatus*, *A. flavus*, and *C. neoformans*, compared with the antifungal drugs individually using disk diffusion assay (Singh et al. 2013, Yosri et al. 2021).

Antibacterial activity against *S. aureus*, *Ps. aeruginosa*, and *E. coli* was shown by AgNPs synthesized from *U. lactuca* (Raja et al. 2012). Moreover, metallic AgNPs were produced (Rajesh et al. 2012) using crude ethyl acetate extract from the marine seaweed *Ulva fasciata*. AgNPs restricted *Xanthomonas campestris* pv. *malvacearum*, displaying a maximum inhibitory concentration (MIC) of 40.0 ± 5.77 μg/ml. Extracts of the green seaweed *Urospora* spp. were employed for AgNP biosynthesis, and the antibacterial effect against *E. coli*, *Ps. aeruginosa*, and *S. aureus* was discussed (Suriya et al. 2012). Fresh and dry extracts of *Codium capitatum* were also used in the biosynthesis of AgNPs (Kannan et al. 2013). Yousefzadi et al. (2014) synthesized AgNPs using *Enteromorpha flexuosa* (a green alga) and experimentally reported potent antimicrobial activity.

15.4.2 ANTITUMOR

Nanoparticles have been at the forefront of cancer treatment, attributed to their distinct properties. The unique properties of NPs, including biocompatibility, enhanced permeability, and improved stability, among others, have aided the development of NP-based therapeutics in recent years. In a search among metal NPs, AgNPs are excellent examples of the best anticancer or antitumor therapeutic agents, found to be potent and effective against multiple cancers. AgNPs' anticancer activity has already been reported against various types of cancer, including cervical, ovarian, breast, lung, and pancreatic ductal adenocarcinoma, etc., and the NPs display different anticancer activities. Marine seaweed–derived NPs are potential alternatives to plant-derived NPs (Raja et al. 2012) in terms of increased cytotoxic effects via enhanced endocytosis. The biosynthesis of potent anticancer AgNPs was achieved using the red seaweed *Pterocladiella capillacea* (El-Kassas and Attia 2014). Another AuNP-based anticancer drug for breast cancer was obtained and synthesized from the extract of the red seaweed *Corallina officinalis* (El-Kassas and El-Sheekh 2014). Cubical-shaped AgNPs (30 nm) were derived from brown seaweed (*Sargassum longifolium*). Another spherical-shaped (50–100 nm) AgNP having strong anticancer activity was synthesized by the seaweed *Sargassum polycystum* (Thangaraju et al. 2012, Devi et al. 2013). Spherical AuNPs (53 nm in size) were synthesized from a brown marine alga, *Padina gymnospora*. Fucoxanthin or flavonoids have good anticancer activity (Singh et al. 2013). Seaweed (*Sargassum muticum*)-derived 18-nm cubic magnetic iron–oxide–sulfated polysaccharides, hydroxyl, and aldehyde groups have anticancer properties (Mahadavi et al. 2013, Namvar et al. 2014). *Sargassum muticum* spherical AuNPs and other biomolecules showed anticancer activity (Namvar et al. 2014, Namvar et al. 2015). *Sargassum swartzii* AgNP, $AuCl_4$, and alcohol, carboxylic, and amide I groups display anticancer functions (Dhas et al. 2014). A study by Inbakandan et al. (2016) discussed *Haliclona exigua* extract–mediated synthesis of silver nanocolloids and their activity against oral biofilm bacteria and cancer cell lines.

15.4.3 Antidiabetic

Nanotechnology has significantly contributed to diabetes treatment in many ways, including therapeutic delivery, the creation of glucose biosensors to monitor blood glucose level changes, and smart molecular imaging to visualize β -cells (Mukhopadhyay and Prosenjit 2018). Bhardwaj and coworkers (2020) discussed the key contributions of bio-based nanotechnological methods in providing key insights into treatment approaches in diabetes, and biochemical and histological parameters, including other benefits (Bhardwaj et al. 2020). The inhibition of α-amylase and α-glucosidase enzymes that catalyze starch hydrolysis in the intestine highlights a prospective strategy for the regulation of hyperglycemia (Lordan et al. 2013). Deepak et al. (2018) studied the antidiabetic properties of bio-based NPs by assessing the α-glucosidase and α-amylase inhibitory function. The biosynthesis of AuNPs from *Sargassum swartzii* (a marine brown alga) and its positive regulation of glucose levels and insulin resistance were studied (Dhas et al. 2016). Other studies showed beneficial effects of AuNPs, namely, a decrease in glycosylated hemoglobin, hemoglobin, fasting blood glucose, and serum insulin levels, in diabetic rats compared with control (Singaravelu et al. 2007, Dhas et al. 2016).

15.4.4 Antifouling

The attachment of unwanted marine organisms is prevented by the use of an antifouling system (a coating or surface treatment device) and acts via the secretion of bioactive substances for the prevention of seaweeds, slime, and barnacles from attaching to the ships. The antifouling coatings (comprising organic and inorganic biocides) restrict the organisms (Yebra 2004). Recent advances in nano-based approaches have proved to be highly effective in developing eco-friendly and less-toxic antifouling agents (Kumar et al. 2021).

The "nano-functionalization" technique facilitates the creation of effective antimicrobials that inhibit bacterial colonization and the formation of biofilms via coating and embedding nanomaterials (Lellouche 2012). Kumar et al. (2021) demonstrated that seaweed-based NPs have bactericidal, fungicidal, and algicidal properties. AgNPs and AuNPs interact with microbial biofilms and exert an antimicrobial action. Moreover, the characteristics and properties of NPs, namely, charge, surface chemistry, and biofilm hydrophobicity, promote the NPs' interaction with the proteins, lipopolysaccharide (LPS), and lipids present in bacterial membranes, enabling them to access biofilms, interact with cellular components, and exert their antimicrobial action (Ikuma et al. 2015, Joshi et al. 2020). In this direction, extracts from marine seaweeds demonstrate potent antibiofilm activity; in particularly, the sulfated polysaccharides are key antibacterials and anti-biofilm agents. A key example is fucoidan (a sulfated polysaccharide) from *Fucus vesiculosus*, which restricts biofilm formation in dental plaque and cell growth in *Streptococcus sobrinus* and *S. mutans* (Jun et al. 2018). Highly potent antifouling agents have been generated via CRISPR/Cas9-based genome editing in algae (Nag et al. 2021).

15.4.5 Other Applications

Bio-fabrication of AuNPs from the marine endophytic fungus *Penicillium citrinum* (Manjunath et al. 2017), antioxidant production, and bio-products from marine yeasts, etc., have been attempted. Zirconia (ZrO_2) nanoparticles were synthesized using *Sargassum wightii* seaweeds and characterized in terms of mechanical strength, biocompatibility, and anti-corrosive properties. The NPs find broad applications in the biomedical sector (implants, dentistry, biosensors, etc.), refractories, ceramics, and others (Ye and Shi 2018). *Sargassum wightii* was successfully used as a reducing and stabilizing agent for the green synthesis of ZrO_2 NPs (Kumaresan et al. 2018).

15.5 CONCLUSIONS AND FUTURE PERSPECTIVES

The global synthesis and utilization of NPs have witnessed enormous popularity, attributed to their unique physicochemical and biological characteristics. The usual drawback associated with the traditional sources of nanoparticles is the generation of toxic metabolites, which can be effectively avoided through the biosynthesis of nanoparticles using marine seaweeds. Currently, the production of nanoparticles from biogenic sources has gained considerable momentum (Fawcett et al. 2017). The biological techniques of nanoparticle production are usually preferred on account of eco-safety, cost-effectiveness, a single-step synthetic process, and the convenience of large-scale manufacturing (Cassani et al. 2021). Eco-friendly and reliable manufacturing of nanoparticles represents a significant aspect of nanotechnological research and development (Cassani et al. 2021). Green synthesis of nanoparticles from marine sources has gained considerable attention in recent decades. Marine seaweeds are considered relatively under-explored nano-factories for the manufacturing of nanomaterials (González-Ballesteros and Rodríguez-Argüelles 2020). Several beneficial substances of potential biomedical and industrial applications, including amino acids, peptides, polysaccharides such as chitosan, agar, carrageenan, and alginate, chitin, and silica have been previously isolated from marine organisms (Asmathunisha and Kathiresan 2013, Singh et al. 2015). Contrary to traditional synthetic processes, marine seaweeds can provide non-toxic nanoparticles. Several bioactive products, including proteins and polysaccharides, that are found in the cell membrane of marine algae can be efficiently used as capping and reducing agents during the biosynthesis of nanoparticles. The eco-friendly, stable, and non-toxic nanomaterials obtained from marine seaweeds can be effectively used as coating agents in biomedical, pharmaceutical, marine, and aquaculture systems (Vijayan et al. 2016). Fucoidan, carrageenan, chitosan, and alginate are among the well-known seaweed polysaccharides that are quite frequently exploited for nanoparticle biosynthesis. Being non-toxic, biodegradable, and biocompatible, the anionic and chitosan-based nanoparticles impart protection to encapsulated materials and facilitate their sustained release in living systems (Venkatesan et al. 2016). The sustained drug-releasing potential, biocompatibility, and stability of seaweed-derived nanoparticles have been demonstrated by many *in vivo* and *in vitro* studies during the last few decades. Sustained-release formulations of anticancer drugs and peptides can be efficiently delivered by means of NPs derived from seaweed polysaccharides (Venkatesan et al. 2016).

The enormous potential of the marine ecosystem to provide diverse bioactive substances and metabolic products has been partially exploited. So far, a number of *in vivo* and *in vitro* studies have primarily targeted fucoidan, carrageenan, and alginate for nanoparticle production (Venkatesan et al. 2016). Accordingly, green chemistry–based biological synthetic pathways should be explored through future research for the manufacturing of metals and metal oxide nanoparticles using marine seaweeds (Fawcett et al. 2017). Sufficient amounts of nanoparticles can be generated using natural polymeric substances by means of polyelectrolyte complexation and ionic gelation techniques (Venkatesan et al. 2016). Nevertheless, the temperature, pH, speed, time of addition, stirring, and molecular weight should be taken into consideration for converting natural polymers into nanoparticles (Venkatesan et al. 2016). The cellular and molecular mechanisms of nanoparticle biosynthesis from marine seaweeds necessitate an in-depth explication (González-Ballesteros and Rodríguez-Argüelles 2020). The bioavailability and effectiveness of marine seaweed–derived nanoparticles should be explored using appropriate *in vitro* assays and *in vivo* models (González-Ballesteros and Rodríguez-Argüelles 2020). Besides, the safety profile of seaweed-derived nanoparticles should be thoroughly investigated by means of carefully conducted nanotoxicity studies. Similarly, risk assessment focusing on the manufacturing, handling, transportation, storage, utilization, and disposal of seaweed-derived nanoparticles is also recommended (Cassani et al. 2021). There is a need to scale up the previously established biosynthetic methods for harnessing the nanoparticle-producing capacity of marine seaweeds at the industrial level (Pallavi et al. 2015). Another limitation is the lack of appropriate methods for the purification of biosynthesized nanoparticles. Multidisciplinary and collaborative research projects should be devised for developing rapid, accurate, and cost-effective techniques for nanoparticle biosynthesis using marine seaweeds.

ACKNOWLEDGMENTS

The authors thank their respective organizations for their encouragement and support. No conflict of interest was declared.

REFERENCES

Abdel-Raouf, N, NM Al-Enazi, and IBM Ibraheem. 2013. "Green biosynthesis of gold nanoparticles using *Galaxaura elongata* and characterization of their antibacterial activity." *Arabian Journal of Chemistry*. doi:10.1016/j.arabjc.2013.11.044.

Ahila, NK, VS Ramkumar, S Prakash, B Manikandan, J Ravindran, PK Dhanalakshmi, and E Kannapiran. 2016. "Synthesis of stable nanosilver particles (AgNPs) by the proteins of seagrass *Syringodium isoetifolium* and its biomedicinal properties." *Biomedicine and Pharmacotherapy* 84:60–70.

Ai, J, E Biazar, M Jafarpour et al. 2011. "Nanotoxicology and nanoparticle safety in biomedical designs." *International Journal of Nanomedicine* 6:1117–1127.

Asmathunisha, N, and K Kathiresan. 2013. "A review on biosynthesis of nanoparticles by marine organisms." *Colloids and Surfaces B: Biointerfaces* 103:283–287.

Bansal, V, R Ramanathan, and SK Bhargava. 2011. "Fungus-mediated biological approaches towards "green" synthesis of oxide nanomaterials." *Australian Journal of Chemistry* 64(3):279–293.

Barros, FCN, DC da Silva, VG Sombra et al. 2013. "Structural characterization of polysaccharide obtained from red seaweed *Gracilaria caudata* (J Agardh)." *Carbohydrate Polymers* 92(1):598–603.

Bhardwaj, M, P Yadav, S Dalal, and SK Kataria. 2020. "A review on ameliorative green nanotechnological approaches in diabetes management." *Biomedicine and Pharmacotherapy* 127:110198.

Cassani, L, NE Marcovich, and A Gomez-Zavaglia. 2021. "Seaweed bioactive compounds: Promising and safe inputs for the green synthesis of metal nanoparticles in the food industry." *Critical Reviews in Food Science and Nutrition*: 1–24.

Chanda, S, R Dave, M Kaneria, and K Nagani. 2010. "Seaweeds: A novel, untapped source of drugs from sea to combat infectious diseases." In A. Mendez-valas Ed., *Current Research, Technology, and Education Topics in Applied Microbiology and Microbial Biotechnology*, Formatex Research Center, Badazoj, pp. 473–480.

Da Gama, BAP, RC Pereira, AGV Carvalho, R Coutinho. 2002. "The effects of seaweed secondary metabolites on biofouling." *Biofouling* 18(1):13–20.

Deepak, P, V Amutha, R Birundha, R Sowmiya, C Kamaraj, V Balasubramanian, et al. 2018. "Facile green synthesis of nanoparticles from brown seaweed *Sargassum wightii* and its biological application potential." *Advances in Natural Sciences: Nanoscience and Nanotechnology* 9(3):035019.

De Matteis, V, and R Rinaldi. 2018. "Toxicity assessment in the nanoparticle era." *Cellular and Molecular Toxicology of Nanoparticles*, 1048:1–19.

Devi, JS, BV Bhimba, and DM Peter. 2013. "Production of biogenic silver nanoparticles using *Sargassum longifolium* and its application." *Indian Journal of Marine Sciences* 42:125–130.

Dhas, TS, VG Kumar, V Karthick, K Govindaraju, and T Shankara Narayana. 2014. "Biosynthesis of gold nanoparticles using *Sargassum swartzii* and its cytotoxicity effect on HeLa cells." *Spectrochimica Acta, Part A: Molecular and Biomolecular Spectroscopy* 133:102–106.

Dhas, TS, VG Kumar, V Karthick, K Vasanth, G Singaravelu, and K Govindaraju. 2016. "Effect of biosynthesized gold nanoparticles by *Sargassum swartzii* in alloxan-induced diabetic rats." *Enzyme and Microbial Technology* 95:100–106.

Ealia, SA, MP Saravanakumar. 2017. "A review on the classification, characterization, synthesis of nanoparticles and their application." IOP Conference Series. *Materials Science and Engineering* 263:Article ID 032019.

El-Kassas, HY, and AA Attia. 2014. "Bactericidal application and cytotoxic activity of biosynthesized silver nanoparticles with an extract of the red seaweed *Pterocladiella capillacea* on the HepG2 cell line." *Asian Pacific Journal of Cancer Prevention* 15(3):1299–1306.

El-Kassas, HY, and MM El-Sheekh. 2014. "Cytotoxic activity of biosynthesized gold nanoparticles with an extract of the red seaweed *Corallina officinalis* on the MCF-7 human breast cancer cell line." *Asian Pacific Journal of Cancer Prevention* 15(10):4311–4317.

Fawcett, D, JJ Verduin, M Shah, SB Sharma, and GEJ Poinern. 2017. "A review of current research into the biogenic synthesis of metal and metal oxide nanoparticles via marine algae and seagrasses." *Journal of Nanoscience* 2017:1–15.

González-Ballesteros, N, and MC Rodríguez-Argüelles. 2020. "Seaweeds: a promising bionanofactory for ecofriendly synthesis of gold and silver nanoparticles." In *Sustainable seaweed technologies* Elsevier, pp. 507–541.

Gracia-Sevillano, MA, T Garcia-Barrera, and JL Gomex-Ariza. 2015. "Environmental metabolomics: Biological markers for metal toxicity." *Electrophoresis* 36(18):2348–2365.

Holdt, SL, and S Kraan. 2011. "Bioactive compounds in seaweed: Functional food applications and legislation." *Journal of Applied Phycology* 23(3):543–597.

Ikuma, K, AW Decho, and BL Lau. 2015. "When nanoparticles meet biofilms-interactions guiding the environmental fate and accumulation of nanoparticles." *Frontiers in Microbiology* 6:591.

Inbakandan, D, C Kumar, M Bavanilatha, DN Ravindra, R Kirubagaran, and SA Khan. 2016. "Ultrasonic-assisted green synthesis of flower-like silver nano colloids using marine sponge extract and its effect on oral biofilm bacteria and oral cancer cell lines." *Microbial Pathogenesis* 99:135–141.

Javed, R, M Zia, S Naz, SO Aisida, NU Ain, Q Ao. 2020. "Role of capping agents in the application of nanoparticles in biomedicine and environmental remediation: Recent trends and future prospects." *Journal of Nanobiotechnology* 18(1):172.

Jiang, W, A Saxena, B Song, BB Ward, TJ Beveridge, and SC Myneni. 2004. "Elucidation of functional groups on gram-positive and gram-negative bacterial surfaces using infrared spectroscopy." *Langmuir* 20(26):11433–11442.

Joshi, AS, P Singh, and I Mijakovic. 2020. "Interactions of Gold and silver nanoparticles with bacterial biofilms: Molecular interactions behind inhibition and resistance." *International Journal of Molecular Sciences* 21(20):7658.

Jun, JY, MJ Jung, IH Jeong, K Yamazaki, Y Kawai, and BM Kim. 2018. "Antimicrobial and antibiofilm activities of sulfated polysaccharides from marine algae against dental plaque bacteria." *Marine Drugs* 16(9):301.

Kannan, RRR, WA Stirk, and J van Staden. 2013. "Synthesis of silver nanoparticles using the seaweed *Codium capitatum* P.C. Silva (Chlorophyceae)." *South African Journal of Botany* 86:1–4.

Khan, MNA, JS Choi, MC Lee et al. 2008. "Anti-inflammatory activities of methanol extracts from various seaweed species." *Journal of Environmental Biology* 29(4):465–469.

Kumar, P, M Govindaraju, S Senthamilselvi, and K Premkumar. 2013. "Photocatalytic degradation of methyl orange dye using silver (Ag) nanoparticles synthesized from *Ulva lactuca*." *Colloids and Surfaces, Part B: Biointerfaces* 103:658–661.

Kumar, P, S Senthamilselvi, A Lakshmiprabha, K Premkumar, VR Muthukumaran, P Visvanathan, RS Ganeshkumar, and M Govindaraju 2012. "Efficacy of bio-synthesized silver nanoparticles using *Acanthophora spicifera* to encumber biofilm formation." *Digest Journal of Nanomaterials and Biostructures* 7:511–522.

Kumar, S, F Ye, S Dobretsov, and J Dutta. 2021. "Nanocoating is a new way for biofouling prevention." *Frontiers in Nanotechnology* 3:771098.

Kumaresan, M, KV Anand, K Govindaraju, S Tamilselvan, and VG Kumar. 2018. "Seaweed *Sargassum wightii* mediated preparation of zirconia (ZrO2) nanoparticles and their antibacterial activity against gram-positive and gram-negative bacteria." *Microbial Pathogenesis* 124:311–315.

Lellouche, J, A Friedman, R Lahmi, A Gedanken, E Banin. 2012. "Antibiofilm surface functionalization of catheters by magnesium fluoride nanoparticles." *International Journal of Nanomedicine* 7:1175–1188.

Lordan, S, TJ Smyth, A Soler-Vila, C Stanton, and RP Ross. 2013. "The α-amylase and α glucosidase inhibitory effects of Irish seaweed extracts." *Food Chemistry* 141(3):2170–2176.

Mahadavi, M, F Namvar, MB Ahmad, and R Mohamad. 2013. "Green biosynthesis and characterization of magnetic iron oxide (Fe_3O_4) nanoparticles using seaweed (*Sargassum muticum*) aqueous extract." *Molecules* 18(5):5954–5964.

Makarov, VV, AJ Love, OV Sinitsya, SS Makarova, IV Yaminsky ME Taliansky, and NO Kalinina. 2014. "Green nanotechnologies: Synthesis of metal nanoparticles using plants." *Acta Naturae* 6:35–44.

Manjunath, HM, CG Joshi, and NG Raju. 2017. "Biofabrication of gold nanoparticles using marine endophytic fungus - *Penicillium citrinum*." *IET Nanobiotechnology* 11(1):40–44.

Mauricio, MD, S Guerra-Ojeda, P Marchio et al. 2018. "Nanoparticles in medicine: A focus on vascular oxidative stress." *Oxidative Medicine and Cellular Longevity 2018*: 1–20.

Mohamed, S, SN Hashim, and HA Rahman. 2012. "Seaweeds: A sustainable functional food for complementary and alternative therapy." *Trends in Food Science and Technology* 23(2):83–96.

Mohandass, C, AS Vijayaraj, R Rajasabapathy, S Satheeshbabu, SV Rao, C Shiva, et al. 2013. "Biosynthesis of silver nanoparticles from marine seaweed *Sargassum cinereum* and their antibacterial activity." *Indian Journal of Pharmaceutical Sciences* 75(5):606–610.

Mohite, Pallavi, Mugdha Apte, Ameeta R. Kumar, and Smita Zinjarde. 2015. "Marine organisms in nanoparticle synthesis." *Springer Handbook of Marine Biotechnology* 2015:1229–1245.

Mukhopadhyay, A, and M Prosenjit. 2018. "Application of nano-biotechnology for improvement in therapeutic approaches for the treatment of diabetes." *Journal of Clinical and Molecular Endocrinology* 3(2):47.

Murugan, K, CM Samidoss, C Panneerselvam, A Higuchi, M Roni, U Suresh, et al. 2015. "Seaweed synthesized silver nanoparticles: An eco-friendly tool in the fight against *Plasmodium falciparum* and its vector *Anopheles stephensi*?." *Parasitology Research* 114(11):4087–4097.

Murugan, K, G Benelli, S Ayyappan, D Dinesh, C Panneerselvam, M Nicoletti, et al. 2015. "Toxicity of seaweed-synthesized silver nanoparticles against the filariasis vector *Culex quinquefasciatus* and its impact on predation efficiency of the cyclopoid crustacean *Mesocyclops longisetus*." *Parasitology Research* 114(6):2243–2253.

Nag, M, D Lahiri, A Dey, T Sarkar, S Joshi, and RR Ray. 2021. "Evaluation of algal active compounds as a potent antibiofilm agent." *Journal of Basic Microbiology* 62(9):1–12.

Namvar, F, S Azizi, MB Ahmad, K Shameli, R Mohamad, M Mahadavi, and PMd Tahir. 2014. "Green synthesis and characterization of gold nanoparticles using the marine macroalgae *Sargassum muticum*." *Research on Chemical Intermediates* 41(8):5723–5730.

Namvar, F, S Mohamed, SG Fard et al. 2012. "Polyphenol-rich seaweed (*Eucheuma cottonii*) extract suppresses breast tumor via hormone modulation and apoptosis induction." *Food Chemistry* 130(2): 376–382.

Namvar, F, HS Rahman, R Mohamad, A Rasedee, SK Yeap, MS Chartrand, S Azizi, and PM Tahir. 2015. "Apoptosis induction in human leukemia cell lines by gold nanoparticles synthesized using the green biosynthetic approach." *Nanomaterial Journal*: 1–10.

Panchanathan, M, and K Se-Kwon. 2015. Introduction to marine bioenergy. In *Marine Bioenergy*, CRC Press, Boca Raton, FL, pp. 3–12.

Pereira, RC, and LV Costa-Lotufo. 2012. "Bioprospecting for bio actives from seaweeds: Potential, obstacles and alternatives." *Revista Brasileira de Farmacognosia* 22(4):894–905.

Prasad, NVKV, S Rao, V Kambala, and R Naidu. 2011. "A critical review on biogenic silver nanoparticles and their antimicrobial activity." *Current Nanoscience* 7(4):531–544.

Raja, SB, J Suriya, V Sekar, and R Rajasekaran. 2012. "Biomimetic of silver nanoparticles by *Ulva lactuca* seaweed and evaluation of its antibacterial activity." *International Journal of Pharmacy and Pharmaceutical Sciences* 4:139–143.

Rajesh, S, DP Raja, JM Rathi, and K Sahayaraj. 2012. "Biosynthesis of silver nanoparticles using *Ulva fasciata* (delile) ethyl acetate extract and its activity against *Xanthomonas campestris pv. malvacearum*." *Journal of Biopesticides* 5:119–128.

Rajeshkumar, S, C Malarkodi, G Gnanajobitha, K Paulkumar, M Vanaja, C Kannan, and G Annadurai. 2013. "Seaweed-mediated synthesis of gold nanoparticles using *Turbinaria conoides* and its characterization." *Journal of Nanostructure in Chemistry* 3(1):44.

Rana, A, K Yadav, and S Jagadevan. 2020. "A comprehensive review on green synthesis of nature-inspired metal nanoparticles: Mechanism, application and toxicity." *Journal of Cleaner Production* 272:122880.

Safiyev, ES, and MY Abdullayeva. 2021. "Nanotechnologies and carbon nanoparticles." *World Science* 5(66), https://doi.org/10.31435/rsglobal_ws/30052021/7585.

Shanmugam, N, P Rajkamal, S Cholan, K Kannadasan, K Sathiskumar, G Viruthagiri, and A Sundaramanickam. 2013. "Biosynthesis of silver nanoparticles from the marine seaweed *Sargassum wightii* and their antibacterial activity against some human pathogens." *Applied Nanoscience* 4(7):881–888.

Shaw, AJ. 1990. "Heavy metal tolerance in plants: Evolutionary aspects." *Nordic Journal of Botany* 13:330.

Singh, M, M Kumar, R Kalaivani, S Manikandan, and AK Kumaraguru. 2013. "Metallic silver nanoparticle: A therapeutic agent in combination with the antifungal drug against human fungal pathogen." *Bioprocess and Biosystems Engineering* 36(4):407–415.

Singaravelu, G, JS Arockiamary, VG Kumar, and K Govindaraju. 2007. "A novel extracellular synthesis of monodisperse gold nanoparticles using marine alga, *Sargassum wightii grevillea*." *Colloids and Surfaces, B: Biointerfaces* 57(1):97–101.

Stalin, DT, V Ganesh Kumar, LS Abraham, V Karthick, and K Govindaraju. 2012. "*Sargassum myriocystum* mediated biosynthesis of gold nanoparticles." *Spectrochimica Acta. Part A: Molecular and Biomolecular Spectroscopy* 99:97–101.

Suriya, J, S Bharathiraja, V Sekar, and R Rajasekaran. 2012. "Biosynthesis of silver nanoparticles and its antibacterial activity using seaweed *Urospora* sp.." *African Journal of Biotechnology* 11:12192–12198.

Thangaraju, N, RP Venkatalakshmi, A Chinnasamy, and P Kannaiyan. 2012. "Synthesis of silver nanoparticles and the antibacterial and anticancer activities of the crude extract of *Sargassum polycystum C. Agardh*." *Nano Biomedicine and Engineering* 4(2):89–94.

Tiwari, P. 2022. "Sustainable agriculture and nanotechnologies for food and nutraceutical production - An update." In J. Chen Ed., *Plant and Nanoparticles*, Springer book series, 2022, ISBN 978-981-19-2502-3.

Tiwari, P, T Khare, V Shriram, H Bae, and V Kumar. 2021. "Exploring synthetic biology strategies for producing potent antimicrobial phytochemicals." *Biotechnology Advances* 48:107729. doi:10.1016/j.biotechadv.2021.107729.

Tiwari, P, M Bajpai, and A Sharma. 2023. "Antimicrobials from medicinal plants: Key examples, success stories and prospects in tackling antibiotic resistance." *Letters in Drug Design and Discovery* 20(4):420–438.

Singh, CR, K Kathiresan,and S Anandhan. 2015. "A review on marine based nanoparticles and their potential applications." *African Journal of Biotechnology* 14(18):1525–1532.

Sri Ramkumar, V, P Santhiyagu, R Ramasamy, P Arivalagan, G Kumar, K Ethiraj, and BR Ramaswamy. 2016. "Seaweeds: A resource for marine bionanotechnology." *Enzyme and Microbial Technology* 95:45–57.

Venkatesan, J, S Anil, SK Kim, and MS Shim. 2016. "Seaweed polysaccharide-based nanoparticles: Preparation and applications for drug delivery." *Polymers* 8(2):1–25.

Venkatesan, J, I Bhatnagar, P Manivasagan, KH Kang, and SK Kim. 2015. "Alginate composites for bone tissue engineering: A review." *International Journal of Biological Macromolecules* 72:269–281.

Vijayan, SR, P Santhiyagu, M Singamuthu, N Kumari Ahila, R Jayaraman, and K Ethiraj. 2014. "Synthesis and characterization of silver and gold nanoparticles using aqueous extract of seaweed, *Turbinaria conoides*, and their antimicrofouling activity." *Scientific World Journal* 2014:1–10.

Vijayan, SR, P Santhiyagu, R Ramasamy, P Arivalagan, G Kumar, K Ethiraj, and BR Ramaswamy. 2016. "Seaweeds: A resource for marine bionanotechnology." *Enzyme and Microbial Technology* 95:45–57.

Vivek, M, PS Kumar, S Steffi, and S Sudha. 2011. "Biogenic silver nanoparticles by *Gelidiella acerosa* extract and their antifungal effects." *Avicenna Journal of Medical Biotechnology* 3(3):143–148.

Wada, K, K Nakamura, Y Tamai et al. 2011. "Seaweed intake and blood pressure levels in healthy pre-school Japanese children." *Nutrition Journal* 10:83.

Xu, J, Y Huang, S Zhu, N Abbes, X Jing, and L Zhang. 2021. "A review of the green synthesis of ZnO nanoparticles using plant extracts and their prospects for application in antibacterial textiles." *Journal of Engineered Fibers and Fabrics* 16:1–14.

Yadav, A, S Singh, H Sohi, and S Dang. 2022. "Advances in delivery of chemotherapeutic agents for cancer treatment." *AAPS PharmSciTech* 23(1):25.

Ye, M, and B Shi. 2018. "Zirconia nanoparticles-induced toxic effects in osteoblast-like 3T3-E1 cells." *Nanoscale Research Letters* 13(1):353.

Yebra, DM, S Kiil, and K Dam-Johansen. 2004. "Antifouling technology-past, present and future steps towards efficient and environmentally friendly antifouling coatings." *Progress in Organic Coatings* 50(2):75–104.

Yosri, N, SAM Khalifa, Z Guo, B Xu, X Zou, and HR El-Seedi. 2021. "Marine organisms: Pioneer natural sources of polysaccharides/proteins for green synthesis of nanoparticles and their potential applications." *International Journal of Biological Macromolecules* 193(B):1767–1798.

Yousefzadi, M, Z Rahimi, and V Ghafori. 2014. "The green synthesis, characterization and antimicrobial activities of silver nanoparticles synthesized from green alga *Enteromorpha flexuasa* (Wulfen) J. Agardh." *Material Letters* 137:9–16.

16 Recent Methods for Producing Alginate Nanoformulations to Alleviate Climate Change Effects

Fatima El Amerany and Mohammed Rhazi

16.1 INTRODUCTION

Recently, the world has faced serious global epidemics, such as COVID-19 and Monkeypox, that have claimed millions of lives. And, the most important factors that lead to increased mortality in the world are having a weak immune system and suffering from severe malnutrition. Therefore, providing nutritious food can have a therapeutic and preventive effect to reduce the impact of current and future pandemics.

The increase in climate change over time has highlighted the depletion of natural resources, such as water and mineral elements, and the need to use innovative methods and tools to increase the productivity of cereals, crops, and herbs. Over the past two decades, much research has been devoted to adapting to climate change and maintaining a sustainable food system. Among the methods proposed is the use of biofertilizers, biostimulants, and biopesticides of microbial, plant, and marine origin, and these groups can be molecules, materials, microorganisms, or a mixture of these (Singh et al., 2020; Fenibo et al., 2021; Bhupenchandra et al., 2022).

A number of research studies have revealed the beneficial effect of polysaccharides, particularly sodium alginate and its oligomers, isolated from marine and microbial sources, on shoot development, root architecture, seed germination, and plant defense (Saberi Riseh et al., 2022). The promising effect of sodium alginate and its oligomers might be due to their ability to induce the expression of auxin response genes that are involved in cell division and elongation and to enhance the activity of antioxidant enzymes essential for reducing oxidative stress (Zhang et al., 2014; Salachna et al., 2018).

With diligent work, it has been shown that using these materials at nano-size would have a potential effect on agricultural products compared with bulk materials (Bijali and Acharya, 2020). The positive effect of the nano-sized materials is due to not only the physico-chemical characteristics of the materials used but also their unique properties, such as large surface area and high encapsulation capacity of various kinds of bioactive molecules (Mali et al., 2020). However, the exploitation of alginate NPs in the agricultural field has not received much attention in comparison to other nanomaterials, especially chitosan. Also, their mechanisms and their impact on plants exposed to environmental stress caused by climate change are not well studied.

In this chapter, we discuss the recent methods used for the extraction of alginate from microbial and marine algal sources, the main steps and conditions for producing alginate NPs, and the effect of alginate NPs on plant systems, with a focus on the reasons why they are portrayed as a good component for reducing the impact of climate change.

DOI: 10.1201/b23308-16

16.2 ALGINATE SOURCES

Alginate is an exopolysaccharide produced by some bacteria, especially nitrogen-fixing bacterial species (i.e., *Azotobacter* spp., *Pseudomonas* spp., *Paenibacillus riograndensis*), to increase their resistance to environmental stress, specifically desiccation, as well as to promote the formation of biofilms (Hay et al., 2014). It can also be found in the cell wall of brown seaweeds (i.e., *Sargassum* spp.), and it is the compound responsible for seaweed's flexibility and rigidity (Gao et al., 2018). Alginate is composed of two units (D-mannuronic acid and L-guluronic acid) linked via (1→4) bonds (Mathieu et al., 2016). The flexibility of seaweeds is attributed to the abundance of D-mannuronic acid, while the rigidity in marine algae is due to the abundance of L-guluronic acid (Singh and Gross, 2001).

Seaweeds are usually used to produce sodium alginate because of its high proportion of alginate in comparison with bacteria. For instance, Table 16.1 shows that the levels of alginate vary from 6 to 52% dry weight (DW) in marine species and from 0.04 to 100 g in the viscous liquid produced by bacteria. In addition, Yu et al. (2011) demonstrated that manipulating strain genes, especially *AlgU*, can increase the levels of alginate produced.

Alginate is characterized by different properties, e.g., low toxicity, biocompatibility, and gel-forming capacity, which make alginate useful in the food and pharmaceutical industry (Lee and Mooney, 2012). It can be applied as a natural ingredient, a stabilizer, a food additive, and a source of dietary fiber (Brownlee et al., 2005; Gheorghita Puscaselu et al., 2020; Bennacef et al., 2021).

16.3 METHODS FOR ISOLATION OF ALGINATE

To date, various methods have been proposed to prepare sodium and calcium alginate, and each method's conditions are dependent on the chemical composition of marine macroalgae and liquids produced by bacteria. Alginate salts can be produced from different seaweed species based on four main steps: depigmentation, acidic pretreatment, alkaline extraction, and alginate salt precipitation (Mohammed et al., 2018). As shown in Table 16.1, before the isolation of alginic acid, seaweeds are treated with a 2% formaldehyde solution to remove pigments and phenolic acids that are localized in the cell wall. This step can be postponed to a later time by adding sodium hypochlorite or hydrogen peroxide (Table 16.1). The depigmented powder can then be treated with either hydrochloric acid (2.5%, *v/v*) or sulfuric acid (0.1–1 M, *v/v*) to remove minerals (Table 16.1). Afterward, the resulting residue is immersed in an alkaline solution to produce soluble or insoluble alginate salts (Table 16.1). The solubility of alginate salts is dependent on the type of alkaline solution used, the solvent employed for solubilization, and the molecular weight of alginate. For instance, the use of sodium carbonate and sodium hydroxide solutions to make alginate salt increases its solubility in water and organic solvents, whereas the application of calcium chloride restricts its solubility in these solvents (Cebrián-Lloret et al., 2022; Frent et al., 2022). Alginate can be precipitated from the solution as sodium alginate, calcium alginate, or alginic acid by adding ethanol, calcium chloride, or hydrochloric acid, respectively (Arvizu-Higuera et al., 1997).

Moreover, the production of highly viscous fluid, containing exopolysaccharide alginate, by nitrogen-fixing bacteria can be promoted in the presence of yeast extracts and carbohydrates, in particular glycerol, molasses, maltose, starch, and sucrose (Table 16.1; Yoshida et al., 2022). Bacterial alginate may be isolated through four steps: filtration, deacetylation, alginic acid precipitation, and the production and precipitation of alginate salts. Regarding the first step, the viscous liquid is filtered through a membrane filter (0.45 μm) to remove microorganisms and gross molecules. Microorganisms can also be eliminated by adding an antibiofilm agent, especially tetrasodium ethylenediaminetetraacetic acid. The resulting supernatant is then deacetylated by adding sodium hydroxide. Afterward, alginic acid is formed and precipitated by decreasing the pH of the supernatant. Finally, alginic acid is converted to salts and precipitated as previously described.

TABLE 16.1
Different Sources for the Biosynthesis of Alginate

	Species	Alginate yield	Raw material	Isolation conditions	References
Alginate isolated from brown seaweed species	*Sargassum vulgare*	10–17% DW	**Brown seaweed**	2% (*v/v*) formaldehyde for 24 h, 0.2 M HCl for 24 h, 2% (*w/v*) sodium carbonate, and precipitation with ethanol	Torres et al. (2007)
	Laminaria digitata	52% DW		2% (*v/v*) formaldehyde for 24 h, 0.2 M HCl for 24 h, 2% (*w/v*) sodium carbonate for 5 h, and precipitation with ethanol, methanol, and acetone	Fertah et al. (2017)
	Sargassum natans	19% DW		2% (*w/v*) formaldehyde overnight, 0.1–1 M sulfuric acid for 1 h at 25–60 °C, and 5–10% (*w/v*) sodium carbonate for 0.5–6 h at 25–75 °C	Mohammed et al. (2018)
	Sargassum polycystum	6–19% DW		Immersion in calcium chloride, extraction using sodium carbonate and sodium hydroxide, and blanching using sodium hypochlorite or hydrogen peroxide	Dharmayanti et al. (2019)
	Cystoseira barbata	19% DW		70% (*v/v*) ethanol for 24 h, 0.1 M HCl for 2 h at 60 °C, 3% (*w/v*) sodium carbonate for 2 h at 60 °C, and precipitation with acetone	Trica et al. (2019)
	Dictyota dichotoma	9% DW		No data	Osman et al. (2020)
	Hormophysa cuneiformis	19% DW			
	Padina pavonica	8% DW			
	Turbinaria ornata	14% DW			
	Laminaria ochroleuca	28% DW		2% (*v/v*) formaldehyde for 24 h, 0.2 M HCl for 24 h, 2% (*w/v*) sodium carbonate overnight, 2.5% (*v/v*) sodium hypochlorite for 15 s, and precipitation by ethanol	Belattmania et al. (2020)
	Cystoseira humilis	19% DW			Rashedy et al. (2021)
	Fucus guiryi	14% DW			
	Padina boergesenii	17% DW			
	Turbinaria triquetra	22% DW			
	Hormophysa cuneiformis	13% DW			
	Dictyota ciliolata	20% DW			
	Sargassum aquifolium	18% DW			
	Sargassum latifolium	18% DW		Hot water for 30 min, 0.5% (*w/v*) calcium chloride for 30 min, 0.5% (*w/v*) sodium chloride for 1 h, and 5% (*w/v*) sodium carbonate for 30 min	Dalal et al. (2021)
	Sargassum wightii	32 % DW		1% (*w/v*) calcium chloride for 2 h, 1% (*v/v*) formaldehyde for 6 h, 5% HCl for 30 min, 2.5% (*w/v*) sodium carbonate for 3 h, and precipitation by 95% (*v/v*) ethanol	Jayasinghe et al. (2022)

(Continued)

TABLE 16.1 (CONTINUED)
Different Sources for the Biosynthesis of Alginate

	Species	Alginate yield	Raw material	Isolation conditions	References
Alginate produced by nitrogen-fixing bacteria species	*Azotobacter vinelandii* *Azotobacter chroococcum* *Pseudomonas putida* *Pseudomonas mendocina* *Pseudomonas fluorescens* *Pseudomonas maltophilia* *Pseudomonas cepacia* *Pseudomonas mallei* *Pseudomonas pseudomallei* *Pseudomonas aeruginosa* *Pseudomonas alcaligenes* *Pseudomonas acidovorans* *Pseudomonas diminuta* *Pseudomonas putrefaciens*	2–100 g	Unknown	Calcium chloride + ethanol or propan-2-ol	Yu et al. (2011)
	Paenibacillus riograndensis	0.262 g	Yeast extract, sucrose, and inorganic compounds	0.5 M tetrasodium ethylenediaminetetraacetic acid + 5 M sodium chloride. Precipitation by ice cold isopropyl alcohol	Shroff et al. (2018)
	Azotobacter vinelandii	0.018 g (per 1 L crude glycerol)	Crude glycerol	0.5 M ethylenediaminetetraacetic acid, 5 M sodium chloride, isopropanol (4 °C) for 10 min, and 70% (*v/v*) ethanol	Yoshida et al. (2022)
	Azotobacter vinelandii 12	0.04–0.18 g (free alginate)	Sucrose and phosphate	Centrifugation + ethanol (−20 °C)	Dudun et al. (2021)
		0.15–0.42 g (capsular alginate)		1 M sodium chloride + 0.1 M ethylenediaminetetraacetic acid for 1 h at 60 °C, and centrifugation + ethanol (−20 °C)	
	Azotobacter vinelandii ATCC 9046	0.131 g	Molasses, maltose, and starch	0.1 M tetrasodium EDTA + 1.5% (*w/v*) sodium chloride for 10 min, 0.1 M sodium hydroxide (*w/v*) for 1 h, 0.1 N HCl (4 °C) for 30 min, and precipitation by ethanol + 0.2% (*w/v*) sodium chloride	Kıvılcımdan Moral and Yıldız (2016)
	Pseudomonas stutzeri	15.25 g	Yeast extract, sucrose, MOPS, ferrous sulfate, iron chloride	95% (*v/v*) ethanol	Meena et al. (2020)

16.4 METHODS FOR PREPARATION OF ALGINATE NPS

In recent decades, alginate has been used as a matrix to make different derivatives and formulations appropriate for protecting or carrying molecules and microorganisms that need to be treated with great caution, especially in certain unsuitable conditions like the pH of a solution and the temperature of a medium. These new products based on alginate are categorized as hydrogels, composites, oligomers, microparticles, nanoparticles, and alkyl alginate ester (He et al., 2015; Grillo et al., 2010; Liu et al., 2019; Chen et al., 2021; Frent et al., 2022; Twizeyimana et al., 2022). And, the synthesis of each category depends on four important parameters: molecular weight of alginate, its solubility, associated molecules (grafted and crosslinkers), and the physico-chemical treatments employed (i.e., applying enzymes and gamma radiation).

Alginate NPs are particles with organized and disorganized shapes made from polymeric alginate (Figure 16.1). The size of one or all dimensions of these particles is less than 1000 nm. The production of alginate NPs can be based on two approaches: complexation and emulsification methods (Pâques et al., 2014). The first approach requires mixing alginate with an oppositely charged molecule called a crosslinker, and this reaction can lead to the synthesis of nano-aggregates and nanocapsules if it occurs in a watery fluid and at the oil droplet interface, respectively. In the case of nanocapsule formation, the addition of surfactants is necessary to ensure the stability of the NPs' external core (Miyazawa et al., 2021). Concerning the second approach, firstly, the desired substances that are to be enclosed within particles are dispersed in an alginate solution; secondly, the mixture is emulsified in an oil liquid, resulting in the formation of alginate droplets; and thirdly, these emulsion droplets are gelified and demixed by adding a crosslinker.

In the last 20 years, only a few reports on alginate-based NP production have been released (Table 16.2). The particles are made by using sodium alginate, especially with low molecular weight, as a precursor, and ten kinds of crosslinkers (calcium chloride, chitosan, cobalt chloride, ferrous chloride, magnesium chloride, manganous chloride, methylene chloride, propylamine, sodium tripolyphosphate, and zinc chloride) (Table 16.2). In addition, these NPs were used to carry drugs (i.e., insulin, bupivacaine, miltefosine, rifampicin, lovastatin, ofloxacin), active compounds (i.e., cinnamaldehyde, meta-bromo-thiolactone, glycomacropeptide), pesticides (i.e., chloroinconazide), oils (i.e., turmeric oil), and plant extract (Table 16.2).

Recently, Belguesmia et al. (2020) demonstrated that alginate NPs can also be produced by grinding alginate with zirconium oxide beads using a Retsch ball mill for 10 h. The produced NPs can then be dissolved in purified water and used to carry antibiotics (i.e., bacteriocins). In addition, nanoemulsions from sodium alginate can be made by using a particular homogenizer (i.e., microfluidizer) and an organic phase composed of oil (Gutiérrez-Jara et al., 2021).

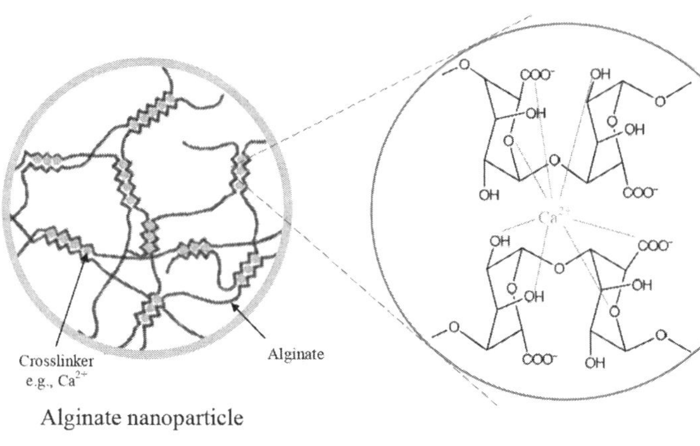

FIGURE 16.1 Structure of calcium alginate nanoparticle.

TABLE 16.2
Conditions for Alginate Nanoparticle Production

Loading substances	Substances carried	Sodium alginate characteristics: Concentrations/ quantities	Sodium alginate characteristics: MW (kDa)	Sodium alginate characteristics: Pre-treatments	Requirement for NP production: Crosslinkers	Requirement for NP production: Surfactants	Requirement for NP production: Treatments needed during NP production	Size of NPs (nm)	References
Without	–	0.06%	12 to 40	Hydrolysis by using 3 M HCl at 50 °C for 20 to 140 min	18 mM calcium chloride 36 mM calcium chloride	–	–	60 to 70 40 to 50	Daemi and Barikani (2012)
	–	Unknown	126	Purification using 100 mM NaCl	Adding two calcium chloride solutions: with 500 mM and then 100 mM	Decane	40 °C for 1 h and then at room temperature for 2 h	100 to 160	Machado et al. (2012)
	–	2%	ND	–	1.5% (*w/v*) calcium chloride	Cetyltrimethylammonium bromide	–	223 to 485	Kafshgari et al. (2012)
	–	0.05%	ND	Dialysis	Zinc chloride	–	–	<300	Pistone et al. (2015)
	–	1%	ND	–	–	1% (*v/v*) Tween 80	Using an organic phase composed of soybean oil and a homogenizer (microfluidizer)	377	Gutiérrez-Jara et al. (2021)
	–	0.2%	120 to 190	–	Manganous chloride Ferrous chloride Cobalt chloride	Span 20	–	275 120 8	Saxena et al. (2022)
With	Insulin	0.063%	50	–	18 mM calcium chloride and 0.05 to 0.09% (*w/v*) chitosan	–	–	764 to 2209	Sarmento et al. (2006)
	Turmeric oil	0.6 mg mL^{-1}	80 to 120	–	0.67 mg mL^{-1} calcium chloride	1% (*v/v*) Tween 80®	–	100 to 600	Lertsutthiwong et al. (2008)
	0.5% (*w/v*) bupivacaine	0.063%	ND	–	18 mM calcium chloride and 0.07% (*w/v*) chitosan	–	–	944	Grillo et al. (2010)
	2% (*w/v*) bupivacaine	1%		–	5% (*w/v*) methylene chloride and 60% (*w/v*) calcium chloride	Bis(2-ethylhexyl) sulfosuccinate	Ice bath for 60 s and 18 h at room temperature	926	

(Continued)

TABLE 16.2 (CONTINUED)
Conditions for Alginate Nanoparticle Production

Loading substances	Substances carried	Sodium alginate characteristics			Requirement for NP production			Size of NPs (nm)	References
		Concentrations/ quantities	MW (kDa)	Pre-treatments	Crosslinkers	Surfactants	Treatments needed during NP production		
	Peppermint oil	1.5%	ND	–	2% (*w/v*) Calcium chloride	0.02% (*v/v*) Tween 20	Sonication for 30 min	63	Ghayempour and Mortazavi (2015)
	Cinnamaldehyde	0.06%	10 to 100	–	0.2% (*w/v*) calcium chloride and 0.05% (*w/v*) chitosan (DA <10%)	–	–	166	Loquercio et al. (2015)
	1 mg mL^{-1} glycomacropeptide 1 mg mL^{-1} 5-aminosalycic acid	1 mg mL^{-1}	ND	–	1 mg mL^{-1} chitosan (DA <5%)	–	Sonication for 30 min	220 to 540 260 to 560	Rivera et al. (2015)
	0.25 mg mL^{-1} acetamiprid	0.06%	ND	–	0.05% (*w/v*) chitosan and 0.067% (*w/v*) calcium chloride	–	–	202	Kumar et al. (2015)
	0.57 mg mL^{-1} curcumin	1.15 mg mL^{-1}	50 to 160	–	–	Cetyltrimethylammonium-ammonium bromide, Cremophor EL, and poloxamer 188	–	178	Hernandez-Patlan et al. (2019)
	Miltefosine	1%	ND	–	220 mM calcium chloride	3% (*v/v*) Span 80	–	279 to 320	Spadari et al. (2019)
	1 mg mL^{-1} ofloxacin	0.05 mg mL^{-1}	ND	–	1.5 mg mL^{-1} chitosan and 0.75 mg mL^{-1} sodium tripolyphosphate	–	–	228	Taghe et al. (2019)
	Metabolites of plant growth–promoting rhizobacteria	2%	ND	–	Calcium chloride NPs	Span 80	Adding gelatin, metal and carbonic NPs, and soybean oil and mixing the whole for 15 min	ND	Moradipour et al. (2019)

(Continued)

TABLE 16.2 (CONTINUED)
Conditions for Alginate Nanoparticle Production

Loading substances	Substances carried	Sodium alginate characteristics: Concentrations/ quantities	MW (kDa)	Pre-treatments	Requirement for NP production: Crosslinkers	Surfactants	Treatments needed during NP production	Size of NPs (nm)	References
	Silver NPs	5%	ND	Oxidation	2% (*v/v*) ammonia solution	–	Hot bath (55 to 85 °C)	6 to 40	Xiang et al. (2019)
	0.5% (*w/v*) rifampicin	1%	ND	Adding honey as a stabilizer	1% (*w/v*) calcium chloride	–	–	64	Thomas et al. (2020)
	60 mg L^{-1} bacteriocins	2 g	ND	–	–	–	Mixing with zirconium oxide beads using a ball mill	124	Belguesmia et al. (2020)
	9% lovastatin	62%	ND	–	Calcium chloride and 29% (*w/v*) chitosan (DA <25%)	–	–	86	Thai et al. (2020)
	5 mg L^{-1} *Artemisia ciniformis* extract	1%	ND	–	Propylamine	–	Sonication for 30 min in a cold bath	150 to 400	Rahimivand et al. (2020)
	3.6 mg mL^{-1} chloroinconazide	1%	ND	–	1% (*w/v*) magnesium chloride and 1% (*w/v*) calcium chloride	Span 80 and Tween 80	–	<100	Lv et al. (2021)
	0.4% (*w/v*) meta-bromo-thiolactone	2%	ND	–	2% (*w/v*) calcium chloride	Tween 20	–	175	Eltayb et al. (2022)
	0.36 mg mL^{-1} atrazine	0.063%	ND	–	18 mM calcium chloride and 0.05% (*w/v*) chitosan	–	–	<300	Cordon et al. (2022)
	Chloridazon	0.06%	ND	–	0.05% (*w/v*) chitosan and 0.067% (*w/v*) calcium chloride	–	Sonication for 15 min	253	Babaei et al. (2022)
	20% (*w/v*) *Phyllanthus niruri* extract	1%	ND	–	–	Tween 80	The organic phase used is composed of oil extracted from seeds of *Citrullus lanatus*	192	Pathania et al. (2022)

16.5 REASONS WHY ALGINATE NPS ARE BENEFICIAL FOR REDUCING THE IMPACT OF CLIMATE CHANGE

16.5.1 Effects of Alginate NPs on Plants

Alginate has been employed in agriculture for many reasons: for purifying water, absorbing metals, increasing seed germination, eliciting plant defense, maintaining the quality of harvested fruits, and carrying fertilizers and beneficial microorganisms (He et al., 2015; Üçüncü Tunca et al., 2017; He et al., 2018; Dey et al., 2019; Wang et al., 2020; van der Merwe et al., 2022; Vu and Wu, 2022). Regarding the application of alginate nanoformulations (nanogels, nanocapsules, and nanoemulsions) in the agricultural field, as far as we know, only six studies have been conducted to improve the immune system of plants and fruits, to reduce the toxicity of pesticides (i.e., atrazine, chloroinconazide, and silver NPs), and to control the release of plant growth–promoting bacterial metabolites (i.e., auxins) (Table 16.3). These nanoformulations were applied to leaves, plant medium, and wounded tissues, and for covering fruits.

To date, the NPs produced are composed of not only alginate but also minerals and molecules that are very important for the plant system (Table 16.2). Moreover, releasing one of them, or even more, might improve and help plants that are suffering from stress.

16.5.2 Releasing Ions and Crosslinkers Attached to the Alginic Acid Backbone

The stability of some alginate NPs, specifically nanogel formulations, cannot last long because of their high level of D-mannuronic acid residue. And when the unstable alginate nanogels dissolve in water, they release divalent ions and crosslinkers (Lee and Mooney, 2012).

The release of alginate divalent ions, particularly calcium ion (Ca^{2+}), which is a necessary mineral element for plant development in normal conditions, can protect the plant from environmental stresses. It ensures the cell's membrane stability and acts as a second messenger involved in the biochemical and physiological response of the plant to stress (Thor, 2019). For instance, Zhao and Tan (2005) reported that pretreatment of wheat leaves with Ca^{2+} ions could reduce the oxide production and lipid degradation that occurred during heat and radiation stress while inducing photosynthetic activity. In rice, Alam et al. (2002) showed that the application of Ca^{2+} ions to saline soil reduced the absorption of sodium ions, prevented chlorosis and cell damage, and improved nutrient (i.e., nitrogen, phosphorus, potassium) absorption. Similar effects were observed in forage crops exposed to salt stress (Sharavdorj et al., 2022).

Crosslinker-based polysaccharides, such as chitosan, are known to be excellent biostimulants and elicitors that have been used for decades. Studies have shown that chitosan, a deacetylated chitin, can induce a plant's growth process due to increasing the aperture of stomata, energy transfer via chlorophyll molecules, and the size of xylem vessels, and releasing nitrogen (Mulaudzi et al., 2022; El Amerany et al., 2022). It can also deal with pathogens and abiotic stress by employing different strategies, and the most efficient strategies are the ability of chitosan to either interact with other molecules or stimulate the plant's defense mechanisms. Chitosan can interact with negatively charged molecules, especially the lipopolysaccharides of bacteria and fungi; when this interaction occurs, the microorganisms' cell walls are disrupted, and micropores are formed, which thereafter lead to the loss of cytoplasmic content and eventually, the death of cells (Ardean et al., 2021). It can also stimulate the defense of plants by controlling the aperture of guard cells, inducing the levels of defense-mediated phytohormones (i.e., jasmonic acid-isoleucine), enzymes (i.e., chitinase), and antioxidants (i.e., ascorbic acid, phenolic compounds) (Lee et al., 1999; El Amerany et al., 2022; Pongprayoon et al., 2022; Hernández et al., 2022). The release of chitosan may be beneficial or harmful to plants and soil microorganisms depending on its molecular weight, its size, its dose, and the plant's condition (El Amerany et al., 2020a; El Amerany et al., 2020b; El Amerany et al., 2023).

TABLE 16.3
Effects of Alginate Nanoformulations Applied in the Agricultural Field

Nano-alginates	Carrying substance	Plant species	Stress	Application mode	Impacts	References
Nanoemulsions	PGPR (*Pseudomonas fluorescens*) metabolites	Pistachio (*Pistacia vera* L.)	Without	Application on the medium	-Increase in the number and weight of leaves, stem length, and root system	Moradipour et al. (2019)
	–	Sweet cherries (cv. Bing)	Storage conditions	Coating fruits	-Reduction in cell damage, water evaporation, weight loss, and color change. Increase in the total soluble sugars and the levels of antioxidants (anthocyanins, phenolic compounds, and 2,2-diphenyl-1-picrylhydrazyl)	Gutiérrez-Jara et al. (2021)
Nanogels	Pesticide (chloroinconazide)	Tobacco (*Nicotiana benthamiana* L.)	Without	Foliar spray	-Maintaining the release of pesticides and magnesium and calcium ions for the longest period. Increase in antioxidant activity (catalase, superoxide dismutase, and peroxidase), salicylic acid content, total chlorophyll, plant length, and fresh and dry weight	Lv et al. (2021)
	–	Tobacco (*Nicotiana benthamiana* L.)	Wildfire disease	Foliar spray	-Increase in the concentration of calcium ion. Alteration of pathogen's cell wall	Ma et al. (2022)
Nanocapsules	Pesticide (silver nanoparticles)	Citrus	Plant pathogenic fungi	Application to the wounded tissues	-Alteration of the fungal cell's membrane and DNA structure. Reduction in the activity and hyphae growth of *Fusarium solani*, *Sclerotinia sclerotiorum*, and *Colletotrichum gloeosporioides*	Xiang et al. (2019)
	Pesticide (atrazine)	Chicory (*Cichorium intybus*) and corn (*Zea mays*)	Herbicide toxicity	Foliar spray	-Reduction in toxicity of herbicide. Slight reduction in plant growth as well as the maximal quantum yield of PSII photochemistry	Cordon et al. (2022)

DA, acetylation degree; ND, not determined; PSII, photosystem II.

Crosslinker-based propylamines are molecules found in terrestrial and small marine plants (i.e., *Peucedanum palustre, Hippospongia communis*) and are also known as 1-aminopropane. Even though the role of these compounds in the plant is unknown, they likely have a positive effect on photosynthetic compounds due to their nitrogen content.

16.6 CONCLUSION AND PERSPECTIVES

Alginate is a polymer that shows great promise for developing sustainable agriculture. This polymer is extensively used in nanobiotechnology to reduce the toxicity of phytochemicals and to control the release of phytohormones that are essential for cell division and elongation. Until now, alginate NPs have received less attention, despite their beneficial effect on plant growth and recovery from shock and injury. Also, alginate NPs appear to increase the tolerance of plants to various environmental stresses by releasing divalent ions and cationic polysaccharides, such as chitosan, and propylamine. Therefore, producing new alginate nanoformulations by using pullulan and starch as crosslinkers might improve agricultural productivity.

REFERENCES

Alam, S., Huq, S. I., Kawai, S., &Islam, A. (2002). Effects of applying calcium salts to coastal saline soils on growth and mineral nutrition of rice varieties. *Journal of Plant Nutrition, 25*(3), 561–576.

Ardean, C., Davidescu, C. M., Nemeş, N. S., Negrea, A., Ciopec, M., Duteanu, N., Negrea, P., Duda-Seiman, D., &Musta, V. (2021). Factors influencing the antibacterial activity of chitosan and chitosan modified by functionalization. *International Journal of Molecular Sciences, 22*(14), 7449.

Arvizu-Higuera, D. L., Hernández-Carmona, G., &Rodriguez-Montesino, E. (1997). Effect of the type of precipitation on the process Toobtain sodium alginate: Calcium alginate method and alginic acid method. *Ciencias Marinas, 23*(2), 195–207.

Babaei, S., Kahrizi, D., Nosratti, I., Karimi, N., Arkan, E., &Tahir, M. B. (2022). Preparation and characterization of chloridazon-loaded alginate/chitosan nanocapsules. *Cellular and Molecular Biology, 68*(3), 34–42.

Belattmania, Z., Kaidi, S., El Atouani, S., Katif, C., Bentiss, F., Jama, C., Reani, A., Sabour, B., &Vasconcelos, V. (2020). Isolation and FTIR-ATR and 1H NMR characterization of alginates from the main alginophyte species of the Atlantic coast of Morocco. *Molecules, 25*(18), 4335.

Belguesmia, Y., Hazime, N., Kempf, I., Boukherroub, R., &Drider, D. (2020). New bacteriocins from Lacticaseibacillus paracasei CNCM I-5369 adsorbed on alginate nanoparticles are very active against Escherichia coli. *International Journal of Molecular Sciences, 21*(22), 8654.

Bennacef, C., Desobry-Banon, S., Probst, L., &Desobry, S. (2021). Advances on alginate use for spherification to encapsulate biomolecules. *Food Hydrocolloids, 118*, 106782.

Bhupenchandra, I., Chongtham, S. K., Devi, E. L., Ramesh, R., Choudhary, A. K., Salam, M. D., Sahoo, M. R., Bhutia, T. L., Devi, S. H., Thounaojam, A. S., Behera, C., Harish, M. N., Kumar, A., Dasgupta, M., Devi, Y. P., Singh, D., Bhagowati, S., Devi, C. P., Singh, H. R., &Khaba, C. I. (2022). Role of biostimulants in mitigating the effects of climate change on crop performance. *Frontiers in Plant Science, 13*, 967665.

Bijali, J., &Acharya, K. (2020). Current trends in nano-technological interventions on plant growth and development: A review. *IET Nanobiotechnology, 14*(2), 113–119.

Brownlee, I. A., Allen, A., Pearson, J. P., Dettmar, P. W., Havler, M. E., Atherton, M. R., &Onsøyen, E. (2005). Alginate as a source of dietary fiber. *Critical Reviews in Food Science and Nutrition, 45*(6), 497–510.

Cebrián-Lloret, V., Metz, M., Martínez-Abad, A., Knutsen, S. H., Ballance, S., López-Rubio, A., &Martínez-Sanz, M. (2022). Valorization of alginate-extracted seaweed biomass for the development of cellulose-based packaging films. *Algal Research, 61*, 102576.

Chen, X., Zhu, Q., Liu, C., Li, D., Yan, H., &Lin, Q. (2021). Esterification of alginate with alkyl bromides of different carbon chain lengths via the bimolecular nucleophilic substitution reaction: Synthesis, characterization, and controlled release performance. *Polymers, 13*(19), 3351.

Cordon, G., Valiño, I. L., Prieto, A., Costa, C., Marchi, M. C., &Diz, V. (2022). Effects of the nanoherbicide made up of atrazine-chitosan on the primary events of photosynthesis. *Journal of Photochemistry and Photobiology, 12*, 100144.

Daemi, H., &Barikani, M. (2012). Synthesis and characterization of calcium alginate nanoparticles, sodium homopolymannuronate salt and its calcium nanoparticles. *Scientia Iranica*, *19*(6), 2023–2028.

Dalal, S. R., Hussein, M. H., El-Naggar, N. E. A., Mostafa, S. I., &Shaaban-Dessuuki, S. A. (2021). Characterization of alginate extracted from Sargassum latifolium and its use in Chlorella vulgaris growth promotion and riboflavin drug delivery. *Scientific Reports*, *11*(1), 1–17.

Dey, P., Ramanujam, R., Venkatesan, G., &Nagarathnam, R. (2019). Sodium alginate potentiates antioxidant defense and PR proteins against early blight disease caused by Alternaria solani in Solanum Lycopersicum Linn.*PLOS ONE*, *14*(9), e0223216.

Dharmayanti, N., Supriatna, J., Abinawanto, A., &Yasman, Y. (2019). Isolation and partial characterization of alginate extracted from Sargassum polycystum collected from three habitats in Banten, Indonesia. *Biodiversitas Journal of Biological Diversity*, *20*(6), 1776–1785.

Dudun, A. A., Akoulina, E. A., Zhuikov, V. A., Makhina, T. K., Voinova, V. V., Belishev, N. V., Khaydapova, D. D., Shaitan, K. V., Bonartseva, G. A., &Bonartsev, A. P. (2021). Competitive biosynthesis of bacterial alginate using Azotobacter vinelandii 12 for tissue engineering applications. *Polymers*, *14*(1), 131.

El Amerany, F., Meddich, A., Wahbi, S., Porzel, A., Taourirte, M., Rhazi, M., &Hause, B. (2020a). Foliar application of chitosan increases tomato growth and influences mycorrhization and expression of endochitinase-encoding genes. *International Journal of Molecular Sciences*, *21*(2), 535.

El Amerany, F., Rhazi, M., Wahbi, S., Taourirte, M., &Meddich, A. (2020b). The effect of chitosan, arbuscular mycorrhizal fungi, and compost applied individually or in combination on growth, nutrient uptake, and stem anatomy of tomato. *Scientia Horticulturae*, *261*, 109015.

El Amerany, F., Meddich, A., Wahbi, S., Taourirte, M., &Rhazi, M. (2023). Application of Compost, as a Peat Substitute, and Foliar Spray of chitosan and/or Roots Inoculation with Beneficial Fungus Increases Tomato Plants Productivity and Fruit Size. *Gesunde Pflanzen*.

El Amerany, F., Rhazi, M., Balcke, G., Wahbi, S., Meddich, A., Taourirte, M., &Hause, B. (2022). The effect of chitosan on plant physiology, wound response, and fruit quality of tomato. *Polymers*, *14*(22), 5006.

Eltayb, E. K., Aleanizy, F. S., Alqahtani, F. Y., Alkahtani, H. M., Ansari, S. A., &Alsarra, I. (2022). Preparation and characterization of meta-bromo-thiolactone calcium alginate nanoparticles. *Saudi Pharmaceutical Journal*, *30*(7), 946–953.

Fenibo, E. O., Ijoma, G. N., &Matambo, T. (2021). Biopesticides in sustainable agriculture: A critical sustainable development driver governed by green chemistry principles. *Frontiers in Sustainable Food Systems*, *5*, 619058.

Fertah, M., Belfkira, A., Taourirte, M., Brouillette, F. (2017). Extraction and characterization of sodium alginate from Moroccan Laminaria digitata brown seaweed. *Arabian Journal of Chemistry*, *10*, S3707–S3714.

Frent, O. D., Vicas, L. G., Duteanu, N., Morgovan, C. M., Jurca, T., Pallag, A., Muresan, M. E., Filip, S. M., Lucaciu, R.-L., &Marian, E. (2022). Sodium alginate—Natural microencapsulation material of polymeric microparticles. *International Journal of Molecular Sciences*, *23*(20), 12108.

Gao, F., Liu, X., Chen, W., Guo, W., Chen, L., &Li, D. (2018). Hydroxyl radical pretreatment for low-viscosity sodium alginate production from brown seaweed. *Algal Research*, *34*, 191–197.

Ghayempour, S., &Mortazavi, S. M. (2015). Preparation and investigation of sodium alginate nanocapsules by different microemulsification devices. *Journal of Applied Polymer Science*, *132*(17), 1–8.

Gheorghita Puscaselu, R., Lobiuc, A., Dimian, M., &Covasa, M. (2020). Alginate: From food industry to biomedical applications and management of metabolic disorders. *Polymers*, *12*(10), 2417.

Grillo, R., de Melo, N. F., De Araujo, D. R., de Paula, E., Rosa, A. H., &Fraceto, L. F. (2010). Polymeric alginate nanoparticles containing the local anesthetic bupivacaine. *Journal of Drug Targeting*, *18*(9), 688–699.

Gutiérrez-Jara, C., Bilbao-Sainz, C., McHugh, T., Chiou, B. S., Williams, T., &Villalobos-Carvajal, R. (2021). Effect of cross-linked alginate/oil nanoemulsion coating on cracking and quality parameters of sweet cherries. *Foods*, *10*(2), 449.

Hay, I. D., Wang, Y., Moradali, M. F., Rehman, Z. U., &Rehm, B. H. (2014). Genetics and regulation of bacterial alginate production. *Environmental Microbiology*, *16*(10), 2997–3011.

He, J., Li, R., Sun, X., Wang, W., Hu, J., Xie, H., &Yin, H. (2018). Effects of calcium alginate submicroparticles on seed germination and seedling growth of wheat (Triticum aestivum L.). *Polymers*, *10*(10), 1154.

He, Y., Wu, Z., Tu, L., Han, Y., Zhang, G., &Li, C. (2015). Encapsulation and characterization of slow-release microbial fertilizer from the composites of bentonite and alginate. *Applied Clay Science*, *109*, 68–75.

Hernández, V., Botella, M. Á., Hellín, P., Fenoll, J., &Flores, P. (2022). Dose-dependent potential of chitosan to increase yield or bioactive compound content in tomatoes. *Horticulturae*, *8*(12), 1152.

Hernandez-Patlan, D., Solis-Cruz, B., Cano-Vega, M. A., Beyssac, E., Garrait, G., Hernandez-Velasco, X., Lopez-Arellano, R., &Rivera-Rodriguez, G. R. (2019). Development of chitosan and alginate nanocapsules to increase the solubility, permeability and stability of curcumin. *Journal of Pharmaceutical Innovation*, *14*(2), 132–140.

Jayasinghe, G. D. T. M., Jinadasa, B. K. K. K., &Sadaruwan, N. A. G. (2022). Pathway of sodium alginate synthesis from marine brown algae, Sargassum wightii from Sri Lanka. *Discover Food*, *2*(1), 1–7.

Kafshgari, M. H., Khorram, M., Mansouri, M., Samimi, A., &Osfouri, S. (2012). Preparation of alginate and chitosan nanoparticles using a new reverse micellar system. *Iranian Polymer Journal*, *21*(2), 99–107.

Kıvılcımdan Moral, Ç., &Yıldız, M. (2016). Alginate production from alternative carbon sources and use of polymer based adsorbent in heavy metal removal. *International Journal of Polymer Science*, *2016*, 1–8.

Kumar, S., Chauhan, N., Gopal, M., Kumar, R., &Dilbaghi, N. (2015). Development and evaluation of alginate–chitosan nanocapsules for controlled release of acetamiprid. *International Journal of Biological Macromolecules*, *81*, 631–637.

Lee, K. Y., &Mooney, D. J. (2012). Alginate: Properties and biomedical applications. *Progress in Polymer Science*, *37*(1), 106–126.

Lee, S., Choi, H., Suh, S., Doo, I. S., Oh, K. Y., Choi, E. J., Schroeder Taylor, A. T., Low, P. S., &Lee, Y. (1999). Oligogalacturonic acid and chitosan reduce stomatal aperture by inducing the evolution of reactive oxygen species from guard cells of tomato and Commelina communis. *Plant Physiology*, *121*(1), 147–152.

Lertsutthiwong, P., Noomun, K., Jongaroonngamsang, N., Rojsitthisak, P., &Nimmannit, U. (2008). Preparation of alginate nanocapsules containing turmeric oil. *Carbohydrate Polymers*, *74*(2), 209–214.

Liu, J., Yang, S., Li, X., Yan, Q., Reaney, M. J., &Jiang, Z. (2019). Alginate oligosaccharides: Production, biological activities, and potential applications. *Comprehensive Reviews in Food Science and Food Safety*, *18*(6), 1859–1881.

Loquercio, A., Castell-Perez, E., Gomes, C., &Moreira, R. G. (2015). Preparation of chitosan-alginate nanoparticles for trans-cinnamaldehyde entrapment. *Journal of Food Science*, *80*(10), N2305–N2315.

Lv, X., Yuan, M., Pei, Y., Liu, C., Wang, X., Wu, L., Cheng, D., Ma, X., &Sun, X. (2021). The enhancement of antiviral activity of chloroinconazide by aglinate-based nanogel and its plant growth promotion effect. *Journal of Agricultural and Food Chemistry*, *69*(17), 4992–5002.

Ma, X., Zhu, X., Mu, Y., Gao, C., He, W., Ran, M., Cai, L., Fan, G., Ma, G., &Sun, X. (2022). Fabrication of polydopamine reduced CuO nanoparticle-alginate composite nanogels for management of *Pseudomonas synringae* pv. tabaci in tobacco. *Pest Management Science*, *79*(3), 1213–1224.

Machado, A. H., Lundberg, D., Ribeiro, A. J., Veiga, F. J., Lindman, B., Miguel, M. G., &Olsson, U. (2012). Preparation of calcium alginate nanoparticles using water-in-oil (W/O) nanoemulsions. *Langmuir*, *28*(9), 4131–4141.

Mali, S. C., Raj, S., &Trivedi, R. (2020). Nanotechnology a novel approach to enhance crop productivity. *Biochemistry and Biophysics Reports*, *24*, 100821.

Mathieu, S., Henrissat, B., Labre, F., Skjåk-Bræk, G., &Helbert, W. (2016). Functional exploration of the polysaccharide lyase family PL6. *PLOS ONE*, *11*(7), e0159415.

Meena, S., Tripathi, A. D., &Ts, R. L. (2020). Optimization and characterization of Alginic acid synthesized from a novel strain of Pseudomonas stutzeri. *Biotechnology Reports*, *27*, e00517.

Miyazawa, T., Itaya, M., Burdeos, G. C., Nakagawa, K., &Miyazawa, T. (2021). A critical review of the use of surfactant-coated nanoparticles in nanomedicine and food nanotechnology. *International Journal of Nanomedicine*, *16*, 3937.

Mohammed, A., Bissoon, R., Bajnath, E., Mohammed, K., Lee, T., Bissram, M., John, N., Jalsa, N. K., Lee, K.-L., &Ward, K. (2018). Multistage extraction and purification of waste Sargassum natans to produce sodium alginate: An optimization approach. *Carbohydrate Polymers*, *198*, 109–118.

Moradipour, M., Saberi-Riseh, R., Mohammadinejad, R., &Hosseini, A. (2019). Nano-encapsulation of plant growth-promoting rhizobacteria and their metabolites using alginate-silica nanoparticles and carbon nanotube improves ucb1 pistachio micropropagation. *Journal of Microbiology and Biotechnology*, *29*(7), 1096–1103.

Mulaudzi, T., Nkuna, M., Sias, G., Doumbia, I. Z., Njomo, N., &Iwuoha, E. (2022). Antioxidant capacity of chitosan on sorghum plants under salinity stress. *Agriculture*, *12*(10), 1544.

Osman, N., Suliman, T., &Osman, K. (2020). Characterization of native alginates of common alginophytes from the red sea coast of Sudan. *International Journal of Secondary Metabolite*, *7*(4), 266–274.

Paques, J. P., van der Linden, E., van Rijn, C. J., &Sagis, L. M. (2014). Preparation methods of alginate nanoparticles. *Advances in Colloid and Interface Science*, *209*, 163–171.

Pathania, R., Najda, A., Chawla, P., Kaushik, R., &Khan, M. A. (2022). Low-energy assisted sodium alginate stabilized Phyllanthus niruri extract nanoemulsion: Characterization, in vitro antioxidant and antimicrobial application. *Biotechnology Reports, 33*, e00711.

Pistone, S., Qoragllu, D., Smistad, G., &Hiorth, M. (2015). Formulation and preparation of stable cross-linked alginate–zinc nanoparticles in the presence of a monovalent salt. *Soft Matter, 11*(28), 5765–5774.

Pongprayoon, W., Siringam, T., Panya, A., &Roytrakul, S. (2022). Application of chitosan in plant defense responses to biotic and abiotic stresses. *Applied Science and Engineering Progress, 15,* 3865.

Rahimivand, M., Tafvizi, F., &Noorbazargan, H. (2020). Synthesis and characterization of alginate nanocarrier encapsulating Artemisia ciniformis extract and evaluation of the cytotoxicity and apoptosis induction in AGS cell line. *International Journal of Biological Macromolecules, 158*, 338–357.

Rashedy, S. H., Abd El Hafez, M. S., Dar, M. A., Cotas, J., &Pereira, L. (2021). Evaluation and characterization of alginate extracted from brown seaweed collected in the Red Sea. *Applied Sciences, 11*(14), 6290.

Rivera, M. C., Pinheiro, A. C., Bourbon, A. I., Cerqueira, M. A., &Vicente, A. A. (2015). Hollow chitosan/alginate nanocapsules for bioactive compound delivery. *International Journal of Biological Macromolecules, 79*, 95–102.

Saberi Riseh, R., Gholizadeh Vazvani, M., Ebrahimi-Zarandi, M., &Skorik, Y. A. (2022). Alginate-induced disease resistance in plants. *Polymers, 14*(4), 661.

Salachna, P., Grzeszczuk, M., Meller, E., &Soból, M. (2018). Oligo-alginate with low molecular mass improves growth and physiological activity of Eucomis autumnalis under salinity stress. *Molecules, 23*(4), 812.

Sarmento, B., Ferreira, D., Veiga, F., &Ribeiro, A. (2006). Characterization of insulin-loaded alginate nanoparticles produced by ionotropic pre-gelation through DSC and FTIR studies. *Carbohydrate Polymers, 66*(1), 1–7.

Saxena, A., Sharda, S., Kumar, S., Kumar, B., Shirodkar, S., Dahiya, P., &Sahney, R. (2022). Synthesis of alginate nanogels with polyvalent 3D transition metal cations: Applications in urease immobilization. *Polymers, 14*(7), 1277.

Sharavdorj, K., Byambadorj, S. O., Jang, Y., &Cho, J. W. (2022). Application of magnesium and calcium sulfate on growth and physiology of forage crops under long-term salinity stress. *Plants, 11*(24), 3576.

Shroff, P., Parikh, S., &smt, S. M. (2018). Production and characterization of alginate extracted from *Paenibacillus riograndensis. Research Journal of Life Sciences, Bioinformatics, Pharmaceutical and Chemical Sciences, 4*(6), 560–575.

Singh, B., Upadhyay, A. K., Al-Tawaha, T. W., Al-Tawaha, A. R., &Sirajuddin, S. N. (2020, April). Biofertilizer as a tool for soil fertility management in changing climate. In *IOP Conference Series: Earth and Environmental Science* (Vol. 492, No. 1, p. 012158). IOP Publishing.

Singh, S. K., &Gross, R. A. (2001). Overview: Introduction to polysaccharides, agroproteins, and poly (amino acids) In: R. A. Gross & C. Scholz (eds) *Biopolymers from Polysaccharides and Agroproteins ACS Sumposium Series.* American Chemical Society, Washington, DC.

Spadari, C. D. C., de Bastiani, F. W. M. D. S., Lopes, L. B., &Ishida, K. (2019). Alginate nanoparticles as non-toxic delivery system for miltefosine in the treatment of candidiasis and cryptococcosis. *International Journal of Nanomedicine, 14*, 5187–5199.

Taghe, S., &Mirzaeei, S. (2019). Preparation and characterization of novel, mucoadhesive ofloxacin nanoparticles for ocular drug delivery. *Brazilian Journal of Pharmaceutical Sciences, 55*, e17105.

Thai, H., Thuy Nguyen, C., Thi Thach, L., Thi Tran, M., Duc Mai, H., Thi Thu Nguyen, T., Duc Le, G., Van Can, M., Dai Tran, L., Long Bach, G., Ramadass, K., Sathish, C. I., &Van Le, Q. (2020). Characterization of chitosan/alginate/lovastatin nanoparticles and investigation of their toxic effects in vitro and in vivo. *Scientific Reports, 10*(1), 1–15.

Thomas, D., KurienThomas, K., &Latha, M. S. (2020). Preparation and evaluation of alginate nanoparticles prepared by green method for drug delivery applications. *International Journal of Biological Macromolecules, 154*, 888–895.

Thor, K. (2019). Calcium—Nutrient and messenger. *Frontiers in Plant Science, 10*, 440.

Torres, M. R., Sousa, A. P., Silva Filho, E. A., Melo, D. F., Feitosa, J. P., de Paula, R. C., &Lima, M. G. (2007). Extraction and physicochemical characterization of Sargassum vulgare alginate from Brazil. *Carbohydrate Research, 342*(14), 2067–2074.

Trica, B., Delattre, C., Gros, F., Ursu, A. V., Dobre, T., Djelveh, G., Michaud, P., &Oancea, F. (2019). Extraction and characterization of alginate from an edible brown seaweed (*Cystoseira barbata*) harvested in the Romanian Black Sea. *Marine Drugs, 17*(7), 405.

Twizeyimana, E., Zhang, S., Mukerabigwi, J. F., &Ge, Z. (2022). Oxidized alginate hydrogel-based derivatives with optimized features for cell culture scaffold. *Macromolecular Research, 30*(4), 238–244.

Üçüncü Tunca, E., Terzioğlu, K., &Türe, H. (2017). The effects of alginate microspheres on phytoremediation and growth of Lemna minor in the presence of Cd. *Chemistry and Ecology*, *33*(7), 652–668.

van der Merwe, R. D. T., Goosen, N. J., &Pott, R. W. M. (2022). Macroalgal-derived alginate soil amendments for water retention, nutrient release rate reduction, and soil ph control. *Gels*, *8*(9), 548.

Vu, C. T., &Wu, T. (2022). MgAl-layered double hydroxides/sodium alginate beads for nitrate adsorption from groundwater and potential use as a slow-release fertilizer. *Journal of Cleaner Production*, *379*, 134508.

Wang, A., Siddique, B., Wu, L., Ahmad, I., &Liu, X. (2020). Sodium alginate edible coating augmented with essential oils maintains fruits postharvest physiology during preservation: A review. *International Journal of Multidisciplinary Research and Development*, *7*(8), 135–140.

Xiang, S., Ma, X., Shi, H., Ma, T., Tian, C., Chen, Y., Chen, H., Chen, X., Luo, K., Cai, L., Wang, D., Xue, Y., Huang, J., &Sun, X. (2019). Green synthesis of an alginate-coated silver nanoparticle shows high antifungal activity by enhancing its cell membrane penetrating ability. *ACS Applied Bio Materials*, *2*(9), 4087–4096.

Yu, H. D., Qiu, D., &Damron, F. H. (2011). *Methods of Producing Bacterial Alginates*. US 2011/0184157 A1, United States.

Yoshida, N., Takase, R., Sugahara, Y., Nambu, Y., &Hashimoto, W. (2022). Direct production of polyhydroxybutyrate and alginate from crude glycerol by Azotobacter vinelandii using atmospheric nitrogen. *Scientific Reports*, *12*(1), 1–9.

Zhang, Y., Yin, H., Zhao, X., Wang, W., Du, Y., He, A., &Sun, K. (2014). The promoting effects of alginate oligosaccharides on root development in Oryza sativa L. mediated by auxin signaling. *Carbohydrate Polymers*, *113*, 446–454.

Zhao, H. J., &Tan, J. F. (2005). Role of calcium ion in protection against heat and high irradiance stress-induced oxidative damage to photosynthesis of wheat leaves. *Photosynthetica*, *43*(3), 473–476.

17 The Application of Nanoparticles in Phytoremediation of Soil and Water Pollution

Amna Mustafa, Hajra Abid, Khawaja Shafique Ahmad, Ansar Mehmood, Fahim Nawaz, Muhammad Asif Shehzad, Saiqa Nazir, and Sajjad Hussain

17.1 INTRODUCTION

Anthropogenic activities such as rapid industrialization, metal processing, excessive pesticide utilization, burning of fossil fuels, urbanization, and synthetic chemical production result in the release of toxic metals and their accumulation in the environment (Park et al., 2021). These toxic metals, collectively called heavy metals (HMs), include different elements like mercury (Hg), cadmium (Cd), chromium (Cr), arsenic (As), nickel (Ni), lead (Pb), zinc (Zn), and copper (Cu). Heavy metals are sourced from both human and natural sources such as sewage sludge, water obtained from the gas and oil industry (Neff et al., 2011; Pichtel, 2016), utilization of phosphate fertilizer in agriculture (Rafique and Tariq, 2016; Hamzah et al., 2016), mining and smelting of metals (Chen et al., 2016), use of pesticides (Iqbal et al., 2016), combustion of fossil fuels, and electroplating (Muradoglu et al., 2015).

Human activities and nature both contribute to the presence of metalloids and heavy metals in the ecosystem. Among the man-made causes are water produced by the oil and gas industry, the usage of phosphate fertilizers in agriculture, sewage sludge, metal mining, and smelting sources (Hamzah et al., 2016; Neff et al., 2011; Pichtel, 2016; Rafique and Tariq, 2016; Muradoglu et al., 2015; Farahat and Linderholm, 2015). Due to industrialization and urbanization, the amount of heavy metals in the environment has dramatically increased recently, creating concern on a global scale (Ashraf et al., 2019; Suman et al., 2018). Furthermore, excessive mining, waste management, rapid urbanization and industrialization, and continuous use of fertilizers and agrochemicals have contributed to the escalation of heavy metal pollution, making it an increasingly urgent issue (Zhai et al., 2018).

Over 10 million sites worldwide are contaminated, with heavy metal pollution affecting more than 50% of them, according to Khalid et al. (2017). Heavy metal contamination has become a major concern due to the frequent detection of HMs in blood and food. The impact of the stress of heavy metals on the growth of plants has a ripple effect on human health through the food chain, as highlighted by Arif et al. (2019).

Phytoremediation is also referred to as natural green biotechnology. Compared with other bioremediation techniques, phytoremediation is more adaptable, causes less harm to the soil structure, has an attractive appearance, and has a higher level of public acceptance (Costello, 2003). It is one of the most popular bioremediation techniques, and it involves the absorption of ionic compounds in soil by plant root systems, equivalent at maximum concentration less than 0.05 mg s (Gong et al., 2018). To restore polluted soil and control soil fertility, the root system extends into the matrix of soil

DOI: 10.1201/b23308-17

and develops a rhizosphere ecosystem to control and collect bioavailable heavy metals (DalCorso et al., 2019; Jacob et al., 2018).

Nanoparticles (NPs) are materials that possess at least one dimension less than 100 nanometers and can be synthesized with various particle sizes, geometries, and functions, as per the requirements (Adeel et al., 2020). The development of nanoscience has opened up a new direction for the establishment of soil remediation due to the unique characteristics of NPs, which not only have a better impact on repairs than conventional materials but can also add new functions Moharem et al., 2019; Liu et al., 2020).

Nanomaterials (NMs) have gained prominence as a significant research area for heavy metal soil treatment in recent years, and the US nanotechnology initiative has categorized them as one of the eight cross sectors in the field of environmental development, due to their large specific surface area, high activity point, and impressive photoelectric characteristics (Tratnyek and Johnson, 2006). The advancement of phytoremediation of contaminated soil has also been facilitated by the development of nanoscience (Liu et al., 2020).

Numerous studies have shown that nanoparticles have many advantages and are rapidly being used in different developing techniques. One promising application is the utilization of NPs for remediating surface water, soil contamination, and groundwater (Nutt et al., 2006; Thatai et al., 2014; Wang et al., 2014). Another emerging technique is nano-phytoremediation, which involves using plants with high metal uptake efficiency to remove polluted water and soil. Nanoparticles can help reduce toxicity and increase phytoremediation by improving the plant's ability to withstand stress (Vázquez-Nez, 2020). The unique properties of nanoparticles, such as their increased reactivity, high surface area, and properties emerging from their nanoscale size, make them suitable materials for contaminant removal (Fan et al., 2013). Nano-phytoremediation, which involves incorporating nanomaterials into phytoremediation, is a green technology that can store polluting materials that have been degraded but not completely removed from the environment. The utilization of nanomaterials in phytoremediation has been shown to increase the decontamination efficiency and regeneration rate compared with other phytoremediation methods alone (Akansha et al., 2018).

Soil is under threat due to human activities. Rapid industrialization and various human practices have led to an increase in the buildup of hazardous compounds in soil and water bodies in the past few decades (Amundson et al., 2015). These compounds are harmful to the environment and human health, posing an important threat to sustainable development (Cachada et al., 2012; Li et al., 2014).

Therefore, it is critical to remediate contaminated water and soil. This process includes pollutant removal or reduction to safe levels, making the environment suitable for human use, and protecting natural resources. Remediation methods include physical, chemical and biological techniques, which can be applied alone or in combination, depending on the nature and extent of contamination (US EPA, 2017) In conclusion, the remediation of contaminated soil and water is vital for sustainable development and human well-being. It is crucial to adopt appropriate remediation techniques to ensure the safety and protection of our environment.

17.2 PHYTOREMEDIATION OF CONTAMINATED SOIL AND WATER

Plants can absorb, degrade, and/or confine pollutants from the soil, water, air, and sediment. Several phytoremediation techniques can be used for affected soils. The plants act as a barrier, preventing the heavy metals from being absorbed by other organisms and reducing the risk of their accumulation in the food chain (Chandra & Yadav, 2010). Phytostabilization involves using plants to decrease the bioavailability of heavy metals in soil. Phytoextraction is another phytoremediation technique that uses plants to extract and remove heavy metals from soil (Eapen et al., 2003). Phytovolatilization involves using plants to absorb heavy metals from soil and discharge them as volatile compounds into the atmosphere. This technique is useful for remediating soils contaminated with volatile organic compounds (VOCs) and heavy metals such as mercury (Jadia & Fulekar, 2009).

Phytofiltration is a technique of phytoremediation that involves the absorption or adsorption of heavy metal ions through hydroponically cultured plants. This technique is particularly useful for treating water (Marques et al., 2009). Rhizodegradation and phytodegradation are two other phytoremediation techniques that break down organic contaminants. Rhizodegradation involves the degradation of contaminants in the root zone of plants, while phytodegradation involves the degradation of contaminants within the plant tissue itself (Pilon-Smits and Pilon, 2005). Figure 17.1 shows the different phytoremediation techniques for remediating heavy metal–contaminated soils.

17.3 PHYTOEXTRACTION

Phytoextraction is a technique of phytoremediation that entails employing plants to extract pollutants, such as heavy metals and metalloids, from contaminated soil and water. It is a more economically viable way of phytoremediation, since it is a long-term means of eliminating heavy metals from contaminated soil. Hyperaccumulator plants, which can accumulate pollutants to levels 100 times higher than most plants, are commonly used for phytoextraction (Reeves et al., 2018; Shaheen and Rinklebe, 2015).

Phytoextraction involves the absorption of pollutants by plant roots. They are then transported into the aboveground portion of the plant and eliminated during cultivation. The contaminated plant material is then harvested, removing the pollutant from the soil or water. Phytoextraction is most commonly applied to soil or water contaminated with toxic metals (Jacob et al., 2018). This technique is an efficient and sustainable method for removing pollutants from contaminated environments. It offers an environmentally friendly, cost-effective alternative to traditional remediation methods and has the potential to address some of the most challenging environmental problems we face today.

17.4 PHYTOSTABILIZATION

Phytostabilization is a technique that uses metal-tolerant plant species to restrict heavy metals in contaminated soil and prevent them from migrating into the environment (Marques et al., 2009). Several mechanisms are involved in phytostabilization, including heavy metal deposition (Gerhardt et al., 2017). This process has been shown to effectively treat toxic metals, including lead, cadmium, and zinc (Jadia & Fulekar, 2009). However, the primary limitation of phytostabilization is that it does not remove pollutants from the soil. Instead, it immobilizes the metals, preventing their

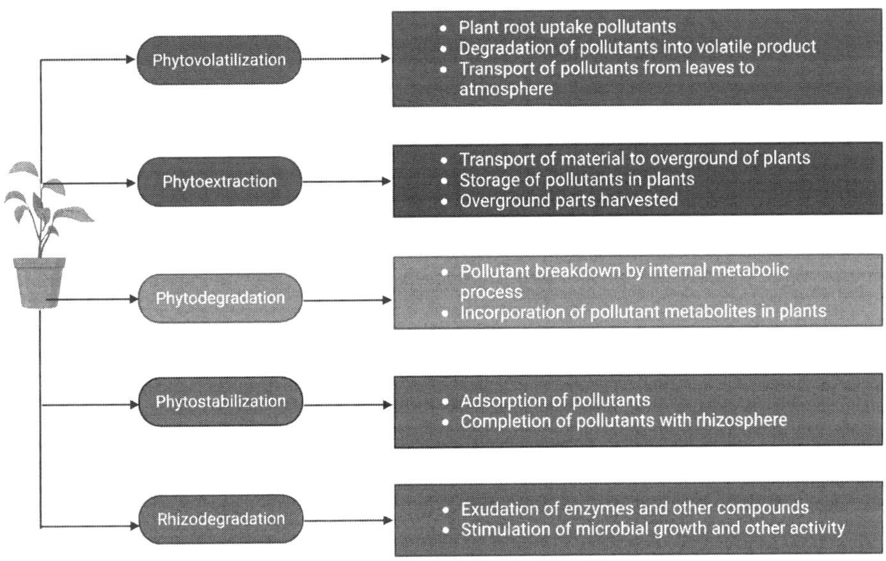

FIGURE 17.1 Technologies of phytoremediation for reduction of soil pollution.

movement into the environment. As a result, phytostabilization is typically used with other methods to achieve a more comprehensive cleanup of contaminated sites (Huang et al., 2019). Overall, phytostabilization is a promising technique for managing heavy metal contamination in soil. However, its limitations mean that it is best used as part of a comprehensive remediation strategy, rather than as a standalone approach.

17.5 PHYTOVOLATILIZATION

Phytovolatilization is a sustainable and cost-effective method that has gained considerable attention in recent years. The process involves the absorption of pollutants by plants from the soil, transforming them into less toxic volatile forms, and releasing them into the atmosphere by transpiration through the foliage or leaf system. Research has shown that phytovolatilization can effectively remediate HMs, such as Hg, Se, and As, as well as organic contaminants. Phytovolatilization works for the remediation of HMs, such as selenium, mercury, and arsenic (Mahar et al., 2016). The study found that phytovolatilization can reduce the metal concentration in soil, and released volatile compounds are less toxic to the environment. Another study by Sakakibara et al. (2010) explored the utilization of phytovolatilization for the remediation of soil contaminated by mercury. The authors reported that plants can transform mercury into gaseous species, which can be released into the atmosphere through transpiration. The study found that the phytovolatilization technique was effective for the remediation of contaminated soil of mercury, and volatile compounds released by plants were less toxic to the environment.

17.6 PHYTOFILTRATION

Phytofiltration is a form of phytoremediation, which involves using plants to eliminate pollutants from soil, water, or air (Mesjasz-Przybyłowicz et al., 2004). Heavy metals, one of the most prevalent types of contaminants, can be removed using phytofiltration. Root exudates alter the pH of the soil surrounding the roots (known as the rhizosphere) and may cause heavy metals to precipitate on plant roots, decreasing their ability to migrate to groundwater (Javed et al., 2019). Overall, phytofiltration is a promising and sustainable approach to removing contaminants from polluted water sources.

17.7 RHIZODEGRADATION

The process of breaking down contaminants in soil is known as biodegradation. This process can be made more effective by the occurrence of an active root zone, which is referred to as increased rhizosphere biodegradation. In the degradation of pollutants, microorganisms show a crucial role in metabolic methods. They break down complex organic compounds into simpler ones that are easier to utilize. Plants play a significant part in the process of biodegradation. Their roots break up the soil, providing microbial growth surfaces for attachment, and excrete saccharides, amino acids, enzymes, and other substances that increase metabolic microbial activity. Some grasses with large root systems, such as tall fescue and ryegrass, are commonly used for rhizodegradation because the rhizospheric microorganisms and root interaction are essential to success. In summary, the biodegradation of contaminants in soil is a complex mechanism that includes the metabolic activity of microorganisms and the active participation of plants. The success of this process can be enhanced through the plant's utilization of a large system of roots and an active root zone (Lin et al., 2011).

17.8 PHYTODEGRADATION

Plants absorb an organic contaminant, which is then decomposed by intracellular biochemical processes. The plant roots can absorb organic pollutants, and these pollutants are transformed into easy-to-integrate compounds that can be used in plant tissues through internal metabolic processes. By producing a variety of enzymes with specialized uses, such as nitroreductase, peroxidases, and dehalogenases, plants can facilitate the degradation of organic contaminants (Lee, 2013). Different plant species have specific degradation pathways that are still poorly understood.

17.9 FACTORS AFFECTING PHYTOREMEDIATION

Inaccessible heavy metals in soil can hinder metal uptake by plants through bioaccumulation (Blaylock and Huang, 2000). Only a small portion of heavy metals present in soil are in soluble forms that can be absorbed by plants, with Zn and Cd being more easily absorbed than other heavy metals (Lasat, 1999; Prasad, 2003; Wang et al., 2006).

Despite the typically low accessibility and uptake of HMs in affected water (Ashraf and Asghar, 2019; Ke and Wang, 2018), the associated soil microorganism and plant utilization in phytoremediation has shown promising results in reducing pollution levels and mitigating negative environmental consequences (Sharma et al., 2021c; Zhang et al., 2018). Phytoremediation is dependent on a variety of factors, including metal speciation, environmental conditions, and plant-related variables (Kang et al., 2017). The restoration of agricultural regions and the productive growth of native plants in contaminated areas also suggest the efficacy of phytoextraction in concentrating various harmful substances (Fernandez et al., 2017).

The criteria used to choose plants for phytoremediation include rapid population development in polluted areas, significant surface area relative to the water body, and high translocation potential (Susarla et al., 2002). These criteria are essential, because plants with these characteristics can efficiently remove contaminants from the environment. One of the most effective phytoremediation approaches is the use of brake ferns (*Pteris vittata*). These ferns have high rates of root-to-shoot metal transduction, which makes them excellent for heavy metal phytoextraction. The translocation potential (TP) or BCF (bioconcentration factor) affects plants' susceptibility to phytoremediation, and the BCF value is greater than 1 for brake ferns (*P. vittata*), indicating their effectiveness in metal uptake (Ali et al., 2013). Phytoextraction, which relies on metal hyperaccumulation throughout the entire plant, is the most effective phytoremediation method. To be effective, a plant should have a strong root system for phytoextraction of metals and be fast-growing with a high biomass output per hectare. After harvesting is finished for the season, the contaminated biomass-rich product needs to be further processed to remove the secondary metabolic waste it produces. This can be done either naturally or artificially. Additionally, the phytoextraction process is a cost-effective method that can be used to produce yield biomass, as the energy can be recovered from burning the biomass (Kramer, 2010). *Salix* and *Populus* species are also commonly utilized in phytoremediation technologies due to their high biomass production, fast growth rates, and tolerance of heavy metals. These plants have been used in numerous studies and are effective in removing contaminants from soil and water bodies (Zhao et al., 2020; Adesodun et al., 2010).

17.10 REMOVING POLLUTANTS THROUGH PHYTOREMEDIATION ASSISTED WITH NANOMATERIALS

Nano-bioremediation has gained popularity as a promising approach to address environmental contamination. Various studies have reported successful remediation of contaminants, including inorganic and organic pollutants, heavy metals, and e-waste using nanoparticles produced through bacteria, algae, and fungi as well as higher plants with nanotechnology usage. Several examples of nanoparticle-supported phytoremediation are shown in Table 17.1 (Sarkar et al., 2020; Siddiqui et al., 2017).

The variety of nano-phytoremediation processes include nano-phytoaccumulation, nano-phytodegradation, nano-phytovolatilization, nano-phytohydraulics, nano-phytostabilization, and nano-rhizofiltration, similar to phytoremediation (Ezhil Vilian et al., 2020). By enhancing the functions

TABLE 17.1
Latest Examples of Assisted Phytoremediation with Nanomaterials

Pollutants	Plant species	Nanomaterials	Role of nanomaterials	Main results	Reference
Trinitrotoluene	Panicum (*Panicum maximum* Jacq.)	nZVI	Removal by nZVI particles by direct trinitrotoluene	Nanoparticles of zero-valent iron enhanced the efficiency removal from 85.7% to 100% after 20 days	Jiamjitrpanich et al., 2012
Endosulfan	Tulsi (*Ocimum sanctum* L.), Chittaratha (*Alpinia calcarata* R.)	nZVI	Endosulfan direct removal through particles of nZVI	The rate of removal from soil was enhanced from 81.2% to 100%	Pillai and Kottekottil, 2016
Pb	Wood (*Populus deltoides* Bartr.)	Nanoparticles (carbon black nano)	Nanomaterials directly stabilize lead (Pb), thereby alleviating phytotoxicity and stimulating plant growth.	Rate of removal after 12-months increased from 31.7% to 46.5%	Liang et al., 2017
Cr	Pea (*Pisum sativum* L.)	Silicon NPs	In Cr stress tolerance, plant growth promoted by alleviating phytotoxicity	The accumulation concentration of Cr in the shoot and root decreased from 14 to 2.6	Tripathi et al., 2015
As	*Isatis cappadocica* (Desv.)	Salicylic acid NPs	Increasing rate of absorption and utilization for plant growth	Accumulation concentration in root and shoot reached 705–1188 mg/kg	Souri et al., 2017
As, Cd, Pb, and Zn	Sunflower (*Helianthus annus* L.)	nZVI	nZVI particles are direct pollutant	It is used for absorption of Pb, As, Cd, and Zn in shoot and root lessened through 50–60%	Gong et al., 2017
Pb and Cd	Rye (*Secale monta-num* Guss.)	Nano-silica	Phytoavailability increased and growth promoted	Increased concentration of Cd (208.6 mg/kg) and Pb (533.6 mg/kg)	Venkatachalam et al., 2017

of microorganisms and plants through adding nanoencapsulated enzymes, nano-phytoremediation can help break down complex organic chemicals that are resistant to degradation into simpler ones. Nanoparticles with high affinity for metal/metalloid absorption, such as nano-sized zero-valent iron owing to its high reactivity to effectively remove a wide spectrum of water contaminants (Kumar et al., 2021; Huang et al., 2019). Nanotechnology can accelerate phytoremediation by enabling access to previously inaccessible locations and encouraging in-situ repair, among other things. In summary, nano-bioremediation and nano-phytoremediation hold great promise for addressing environmental contamination sustainably and effectively.

17.11 PHYTOREMEDIATION FACILITATED BY NANOMATERIALS FOR REMOVAL OF HEAVY METALS

Soil and water pollution by HMs creates a severe hazard to food and the health of humans on a global scale. The in-situ cleanup of HMs and phytoremediation are frequently applied (Liang et al., 2017). According to reports (Khan & Bano, 2016; Vtkova et al., 2018), the phytoremediation of water and soil contaminated by Cr, Pb, Cd, Zn, and Ni can be increased by the application of nanomaterials. Due to their prevalence in contaminated areas, the heavy metals lead and cadmium have undergone the most research.

17.11.1 Lead

Lead is a commonly used industrial metal with a variety of applications, including in solder, ammunition, fuel additives, and storage batteries. However, as a well-known soil and water contaminant, it poses serious hazards to health (Yu et al., 2001). In phytoremediation, removing pollutants by using plants is a promising technique for lead removal from contaminated water and soil. The most well-known phytoextraction technique is the phytoremediation method for removing lead from polluted water and soil (Ali et al., 2013). Ryegrass (*Lolium perenne* L.) is a plant species that grows quickly, has a high tolerance to Pb, and is inexpensive, making it a popular choice for experimental research in phytoextraction of Pb from polluted soil and water. Ryegrass is more effective at extracting Pb when nanomaterials are used. For instance, Liang et al. (2017) evaluated the effectiveness of remediation over different periods. The findings demonstrated that adding nano-hydroxyapatite (0.2% w/w) after 1.5 months greatly enhanced the Pb accumulation in plants over the ground area. After one month, ryegrass treated with nano-hydroxyapatite removed more than 30% of the Pb contained in the soil, and after three months, the most effective treatment time, removal rates reached 44.39%. After a year, the removal rate rose to 46.55%.

7.11.2 Cadmium

Industrial activities, for example, storage batteries, mining, smelting, electroplating, phosphate fertilizers, and color pigments, are known to discharge the toxic metal cadmium (Cd) into water and soil, creating a severe danger to the environment and health of humans (Godt et al., 2006). Phytoremediation using hyperaccumulators is a promising method for Cd removal from contaminated water and soil, but the limited number of Cd hyperaccumulator species restricts its application (Kirkham, 2006). Recent research has shown that nanomaterials can improve the efficiency of Cd phytoextraction from polluted soil and water. For instance, TiO_2 nanoparticles have been found to increase Cd accumulation in soybean significantly (Singh and Lee, 2016). The effects of TiO_2 nanoparticles on the dispersion and accumulation of Cd in plant and contaminated soil with TiO_2 nanoparticles at different concentrations of 100, 200, 300 mg/kg, and detected a significant enhance Cd accumulation in soybean roots and shoots (Gong et al., 2017). The contaminated soil was treated with TiO_2 nanoparticles at different concentrations of 100,

200, and 300 mg/kg, and a significantly enhanced Cd accumulation was detected in soybean roots and shoots (Gong et al., 2017). At the highest concentration of TiO_2 nanoparticles (300 mg/kg), the maximum Cd accumulation was 1534.7 mg/g. Similarly, starch-stabilized nano zero-valent iron (nZVI) particles that provide reducing power for reactions with pollutants have been shown to enhance Cd accumulation in ramie seedlings grown in Cd-contaminated sediment (Yang et al., 2022).

7.11.3 Arsenic

Arsenic pollution in the soil is a growing concern due to its high toxicity and carcinogenicity, which may result from human activities such as pesticide and herbicide usage, phosphate fertilizers, wood preservatives, and industrial processes (Vithanage et al., 2012). Phytoremediation techniques, such as phytoextraction and phytostabilization, are commonly employed to mitigate soil arsenic pollution. Phytoextraction is the preferred method, since it involves arsenic accumulation and uptake via plants from contaminated soil. According to Lei et al. (2018), phytoextraction has shown promising results in removing arsenic from soil.

Nanotechnology can play a critical role in enhancing the efficacy of phytoremediation. Salicylic acid nanoparticles can increase the phytoextraction of arsenic through *Isatis cappadocica*. Salicylic acid nanoparticles were added to the arsenic phytoextraction system after pretreating the plant seedlings with salicylic acid 250 mM nanoparticles for ten days (Souri et al., 2017). The researchers found that the addition of salicylic acid nanoparticles significantly increased the efficacy of phytoremediation, as well as improving plant growth. The shoots and roots had maximum arsenic accumulations of 705 and 1188 mg/kg, respectively. Vtkova et al. (2018) found that the use of nZVI particles improved the stability of arsenic in the sunflower rhizosphere. After a five-week growth period, the level of As in pore soil water decreased by nearly 80% as compared with control, from 5.55 to 0.95 mg/L, and the As buildup in shoot and root decreased by 47 and 27%, respectively. These findings demonstrate the potential of nZVI particles in increasing phytostabilization systems.

7.12 PHYTOREMEDIATION FACILITATED BY NANOMATERIALS FOR ORGANIC POLLUTANT REMOVAL

Organic pollutants, such as organo-chlorine pesticides, chlorinated hydrocarbons, polychlorinated biphenyls, polycyclic aromatic hydrocarbons, phenols, and their derivatives, are commonly found in water and soil and are primarily produced by human activities (Jones & de Voogt, 1999). Due to their toxicity, persistence, and bioaccumulation, these pollutants must be eliminated. A green technology, phytoremediation, to remove pollutants from the environment using plants has been successful in treating water and soil affected by the organic contaminants polycyclic aromatic hydrocarbons, polycyclic aromatic hydrocarbons, pesticides, petroleum, and explosives (Kang, 2014). To improve the effectiveness of phytoremediation, nanomaterials have been applied (Pillai and Kottekottil, 2016; Ma and Wang, 2010; Jiamjitrpanich et al., 2012)

7.13 THE ROLE OF NANOPARTICLES IN PROMOTING PLANT GROWTH

To encourage plant growth in phytoremediation, various tactics are employed (Table 17.2), including the use of transgenic plants, regulators of plant growth, and plant growth-promoting rhizobacteria (PGPR) (Yadu et al., 2018; Nahar et al., 2017).

In recent years, nanomaterials have emerged as a promising tool to increase plant growth. Several nanomaterials, including carbon nanotubes, graphene quantum dots, silver NPs, zinc oxide NPs, iron NPs, and up-conversion NPs, have been reported to enhance the growth of plants (Chakravarty et al., 2015; Khodakovskaya et al., 2013; Tripathi et al., 2015; Praveen et al., 2018;

TABLE 17.2
Several Illustrative Studies about the Effects of Nanomaterials on the Growth of Plants

Nanomaterials	Species of plant	Outcome	Mechanism	Reference
Quantum dots Graphene	Garlic (*Allium sativum* L.), Coriander (*Coriandrum sativum* L.)	Increased	Nanomaterials functioned as pesticides and fertilizer	Chakravarty et al., 2015
Carbon nanotubes	Tomato (*Solanum lycopersicum* L.)	Increased	Activated the plant reproductive system and increased fruit production	Khodakovskaya et al., 2013
Graphene	Tomato (*Solanum lycopersicum* L.)	Decreased	Penetrated vacuoles and reduced biomass production	Zhang et al., 2015
Water-soluble C_{70} fullerenes	Cress (*Arabidopsis thaliana* L.)	Decreased	Caused abnormal cell division, auxin disruption, and retarded growth	Liu et al., 2010
Metal-based nanomaterials, Ag nanoparticles	Kidney beans (*Phaseolus vulgaris* L.)	Increased	Stimulated plant metabolism and nitrogen uptake and accumulated biomass	Das et al., 2018
TiO_2 nanoparticles	Duck meat (*Spirodela polyrrhiza* L.)	Decreased	Inhibited the synthesis of protein, nitrogen fixation, and photosynthesis	Movafeghi et al., 2018
ZnO nanoparticles	Cotton (*Gossypium hirsutum* L.)	Increased	Increased the utilization of zinc and phosphorus and caused phytotoxicity	Venkatachalam Prinyanka et al., 2017
CuO nanoparticles	Indian mustard (*Brassica juncea* L.)	Decreased	Enhanced lignification and reduced growth	Nair and Chung, 2015
nZVI nanoparticles	Cress (*Arabidopsis thaliana* L.)	Enhanced	Caused elongation of the root and induced cell wall loosening	Kim et al., 2014
Upconversion nanoparticles	Mung beans (*Phaseolus radiatus* L.)	Increased	Mechanism not clear	Peng et al., 2012

Ding et al., 2017; Souri et al., 2017; Singh & Lee, 2016). By promoting plant development, nanomaterials can increase the efficacy of phytoremediation systems, lowering phytotoxicity, and enhancing plant resilience to pollution through the genetic expression of enzyme control (Praveen et al., 2018). Similarly, silicon nanoparticles were found to reduce Cr(VI) phytotoxicity against peas by improving antioxidant activities and photosynthetic performance (Tripathi et al., 2015).

Nano-hydroxyapatite, for instance, has been used to enhance the phytoremediation of lead by ryegrass through increasing phosphorus concentration in the soil, thereby improving plant growth (Ding et al., 2017). Also, it has been discovered that TiO_2 nanoparticles can help promote plant growth by accelerating the photosynthetic rate in a Cd phytoextraction technique by soybean (Singh & Lee, 2016) and improving PGPR performance (Timmusk et al., 2018).

7.14 INCREASING PHYTOAVAILABILITY OF POLLUTANTS

The effectiveness of phytoremediation, particularly for extraction, is significantly influenced by the phytoavailability of contaminants. Only accessible forms of contaminants are absorbed by plants. The chemical speciation and distribution of contaminants in soil and water have a significant impact on their phytoavailability. For instance, a phytoavailability study on cadmium in various binding forms revealed that when compared with other oxide minerals in soil and water (alumina, goethite, magnetite, and manganese oxide), adsorbed cadmium on gibbsite is most accessible to reed (Wang et al., 2009; Sheoran et al., 2016; Liang et al., 2017). The phytoremediation process is frequently restricted by low phytoavailability. For instance, lead typically resides in insoluble forms in the soil as a result of adsorption, complexation, and precipitation, making phytoextraction problematic (Zaier et al., 2014; Ren et al., 2018; Habiba et al., 2015; Jacobs et al., 2018).

Two divergent influences can be found on nanomaterials for phytoavailability in affected water and soil (Table 17.3). On the one hand, when contaminants reach cells, nanomaterials can act as a transporter, enhancing the bioavailability of the substance (Su et al., 2013; Wild & Jones, 2009). On the other hand, the bioavailability of contaminants may decrease as a result of their adsorption onto nanomaterials outside living things (Glomstad et al., 2016). Based on this, two essential

TABLE 17.3
Effects of Nanomaterials on the Uptake and Accumulation of Pollutants

Pollutant	Plant species	Nanomaterials	Main effects	References
Dichlorodiphenyldichloroethylene	Soybean (*Glycine max*) and tomato (*Solanum lycopersicum*)	C_{60} fullerenes	Increased uptake by 30%–65%	Torre-Roche et al., 2012
Chlordane, dichlorodiphenyltrichloroethane	Soybean (*Glycine max*) and corn (*Zea mays*)	Multiwalled carbon nanotubes	Decreased the accumulation of pollutants and increased the accumulation of chlordane	Deng et al., 2017
Pb and Cd	Fava bean (*Vicia faba* L.)	Carbon nanotubes	Facilitated the enhancement of Cd and Pb in leaf	Wang et al., 2014
Chlordane and DDE	Lettuce (*Lactuca sativa* L.)	Carbon nanotubes	Decreased pesticides content in roots and shoots	Wu et al., 2018
Tetracycline	Rice (*Oryza sativa* L.)	TiO_2 nanoparticles	Lowered cycline increase in seedlings of rice	Ma et al., 2017
Cd	Soyabean (*Glycine max*)	Carbon nanotubes	Reduced the concentrations of pyrene and suppressed translocation	Zhang et al., 2017
Anthracene, phenanthrene, pyrene	Cucumber (*Cucumis sativus*) and corn (*Zea mays*)	TiO_2, Ag nanoparticles	Increased the accumulation of pollutants	Wu et al., 2018

requirements must be fulfilled to increase the phytoavailability of contaminants. The utilized nanomaterial has the following properties: (1) it may mix with the pollutant (mostly by adsorption) and (2) it is phytoavailable.

To help remove trichloroethylene from contaminated soil and water, Ma and Wang (2010) employed a phytoremediation system with eastern cottonwood and C60 fullerene nanoparticles. Their findings demonstrated that fullerenes improved the plant's absorption of trichloroethylene. Trichloroethylene and fullerene nanoparticles were co-transported, and this allowed the authors to explain the experimental result. With the uptake of NPs, the trichloroethylene adsorbed on fullerene reached the plant. Torre-Roche et al. (2012) indicated that the presence of C60 fullerene enhanced the accumulation of dichlorodiphenyldichloroethylene (p,p′-DDE) in soybean, tomato, and zucchini. The effect was most noticeable in roots, where the fullerene addition boosted the uptake of pollutants by 30%–65%. A few additional nanomaterials, including Fe_3O_4 nanoparticles, short carbon nanotubes, quantum dots, TiO_2 nanoparticles, and silica nanoparticles, have also been shown to be directly ingested by plants (Wang et al., 2016). This makes it possible to use more nanomaterials in the future to increase the phytoavailability of contaminants in phytoremediation systems.

17.15 EFFICIENCY OF PHYTOREMEDIATION

The amount of highly toxic and bioavailable (phytostabilization) pollutant, or pollutant accumulation in the tissue of plants (phytoextraction), is usually an index for a judgment of subsequent remedial treatment and to measure the success of phytoremediation (Liang et al., 2017). So, to identify the positive effects of NPs on remediation efficiency, when assessing the efficiency of phytoremediation by nanoparticles, an experimental group without nanomaterials is often used as a control. In condensed examples of nanomaterial-facilitated phytoremediation, the phytoremediation effectiveness is summarized by the RAS (residual amount of pollutant in soil), AAP (accumulated amount of pollutant in the tissue of plants), BCF (bioconcentration factor), TF (translocation factor), and RF (remediation factor). The RAS can accurately reflect the increased degradation efficiency achieved by the use of nanomaterials for organic contaminants. In their article, Pillai and Kottekottil (2016) provide a common example. Three control groups were used in the authors' experiments: one that received no treatment at all, one that also received nZVI, and one that received only phytoremediation (Souri et al., 2017),

Processes for phytoextraction are assessed using the TF, BCF, and RF. The following shows the calculation method.

TF ¼ Cshoots = Croots
BCF ¼ Cplant = Csoil
RF ¼ Cplant = Csoil _Wplant = Wsoil

The concentrations of pollutants in harvested plant biomass, shoots of the plant, roots of the plant, and soil are denoted by the letters Cshoots, Croots, Cplant, and Csoil, respectively. The weight of harvested plant biomass and soil is Wsoil and Wplant, respectively. The value of BCF is used to classify the plant as an accumulator (1BCF=10), an excluder (BCF<1), or a superaccumulator (BCF >10) (Lam et al., 2018). To assess the capacity of phytoextraction for affected soil, the RF is computed, and the TF value reflects the capacity of plants to translocate pollutants to the harvestable section (Liang et al., 2017; Moameri et al., 2017). To Liang et al. (2017), who assessed the effectiveness of nano-hydroxyapatite-assisted Pb phytoextraction, the TF could be greatly increased by the nanoparticles after a two-month treatment period. However, the nano-silica level of 500 mg/kg in the phytoremediation study by Moameri et al. (2017) boosted the BCF of Cd in the rye but reduced TF. As a result of a reduction in biomass and growth of plants, the researchers further evaluated the value of RF and discovered no discernible change when compared with control. Both the pollutant and plant change nanomaterials assisted phytoremediation techniques this scenario implies that the RF is more appropriate for assessing the efficacy of phytoremediation.

17.16 CHALLENGES AND RECOMMENDATIONS

In recent years, utilizing nanomaterials in the phytoremediation process has gained increasing attention due to their potential to enhance efficiency. However, there are concerns about the potential environmental risks of nanomaterials. Many nanomaterials are toxic to plants, animals, and soil-based microbial ecosystems (Maurer-Jones et al., 2013). Therefore, it is essential to assess the environmental risks of nanomaterials before using them in phytoremediation. Moreover, nanomaterial phytotoxicity is a significant concern in phytoremediation. Further investigation is needed into the environmental risk posed by NPs to fully comprehend toxicity. To reduce hazards and enhance benefits, the utilization of nanomaterials in phytoremediation must be regulated. The utilization of nanomaterials is still in the experimental stages despite some encouraging outcomes. To ascertain the continuing effectiveness of NMs and their impact on the soil ecosystem, more research is required.

Among various nanomaterials, nZVI has been commonly used for remediating soil and water at a large scale (Mueller & Nowack, 2010). These are challenges associated with its use, such as oxidative corrosion, particle aggregation, and intrusion from water and soil constituents (Crane & Scott, 2012). Therefore, understanding the structure and factors of nanomaterials that is crucial for successful phytoremediation using nZVI. To improve remediation efficacy, additional techniques like nanomaterial-assisted phytoremediation can be used, which involve treatment with chemical additives, agronomic management, and genetic engineering and inoculation of rhizospheric bacteria. To determine the general applicability of nanomaterials in phytoremediation systems for different species of plants, types of soil, pollutants, and climates, further research is necessary.

17.17 CONCLUSION

Phytoremediation, a green biotechnology with numerous benefits over other methods of soil and water contamination remediation, faces limitations such as prolonged remediation periods, unpredictable weather patterns, and phytotoxicity resulting from high levels of contaminants. Enhancing the efficiency of phytoremediation has required extensive efforts. However, recent research has shown that incorporating nanoparticles into phytoremediation can be promising in boosting its effectiveness. This chapter provides an overview of the latest research and discoveries regarding the utilization of nanomaterials to accelerate the phytoremediation of polluted water and soil. In conclusion, nanomaterials have the potential to eliminate pollutants, promote the growth of plants, improve the phytoavailability of pollutants, and eventually, simplify the phytoremediation process. Although still in the experimental stage, employing nanomaterials in phytoremediation presents a novel approach to improving the efficacy of the entire strategy. To bolster these findings and deepen our knowledge, additional research is necessary for developing and then optimizing nanomaterial-assisted phytoremediation.

REFERENCES

Adeel, M., Tingting, J., Hussain, T., He, X., Muhammad Ahmad, M., Irshad, H. M. K., Shakoor N., Xie C., Hao Y., Zhang P., et al. (2020). Bioaccumulation of ytterbium oxide nanoparticles insinuate oxidative stress, inflammatory, and pathological lesions in ICR mice. *Environmental Science and Pollution Research International*, 27(26): 32944–32953.

Adesodun, J. K., Atayese, M. O., Agbaje, T., Osadiaye, B. A., Mafe, O., Soretire, A. A. (2010). Phytoremediation potentials of sunflowers (Tithonia diversifolia and Helianthus annuus) for metals in soils contaminated with zinc and lead nitrates. *Water Air Soil Poll*, 207: 195–201.

Akansha, S., Krishna, K. Y., Sunita, Y., Gupta, N., Jitendra, K. S., Ravi, K., & Vinit, K. (2018). Nanophytoremediation of pollutants from contaminated soil environment: Current scenario and future prospects. *Phytoremediation*, 16, 383–401.

Ali, H., Khan, E., & Sajad, M. A. (2013). Phytoremediation of heavy metals-concepts and applications. *Chemosphere*, 91(7), 869–881.

Amundson, R., Berhe, A. A., Hopmans, J. W., Olson, C., Sztein, A. E., & Sparks, D. L. (2015). Soil science soil and human security in the 21st century. *Science*, 348(6235), 1261071.

Arif, N., Sharma, N. C., Yadav, V., Ramawat, N., Dubey, N. K., Tripathi, D. K., Chauhan D. K., Sahi S. Understanding heavy metal stress in a rice crop: Toxicity, tolerance mechanisms, and amelioration strategies. *Journal of Plant Biology*, 62(4), 239–253.

Ashraf, S., Ali, Q., Zahir, Z. A., Ashraf, S., & Asghar, H. N. (2019). Phytoremediation: Environmentally sustainable way for reclamation of heavy metal polluted soils. *Ecotoxicology and Environment Safety*, 174, 714–727. doi:10.1016/j.ecoenv.2019.02.068.

Ashraf, S., & Asghar, H. N. (2019). Phytoremediation: Environmentally sustainable way for reclamation of heavy metal polluted soils. *Ecotoxicology and Environment Safety*, 15, 714–727.

Blaylock, M., & Huang, J. (2000). Phytoextraction of metals. In I. Raskin & B. D. Ensley (Eds.), *Phytoremediation of Toxic Metals: Using Plants to Clean-Up the Environment* (p. 303). New York: JohnWiley & Sons, Inc.

Cachada, A., Pato, P., Rocha-Santos, T., Ferreira da Silva, E., & Duarte, A. C. (2012). Levels, sources and potential human health risks of organic pollutants in urban soils. *Science of the Total Environment*, 430, 184–192. doi:10.1016/j.scitotenv.2012.04.075.

Chakravarty, D., Erande, M. B., & Late, D. J. (2015). Graphene quantum dots as enhanced plant growth regulators: Effects on coriander and garlic plants. *Journal of the Science of Food and Agriculture*, 95(13), 2772–2778. doi:10.1002/jsfa.7106.

Chandra, R., & Yadav, S. (2010). Heavy metal tolerance in plants: Role of transcriptomics, proteomics, metabolomics, and ionomics. *Frontiers in Environmental Science*, 4, 1–21.

Chen, B., Stein, A. F., Castell, N., Gonzalez-Castanedo, Y., De La Campa, A. S., & De La Rosa, J. (2016). Modeling and evaluation of urban pollution events of atmospheric heavy metals from a large Cu-smelter. *Science of the Total Environment*, 539, 17–25. doi:10.1016/j.scitotenv.2015.08.117.

Costello, C. (2003). Acid mine drainage: Innovative treatment technologies U.S. environmental protection agency, office of solid waste and emergency response technology innovation office, Washington, DC.

Crane, R. A., & Scott, T. B. (2012). Nanoscale zero-valent iron: Future prospects for an emerging water treatment technology. *Journal of Hazardous Materials*, 211–212, 112–125. doi:10.1016/j.jhazmat.2011.11.073.

DalCorso, G., Fasani, E.,Manara, A., Visioli, G., & Furini, A. (2019). Heavy metal pollutions: State of the art and innovation in phytoremediation. *International Journal of Molecular Sciences*, 20(14), 3412.

Das, P., Barua, S., Sarkar, S., Karak, N., Bhattacharyya, P., Raza, N., Bhattacharya, S. S. (2018). Plant extract–mediated green silver nanoparticles: Efficacy as soil conditioner and plant growth promoter. *Journal of Hazardous Materials*, 346, 62–72. doi:10.1016/ j.jhazmat.2017.12.020.

Deng, Y., Eitzer, B., White, J. C., & Xing, B. (2017). Impact of multiwall carbon nanotubes on the accumulation and distribution of carbamazepine in collard greens (*Brassica oleracea*). *Environmental Science: Nano*, 4(1), 149–159. doi:10.1039/C6EN00419A.

Ding, L., Li, J., Liu, W., Zuo, Q., & Liang, S. (2017). Influence of nano-hydroxyapatite on the metal bioavailability, plant metal accumulation and root exudates of ryegrass for phytoremediation in lead-polluted soil. *International Journal of Environmental Research and Public Health*, 14(5), 532. doi:10.3390/ijerph14050532.

Eapen, S., D'Souza, S. F., & Singh, S. (2003). Advances in development of transgenic plants for remediation of xenobiotic pollutants. *Biotechnology Advances*, 21(5), 383–395.

Ezhil Vilian, A. T., Wilson, L. J., & Gopinath, P. M. (2020). Engineered nanoparticles for the environment: Bioremediation and phytoremediation prospects. *Journal of Environmental Management*, 272, 111095.

Fan, G., Cang, L., Qin, W., Zhou, C., Gomes, H., & Zhou, D. (2013). Surfactants-enhanced electrokinetic transport of xanthan gum stabilized nanoPd/Fe for the remediation of PCbs contaminated soils. *Separation and Purification Technology*, 114, 64.

Farahat, E., & Linderholm, H. W. (2015). The effect of long-term wastewater irrigation on accumulation and transfer of heavy metals in Cupressus sempervirens leaves and adjacent soils. *Science of the Total Environment*, 51, 1–7. doi:10.1016./j.scitotenv.2015.01.032.

Fernandez, S., Poschenrieder, C., Marcenò, C., Gallego, J. R., Jiménez-Gámez, D., Bueno, A., & Afif, E. (2017). Phytoremediation capability of native plant species living on Pb-Zn and Hg-As mining wastes in the Cantabrian range, north of Spain. *Journal of Geochemical Exploration*, 174, 10–20.

Gerhardt, K. E., Gerwing, P. D., & Greenberg, B. M. (2017). Opinion: taking phytoremediation from proven technology to accepted practice. *Plant Science: An International Journal of Experimental Plant Biology*, 256, 170–185.

Glomstad, B., Altin, D., Sørensen, L., Liu, J., Jenssen, B. M., & Booth, A. M. (2016). Carbon nanotube properties influence adsorption of phenanthrene and subsequent bioavailability and toxicity to Pseudokirchneriella subcapitata. *Environmental Science and Technology*, 50(5), 2660–2668. doi:10.1021/acs.est.5b05177.

Godt, J., Scheidig, F., Grosse-Siestrup, C., Esche, V., Brandenburg, P., Reich, A., & Groneberg, D. A. (2006). The toxicity of cadmium and resulting hazards for human health. *Journal of Occupational Medicine and Toxicology*, 1(1), 22. doi:10.1186/1745-6673-1-22.

Gong, X., Huang, D., Liu, Y., Peng, Z., Zeng, G., Xu, P., Wan, J. (2018). Remediation of contaminated soils by biotechnology with nanomaterials: Bio-behavior, applications, and perspectives. *Critical Reviews in Biotechnology*, 38(3), 455–468.

Gong, X., Huang, D., Liu, Y., Zeng, G., Wang, R., Wan, J., Xue, W. (2017). Stabilized nanoscale zerovalent iron mediated cadmium accumulation and oxidative damage of Boehmeria nivea (L.) Gaudich cultivated in cadmium contaminated sediments. *Environmental Science and Technology*, 51(19), 11308–11316. doi:10.1021/acs.est.7b03164.

Habiba, U., Ali, S., Farid, M., Shakoor, M. B., Rizwan, M., Ibrahim, M., Ali, B. (2015). EDTA enhanced plant growth, antioxidant defense system, and phytoextraction of copper by Brassica napus L. *Environmental Science and Pollution Research International*, 22(2), 1534–1544. doi:10.1007/s11356-014-3431-5.

Hamzah, A., Hapsari, R. I., & Wisnubroto, E. I. (2016). Phytoremediation of Cadmium-contaminated agricultural land using indigenous plants. *International Journal of Agriculture and Environmental Research*, 2, 8–14.

Huang, Y., Zhang, H., Wang, Y., Liu, Z., He, S., Li, X., & Li, S. (2019). Application of nanoparticles in phytoremediation: A new era for environmental sustainability. *Science of the Total Environment*, 651, 2114–2125.

Iqbal, M., Iqbal, N., Bhatti, I. A., Ahmad, N., & Zahid, M. (2016). Response surface methodology application in optimization of cadmium adsorption by shoe waste: A good option of waste mitigation by waste. *Ecological Engineering*, 88, 265–275. doi:10.1016/j.ecoleng.2015.12.041.

Jacobs, A., De Brabandere, L., Drouet, T., Sterckeman, T., & Noret, N. (2018). Phytoextraction of Cd and Zn with Noccaea caerulescens for urban soil remediation: Influence of nitrogen fertilization and planting density. *Ecological Engineering*, 116, 178–187. doi:10.1016/j.ecoleng.2018.03.007.

Jadia, C. D., & Fulekar, M. H. (2009b). Phytoremediation of heavy metals: Recent techniques. *African Journal of Biotechnology*, 8, 921–928.

Javed, M. T., Tanwir, K., Akram, M. S., Shahid, M., Niazi, N. K., & Lindberg, S. (2019). Chapter 20 – Phytoremediation of cadmium-polluted water/sediment by aquatic macrophytes: role of plant-induced pH changes. In M. Hasanuzzaman, M. N. V. Prasad, and M. Fujita (Eds.), *Cadmium Toxicity and Tolerance in Plants* (pp. 495–529). London: Academic Press.

Jiamjitrpanich, W., Parkpian, P., Polprasert, C., & Kosanlavit, R. (2012). Enhanced phytoremediation efficiency of TNT-contaminated soil by nanoscale zero valent iron. *2nd International Conference on Environment and Industrial Innovation* (Vol. 35, pp. 82–86). IACSIT Press.

Jones, K. C., & de Voogt, P. (1999). Persistent organic pollutants (POPs): State of the science. *Environmental Pollution*, 100(1–3), 209–221. doi:10.1016/S0269-7491(99)00098-6.

Kang, J. W. (2014). Removing environmental organic pollutants with bioremediation and phytoremediation. *Biotechnology Letters*, 36(6), 1129–1139. doi:10.1007/s10529-014-1466-9.

Kang, X., Song, J., Yuan, H., Duan, L., Li, X., Li, N., Liang, X., Qu, B. (2017). Speciation of heavy metals in different grain sizes of Jiaozhou Bay sediments: Bioavailability, ecological risk assessment, and source analysis on a centennial timescale. *Ecotoxicology and Environmental Safety*, 143, 296–306.

Ke, Y., & Wang, W. X. (2018). Metal accumulation, growth, and reproduction of razor clam *Sinonovacula constricta* transplanted in a multi-metal contaminated estuary. *Science of the Total Environment*, 636, 829–837.

Khalid, S., Shahid, M., Niazi, N. K., Murtaza, B., Bibi, I., & Dumat, C. (2017). A comparison of technologies for remediation of heavy metal contaminated soils. *Journal of Geochemical Exploration*, 182, 247–268.

Khan, N., & Bano, A. (2016). Role of plant growth promoting rhizobacteria and Ag-Nano particle in the bioremediation of heavy metals and maize growth under municipal wastewater irrigation. *International Journal of Phytoremediation*, 18(3), 211–221. doi:10.1080/15226514.2015.1064352.

Khodakovskaya, M. V., Kim, B. S., Kim, J. N., Alimohammadi, M., Dervishi, E., Mustafa, T., & Cernigla, C. E. (2013). Carbon nanotubes as plant growth regulators: Effects on system, and soil microbial community. *Small*, 9(1), 115–123. doi:10.1002/smll.201201225.

Kim, J. H., Lee, Y., Kim, E. J., Gu, S., Sohn, E. J., Seo, Y. S., Chang, Y. S. (2014). Exposure of iron nanoparticles to Arabidopsis thaliana enhances root elongation by triggering cell wall loosening. *Environmental Science and Technology*, 48(6), 3477–3485. doi:10.1021/es4043462.

Kirkham, M. B. (2006). Cadmium in plants on polluted soils: Effects of soil factors, hyperaccumulation, and amendments. *Geoderma*, 137(1–2), 19–32. doi:10.1016/j.geoderma.2006.08.024.

Kramer, U. (2010). Metal hyperaccumulation in plants. *Annual Review of Plant Biology*, 6(1): 517–534.

Kumar, P., Singh, P., & Singh, R. (2021). Nanoparticles in phytoremediation: A promising tool for removal of organic and inorganic pollutants. In Avnesh Kumari, Rajni Garg and Rishav Garg Plants (eds) *Nanotechnology for Sustainable Agriculture* (pp. 309–326). Cham: Springer.

Lam, E. J., Cánovas, M., Montofré, Í. L., & Keith, B. F. (2018). Assessment of the adaptive capacity of plant species in coppers mine tailings in arid and semiarid environments. *Journal of Soils and Sediments*, 18(6), 2203–2216. doi:10.1007/s11368-017-1835-9.

Lasat, M. (1999). Phytoextraction of metals from contaminated soil: A review of plant/soil/metal interaction and assessment of pertinent agronomic issues. *Journal of Hazardous Substance Research*, 2(1), 5.

Lee, J. H. (2013). An overview of phytoremediation as a potentially promising technology for environmental pollution control. *Biotechnology and Bioprocess Engineering*, 18(3), 431–439.

Lei, M., Wan, X., Guo, G., Yang, J., & Chen, T. (2018). Phytoextraction of arsenic contaminated soil with Pteris vittata in Henan Province, China: Comprehensive evaluation of remediation efficiency correcting for atmospheric depositions. *Environmental Science and Pollution Research*, 25(1), 124–131. doi:10.1007/s11356-016-8184-x.

Li, Z., Ma, Z., van der Kuijp, T., Yuan, Z., & Huang, L. (2014). A review of soil heavy metal pollution from mines in China: Pollution and health risk assessment. *Science of the Total Environment*, 468–469, 843–853. doi:10.1016/j.scitotenv.2013.08.090.

Liang, J., Yang, Z., Tang, L., Zeng, G., Yu, M., Li, X., … Luo, Y. (2017). Changes in heavy metal mobility and availability from contaminated wetland soil remediated with combined biochar-compost. *Chemosphere*, 181, 281–288. doi:10.1016/j.chemosphere. 2017.04.081.

Lin, C., Lerch, R. N., Kremer, R. J., & Garrett, H. E. (2011). Stimulated rhizodegradation of atrazine by selected plant species. *Journal of Environment Quality*, 40(4), 1113–1121.

Liu, Y., Wu, T., White, J. C., & Lin, D. (2020). A new strategy using nanoscale zero-valent iron to simultaneously promote remediation and safe crop production in contaminated soil. *Nature Nanotechnology*, 16: 197–205

Liu, Q., Zhao, Y., Wan, Y., Zheng, J., Zhang, X., Wang, C., … Lin, J. (2010). Study of the inhibitory effect of water-soluble fullerenes on plant growth at the cellular level. *ACS Nano*, 4(10), 5743–5748. doi:10.1021/nn101430g.

Ma, C., Liu, H., Chen, G., Zhao, Q., Eitzer, B., Wang, Z., … Dhankher, O. P. (2017). Effects of titanium oxide nanoparticles on tetracycline accumulation and toxicity in Oryza sativa (L.). *Environmental Science: Nano*, 4(9), 1827–1839. doi:10.1039/C7EN00280G.

Ma, X., & Wang, C. (2010). Fullerene nanoparticles affect the fate and uptake of trichloroethylene in phytoremediation systems. *Environmental Engineering Science*, 27(11), 989–992. doi:10.1089/ees.2010.0141.

Mahar, A., Wang, P., Ali, A., Awasthi, M. K., Lahori, A. H., Wang, Q., et al. (2016). Challenges and opportunities in the phytoremediation of heavy metals contaminated soils: A review. *Ecotoxicology and Environmental Safety*, 126, 111–121.

Marques, A. P., Rangel, A. O., & Castro, P. M. (2009). Remediation of heavy metal contaminated soils: Phytoremediation as a potentially promising clean-up technology. *Critical Reviews in Environmental Science and Technology*, 39(8), 622–654.

Maurer-Jones, M. A., Gunsolus, I. L., Murphy, C. J., & Haynes, C. L. (2013). Toxicity of engineered nanoparticles in the environment. *Analytical Chemistry*, 85(6), 3036–3049. doi:10.1021/ac303636s.

Mesjasz-Przybyłowicz, J., Nakonieczny, M., Migula, P., Augustyniak, M., Tarnawska, M., Reimold, U., et al. (2004). Uptake of cadmium, lead nickel and zinc from soil and water solutions by the nickel hyperaccumulator Berkheya coddii. *Acta Biologica Cracoviensia Series Botanica*, 46, 75–85.

Moameri, M., & Abbasi Khalaki, M. (2017). Capability of Secale montanum trusted for phytoremediation of lead and cadmium in soils amended with nano-silica and municipal solid waste compost. *Environmental Science and Pollution Research*. doi:10.1007/s11356-017-0544-7.

Moharem, M., Elkhatib, E., & Mesalem, M. (2019). Remediation of chromium and mercury polluted calcareous soils using nanoparticles: Sorption-desorption kinetics, speciation and fractionation. *Environmental Research*, 170, 366–373.

Movafeghi, A., Khataee, A., Abedi, M., Tarrahi, R., Dadpour, M., & Vafaei, F. (2018). Effects of TiO2 nanoparticles on the aquatic plant Spirodela polyrrhiza: Evaluation of growth parameters, pigment contents and antioxidant enzyme activities. *Journal of Environmental Sciences*, 64, 130–138. doi:10.1016/j.jes.2016.12.020.

Mueller, N. C., & Nowack, B. (2010). Nanoparticles for remediation: Solving big problems with little particles. *Elements*, 6(6), 395–400. doi:10.2113/gselements.6.6.395.

Muradoglu, F., Gundogdu, M., Ercisli, S., Encu, T., Balta, F., Jaafar, H. Z., et al. (2015). Cadmium toxicity affects chlorophyll a and b content, antioxidant enzyme activities and mineral nutrient accumulation in strawberry. *Biological Research*, 48(1), 11. doi:10.1186/S40659-015-0001-3.

Nahar, N., Rahman, A., Nawani, N. N., Ghosh, S., & Mandal, A. (2017). Phytoremediation of arsenic from the contaminated soil using transgenic tobacco plants expressing ACR2 gene of Arabidopsis thaliana. *Journal of Plant Physiology*, 218, 121–126. doi:10.1016/ j.jplph.2017.08.001.

Nair, P. M. G., & Chung, I. M. (2015). Study on the correlation between copper oxide nanoparticles induced growth suppression and enhanced lignification in Indian mustard (*Brassica juncea* L.). *Ecotoxicology and Environmental Safety*, 113, 302–313. doi:10.1016/ j.ecoenv.2014.12.013.

Neff, J., Lee, K., & Deblois, E. M. (2011). Produced water: Overview of composition, fates, and effects. In K. Lee & J. Neff (Eds.), *Produced Water: Environmental Risks and Advances in Mitigation Technologies* (pp. 3–54). New York: Springer. doi:10.1007/978-1-4614-0046-2_1.

Nutt, M. O., Heck, K. N., Alvarez, P., & Wong, M. S. (2006). Improved Pd-on-Au bimetallic nanoparticle catalysts for aqueous-phase trichloroethanehydrodechlorination. *Applied Catalysis B: Environmental*, 69(1–2), 115–125.

Park, S. C., Boyanov, M. I., Kemner, K. M., O'Loughlin, E. J., & Kwon, M. J. (2021). Distribution and speciation of Sb and toxic metal (loid)s near an antimony refinery and their effects on indigenous microorganisms. *Journal of Hazardous Materials*, 403, 123625.

Peng, J., Sun, Y., Liu, Q., Yang, Y., Zhou, J., Feng, W., Li, F. (2012). Upconversion nanoparticles dramatically promote plant growth without toxicity. *Nano Research*, 5(11), 770–782. doi:10.1007/s12274-012-0261-y.

Pichtel, J. (2016). Oil and gas production wastewater: Soil contamination and pollution prevention. *Applied and Environmental Soil Science*, 2016, 2707989. doi:10.1155/2016/2707989.

Pillai, H. P., & Kottekottil, J. (2016). Nano-phytotechnological remediation of endosulfan using zero valent iron nanoparticles. *Journal of Environmental Protection*, 7(5), 734–744. doi:10.4236/ jep.2016.75066.

Pilon-Smits, E. A., & Pilon, M. (2005). Phytoremediation of metals using transgenic plants. *Critical Reviews in Plant Sciences*, 24(4), 311–327.

Prasad, M. N. V. (2003). Phytoremediation of metal-polluted ecosystems: Hype for commercialization. *Russian Journal of Plant Physiology*, 50(5), 686–701.

Praveen, A., Khan, E., Ngiimei, D.S., Perwez, M., Sardar, M., & Gupta, M. (2018). Iron oxide nanoparticles as nano-adsorbents: A possible way to reduce arsenic phytotoxicity in Indian mustard plant (Brassica juncea L.). *Journal of Plant Growth Regulation*, 37(2), 612–624. doi:10.1007/s00344-017-9760-0.

Rafique, N., & Tariq, S. R. (2016). Distribution and source apportionment studies of heavy metals in soil of cotton/wheat fields. *Environmental Monitoring and Assessment*, 188(5), 309. doi:10.1007/ s10661-016-5309-0.

Reeves, R. D., Baker, A. J. M., Jaffr_e, T., Erskine, P. D., Echevarria, G., & van der Ent, A. (2018). A global database for plants that hyperaccumulate metal and metalloid trace elements. *New Phytologist*, 218(2), 407–411. doi:10.1111/nph.14907.

Ren, X., Zeng, G., Tang, L., Wang, J., Wan, J., Liu, Y., Deng, R. (2018). Sorption, transport and biodegradation: An insight into bioavailability of persistent organic pollutants in soil. *The Science of the Total Environment*, 610–611, 1154–1163.

Sakakibara, M., Watanabe, A., Inoue, M., Sano, S., & Kaise, T. (2010). Phytoextraction and phytovolatilization of arsenic from As-contaminated soils by Pteris vittata *Proceedings of the Annual International Conference on Soils, Sediments, Water and Energy*, 26: 12–27.

Sarkar, R., Das, S., & Dash, S. S. (2020). Nanoparticle mediated phytoremediation: Recent advances and future prospects. *Environmental Science and Pollution Research*, 27(10), 10339–10353.

Shaheen, S. M., & Rinklebe, J. (2015). Phytoextraction of potentially toxic elements by Indian mustard, rapeseed, and sunflower from a contaminated riparian soil. *Environmental Geochemistry and Health*, 37(6), 953–967.

Sharma, P., Sirohi, R., Tong, Y. W., Kim, S. H., & Pandey, A. (2021c). Metal and metal (loids) removal efficiency using genetically engineered microbes: Applications and challenges. *Journal of Hard Materials*, 416:125855.

Sheoran, V., Sheoran, A. S., & Poonia, P. (2016). Factors affecting phytoextraction: A review. *Pedosphere*, 26(2), 148–166. doi:10.1016/S1002-0160(15)60032-7.

Siddiqui, M.N., Mostofa, M.G., Akter, M.M., Srivastava, A.K., Md, A.S., Hasan, M.S., Tran, L.S.P., (2017). Impact of salt-induced toxicity on growth and yield-potential of local wheat cultivars: oxidative stress and ion toxicity are among the major determinants of salt-tolerant capacity. *Chemosphere* 187: 385–394.

Singh, J., & Lee, B. K. (2016). Influence of Nano-TiO2 particles on the bioaccumulation of Cd in soybean plants (Glycine max): A possible mechanism for the removal of Cd from the contaminated soil. *Journal of Environmental Management*, 170, 88–96. doi:10.1016/ j.jenvman.2016.01.015.

Souri, Z., Karimi, N., Sarmadi, M., & Rostami, E. (2017). Salicylic acid nanoparticles (SANPs) improve growth and phytoremediation efficiency of Isatis cappadocica Desv., under as stress. *IET Nanobiotechnology*, 11(6), 650–655. doi:10.1049/iet-nbt.2016.0202.

Su, Y., Yan, X., Pu, Y., Xiao, F., Wang, D., & Yang, M. (2013). Risks of single-walled carbon nanotubes acting as contaminants-carriers: Potential release of phenanthrene in Japanese medaka (Oryzias latipes). *Environmental Science and Technology*, 47(9), 4704–4710. doi:10.1021/es304479w.

Suman, J., Uhlik, O., Viktorova, J., & Macek, T. (2018). Phytoextraction of heavy metals: A promising tool for clean-up of polluted environment? *Frontiers in Plant Science*, 9, 1476. doi:10.3389/fpls.2018.01476.

Susarla, S., Medina, V. F., & McCutcheon, S. C. (2002). Phytoremediation: An ecological solution to organic chemical contamination. *Ecological Engineering*, 18(2), 647–658.

Thatai, S., Khurana, P., & Boken, J. (2014). Nanoparticles and core–shell nanocomposite based new generation water remediation materials and analytical techniques: A review. *Microchemical Journal*, 116, 62–76.

Timmusk, S., Seisenbaeva, G., & Behers, L. (2018). Titania (TiO2) nanoparticles enhance the performance of growth-promoting rhizobacteria. *Scientific Reports*, 8(1), 617.

Torre-Roche, R. D. L., Hawthorne, J., Deng, Y., Xing, B., Cai, W., Newman, L. A., & White, J. C. (2012). Fullerene-enhanced accumulation of p,p0-DDE in agricultural crop species. *Environmental Science and Technology*, 46(17), 9315–9323. doi:10.1021/es301982w.

Tratnyek, P. G., & Johnson, R. L. (2006). Nanoteehnologies for environmental cleanup. *Nanotoday*, l, 44–48.

Tripathi, D. K., Singh, V. P., Prasad, S. M., Chauhan, D. K., & Dubey, N. K. (2015). Silicon nanoparticles (SiNp) alleviate chromium (VI) phytotoxicity in Pisum sativum (L.) seedlings. *Plant Physiology and Biochemistry*, 96, 189–198. doi:10.1016/j.plaphy.2015.07.026.

USEPA. (2017). Superfund remedial investigation/feasibility study (site characterization). U.S. 1029 Environmental Protection Agency: Washington, DC. 1030 https://www.epa.gov/superfund/superfund-remedial-investigationfeasibility-study-site1031 characterization (accessed 31 January 2018)

Vázquez-Núñez, E., Molina-Guerrero, C. E., Peña-Castro, J. M., Fernández-Luqueño, F., & de la Rosa-Álvarez, M. G. (2020). Use of Nan-otechnology for the bioremediation of contaminants: A review. *Processes*, 8(7), 826.

Venkatachalam, P., Priyanka, N., Manikandan, K., Ganeshbabu, I., Indiraarulselvi, P., Geetha, N., Sahi, S. V. (2017). Enhanced plant growth promoting role of phycomolecules coated zinc oxide nanoparticles with P supplementation in cotton (Gossypium hirsutum L.). *Plant Physiology and Biochemistry*, 110, 118–127. doi:10.1016/j.plaphy.2016.09.004.

Vithanage, M., Dabrowska, B. B., Mukherjee, A. B., Sandhi, A., & Bhattacharya, P. (2012). Arsenic uptake by plants and possible phytoremediation applications: A brief overview. *Environmental Chemistry Letters*, 10(3), 217–224. doi:10.1007/s10311-011-0349-8.

Vtkova, M., Puschenreiter, M., & Komarek, M. (2018). Effect of nano zero-valent iron application on As, Cd, Pb, and Zn availability in the rhizosphere of metal(loid) contaminated soils. *Chemosphere*, 200, 217–226.

Wang, A. S., Angle, J. S., Chaney, R. L., Delorme, T. A., & Reeves, R. D. (2006). Soil pH effects on uptake of Cd and Zn by Thlaspi caerulescens. *Plant and Soil*, 281(1–2), 325–337.

Wang, C., Liu, H., Chen, J., Tian, Y., Shi, J., Li, D., Ma, Q. (2014). Carboxylated multi walled carbon nanotubes aggravated biochemical and subcellular damages in leaves of broad bean (Vicia faba L.) seedlings under combined stress of lead and cadmium. *Journal of Hazardous Materials*, 274, 404–412. doi:10.1016/j.jhazmat.2014.04.036.

Wang, H., Jia, Y., Wang, S., Zhu, H., & Wu, X. (2009). Bioavailability of cadmium adsorbed on various oxides minerals to wetland plant species Phragmites australis. *Journal of Hazardous Materials*, 167(1–3), 641–646. doi:10.1016/j.jhazmat.2009.01.012.

Wang, P., Lombi, E., Zhao, F.-J., & Kopittke, P. M. (2016). Nanotechnology: A new opportunity in plant sciences. *Trends in Plant Science*, 21(8), 699–712.

Wild, E., & Jones, K. C. (2009). Novel method for the direct visualization of in vivo nanomaterials and chemical interactions in plants. *Environmental Science and Technology*, 43(14), 5290–5294. doi:10.1021/es900065h.

Wu, X., Wang, W., & Zhu, L. (2018). Enhanced organic contaminants accumulation in crops: Mechanisms, interactions with engineered nanomaterials in soil. *Environmental Pollution*, 240, 51–59. doi:10.1016/j.envpol.2018.04.072.

Yadu, B., Chandrakar, V., Korram, J., Satnami, M. L., Kumar, M., & Keshavkant, S. (2018). Silver nanoparticle modulates gene expressions, glyoxalase system and oxidative stress markers in fluoride stressed Cajanus cajan L. *Journal of Hazardous Materials*, 353, 44–52. doi:10.1016/j.jhazmat.2018.03.061.

Yang, L., Ren, Q., Ge, S., Jiao, Z., Zhan, W., Hou, R., Ruan, X., Pan, Y., Wang, Y. (2022). Metal (loid) s spatial distribution, accumulation, and potential health risk assessment in soil-wheat systems near a Pb/Zn smelter in Henan province, central China. *Int. J. Environ. Res. Public Health,* 19: 2527.

Yu, B., Zhang, Y., Shukla, A., Shukla, S. S., & Dorris, K. L. (2001). The removal of heavy metals from aqueous solutions by sawdust adsorption: Removal of lead and comparison of its adsorption with copper. *Journal of Hazardous Materials*, 84(1), 83–94. doi:10.1016/ S0304-3894(01)00198-4.

Zaier, H., Ghnaya, T., Ghabriche, R., Chmingui, W., Lakhdar, A., Lutts, S., & Abdelly, C. (2014). EDTA-enhanced phytoremediation of lead-contaminated soil by the halophyte Sesuvium portulacastrum. *Environmental Science and Pollution Research International*, 21(12), 7607–7615. doi:10.1007/ s11356-014-2690-5.

Zhao, X., Li, Z., Huang, B., Luo, N., Huang, M., Zhang, Q., & Zeng, G. (2018). Remediation of multiple heavy metal-contaminated soil through the combination of soil washing and in situ immobilization. *Science of the Total Environment*, 635, 92–99.

Zhang, H., Liu, Y., Shen, X., Zhang, M., Yang, Y., Tao, S., & Wang, X. (2017). Influence of multiwalled carbon nanotubes and sodium dodecyl benzene sulfonate on bioaccumulation and translocation of pyrene and 1-methylpyrene in maize (Zea mays) seedlings. *Environmental Pollution*, 220(B), 1409–1417. doi:10.1016/j.envpol.2016.10.093.

Zhang, M., Gao, B., Chen, J., & Li, Y. (2015). Effects of graphene on seed germination and seedling growth. *Journal of Nanoparticle Research*, 17(2), 78.

Zhang, Y., Zhang, H., Zhang, Z., Liu, C., Sun, C., Zhang, W., & Marhaba, T. (2018). pH effect on heavy metal release from polluted sediment. *Journal of Chemistry*, 2018: 7597640.

Index

Page number followed by *f* indicate figure and **bold** indicate table respectively